Proceedings of Light-Activated Tissue Regeneration and Therapy Conference

Lecture Notes in Electrical Engineering

Advances in Numerical Methods
Mastorakis, Nikos; Sakellaris, John (Eds.)
2008, Approx. 300 p., Hardcover
ISBN: 978-0-387-76482-5, Vol. 11

Embedded Systems Specification and Design Languages
Villar, Eugenio (Ed.)
2008, Approx. 400 p., Hardcover
ISBN: 978-1-4020-8296-2, Vol. 10

Content Delivery Networks
Buyya, Rajkumar; Pathan, Mukaddim; Vakali, Athena (Eds.)
2008, Approx. 400 p., Hardcover
ISBN: 978-3-540-77886-8, Vol. 9

Unifying Perspectives in Computational and Robot Vision
Kragic, Danica; Kyrki, Ville (Eds.)
2008, Approx. 250 p., Hardcover
ISBN: 978-0-387-75521-2, Vol. 8

Sensor and Ad-Hoc Networks
Makki, S.K.; Li, X.-Y.; Pissinou, N.; Makki, S.; Karimi, M.; Makki, K. (Eds.)
2008, Approx. 350 p. 20 illus., Hardcover
ISBN: 978-0-387-77319-3, Vol. 7

Trends in Intelligent Systems and Computer Engineering
Castillo, Oscar; Xu, Li; Ao, Sio-Iong (Eds.)
2008, Approx. 750 p., Hardcover
ISBN: 978-0-387-74934-1, Vol. 6

Advances in Industrial Engineering and Operations Research
Chan, Alan H.S.; Ao, Sio-Iong (Eds.)
2008, XXVIII, 500 p., Hardcover
ISBN: 978-0-387-74903-7, Vol. 5

Advances in Communication Systems and Electrical Engineering
Huang, Xu; Chen, Yuh-Shyan; Ao, Sio-Iong (Eds.)
2008, Approx. 700 p., Hardcover
ISBN: 978-0-387-74937-2, Vol. 4

Time-Domain Beamforming and Blind Source Separation
Bourgeois, J.; Minker, W. (Eds.)
2009, approx. 200 p., Hardcover
ISBN: 978-0-387-68835-0, Vol. 3

Digital Noise Monitoring of Defect Origin
Aliev, T. (Ed.)
2007, XIV, 223 p. 15 illus., Hardcover
ISBN: 978-0-387-71753-1, Vol. 2

Multi-Carrier Spread Spectrum 2007
Plass, S.; Dammann, A.; Kaiser, S.; Fazel, K. (Eds.)
2007, X, 106 p., Hardcover
ISBN: 978-1-4020-6128-8, Vol. 1

Ronald Waynant • Darrell B. Tata

Editors

Proceedings of Light-Activated Tissue Regeneration and Therapy Conference

 Springer

Ronald W. Waynant
The Food and Drug Administration
10903 New Hampshire Avenue
Silver Spring, MD 20993
USA

Darrell B. Tata
The Food and Drug Administration
CDRH/OSEL/Division of Physics
Silver Spring, MD 20993
USA

ISBN 978-0-387-71808-8   e-ISBN 978-0-387-71809-5

Library of Congress Control Number: 2008922319

Printed on acid-free paper.

9 8 7 6 5 4 3 2 1

springer.com

Contents

Forword xi

Contributors xvii

Keynote

Mitochondrial Mechanisms of Laser Phototherapy xxvii
Tiina I. Karu

Part I Mechanisms

1 **Mechanisms** 3
   Darrell B. Tata and Ronald W. Waynant

2 **Near-IR Picosecond Pulsed Laser Induced Suppression
   of Metabolic Activity in Malignant Human Brain Cancer:
   An In-Vitro Study** 11
   Darrell B. Tata and Ronald W. Waynant

Part II Wound Healing

3 **Combined 660 and 880 nm Light Improves Healing
   of Recalcitrant Diabetic Ulcers** 23
   Debora G. Minatel, Marco Andrey C. Frade, Suzelei C. Franca,
   Gil L. Almeida, and Chukuka S. Enwemeka

4 **Blue Light Photo-Destroys Methicillin Resistant
   Staphylococcus aureus (MRSA) In-Vitro** 33
   Chukuka S. Enwemeka, Deborah Williams, Steve Hollosi, and David Yens

5   Photobiomodulation for the Treatment of Retinal Injury
    and Retinal Degenerative Diseases 39
    Janis T. Eells, Kristina D. DeSmet, Diana K. Kirk, Margaret Wong-Riley,
    Harry T. Whelan, James Ver Hoeve, T. Michael Nork, Jonathan Stone,
    and Krisztina Valter

6   Irradiation with a 780 nm Diode Laser Attenuates Inflammatory
    Cytokines While Upregulating Nitric Oxide in LPS-Stimulated
    Macrophages: Implications for the Prevention
    of Aneurysm Progression 53
    Lilach Gavish, Louise S. Perez, Petachia Reissman, and S. David Gertz

7   New Aspects of Wound Healing 59
    A. Lipovsky, Ankri R. Nitzan, Z.A. Landoy, J. Jacobi, and R. Lubart

Part III Photodynamic Therapy

8   An Introduction to Low-Level Light Therapy 67
    Stuart K. Bisland

9   Enhancing Photodynamic Effect Using Low-Level
    Light Therapy 81
    Stuart K. Bisland

10 Light Fractionated ALA-PDT: From Pre-Clinical Models
    to Clinical Practice 89
    D.J. Robinson, H.S. de Bruijn, E.R.M. de Haas, H.A.M. Neumann,
    and H.J.C.M. Sterenborg

11 Combination Immunotherapy and Photodynamic
    Therapy for Cancer 99
    Michael R. Hamblin, Ana P. Castano, and Pawel Mroz

12 Patient-Specific Dosimetry for Photodynamic Therapy 115
    Jarod C. Finlay, Jun Li, Xiaodong Zhou, and Timothy C. Zhu

13 Novel Targeting and Activation Strategies
    for Photodynamic Therapy 127
    Juan Chen, Ian R. Corbin, and Gang Zheng

Part IV Cardiovascular

14 Light Therapy for the Cardiovascular System 151
    Hana Tuby, Lydia Maltz, and Uri Oron

Part V Dentistry

15 **Introduction: Overview** 159
Donald E. Patthoff

16 **Optical Coherence Tomography Imaging for Evaluating
the Photobiomodulation Effects on Tissue Regeneration
in Periodontal Tissue** 173
Craig B. Gimbel

17 **Photobiomodulation Laser Strategies in Periodontal Therapy** 181
Akira Aoki, Aristeo Atsushi Takasaki, Amir Pourzarandian,
Koji Mitzutani, Senarath M.P.M. Ruwanpura, Kengo Iwasaki,
Kazuyuki Noguchi, Shigeru Oda, Hisashi Watanabe, Isao Ishikawa,
and Yuichi Izumi

18 **Combined New Technologies to Improve Dental
Implant Success and Quantitative Ultrasound Evaluation
of NIR-LED Photobiomodulation** 191
Jerry E. Bouquot, Peter R. Brawn, and John C. Cline

19 **Photobiomodulation by Low Power Laser Irradiation
Involves Activation of Latent TGF-β1** 207
Praveen R. Arany

Part VI Diabetes

20 **The Role of Laser in Diabetic Management** 215
Leonardo Longo

21 **He-Ne Laser Irradiation Stimulates Proliferation
and Migration of Diabetic Wounded Fibroblast Cells** 221
Nicolette Houreld and Heidi Abrahamse

22 **The Role of Colostrum Proline-Rich Polypeptides
in Human Immunological and Neurological Health** 233
Andrew Keech, John I. Buhmeyer, and Dick Kolt

Part VII Neuroscience

23 **Phototherapy and Nerve Tissue Repair** 247
Shimon Rochkind

24   Laser Regeneration of Spine Discs Cartilage: Mechanism,
     In-Vivo Study and Clinical Applications 259
     Emil Sobol, Andrei Baskov, Anatoly Shekhter, Igor Borshchenko,
     and Olga Zakharkina

Part VIII FDA Regulations

25   Requirements for FDA Approval 269
     Sankar Basu

Part IX Pain

26   Pain Relief with Phototherapy: Session Overview 273
     Mary Dyson

27   Is Relief of Pain with Low-Level Laser Therapy (LLLT)
     a Clinical Manifestation of Laser-Induced Neural Inhibition? 277
     Roberta Chow

28   Complex Regional Pain Syndrome: A New
     Approach to Therapy 283
     Allan Gardiner, Robert E. Florin, and Constance Haber

Part X Electric Field Interactions

29   Introduction 295
     Martin J.C. van Gernert

30   The Painful Derivation of the Refractive Index
     from Microscopic Considerations 297
     Bernhard J. Hoenders

31   Independent Applications of Near-IR Broadband Light Source
     and Pulsed Electric Potential in the Suppression
     of Human Brain Cancer Metabolic Activity:
     An In-Vitro Study 307
     Darrell B. Tata and Ronald W. Waynant

32   Electroencephalogram Changes Caused by Mobile
     Phones a Protective Device 315
     Mbonu Ngozy, Weiler Elmar, and Schroeter Careen

Appendix 1 A 3D Dose Model for Low Level Laser/LED
Therapy, Biostimulation and Bioinhibition 327
James D. Carroll

Appendix 2 Does Body Contouring Need to Be Painful? 331
Michail M. Pankratov

Appendix 3 Effect of Far Infrared Therapy on Inflammatory
Process Control After Sciatic Crushing in Rats 347
Carolina L.R.B. Nuevo, Renata Amadei Nicolau,
Renato Amaro Zângaro, Aldo Brugnera, Jr.,
and Marcos Tadeu Tavares Pacheco

Appendix 4 Effects of Diode Laser Therapy
on the Acellular Dermal Matrix 357
Lívia Soares, Marília de Oliveira, Sílvia Reis,
and Antônio Pinheiro

Appendix 5 Laser Therapy in Inflammation:
Possible Mechanisms of Action 361
Rodrigo Alvaro B. Lopes Martins

Appendix 6 A Systematic Review of Post-operative Pain
Relief by Low-Level Laser Therapy (LLLT) After
Third Molar and Endodontic Surgery (Slides Only) 393
Jan M. Bjordal

Index 441

Foreword

The contents of this book represent most of the material presented at the second conference on "Light-Activated Tissue Regeneration and Therapy" an Engineering Conference International (ECI). This Conference was held in Tomar, Portugal from June 24 to 29, 2007 and focused on the use of lasers, light and electromagnetic radiation for medical treatments. The methodology was first presented by Finsen, who used light (mostly sunlight), to cure tuberculosis in the late 19th century. He was awarded the Nobel Prize in 1903. His work led to sanatoriums throughout the world, but Finsen's work was mostly forgotten by the 1940s in favor of new work on antibiotic drugs. After the invention of the laser in 1960 medical research became interested in its possible use. Andre Mester in Hungary tried lasers in medicine and his work led to laser use in medicine. This work spread to numerous uses of low powered gas laser applications for treating a broad spectrum of medical illnesses, afflictions, injuries and cosmetic treatments. The major body of early applications used lasers hence the field was called laser therapy, but the exposures were very low hence many other names were sometimes used to make the low exposure obvious. Early workers thought that the coherence of a laser might be necessary, but later work with non-coherent sources such as LEDs, broadband sources, etc. has worked and shows that lasers are not needed. Experiments with sources of longer wavelengths also work and shorter wavelengths may also work, but have yet to be tested.

Since Mester's work in the 1960s the laser therapy field has spread all over the world. The mechanism of how the therapy works has not yet been accepted worldwide although this conference presents some views. Nor has the therapy been accepted by all of the medical professions. Yet the therapy has been applied to nearly 100 applications and successful double-blinded studies showing positive results have been submitted for some of the applications. In addition, light therapy treatments have migrated to cosmetics treatments as well as to physical therapy treatments and to treatments for accident injuries in the chiropractic area. Stories of nearly miraculously recoveries utilizing these therapies are often told. Many new applications come from small companies that find it difficult to fund the development and market preparation of their therapies.

This book begins with a paper from **Tiina Karu**, the lady who has generated a mountain of work in this field including over one hundred publications and at least three books on the subject and who discusses Mitochondrial Mechanisms of laser Therapy here as the keynote speaker. Following Tiina, **Darrell Tata** and I present our conviction that most of the results published indicate that the mechanism of light therapy follows results produced by hydrogen peroxide — the most stable reactive oxygen specie, which fits closely with the suggestions published by Tiina Karu. In addition, we continue to verify the results of catalase as well as to use the facilities at Johns Hopkins School of Public Health with **Michael Trush** to measure the amount of H_2O_2 produced in every cell of our 96 well cell culture plates. Only a health problem now resolved prevented Michael Trush from being at the conference and telling about the Good and the Bad of reactive oxygen species. **Jan Bjordal** gave a paper by he and **Rodrigo Lopes-Martins** produced on laser therapy for inflammatory diseases that gives information also important to mechanisms.

The remainder of the conference focused on applications of laser therapy. **Chukuka Enwemeka** organized the session on Wound Healing and presented papers from his own work. His first paper discusses the use of two wavelengths to improve the healing of diabetic ulcers. His second paper, given by **Steve Hillosi**, deals with blue light to destroy staphylococcus aureus. **Janis Eells** presents use of laser therapy (also called photobiomodulation) for retinal injury and retinal degenerative diseases. **Lilach Gavish** discusses the reduction of inflammatory cytokines, the up-regulation of nitric oxide and the implications for aneurysm progression. **Rachel Lubart** gives new aspects on wound healing to close a terrific session.

To start the second day **Stuart Bisland** offered a comprehensive look at Photodynamic Therapy, a second application of reactive oxygen species with similarities to laser therapy. Stuart gave an excellent introduction to the session including papers by **D.J. Robinson, Michael Hamblin, Jarod Finlay** and Stuart presented a paper authored by **Gang Zheng**. The evening session was a single paper session on Light Therapy for the Cardiovascular System by **Uri Oron**. Uri has done excellent work in this area producing ground-breaking work on heart attack damage reduction as well as stimulation for work by others. Its implication for progress against this number one killer in the western world demanded that we have a session on it here. Hopefully more work in this area will be stimulated. Rumors from the hydrogen peroxide work from late in the 20th century suggest that a method of reducing arterial plaque was possible. Hopefully more work in this area will soon be taken up. **James Carroll** gave a very interesting presentation on the sensitivity of stimulation to dose.

The third day of the conference featured a morning session where **Don Pattoff** introduced a dynamic Dentistry Session. **Craig Gimbel** presented a paper on Optical Coherence Tomography to evaluate Laser Therapy effects in Periodontal Tissue. **Akira Aoki** then also discussed Light Therapy Strategies for Periodontal Therapy. Don also gave **Jerry Bouquot's** (Jerry misplaced his passport) paper on Laser Therapy for improving tooth implant success. Then **Praveen Arany** gave a paper on the activation of TGF-β1 Çomplexes.

The afternoon session chaired by **Leonardo Longo** highlighted the application of laser therapy to reduction of the effects of diabetes. Leonardo discussed work he has done and the importance of lowering blood glucose in preventing heart attack, stroke, blindness and liver failure in ten percent of the world's population. **Nicolette Hourold** and **Heidi Abrahamse** told how He-Ne laser irradiation stimulates proliferation and migration of diabetic-wounded fibroblast cells and **Dick Kolt** discussed the role of Bovine Colostrum in boosting Human Immunological and Neurological Health. Again, these session where major killers have fewer workers were featured in hopes of increasing interest in generating more work to find cheaper, effective means of reducing health problems around the world.

The evening session was organized by **Shimon Rochkind**, who was hospitalized at the last minute and is presenting his paper here on Phototherapy and Nerve Tissue Repair. **Emil Sobel** also presented a paper on Laser Regeneration of Spine Discs Cartilage. Mechanism, in-vivo study and Clinical Applications Although neurological problems occur far less frequently than the major killers, they are severe injuries that have a major impact on those who have them.

Pain was the topic of the morning session of the fourth day and it was chaired and introduced by **Mary Dyson**. Pain is a debilitating thing that can effect an entire family. It can lead to depression and to suicide. **Roberta Chow** asks *"Is Relief of Pain with Low-Level Laser Therapy a Clinical Manifestation of Laser-Induced Neural Inhibition?"* and presents on the research she has done. **Jan Bjordal** discusses his review of post-operative pain relief by laser Therapy after dental surgery and **Allan Gardiner** talks about Complex Regional Pain Syndrome and his approach that has been highly successful.

The evening session chaired by **Martin J.C van Gemert** features a pure Physics derivation of the refractive index by **Bernhard Hoenders** that originated showing the possible changes in refraction without the need for measurable absorption. Also, **Darrell Tata** and I presented results from a variety of novel sources originally suggested by other workers that suggest a very broad spectrum of operation for laser therapy and also that cheap sources can be used to produce results. The final paper by **Careen Schroeter** discussed electroencephalograph changes caused by cell phones and a simple protective device to prevent the interaction. How similar the mechanisms of the interaction of the rf from the cell phone may be to the interaction of light is not yet known.

Finally the Appendix collects papers that didn't fall into a session or may have been presented in sessions, but not expanded into an explanatory form or similarly were presented by poster. While these papers may be less informative, it is felt that there is still valuable information that can be found. **James Carroll** presents an interesting calculation that seems to define the difficulty of getting the exposure correct to produce stimulation with light therapy. **Michail Pankratov** describes body contouring and the role that light therapy plays in it. **Carolina L.R.B. Nuevo** et al. show the effects of far infrared radiation for the control of inflammation of crushed nerves in rats. **Livia Soares** et al. describe the effects of therapy on the acellular dermal matrix. The Appendix also is home for two papers given by **Jan Bjordal** that we have only as the slides of his talk, two talks which are referred to in the Mechanisms and Pain Sessions.

I believe this book summarizes a step forward for laser therapy, perhaps better described as light therapy or maybe electromagnetic therapy. We may be a few steps closer to a consensus on the mechanism, wavelength may not be confined to so narrow a region near the red and infrared, lasers may only have been a convenient way of pin pointing a treatment and regulation may not be so necessary for something produced everyday by the sun shining on our cells. It may be that we can find simple, inexpensive ways of curing or controlling many of our serious diseases.

Ronald W. Waynant
Darrell B. Tata

Tomar Group Photo

Photoidentifier

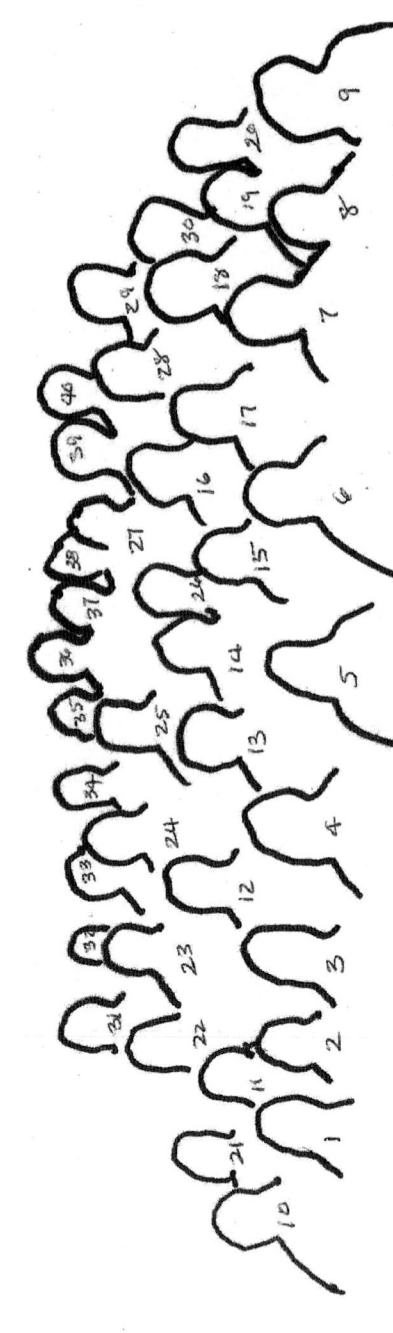

1. James Carroll, UK 2. Lilich Gavish, Israel 3. Chukuka Enwemeka, US 4. Tiina Karu, Russia 5. Roberta Chow, Australia 6. Emil Sobol, Russia 7. Kris Waynant, US 7. Anna-Liisa Nieminen, Netherlands 9. Akira Aoki, Japan 10. Brian Bennett, Canada 11. Allan Gardiner, US 12. Steve Hollosi, US 13. Coreen Schroeter, Netherlands 14. Dick Kolt, US 15. Craig Gimbel, US 16. Richard Fein, US 17. Don Pattoff, US 18. Dominic Robinson, Netherlands 19. Ronald Waynant, US 20. Rachel Lubart, Israel 21. Rita Dunbar 22. Not Identified 23. Michael Hamblin, US 24. Heidi Abrahamse, South Africa 25. Sankar Basu, US 26. Janis Eells, US 27. Jarad Finlay, US 28. Renata Amadei Nicolau, Brazil 29. Martin J.C. van Gemert, Netherlands 30. Josepa Rigau Mas, Spain 31. Arun Danbar, UK 32. Stuart Bisland, Canada 33. Mary Dyson, UK 34. Uri Oron, Israel 35. Gerry Ross, Canada 37. Not Identified 38. Bernhard Hoenders, Netherlands 39. Weidong Yu, Canada 40. Luis De Taboada, US.

Contributors

Heidi Abrahamse
Laser Research Group, Faculty of Health Sciences, University of Johannesburg, Doornfontein, 2028, South Africa

Gil L. Almeida
Department of Physical Therapy, University of Ribeirão Preto, Ribeirão Preto, Brazil

Akira Aoki
Department of Basic Sciences, Faculty of Dental Sciences, University of Peradeniya, Peradeniya, Sri Lanka

Praveen R. Arany
Programs in Oral and Maxillofacial Pathology, Biological Sciences in Dental Medicine and Leder Medical Sciences, Harvard School of Dental Medicine, Boston, MA, USA

Andrei Baskov
Spine and Orthopedic Medical Center, Moscow, Russia

Sankar Basu
Food and Drug Administration, Office of Device Evaluation, 9200 Corporate Boulevard, Rockville, MD 20850, USA

Stuart K. Bisland
Ontario Cancer Institute, University Health Network, University of Toronto, 610 University Avenue, Toronto, ON, Canada M5G 2M9

Jan M. Bjordal
Bergen University College, University of Bergen, Bergen, Norway

Igor Borshchenko
Spine and Orthopedic Medical Center, Moscow, Russia

Jerry E. Bouquot
Department of Diagnostic Services, University of Texas, Dental Branch at Houston, Houston, TX, USA

Peter R. Brawn
Private Practice, Nanaimo, BC, Canada

Aldo Brugnera
Institute of Research and Development, Universidade do Vale do Paraiba (UNI-VAP), Biomodulation Tissue Laboratory and Lasertherapy and Phototherapy Center, São José dos Campos, São Paulo, 12244-000, Brazil

H.S. de Bruijn
Center for Optical Diagnostics and Therapy, University Medical Center Rotterdam, Rotterdam, The Netherlands

John I. Buhmeyer
Sovereign Laboratories, LLC, Sedona, Arizona

James D. Carroll
THO Photomedicine Ltd, Chesham. Castano Wellman Center for Photomedicine, Massachusetts General Hospital, Boston, MA, USA

Juan Chen
Ontario Cancer Institute and University of Toronto, Toronto, ON, Canada

Roberta Chow
Department of Medicine, Central Clinical School, The University of Sydney, Sydney, Australia

John C. Cline
Private Practice, Nanaimo, BC, Canada

Ian R. Corbin
Ontario Cancer Institute and University of Toronto, Toronto, ON, Canada

E.R.M. de Haas
Department of Dermatology and Center for Optical Diagnostics and Therapy, Erasmus MC, Rotterdam, The Netherlands

Kristina D. DeSmet
Department of Clinical Laboratory Sciences, University of Wisconsin, Milwaukee, Milwaukee, WI, USA

Mary Dyson
Emeritus Reader in the Biology of Tissue Repair, Kings College London (KCL), University of London, London, UK

Janis T. Eells
College of Health Sciences, University of Wisconsin, Milwaukee, WI, USA

Weiler Elmar
NeuroNet GmbH, St Annenstr.10, 66606 St. Wendel, Germany

Chukuka S. Enwemeka
School of Health Professions, Behavioral & Life Sciences, New York Institute of
Technology, Old Westbury, NY, USA

Jarod C. Finlay
Department of Radiation Oncology, University of Pennsylvania, Philadelphia, PA,
USA

Robert E. Florin
PhotoMed Technologies, Kensington, CA, USA

Marco Andrey C. Frade
Department of Dermatology, Faculty of Medicine, University of São Paulo,
Ribeirão Preto, Brazil

Suzelei C. Franca
Department of Biotechnology, University of Ribeirão Preto, Ribeirão Preto, Brazil

Allan Gardiner
PhotoMed Technologies, Kensington, CA, USA

Lilach Gavish
Hadassah Medical School, The Hebrew University, Jerusalem, Israel

S. David Gertz
Hadassah Medical School, The Hebrew University, Jerusalem, Israel
and
Shaare Zedek Hospital, The Hebrew University, Jerusalem, Israel

Craig B. Gimbel
Academy of Laser Dentistry, Coral Springs, FL, USA

Constance Haber
PhotoMed Technologies, Kensington, CA, USA

Michael R. Hamblin
Wellman Center for Photomedicine, Massachusetts General Hospital, Boston, MA,
USA

Bernard J. Hoenders
Institute for Theoretical Physics and Zernike Institute for Advanced Materials, University of Groningen, Nijenborgh 4, 9747 AG Groningen, The Netherlands

James Ver Hoeve
Eye Research Institute, University of Wisconsin, Madison, WI, USA

Steve Hollosi
New York College of Osteopathic Medicine, New York Institute of Technology, Old Westbury, NY, USA

Nicolette Houreld
Laser Research Group, Faculty of Health Sciences, University of Johannesburg, P.O. Box 17011, Doornfontein 2028, South Africa

Isao Ishikawa
Institute of Advanced Biomedical Engineering & Science, Tokyo Women's Medical University, Tokyo, Japan

Kengo Iwasaki
Department of Periodontology, Graduate School of Medical & Dental Sciences, Kagoshima University, Kagoshima, Japan

Yuichi Izumi
Section of Periodontology, Department of Hard Tissue Engineering, Tokyo Medical and Dental University, Tokyo, Japan

J. Jacobi
SALT – Swiss Association Laser Therapy, Geneva, Switzerland

Tiina I. Karu
Institute of Laser and Information Technologies, Russian Academy of Sciences, Troitsk, Moscow Region, Russia

Andrew Keech
Advanced Protein Systems, LLC, Phoenix, AZ, USA

Diana K. Kirk
Research School of Biological Sciences, The Australian National University, Canbera, Australia

Dick Kolt
Rejuva-a-Light, Tucson, AZ, USA

Z.A. Landoy
Kaplan Hospital, Rehovot, Israel

Jun Li
Department of Radiation Oncology, University of Pennsylvania, Philadelphia, PA, USA

A. Lipovsky
Departments of Physics, Chemistry & Life Sciences, Bar-Ilan University, Ramat Gan, Israel

Leonardo Longo
Institute for Laser Medicine, Siena University, Firenze, Italy

R. Lubart
Departments of Physics, Chemistry & Life Sciences, Bar-Ilan University, Ramat Gan, Israel

Rodrigo Alvaro B. Lopes
Martins Department of Pharmacology, Institute of Biomedical Sciences, University of Sao Paulo, Ribeirão Preto, Brazil

Lydia Maltz
Department of Zoology, The George S. Wise Faculty of Life Sciences, Tel-Aviv University, Tel-Aviv 69978, Israel

Ngozy C. Mbonu
Department of Laser Therapy, Medical Center Maastricht, 6216 BX Maastricht, The Netherlands

Debora G. Minatel
Department of Biotechnology, University of Ribeirão Preto, Ribeirão Preto, Brazil

Koji Mizutani
Department of Basic Sciences, Faculty of Dental Sciences, University of Peradeniya, Peradeniya, Sri Lanka

Pawel Mroz
Wellman Center for Photomedicine, Massachusetts General Hospital, Boston, MA, USA

H.A.M. Neumann
University of Maastricht, Maastricht, The Netherlands

Renata Amadei Nicolau
Institute of Research and Development, Universidade do Vale do Paraiba (UNIVAP), Biomodulation Tissue Laboratory and Lasertherapy and Phototherapy Center, São José dos Campos, São Paulo, 12244-000, Brazil

Ankri R. Nitzan
Departments of Physics, Chemistry & Life Sciences, Bar-Ilan University, Ramat Gan, Israel

Kazuyuki Noguchi
Department of Periodontology, Graduate School of Medical & Dental Sciences, Kagoshima University, Kagoshima, Japan

T. Michael Nork
Department of Ophthalmology and Visual Sciences, University of Wisconsin, Madison, WI, USA

Carolina L.R.B. Nuevo
Institute of Research and Development – Universidade do Vale do Paraiba (UNI-VAP), Biomodulation Tissue Laboratory and Lasertherapy and Phototherapy Center, São José dos Campos, São Paulo, 12244-000, Brazil

Shigeru Oda
Institute of Advanced Biomedical Engineering & Science, Tokyo Women's Medical University, Tokyo, Japan

Marília de Oliveira
Pontifíca Universidade Católica do Rio Grande do Sul Av. Ipiranga, 6681, Prédio 06, sala 203 Partenon – Porto Alegre, Brazil

Uri Oron
Department of Zoology, The George S. Wise Faculty of Life Sciences, Tel-Aviv University, Tel-Aviv 69978, Israel

Michail M. Pankratov
Elemé Medical, Merrimack, NH 03054, USA

Donald E. Patthoff
300 Foxcroft Ave., Martinsburg, WV, USA

Louise S. Perez
Shaare Zedek Hospital, The Hebrew University, Jerusalem, Israel

Antônio Pinheiro
Faculdade de Odontologia – UFBA, Av. Araújo Pinho, 62, 2o andar, Canela, Cep 40110-150 Salvador, Brazil

Amir Pourzarandian
Department of Basic Sciences, Faculty of Dental Sciences, University of Peradeniya, Peradeniya, Sri Lanka

Sílvia Reis
Faculdade de Odontologia – UFBA, Av. Araújo Pinho, 62, 10o andar, Canela, Cep 40110-150 Salvador, Brazil

Petachia Reissman
Shaare Zedek Hospital, The Hebrew University, Jerusalem, Israel

D.J. Robinson
Department of Radiation Oncology, Center for Optical Diagnostics and Therapy, Rotterdam, The Netherlands

Shimon Rochkind
Division of Peripheral Nerve Reconstruction, Tel Aviv Sourasky Medical Center, Tel Aviv University, Tel-Aviv, Israel

Senarath M.P.M. Ruwanpura
Department of Basic Sciences, Faculty of Dental Sciences, University of Peradeniya, Peradeniya, Sri Lanka

Careen A. Schroeter
Department of Laser Therapy, Medical Center, Maastricht, 6216 BX Maastricht, The Netherlands

Anatoly Shekhter
Sechenov Medical Academy of Moscow, Moscow, Russia

Livia Soares
Pontifica Universidade Católica do Rio Grande do Sul, Porto Alegre, Brazil

Emil Sobol
Institute on Laser and Information Technologies, Russian Academy of Science, Troitsk, Russia

H.J.C.M. Sterenborg
Department of Radiation Oncology, Erasmus University Rotterdam, Rotterdam, The Netherlands

Jonathan Stone
Research School of Biological Sciences, The Australian National University, Canbera, Australia

Marcos Tadeu
Tavares Pacheco Institute of Research and Development, Universidade do Vale do Paraiba (UNIVAP), Biomodulation Tissue Laboratory and Lasertherapy and Phototherapy Center, São José dos Campos, São Paulo, 12244-000, Brazil

Aristo Atsushi Takasaki
Department of Basic Sciences, Faculty of Dental Sciences, University of Peradeniya, Peradeniya, Sri Lanka

Darrell B. Tata
U.S. Food and Drug Administration, Center for Devices and Radiological Health, White Oak, MD, USA

Hana Tuby
Department of Zoology, The George S. Wise Faculty of Life Sciences, Tel-Aviv University, Tel-Aviv 69978, Israel

Krisztina Valter-Kocsi
Clinical Opthamology & Eye Health, Central Clinical School, The University of Sydney, Sydney, Australia

Hisashi Watanabe
Institute of Advanced Biomedical Engineering & Science, Tokyo Women's Medical University, Tokyo, Japan

Martin J.C. van Germert
Laser Center, Academic Medical Center, University of Amsterdam, Amsterdam, The Netherlands

Ronald W. Waynant
U.S. Food and Drug Administration, Center for Devices and Radiological Health, White Oak, MD, USA

Harry T. Whelan
Division of Pediatric Neurology, Medical College of Wisconsin, Milwaukee, WI, USA

Deborah Williams
School of Health Professions, Behavioral & Life Sciences, New York Institute of Technology, Old Westbury, NY, USA

Margaret Wong-Riley
Department of Cell Biology, Neurobiology & Anatomy, Medical College of Wisconsin, Milwaukee, WI, USA

David Yens
New York College of Osteopathic Medicine, New York Institute of Technology, Old Westbury, NY, USA

Olga Zakharkina
Institute on Laser and Information Technologies, Russian Academy of Science, Troitsk, Russia

Renato Amaro Zângaro
Institute of Research and Development, Universidade do Vale do Paraiba (UNIVAP), Biomodulation Tissue Laboratory and Lasertherapy and Phototherapy Center, São José dos Campos, São Paulo, 12244-000, Brazil

Gang Zheng
Ontario Cancer Institute and University of Toronto, Toronto, ON, Canada

Xiaodong Zhou
Department of Radiation Oncology, University of Pennsylvania, Philadelphia, PA, USA

Timothy C. Zhu
Department of Radiation Oncology, University of Pennsylvania, Philadelphia, PA, USA

Mitochondrial Mechanisms of Laser Phototherapy

Tiina I. Karu

Abstract The terminal enzyme of mitochondrial respiratory chain cytochrome c oxidase is considered as a universal photoacceptor in mammalian cells for visible-to-near IR radiation. Two mechanisms occurring in cytochrome c oxidase under irradiation are investigated experimentally. These are an increase of electron flow inside of cytochrome c oxidase and a relieve of NO block in the catalytic center of cytochrome c oxidase. A novel mitochondrial light-activated cellular signaling pathway (retrograde signaling) has been discovered and investigated. Our results evidence that cytochrome c oxidase can work as a signal generator as well as a signal transducer in irradiated cells.

Keywords: Action and absorption spectra, cytochrome c oxidase, Loretzian curve fitting, novel light-activated cellular signaling, relieve of NO block, retrograde mitochondrial signaling.

Cytochrome c Oxidase Is a Universal Photoacceptor in Eukaryotic Cells

The action spectra recorded in HeLa cell culture for processes occurring in the cell nucleus (DNA and RNA synthesis rate) and cell membrane (increase in number of cells attached to a glass matrix) in red to near IR region were analyzed by Lorentzian curve fitting [1]. Red-to-near IR part of one of these spectra is presented in Fig. 1. Insofar as the action spectrum resembles the absorption spectrum of the molecule absorbing the light (photoacceptor), the bands in the action spectra were identified by analogy with the metal-ligand systems absorption spectra characteristic of visible-to-near IR spectral range [2]. This analysis allowed us to conclude that all bands in the action spectra (one maximum at 400 nm with the edge of the envelope near 450 nm and two series of doublet bands in the range 620–680 nm and

T.I. Karu
Institute of Laser and Information Technologies of Russian Academy of Sciences, Troitsk, Moscow Region, Russian Federation

R. Waynant and D.B. Tata (eds.), *Proceedings of Light-Activated Tissue Regeneration and Therapy Conference.*

xxvii

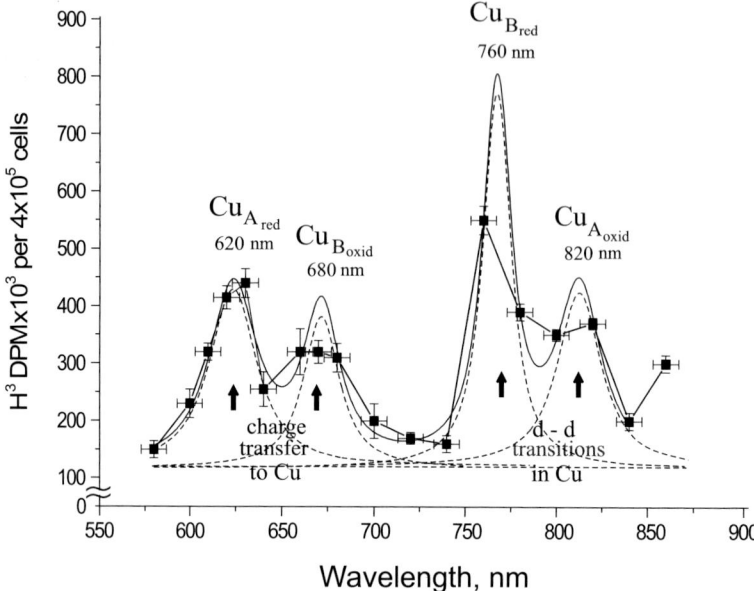

Fig. 1 The action spectrum for stimulation of DNA synthesis rate on cellular level. Suggested absorbing chromophores of the photoacceptor, cytochrome c oxidase, are shown (after [2, 3]). Experimental details are described in [1]

760–895 nm with well-pronounced maxima at 620, 680, 760, and 825 nm) may be related to the cytochrome c oxidase. Cytochrome c oxidase is the terminal enzyme of the respiratory chain in eukaryotic cells, which mediates the transfer of electrons from cytochrome c to molecular oxygen. Bands at 404–420, 680 and 825 nm were attributed to a relatively oxidized form of cytochrome c oxidase. The edge of the blue-violet band at 450 nm and the distinct bands at 620 and 760 nm belong to a relatively reduced form of the enzyme [2].

Figure 1 presents only the red-to-near IR part of the action spectrum. It was suggested that the photoacceptor is one of the intermediate forms of cytochrome c oxidase redox cycle. In the red-to-near IR region the 820 nm band is believed belonging mainly to the relatively oxidized Cu_A chromophore of cytochrome c oxidase, the 760 nm band to the relatively reduced Cu_B, the 680 nm band to the relatively oxidized Cu_B, and the 620 nm band to the relatively reduced Cu_A (Fig. 1).

Comparison of Action and Absorption Spectra: Effect of Irradiation at 830 nm on the Absorption Spectra

Absorption spectra of cellular monolayers were recorded in red-to-near IR region [4] using a sensitive multichannel registration method described in details in [5]. Figure 2a, b present as examples two spectra recorded in (a) enclosed and (b) open cuvettes. Figure 2a$_1$, b$_1$ present the spectra of the same cells after irradiation at 830 nm. Peak intensity ratios of two bands at 760 and 665 nm (I_{760}/I_{665}) were used to characterize every spectrum quantitatively (see gray vertical lanes in Fig. 2). In the case of equal concentrations of the reduced and oxidized forms of the photoacceptor molecule, the ratio I_{760}/I_{665} should be equal to unity. When the reduced forms prevailed, the ratio I_{760}/I_{665} was greater than unity, and it was less than unity in cases where the oxidized forms dominated [4]. Recall that the internal electron transfer within the cytochrome c oxidase molecule causes the reduction of the molecular oxygen via several transient intermediates of various redox states [6].

The magnitude of the I_{760}/I_{665} criterion was 9.5 for spectrum a (Fig. 2a) and 1.0 for spectrum b (Fig. 2b). By this criterion, irradiation of the cells, whose spectrum is marked by a ($I_{760}/I_{665} = 9.5$) caused the reduction of the absorbing molecule (I_{760}/I_{665} for spectrum a$_1$ is equal to 16). Irradiation of the cells characterized by spectrum b also caused the reduction of the photoacceptor, as evidenced by the increase of the I_{760}/I_{665} ratio from 1.0 to 2.5 in spectrum b1. In the spectrum of the cells with an initially more reduced photoacceptor (spectrum a), irradiation caused reduction to a lesser extent ($16/9.5 = 1.7$) than in that of the cells with an initially less reduced photoacceptor (spectrum b). The intensity ratio in this case was $2.5/1 = 2.5$.

So, the irradiation at 830 nm caused changes in the initial absorption spectra of the cellular monolayers, which can be interpreted by the I_{760}/I_{665} band intensity ratio criterion as being due to the reduction of the photoacceptor molecule [4].

The $Cu_A \rightarrow$ heme $a \rightarrow$ [heme $a_3 - Cu_B$] $\rightarrow O_2$ electron transfer within cytochrome c oxidase proceeds rapidly (on a microsecond time scale) between Cu_A and heme a and between the catalytic center [heme a_3-Cu_B] and dioxygen. The only rate-limiting stage in the turnover appears to be the internal electron transfer between heme a and the [heme $a_3- Cu_B$] pair. The reduction of the [$a_3 - Cu_B$] binuclear heme site by the reduced heme a occurs on a millisecond time scale [6]. One can speculate that irradiation intensifies exactly this electron transfer stage within the enzyme. It is quite possible that irradiation makes more electrons available for the reduction of dioxygen in the catalytic center of cytochrome c oxidase (heme $a_3 - u_B$ site). It has long been known that electronic excitation by light stimulates redox processes in organic dyes to intensify electron transfer [7]. This is also true of cytochrome c oxidase [8]. The increase of the availability of electrons can be the crucial result of irradiation in situations when all the four electrons are unavailable for the reduction of dioxygen.

Comparison between the absorption (Fig. 2) and action spectra (one example see in Fig. 1) provided evidence that all bands present in the action spectra were present in the absorption spectra as well [3, 4].

A Discovery of a Novel Light-Activated Mitochondrial Cellular Signaling Pathway

The purpose of these experiments was to demonstrate that a signaling pathway exist between the mitochondria (where the suggested photoacceptor cytochrome c oxidase is located) and cellular membrane. As the experimental approach, we used a

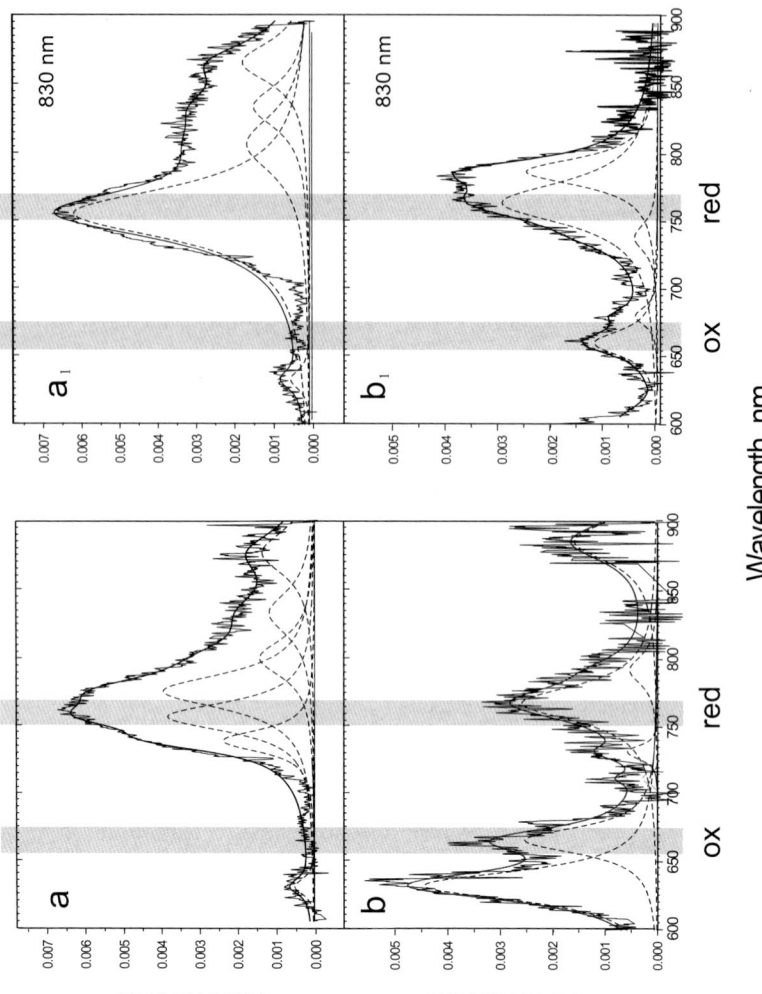

Fig. 2 Absorption spectra of HeLa cell monolayer: (a, b) prior to and (a₁, b₁) after irradiation at 830 nm. a, a₁ – closed cuvette, b, b₁ – open cuvette. Original spectrum, curve fitting (–) and Lorentzian fitting (– – –) are shown as described in [4]. Irradiation procedure is described in [4] in details

modification of the action spectrum associated with the increase of adhesive properties of the cell membrane in the range 600–860 nm. NO donor sodium nitroprusside
(SNP), sodium azide, which both bond to cytochrome c oxidase catalytic center, as
well as ouabain (inhibitor of Na^+, K^+-ATPase in the cell membrane) and amiloride
(inhibitor of Na^+/H^+ antiporter in cell the cell membrane) were added to the cells
before the irradiation. It is in evidence by comparing these spectra (Fig. 3) that these
chemicals have a strong influence on the structure of the intact action spectrum
(Fig. 3a) It was suggested that the putative charge transfer complexes to $Cu_{A_{red}}$
$Cu_{B_{oxid}}$ and (see [9] for explanation) are closed for electron transport in the presence
of azide. There were practically no changes in electron transport connected with the
suggested d–d transitions in $Cu_{B_{red}}$ chromophores (characterized by doublet bands
at 745 and 760 nm), and only a few changes in electron transport occurred,
connected with the suggested d–d transitions in $Cu_{A_{oxid}}$ chromophores in near IR
region (disappearance of the shoulder at 840 nm (Fig. 3c).

Two charge transfer channels putatively to $Cu_{A_{red}}$ and $Cu_{B_{oxid}}$ as well as two
reaction channels putatively connected with d–d transitions in $Cu_{B_{red}}$ and $Cu_{A_{oxid}}$
are reorganized in the presence of NO (Fig. 3b). The action of NO appeared to be
quite different from that of azide, which also reacts directly with the binuclear
catalytic center of cytochrome c oxidase. Azide bridges the heme of cytochrome a_3
and Cu_B permanently, but NO binds to the catalytic center of cytochrome c oxidase
reversibly [10].

The action spectra were recorded also in the presence of two chemicals for which
the plasma membrane is impermeable but react with it (amiloride and ouabain, Fig.
3d, e). Ouabain as an inhibitor of Na^+, K^+-adenosine triphosphatase [Na^+, K^+-
ATPase] and amiloride as an inhibitor of N^+/H^+ exchanger [NHE], both higher
molecular weight substances, cannot react with cytochrome c oxidase (and the
a_3-Cu_B center in particular) in the same way as the small ligands N_3 and NO
radicals. Our action spectroscopy results (Fig. 3) provide evidence that ouabain as
well as amiloride significantly modify the light action spectrum of the increase in
the percentage of attached cells [9–11]. The light action spectrum in the presence of
ouabain was characterized by a single band at 620 nm and by triplet bands in the
near IR region (main peak at 820 nm with shoulders at 800 and 840 nm. Other bands
in the red-to-far red region characteristic of the control spectrum fully disappeared
in the presence of ouabain. This means that a putative charge transfer channel to
$Cu_{A_{red}}$ (characterized by band at 619 nm) and a channel suggested to be connected
with d–d transition in $Cu_{A_{oxid}}$ (band at 820 nm) are working similarly to those of in
the control cells, but both channels to Cu_B (the charge transfer channel characterized by the band at 680 nm and a channel due to d–d transition characterized by
absorption at 760 nm) are closed in the presence of ouabain.

The light action spectrum in the presence of amiloride has the band only in the
near IR region at 831 nm. Noteworthy is the fact that the band at 751 nm in the control
spectrum was not only eliminated but amiloride also caused a slight inhibition of cell
attachment. This result means that only one reaction channel, namely the channel
putatively connected with d-d transition in $Cu_{A_{oxid}}$ chromophore, was working in the

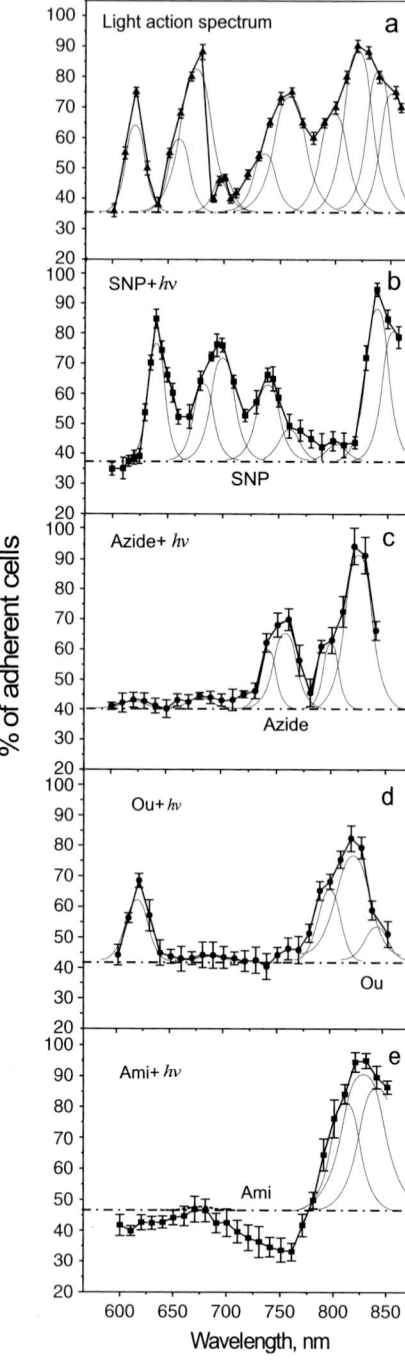

Fig. 3 Action spectra for HeLa cell attachment increase (52 J/m², measurements performed 30 min after irradiation): a – without chemicals added or b, c, d, e – sodium nitroprusside, sodium azide, oubain or amiloride added before the irradiation as described in details in [11]

presence of amiloride. Recall that both ouabain and amiloride in the concentrations used (1×10^{-6} M for ouabain and 1.7×10^{-5} M for amiloride) did not statistically significantly influence cell attachment without irradiation.

The novel light-activated mitochondrial cellular signaling pathway could be classified as a mitochondrial retrograde signaling pathway. Mitochondrial retrograde signaling is a pathway of communication from mitochondria to the nucleus under normal and pathophysiological conditions [12]. Recent experimental results confirm the suggestion that cellular responses to light in red-to-near IR region involve retrograde mitochondrial signaling [13].

Conclusions

Our experiments evidence about existence of a light-activated mitochondrial cellular signaling pathway (mitochondrial retrograde signaling). Cytochrome c oxidase acts as a signal generator after the absorption of light quanta as evidenced by light action spectra. But cytochrome c oxidase can act also as a signal transducer as evidenced by results obtained by using the chemicals (Fig. 3).

One can suggest that nitric oxide, a physiological inhibitor of cytochrome c oxidase that binds to its catalytic center dissociates from the catalytic center when the enzyme is reduced by the irradiation. This event could transiently relieve a block in cytochrome c oxidase that causes a reverse of signaling consequences. First, this suggestion may form a basis for explanation of universal effects of various wavelengths in red-to-near IR region phototherapy as well as of various therapeutic uses of this modality.

References

1. Karu TI, Kolyakov SF (2005) Exact action spectra for cellular responses relevant to phototherapy. Photomed. Laser Surg. 23: 355–361
2. Karu T (1999) Primary and secondary mechanisms of action of visible-to-near IR radiation on cells. J. Photochem. Photobiol. B 49: 1–17
3. Karu T (2007) Ten lectures on basic science of laser phototherapy. Grängesberg Sweden: Prima Books
4. Karu TI, Pyatibrat LV, Kolyakov SF, et al. (2005) Absorption measurements of a cell monolayer relevant to phototherapy: reduction of cytochrome c oxidase under near IR radiation. Photochem. Photobiol. B. 81: 98–106
5. Karu TI, Afanasyeva NI, Kolyakov SF, et al. (2001) Changes in absorbance of monolayer of living cells induced by laser radiation at 633, 670, and 820 nm. IEEE Select. Topics Quantum Elect. 7: 982–988
6. Brunori M, Giuffre A, Sarti P (2005) Cytochrome c oxidase, ligands and electrons. J. Inorg. Biochem. 99: 324–336
7. Terenin AN (1947) Photochemistry of dyes and other organic compounds. Moscow, Leningrad: Acad. Sci. Publ.

8. Marcus RA, Sutin N (1985) Electron transfer in chemistry and biology. Biochim. Biophys. Acta 811: 265–322

9. Karu TI, Pyatibrat LV, Kalendo GS (2004) Photobiological modulation of cell attachment via cytochrome c oxidase. Photochem. Photobiol. Sci. 3: 211–216

10. Karu TI, Pyatibrat LV, Afanasyeva NI (2005) Cellular effects of low power laser therapy can be mediated by nitric oxide. Lasers Surg. Med. 36: 307–314

11. Karu TI, Pyatibrat LV, Afanasyeva NI (2004) A novel mitochondrial signaling pathway activated by visible-to-near infrared radiation. Photochem. Photobiol. 80: 366–372

12. Liu Z, Butow RA (2006) Mitochondrial retrograde signaling. Annu. Rev. Genet. 40: 159–185

13. Schroeder P, Pohl C, Calles C, et al. (2007) Celluar response to infrared radiation involves retrograde mitochondrial signaling. Free Radic. Biol. Med. 43: 128–135

Part I
Mechanisms

Mechanisms

Darrell B. Tata and Ronald W. Waynant

Abstract Our work shows that light exposures of cell cultures results in the production of hydgogen peroxide in direct relation to the dose (J/cm^2) of light given. However, cell growth is stimulated (with respect to unexposed controls) only at low doses, inhibition occurs as dose is increased reaching maximum inhibition near 50 J/cm^2, then at higher doses stimulation of cell growth returns as dose reaches 100 J/cm^2. We believe hydrogen peroxide production is responsible for many of the positive results attributed to laser therapy.

Keywords Light therapy, hydrogen peroxide, stimulation, inhibition, dose, catalase.

This second conference sponsored by the Engineering Conference International (ECI), a group developed by the United Engineering Foundation and Polytechnic University of Brooklyn, on the topic of **Light-Activated Tissue Regeneration and Therapy** was held June 24–29, 2007 in Tomar, Portugal. This topic, also known as "laser therapy" has existed for nearly forty years. It was initially discovered soon after the fabrication of the laser and was practiced using gas lasers for approximately twenty years. "Laser therapy" has acquired a number of additional names, among them "low level light therapy" or (LLLT), "cold laser", "soft laser" and "photobiomodulation." The first of these "Gordon conference style" conferences by this title was held in Hawaii in 2004 (see information on the 2004 conference at http://www.engconfintl.org and http://services.bepress.com/eci/tissue-regen/) and led to a consensus that the mechanism of "laser therapy" was a critical need in order to better focus on the optimization of dosimetry.

At this conference this summer of 2007 our group plans to introduce and discuss a dominant mechanism by which light therapy works. Our group at FDA, USUHS (Dr. Anders) and at Florida Institute of Technology (Dr. Mitra) are in the process of proving that light interacts with the mitochondria of cells and generates. hydrogen

D.B. Tata and R.W. Waynant
U.S. Food and Drug Administration, Center for Devices and Radiological Health, Silver Spring, Maryland

R. Waynant and D.B. Tata (eds.), *Proceedings of Light-Activated Tissue Regeneration and Therapy Conference.*
© Springer Science + Business Media, LLC 2008

peroxide, as implied by Karu [1] and Lubart [2]. They both did not realize, however, that the smallest amount of H_2O_2, approximately 3–15 $\mu mol/10^7$ cells [3], is sufficient to optimize the stimulation of cells leading to the benefit to <u>nearly 100</u> medical problems. Benefits of hydrogen peroxide have been known for hundreds of years, but have largely been abandoned by "modern medicine" in favor of more costly (and profitable) drugs. We believe that this H_2O_2 mechanism does explain, to a large extent, the mystery of laser therapy's success in the treatment of approximately 100 diseases and conditions and suggests an extremely safe way of pinpointing the treatment on the surface of the body. It also calls renewed attention to a cheap, effective drug. A drug naturally produced and used by the body, and that can potentially play a much larger role in health care. It also implies that the use of light (and lasers) is not necessary to utilize this cure. In many cases it can be topically applied with a cotton ball applicator rather than with lasers costing hundreds of dollars. However, using light to generate the drug near the surface of tissue can be done and has some advantages. Light can pinpoint it to the cells that need it. It is safe, non-messy and the curing drug can penetrate deeply through cell layers to seek and destroy diseased cells, stimulate nerve growth and other benefits. Higher concentrations can lead to inhibition of the stimulation, inhibition of cell proliferation or to cell death [4]. Quantification and optimization of light generation of hydrogen peroxide is in progress in our laboratory and may be completed before the conference.

Hydrogen peroxide is well-known in Complementary and Alternative Medicine. The drug is made naturally by the body and is used by bodily defenses. It is cheap, and is claimed to be effective against a wide range of diseases including cancer, diabetes and vascular disease. There are as many similar diseases and ailments for which it has been found effective as there are for light therapy. Of course, we now know why, because they are one and the same. The production of a drug easily transported by the blood explains the systemic observation noted by many observers studying wound healing. When they make two wounds on an animal, one to treat and one for control, they have been confounded by the fact that both respond. The stimulant, H_2O_2 of course, treats both sides when transported to both sides by blood circulation.

Medical uses of hydrogen peroxide have been described in over 7,000 scientific journal articles and popular articles, such as *Hydrogen Peroxide, Medical Miracle* by William Campbell Douglass [5], an attempt to review its benefits for the general public. If we compare the claimed success of hydrogen peroxide therapy, as mentioned by Douglass, and the success of laser (or light) therapy, as documented by Tuner and Hode, as we have shown in Table 1 below, we see that both claim an incredible spectrum of cures not found with many other drugs as well as <u>many similar uses</u>. In fact, if we eliminate the controversial oral and intravenous uses of hydrogen peroxide that light therapy can not easily reach (because of limited light-penetration through skin) to treat, then the similarity of these two treatments is more apparent.

The most prolific researcher on light therapy has been Tiina Karu. Tiina has written over 100 papers on light therapy plus two books already published and a

Table 1 Comparison of diseases cured with hydrogen peroxide with those cured with light theraphy

Cured by hydrogen peroxide	Cured by laser therapy
W.C. Douglass [5]	Tuner and Hode [6]
Illness	*Same illness cured by light therapy – ref in 6*
Allergies	Page 4, 116
Headaches	Pages 86, 146, 182
Herpes simplex	Pages 34, 147
Herpes zoster	Page 215
Asthma	Page 117
Cancer	Pages 130–134
Cerebral vascular disease	Pages 134–135, 136
Periodontal disease	Pages 34, 225
Chronic pain	Page 166
Diabetes type II	Pages 140, 289
Rheumatoid arthritis	Pages 118–125, 414
Shingles	Pages 49, 196
Sinusitis	Page 172
Ulcers	Pages137, 140
Warts	Page 188
Gingivitis	Page 213

third one that will be out soon. She has written a chapter in the book, *Lasers in Medicine* [7] from which I will draw the basic mechanistic pathway.

Light interaction with tissue is important for both the photodynamic treatment of cancer tumors, where exogenous photo-absorbers are added to tissue, and for laser therapy, where the endogenous (natural) photo-absorbers of the mitochondria in almost all cells are used. The photo-absorbers that are used in PDT are described in detail by Marcus in [8]. Karu speculates [7] that cytochrome c oxidase in the mitochondria of cells absorbs visible and near infrared radiation leading to chain enhanced respiratory output of reactive oxygen species (ROS) and ATP production including RNA and protein synthesis. Karu describes four possible pathways following absorption, one of which produces H_2O_2. While we believe this pathway is the one most responsible for the curative effects of light therapy, we have reason to believe that other smaller effects may also take place and be activated by lower energy fields. Experiments to verify the exact path and the range of sources capable of generating hydrogen peroxide and the amount generated per dose are currently underway.

The significance of these results is astounding in several ways. First, after forty years of research, no complete explanation of the mechanism that fits the observations noted by a wide spectrum of researchers has been offered. This is not to discount the tremendous work done by Tiina Karu and Rachel Lubart that have unraveled quite a bit of data. However the work by Burdon [3, 4] and Davies [9] helps solidify the full picture. Now armed with their information, this enables us to focus on the presence of the correct drug, measure its concentration as generated by the correct wavelengths and doses and to observe cell proliferation as produced by weaker concentrations and inhibition as observed by greater concentrations. The presence of a drug is consistent with a systemic effect noticed by many researchers especially in regard to wound healing. The ability to heal difficult wounds with laser therapy is understandable in view of hydrogen peroxide's well-known value as an antiseptic. The ability of hydrogen peroxide to stimulate growth of nerves, such as in spinal cords as observed by Byrnes [10], is understandable due to hydrogen peroxide's ability to penetrate through numerous cell layers thus not requiring light to transmit to great depths, but to simply generate H_2O_2 near the surface in sufficient quantity to soak through to the spinal cord. Similar results might be expected by simply applying H_2O_2 with a cotton swab.

Perhaps the second area of significance is the fact that generating hydrogen peroxide with light is that laser therapy's success in healing a wide spectrum of diseases compares closely to the broad spectrum of success observed by practitioners advocating broad use of hydrogen peroxide for similar illnesses as was mentioned in Table 1. William C. Douglas claims in his book [5] that the conventional practices of using hydrogen peroxide, i.e. either by ingesting very small quantities (one to three drops of H_2O_2 dissolved in a glass of water) by mouth or by dripping dilute quantities of H_2O_2 through the arteries, can produce fantastic success with plaque removal and with other illnesses. Campbell also discusses his own work with AIDS in Africa where the low cost of the drug could allow treatment of many more patients than the more costly western drugs. While the stories told in Campbell's book are for a general lay audience, other doctors have published thousands of success stories that are currently relegated to the complementary and alternative medical field. In view of the successes generated by laser therapy, a much greater number of successes can now be added to those of complementary and alternative medicine. This may mean that another look should be taken at hydrogen peroxide. In fact, each of these successful applications needs to be verified in animal models and taken through the rigorous FDA trials necessary to see if they are safe and effective for human use.

Dr. Waynant and his group have teamed with Juanita Anders, Co-Chair of the conference, for projects over the last five years first directed toward DARPA's Persistence in Combat program which funded studies in how light therapy might aid the warfighter and eventually led to studies of light treatment of spinal cord injuries. While these studies helped to give us experience and knowledge in light therapy, both could have been optimized with our current knowledge of the mechanism. During this initial phase with DARPA we perfected our measurements of dose, made measurements through skin to determine light transmission and determined that 810 nm light seemed to be a good wavelength to do research. This wavelength was close to the cross-over of oxy- and deoxy-hemoglobin where absorption was stable during the respiratory cycle of animals. We also did experiments with spinal cord injuries in rats [9] and found that severed nerves were stimulated to grow beyond the point of the severing when light was applied near the point of the severing.

Later our collaborative research focused on cell cultures. We have done numerous studies of the effects of light on a multi-linage progenitor cell line purchased from Bio-Whitaker. This work made us aware of the effects of light on cell cultures. We also learned that light had an effect similar to some of the factors previously seen to enhance cell growth. From there our work was centered on understanding the effects of light on cell growth and to measurement of the small amounts of H_2O_2 produced. This work continues. Dr. Darrell Tata joined us in these new studies and he will summarize our current work at the conference. Dr. Tata has a strong background in cancer and has studied the slowing of cancer cell proliferation caused by light therapy. Samples of this data are shown in Figs. 1 and 2.

These two figures form a strong indication that hydrogen peroxide is produced by light at 1552 nm. We believe that a broad spectrum of electromagnetic energy produces hydrogen peroxide—from ultraviolet through the mid-infrared, which we have already measured—and likely much broader, e.g. from x-ray through deep IR which we hope to measure by the start of the conference. The use of the catalase enzyme as a means of assessing the production of hydrogen peroxide functions through the reaction below.

$$2H_2O_2 + \text{catalase} \rightarrow 2H_2O + O_2 + \text{catalase} \tag{1}$$

Figure 2 shows a reduced inhibition of cancer cell proliferation than does Fig. 1 due to the addition of the catalase enzyme. Thus we conclude the presence of hydrogen peroxide and believe its stimulatory effects are the cause of the benefits of light therapy.

We believe that our announcement of the mechanism at this conference plus the results of our current research to quantify and verify quantities of H_2O_2 as a function of dose will have a further impact on all researchers. They will now be able to focus

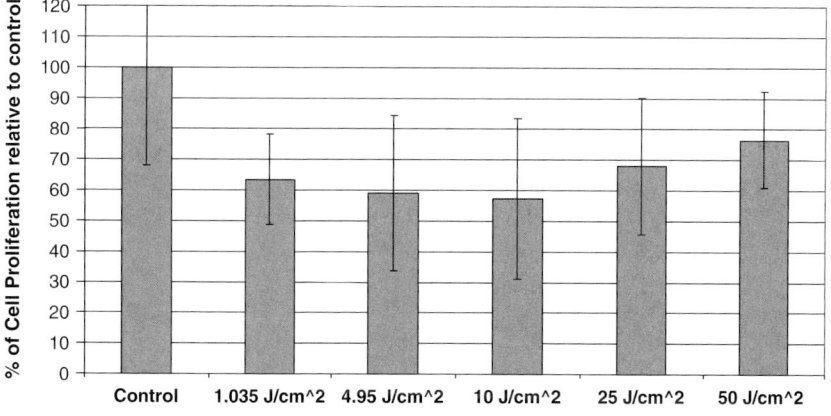

Fig. 1 Therapeutic effects of near IR light on human malignant brain cancer cells

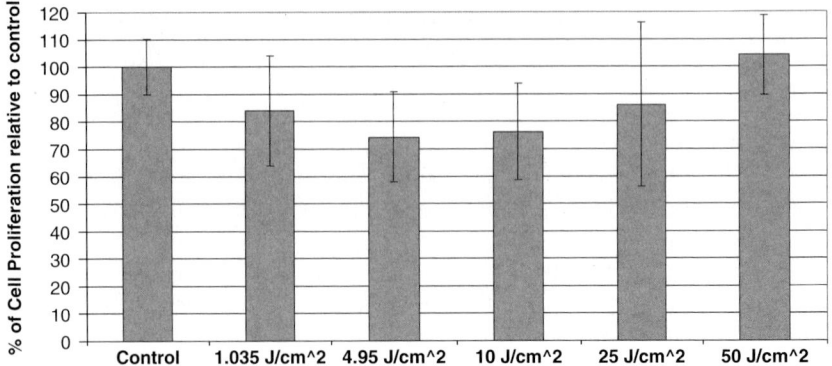

Fig. 2 The same cell line shows less inhibition due to the reduced formation of hydrogen peroxide produced in the presence of catalase

on a known drug pertinent to the disease they are studying. Just as our current results were amplified once the presence of H_2O_2 was confirmed, their results will also be amplified leading to a tremendous advancement in the entire field. Dr. Tata will review more of the studies that we have done with malignant cells in culture.

This is an important time for this conference. Several decisions are being considered that affect the health of the American people. The Committee on Medicare and Medicade is proposing to disallow payment for light therapy as a treatment for diabetes. In addition, the FDA is concerned about inappropriate use of hydrogen peroxide, i.e. the controversial ingestion of small amounts or intravenous use of it. An international conference on the uses of light therapy will draw many attendees from Europe and Asia. Perhaps scientific discussions involving mechanisms of light therapy will result in reasonable procedures leading to trials to exploit the benefits of a light induced drug such as H_2O_2 to bring inexpensive cures to a wide spectrum of diseases.

For this conference to accomplish its goals of significantly advancing effective treatments from light therapy to a large number of illnesses, we have organized eleven sessions. We have invited all of the session chairs and speakers listed with the expectation of providing stimulating talks that will be scientifically beneficial to the advancement of this field. We have high quality session chairs who lead the field and they have invited high quality speakers to the conference. The session chairs have summarized the status of their session and the speakers have provided a summary of their talks that they will discuss with the audience. We hope that the breakthrough with the mechanism to be presented at this conference will enable those attending this meeting. to make rapid progress. We believe a high quality program and a high quality audience will be capable of rapidly advancing this field. We hope to capture the beginning of a new surge of progress in this book and to continue a valuable medical technology in future editions of the book.

References

1. Karu, T., "Primary and secondary mechanisms of action of visible to near-IR radiation on cells," **Journal of Photochemistry Photobiology B: Biology**, **49**, pp. 1–17 (1999).
2. Lubart, R., Wollman, Y., Friedman, H., Rochkind, S., Laulicht, I., "Effects of visible and near–infrared lasers on cell cultures," **Journal of Photochemistry Photobiology**, **12**, pp. 305–310 (1992).
3. Burdon, R.H., "Superoxide and hydrogen peroxide in relation to mammalian cell proliferation," **Free Radical Biology and Medicine**, **18**, pp. 775–794 (1995).
4. Burdon, R.H., "Control of cell proliferation by reactive oxygen species," **Biochemical Society Transactions**, **24**, pp. 2–5 (1996).
5. Douglass, W.C., *Hydrogen Peroxide: Medical Miracle*, Second Opinion Publishing, Box 467939, Atlanta, GA 31146-7939 ISBN 1-885236-07-7 (1996).
6. Tuner, J., Hode, L., *Laser Therapy: Clinical Practice and Scientific Background*, Spjutvagen 11, Prima Books, AB, Grangesberg, Sweden, 77232 (2002).
7. Karu, T. "Low-Power Laser Effects," Chapter 7 in *Lasers in Medicine*, R.W. Waynant, Ed., CRC Press, Boca Raton, FL 33431, ISBN 0-8493-1146-2 (2001).
8. Marcus, S.L., "Lasers in Photodynamic Therapy," Chapter 10 in *Lasers in Medicine*, R.W. Waynant, Ed., CRC Press, Boca Raton, FL 33431, ISBN 0-8493-1146-2 (2001).
9. Davies, K.J.A., "The broad spectrum of responses to oxidants in proliferating cells: A new paradigm for oxidative stress," **IUBMB Life**, **48**, pp. 41–47 (1999).
10. Byrnes, K., Waynant, R., Ilev, I., Wu, X., Barna, L., Smith, K., Heckert, R., Gerst, H. and Anders, J., "Light promotes regeneration and functional recovery and alters the immune response after spinal cord injury," **Lasers in Surgery and Medicine**, **36**, pp. 171.

Near-IR Picosecond Pulsed Laser Induced Suppression of Metabolic Activity in Malignant Human Brain Cancer: An In-Vitro Study

Darrell B. Tata and Ronald W. Waynant

Abstract The role of low light intensity in suppressing metabolic activity of malignant human brain cancer (glioblastoma) cell line was investigated through the application of a 1,552 nm wavelength pulsed picosecond laser. Human glioblastomas were grown in T-75 flasks and were utilized when the cells were 50–70% confluent and thereafter transferred into 96 well plates and exposed in their growth culture medium with serum under various energy doses (i.e., fluence) ranging from 0.115–50 J/cm^2. All exposure doses were reached with an average intensity of 0.115 W/cm^2; 25 kHz repetition rate with 1.6 μJ per pulse; pulse duration = 2.93 ps. The glioblastomas exhibited a maximal decline in the metabolic activity (down 50–60%) relative to their respective sham exposed control counterparts between the fluence dose values of 5.0–10 J/cm^2. The cellular metabolic activities for various treatment doses were measured through the colorimetric MTS metabolic assay 3 days after the laser exposure. Interestingly, the metabolic activity was found to return back to the sham exposed control levels as the fluence of exposure was increased up to 50 J/cm^2. Addition of (the enzyme) Catalase in the growth medium prior to the laser exposure was found to diminish the laser induced metabolic suppression for all fluence treatment conditions, thus suggesting a functional role of H_2O_2 in the metabolic suppression. In view of this evidence, a hypothesis is formulated which attributes the classical biphasic response, in part, to the light induced production of H_2O_2. Furthermore, it was observed that if the glioblastoma cells were allowed to reach 100% confluency within the T-75 flasks the characteristic laser induced metabolic suppression was found to be severely abrogated. Exploratory steps were also undertaken to maximize the suppression in the metabolic activity through repetitive laser dose of exposure every 24 hours for 3 consecutive days. In addition, the efficacy in the metabolic suppression of the 1,552 nm pulsed laser was also compared to a continuous wave broad band continuous wave heating lamp source channeled through a fiber-optic bundle with identical intensity of exposure. Taken

D.B. Tata and R.W. Waynant
U.S. Food and Drug Administration, White Oak, Maryland

R. Waynant and D.B. Tata (eds.), *Proceedings of Light-Activated Tissue Regeneration and Therapy Conference.*
© Springer Science + Business Media, LLC 2008

together, our findings reveal that near-IR low level light exposures could potentially be a viable tool in reducing the metabolic activity of cancers; however, due to the cellular "biphasic" response to the non-ionizing irradiation, further research needs to be undertaken to determine exposure parameters which would optimize metabolic and cellular growth suppression in-vivo.

Keywords: Human brain cancer (glioblastoma), inhibition, cell cultures, catalase, near infrared.

Introduction

Wavelength, fluence and intensity have been noted as important light exposure parameters through previous investigations playing an important role in biomodulations which bring about various biological effects [1]. A feature in low level light exposure, as noted through past research, is a "biphasic" biological response in intensity and most notably in the light energy dose, i.e., the fluence [2]. Consequently, for a specified optical or infra-red wavelength and for a specified low intensity there exists an optimal fluence value for light exposure to produce a maximum modulation for a specified biological response/effect. A few noteworthy bio-effects due to low level light exposure which have been reported in the literature include: (i) modulation in gene expressions [3], (ii) increase in the intracellular calcium levels [4], (iii) increase in the mitochondrial metabolic activity [5] and in the enhanced production levels of ATP [6], (iv) with a concomitant increase in cellular proliferation [7].

Findings from Tina Karu's group on correlations of light irradiated (cellular proliferation) action spectra and the absorption spectra of intact (human cervical cancer) HeLa cells has suggested the hypothesis that absorption of light by certain chromophores within cells do bring about light induced modulations in cellular proliferation. Karu's investigations had found that the intact cell's visible absorption spectrum resembled the absorption spectrum of cytochrome c which is an integral part of the respiratory chain found in the mitochondrion [8]. This observation had led to a hypothesis that selective chromophores within the mitochondria are responsible in light absorption. Additional investigations found that the light absorbed by the mitochondria sped up the electron shuttle within its inner membrane compartments and consequently led to enhancement in the ATP production. As a natural consequence to the enhancement of the respiratory chain activity is the enhancement of the interplay between electrons and molecular oxygen within the mitochondria [9]. Thus, enhancements in the concentration levels of reactive oxygen species (ROS) such as the superoxide anion O_2^-, and H_2O_2 have been observed to be elevated due to light absorption within the mitochondria. Lubart and coworkers have found elevated concentrations of H_2O_2 due to white light exposures from cardiomyocytes [4]. In past literature, enhancement of ROS levels have been shown to activate selective transcription factors which would "turn on" or "turn off" genes and their protein products within the cells [10, 11]. Different cell types have varying degree of responses at their optimal low light level exposure settings.

Reports on low level light bio-effects on cancer cells are sparse and have primarily entailed monitoring proliferation/mitotic rates of cancers with a broad range of discrete optical wavelengths [12]. We report in this communication the biphasic role of near infra-red low light intensity in suppressing the metabolic activity of the human malignant glioblastoma. The cellular metabolic activities for various treatment doses were measured through the colorimetric MTS metabolic assay. Addition of the enzyme catalase in the growth medium prior to the laser exposure was found to partially block the laser induced metabolic suppression for all fluence treatment conditions. This finding has led us to formulate a hypothesis on the role of light induced H_2O_2 in bringing about the light induced biphasic metabolic response.

Materials and Methods

Cell Line Maintenance

Human malignant (brain cancer) glioblastoma was purchased from American Type Culture Collection (Rockville, MD) and grown in monolayer and maintained in T-75 flasks under incubation conditions of 5% CO_2 at 37°C. The glioblastoma cells were maintained in ATCC formulated DMEM/F12 growth medium with 10% of fetal bovine serum (FBS) and 50 units/ml of penicillin and streptomycin (Pen/Strep) antibiotics.

Cell Preparation

When the adherent glioblastoma cells reach 50–70% confluence within the T-75 flasks, the glioblastomas were trypsinized and brought into suspension. The cells were spun down and the (trypsin) supernate was discarded. The cells were re-suspended in fresh growth medium with 10% FBS and 50 units/ml of Penicillin/Streptomycin at an initial working concentration 75,000 cells/ml. The cell suspension was then transferred into single well's of the 96 well plates with a transfer volume of 0.2 ml or 15,000 cells seeded per selected well. The cells were seeded into every other well in order to ensure no possible overlap in the laser light exposure. The cells within the 96 well plates were returned back into the incubator for approximately 16 hours before the laser treatments.

Cell Exposure

Laser exposure set-up is shown in the schematic diagram in Fig. 1 below. A near infra-red 1,552 ± 0.5 nm wavelength pulsed picosecond laser, manufactured by Raydiance, Inc., was utilized in exposing cells within the selected wells of the 96 well polystyrene (Perkin Elmer, Inc. View plate – 96, #6005182) plates (well

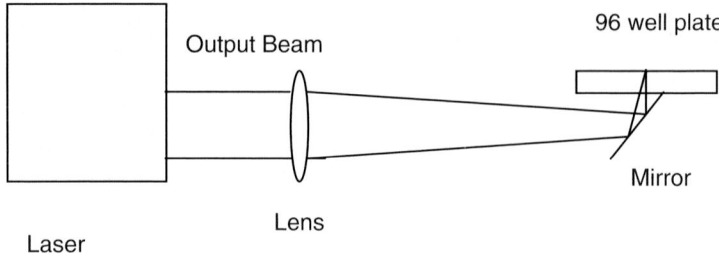

Fig. 1 Schematic diagram of the experimental set-up

diameter of 6.5 mm) with fixed laser parameters of: 2.93 ps pulse width, 25 kHz repetition rate, and with 1.6 μJ pulse energy. The measured average power delivered at the underside of the well = 30.0 mW (25% of laser energy was lost due to external optics) with a spot size of 5.75 mm in diameter (measured through knife – edge technique), yielding an average laser intensity of 115 mW/cm^2.

Each alternating column of the 96 well plate (with the exception of column#1 which served as the control/sham exposed condition) received a pre-determine fluence dose of exposure for those wells within the column which contained cells. Fluence levels were varied by keeping the aforementioned laser parameters fixed and varying the irradiation time of exposure. The selected times of laser exposure ranged from 1 second to 7 minutes and 15 seconds with the corresponding fluence range from 0.115 to 50 J/cm^2. Experiments were done in quadruplicates for each fluence value. All exposures were done at room temperature (~23°C), and the average duration that the 96 well plates were left out at room temperature for laser treatment was approximately 2 hours. Upon completion of the experiment, the 96 well plates were immediately returned to the incubator.

The efficacy in the metabolic suppression of the 1,552 nm pulsed laser was also compared to a continuous wave broad-band lamp source (Tungsram, Inc.). The light emission was channeled through a fiber-optic bundle (5 mm in diameter) and the fiber bundle was position in the center at the underside of the well and the cells were exposed with an identical intensity of exposure of 0.115 W/cm^2 (as measured through the Spectra-Physics, Inc. model #407A power meter). Figure 2 below exhibits the normalized emission characteristics of the broad band lamp source after passing through the fiber-optic bundle (at the site of exposure) measured through Ocean Optics, Inc. high resolution spectrophotometer HR4000 CG UV-NIR.

Measuring Cellular Metabolic Activity

The metabolic response of glioblastoma cells to various laser fluences were assessed with a non-radioactive colorimetric cell metabolic tetrazolium compound (MTS) assay (Promega, Madison, WI) in four independent (control and exposed) replicates, 3 days after the laser treatments. On the day of measurement, the 96-well

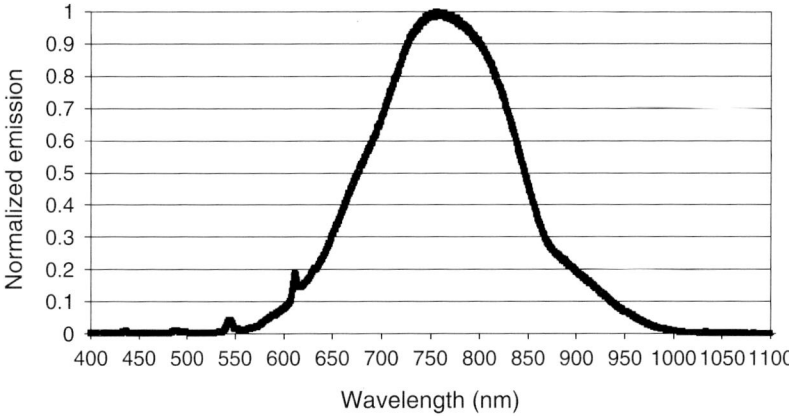

Fig. 2 Normalized emission through a fiber bundle of a broad band lamp source. Location of the central peak: 757 nm

plates were removed from the incubator and 20 μl of the MTS solution was added to each cell containing well. Thereafter, the plates were immediately returned to the incubator for a 2 hour incubation period.

Functionally, the MTS readily permeates through the cell membrane and is metabolized and is converted into formazan by living cells. Conversion into formazan induces a maximum change in absorption at 490 nm wavelength.

Two hours after the addition of MTS, absorption measurements were made at 490 nm with a 96 well plate reader (Perkin Elmer 1420 Multilabel counter: VICTOR[3]), and the average absorbance value at 490 nm of laser treated cell's metabolic activity was computed with standard deviations and compared to the sham laser exposed average absorbance value with its standard deviation. The percentage of laser treated cell's metabolic activity was computed relative to the sham exposed metabolic activity.

Results and Discussion

The human glioblastoma cells exhibited a decline in metabolic activity relative to their control (sham exposed) counterparts between the fluence values of 0.115–10 J/cm^2. Maximal suppression in metabolic activity was noted between 5 to 10 J/cm^2. See Fig. 3 below. As the near infrared laser light dose was further increased beyond 10 J/cm^2 the metabolic activity was found to return towards the control levels. Thus, due to the biphasic response characteristics in the metabolic activity we deduce that a window of opportunity in the fluence level at which maximum suppression occurs. Although the bulk growth medium temperature immediately after the laser irradiation did not appreciably change ($\Delta T \sim 0.5°C$), one plausible and speculative mechanism for the metabolic activity trend to return towards the control condition could be attributed to the intra-cellular temperature

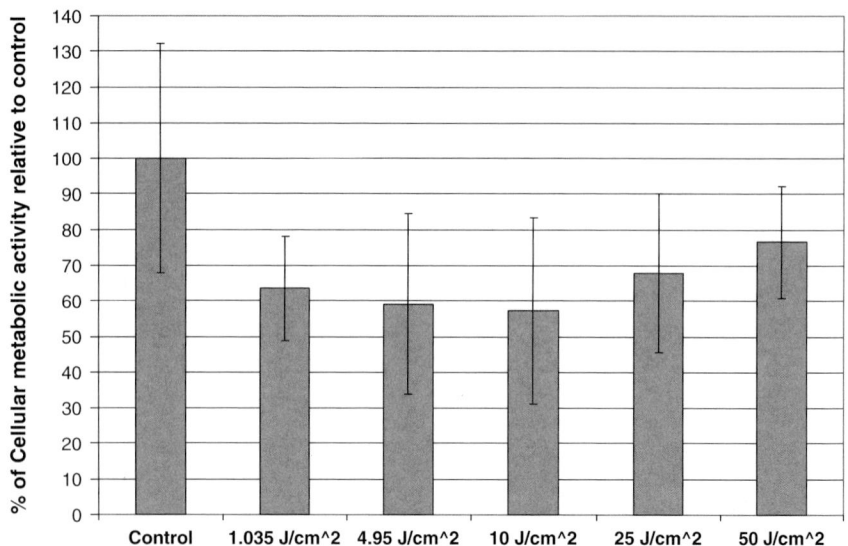

Fig. 3 Malignant human glioblastoma cellular metabolic activity measured 72 hours after laser exposure. Average ± SD, n = 4

levels which may potentially reach hyperthermic levels in which case the cellular constitutive heat shock proteins would be immediately called into action to protect the integrity of cellular and mitochondrial proteins and the DNA from thermal damage. Alternatively, elevated and sustained levels of long lived reactive oxygen species, such as hydrogen peroxide, could be expected to mediate similar pro-active responses from the constitutive heat shock proteins.

In Fig. 4 below, addition of the H_2O_2 scavenger enzyme catalase in the growth medium prior to the laser exposures was found to diminish the extent of the laser induce metabolic suppression, thus implicating a role of H_2O_2 in diminishing cellular metabolic activity. In view of the fact that the catalase scavenger activity must reside outside of the cell membrane (since catalase is a fairly large protein and is not known to permeate through the cell membrane), we hypothesize that the mode of action for H_2O_2 in part, to affect the cellular metabolic activity is likely to be through the H_2O_2 induced oxidation of (as yet unidentified) cellular membrane biochemical signal transduction proteins on the outer side of the cell membrane.

Additionally, minimal near infra-red light induced changes in the metabolic activities were observed if the cells were allowed to be grown fully confluent within the T-75 flasks. Figure 5 below gives an example of the minimal changes in the metabolic activity for the case of the glioblastomas which were grown up to 100% confluent within the T-75 flasks. Thus, the near infra-red light induced changes in the metabolic activities were found to be robust when the cells are not in the static cell cycle (G0) phase.

An important question which needs to be addressed in the pre-clinical/clinical setting is that of the light dosing scheme: Would repetitive light exposures with a

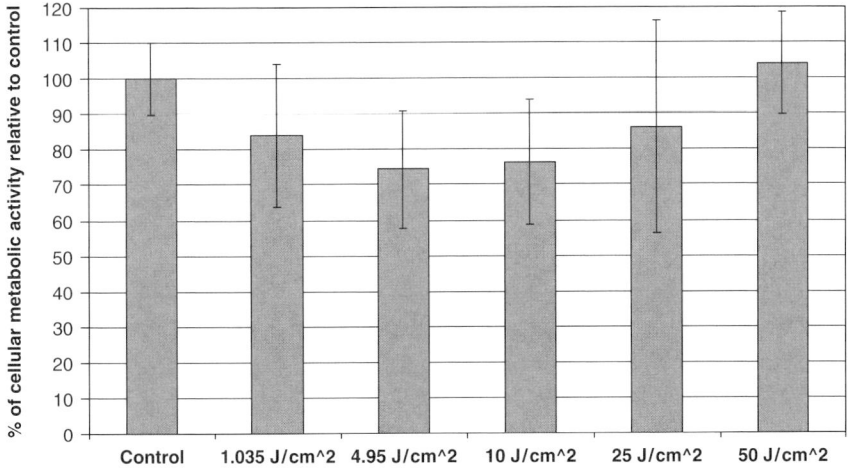

Fig. 4 Metabolic activity measured 72 hours after laser exposure on human glioblastoma in presence of [Catalase] = 2,000 units/well. Average ± SD, n = 4

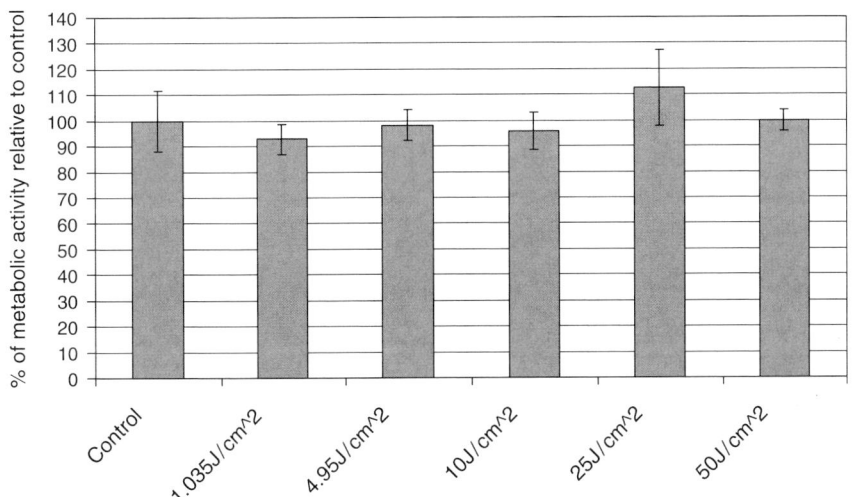

Fig. 5 Fully confluent glioblastoma cellular metabolic activity measured 72 hours after laser exposure. Average ± SD, n = 4

specified fluence dose, separated by a specified time interval yield a greater photo-modulation in a bio-effect than a single exposure dose? To address this question, preliminary experiments were undertaken with a goal to maximize suppression in the metabolic activity through repetitive laser exposures every 24 hours for 3 consecutive days. As evidence in the preliminary findings in Fig. 6, repetitive dosing every 24 hours at a specified fluence value yielded less reduction in the

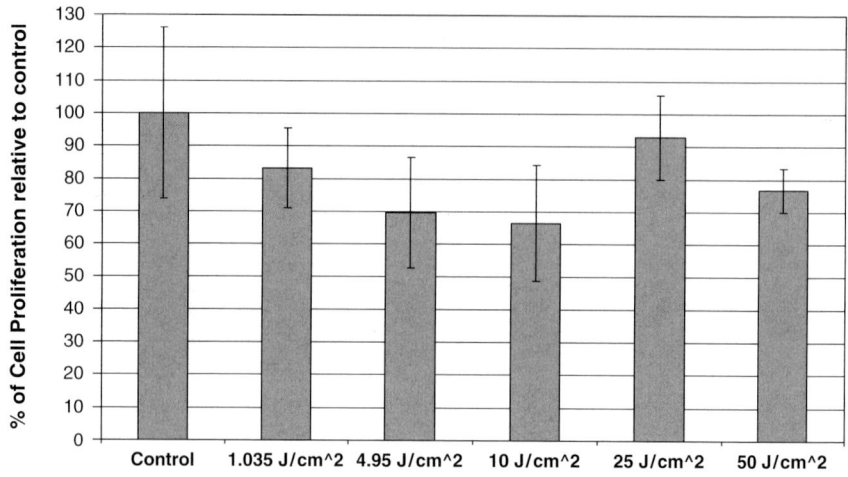

Fig. 6 Glioblastoma growth characteristics relative to control. Measured 72 hours after initial laser dose. Glios were dosed every 24 hours with a total of three doses

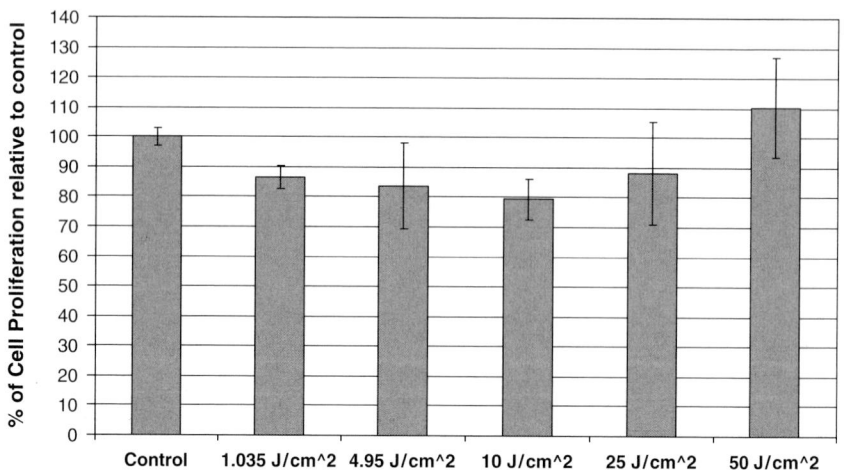

Fig. 7 Malignant human glioblastoma cellular metabolic activity measured 72 hours after the broad band lamp exposure through a fiber bundle. Average ± SD, n = 4

metabolic activity then a single fluence of exposure, however, the timing between the dosing may play a significant role in the outcome of the results, as this has not yet been investigated.

The efficacy in the metabolic suppression of the 1,552 nm pulsed laser was also compared to a continuous wave broad band lamp source (centered near 760 nm) channeled through a fiber-optic bundle with an identical intensity of exposure upon the human glioblastoma cells. Figure 7 below reveals similar characteristic biphasic tendency of suppression in the metabolic activity for the broad band light source, although not as pronounced as the pulsed picosecond laser source operating at 1,552 nm wavelength.

Conclusion

These in-vitro findings have suggested further scrutiny into pre-clinical tumor bearing animal models. Of particular interest are the possible light induced growth modulations in aggressive surface cancers, such as the melanomas, due to their convenient accessibility to sunlight and to the conventional and non-conventional low level light exposures.

References

1. Tuner, J., Hode, L. Laser Therapy – Clinical Practice and Scientific background. Prima Books AB, Grangesberg, Sweden. Chapter 3: Biostimulation, Chapter 4: Medical Indications, 2002.
2. Hamblin, M., Demidova, T.N., Mechanisms of Low Light Therapy. Proc. SPIE. Vol. 6140, pp 1–12, 2006.
3. Zhang, Y., Song, S., Fong, C.C., Tsang, C.H., Yang, Z., Yang M. cDNA Microarray Analysis of Gene Expression Profiles in Human Fibroblast Cells Irradiated with Red Light. J. Invest. Dermatol., Vol. 120, pp 849–857, 2003.
4. Lubart, R., Lavi, R., Friedmann, H., Rochkind, S. Photochemistry and Photobiology of Light Absorption by Living Cells. Photomed Laser Surg., Vol. 24, pp 179–185, 2006.
5. Yu, W., Naim, J.O., McGowan, M., Ippolito, K., Lanzafame, R.J. Photomodulation of oxidative metabolism and electron chain enzymes in rat liver mitochondria. Photochem Photobiol. Vol. 66, pp 866–871, 1997.
6. Passarella, S. Helium – Neon Laser Irradiation of Isolated Mitochondria. J Photochem Photobiol B., Vol. 3, pp 642–643, 1989.
7. Hawkins, D., Abrahamse, H. Biological Effects of Helium – Neon Laser Irradiation on Normal and Wounded Human Skin Fibroblast. Photomed Laser Surg. Vol. 23, pp 251–259, 2005.
8. Karu, T.I., Kolyakov, S.F., Exact Action Spectra for Cellular Responses Relevant to Phototherapy. Photomed Laser Surg. Vol. 23, pp 355–361, 2005.
9. Lubart, R., Eichler, M., Lavi, R., Friedman, H., Shainberg, A. Low – Energy Laser Irradiation Promotes Cellular Redox Activity. Photomed Laser Surg. Vol. 23, pp 3–9, 2005.
10. Davies, K.J.A. The Broad Spectrum of Responses to Oxidants in Proliferating Cells: A New Paradigm for Oxidative Stress. IUBMB Life, Vol. 48, pp 41–47, 1999.
11. Nemoto, S., Takeda, K., Yu, Z., Ferrans, V.J., Finkel, T. Role of Mitochondrial Oxidants as Regulators of Cellular Metabolism. Mol. Cell Biol. Vol. 20, pp. 7311–7318, 2000.
12. Tuner, J., Hode, L. 2002, Laser Therapy - Clinical Practice and Scientific background. Prima Books AB, Grangesberg, Sweden. Biostimulation, Chapter 4.1.7: Cancer 130–134, and see table listing in Chapter 11 on page 349.

Part II
Wound Healing

Combined 660 and 880 nm Light Improves Healing of Recalcitrant Diabetic Ulcers

Initial Pilot Study

Debora G. Minatel, Marco Andrey C. Frade, Suzelei C. Franca, Gil L. Almeida, and Chukuka S. Enwemeka

Abstract Diabetes not under control results in many additional health risks including skin ulcers that are extremely hard to cure. Laser therapy has been seen to be effective. This study shows that two wavelengths, 660 and 880 nm, when used together seems to be more effective than other forms of light therapy. This hypothesis will be tested with larger populations.

Keywords: Diabetic ulcers, clinical, 660 and 880 nm wavelengths of laser light used together.

Introduction

Diabetes poses a significant health concern in many parts of the world, including Brazil and the United States, which have high cases of ulcers resulting from chronic diabetes. In the US alone, more than one billion dollars is expended to treat chronic healing-resistant diabetic ulcers each year [1]. As the incidence of diabetes continues to rise world-wide, other nations are expected to see a similar trend in the cost of treating diabetic ulcers. There have been significant advances in ulcer care within the last 20 years, but, to date, there is no cure for healing-resistant diabetic ulcers. Thus, finding a cure for healing-resistant ulcers associated with diabetes will be of immense benefit to society. In recent times interest has focused on the use of polypeptide growth factors to stimulate wound repair [2, 3]. Whereas, these growth factors have been shown to promote healing of experimental wounds in animal models, their use in human cases of ulcers is relatively new, and results have been

C.S. Enwemeka
School of Health Professions, Behavioral and Life Sciences, New York Institute of Technology, Old Westbury, NY 11568–8000, USA, e-mail: Enwemeka@nyit.edu

R. Waynant and D.B. Tata (eds.), *Proceedings of Light-Activated Tissue Regeneration and Therapy Conference.*
© Springer Science + Business Media, LLC 2008

disappointing [4]. The cost of growth factor therapy is prohibitive, and their potential side effects remain little known.

Emerging data suggest that certain wavelengths of light, $\lambda = 600–1,000$ nm [5, 6], promote the repair of skin [7–16], ligament [17–20], tendon [21–24], bone [17, 25–28], and cartilage [17, 29–33] in experimental animals, as well as human wounds and ulcers of varying etiologies [34–37]. Phototherapy is thought to enhance wound healing by promoting cell proliferation [38–48] collagen synthesis and the formation of granulation tissue [49–52], advancing the formation of type I and type III procollagen specific pools of mRNA [53], enhancing ATP synthesis, and by activating lymphocytes, and increasing their ability to bind pathogens [9, 54–55].

However, other reports [56–62] suggest the contrary. Consequently, clinicians are skeptical to embrace laser phototherapy as an evidence-based treatment. Since phototherapy is relatively safe, inexpensive, non-invasive and has the potential to promote ulcer repair, minimize infection, and restore function in patients with healing resistant wounds, we designed a completely randomized double-blind clinical study to ascertain its true effect on healing of diabetic ulcers that failed to respond to any other available therapies. We selected as our initial test cases, patients with diabetic ulcers that have been slow- or non-healing for at least four weeks. This initial report is prompted by the vivid beneficial effects obtained from our first set of patients.

Methods

We used a double-blind completely randomized clinical trial to determine if treatment with a combination of 660 and 880 nm light improves healing of recalcitrant slow or non-healing diabetic leg ulcers. Subjects were recruited from the ulcer clinic of the Department of Dermatology, University of São Paulo Medical School, Ribeirão Preto, São Paulo State, Brazil. Eight of ten patients that met our inclusion and exclusion criteria complied fully with the protocol and were treated either with Probe One (placebo treatment) or Probe Two (real treatment) until their ulcers healed fully or for a maximum of 12 weeks.

To achieve treatment with real or placebo probe, a Dynatron Solaris® 705 light therapy (LT) machine (Dynatronics, Corporation, Salt Lake City, UT) was customized to permit the use of two identical applicators, labeled Probes One and Two. One of the probes was designed to emit the full compliment of light from thirty-two 15 mW superluminous 880 nm diodes (SLDs) and four 5 mW 660 nm SLDs, making a total of 36 diodes interspersed in a 5 cm^2 cluster. Because 660 nm light is visible red light and the infrared 880 nm light, invisible infrared radiation, the 880 nm SLDs in one probe were disabled so that it produced just 660 nm light. To further reduce the amount of light emitted by this probe, the outer three diodes were also disabled so that only the centrally located red diode produced light.

Ulcers were cleaned with physiological saline, dried and then treated twice a week. Following phototherapy, the sores were dressed with sulfadiazine covered with gauze and bandage. Given the 100 mW cm^{-2} irradiance of the treatment device, each spot size was treated for 30 seconds, yielding a fluence of 3.0 J/cm^2. Larger ulcers, having more square centimeters, required more time as treatment was applied in the contact mode, sequentially from one spot size to adjoining areas, until the entire ulcer area was completely treated. Treatment was delivered in the contact mode; therefore, to avoid cross-contamination, each ulcer was covered with a clear transparent thin film of plastic verified to have 100% light transmission.

Healing of each ulcer was evaluated by digital photography, using a Sony$^®$ DSC-P100. The camera was custom fixed to an aluminum base with a 30 cm ruler attached perpendicularly to one end of the fixture. The fixture enabled standardized photography, making it easy for pictures to be taken vertically from a regular distance of 30 cm every week (Fig. 3). Ulcer area and the area of granulation in each photograph were then measured using Image J$^®$ software. For this analysis, the edge of each ulcer was first delineated; then, the computer automatically computed the area of each ulcer. The computed area was then ratioed to the scale of the metric ruler that was photographed with each ulcer, Fig. 1.

Results

Ten patients with healing resistant ulcers associated with type II diabetes consented to participate in the study. Four each were assigned to treatment with Probe One or Two. The remaining two patients had multiples ulcers. One ulcer in each patient

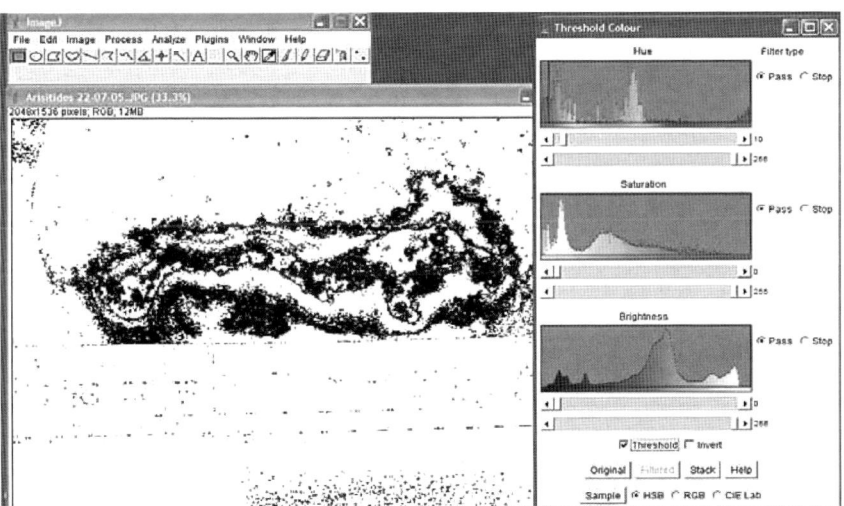

Fig. 1 An example of a computer generated image of an ulcer obtained with the Image J software software

was treated with Probe One; the others were treated with Probe Two (as detailed below; see Patients Four and Eight) in order to pilot test the potential effect of treating the same patient with the two probes simultaneously. Two patients assigned to treatment with Probe One were lost to follow up as they stopped showing up for treatment after two and four weeks respectively. Thus, eight patients completed the study one of whom (Patient Number Six) developed complications and was referred back to the attending physician as detailed below. Table 1 is a summary of the clinical characteristics of the patients.

Absolute (Ar) and percent absolute (%Ar) reduction of ulcer size were 2.9 + 1.6 and 54.9 + 31.7 respectively for treatment with Probe One, and 12.2 + 3.7 and 89.0 + 26.8 respectively for five treatment with Probe Two. Similarly, absolute tissue granulation (Ga) and percent tissue granulation (%Ga) were 2.2 + 1.2 and 65.0 + 37.9 respectively for Probe One. The corresponding values for Probe Two were 10.9 + 3.3 and 91.8 + 27.7. The mean rate of healing and granulation of ulcers treated with Probe One were 4.4 + 2.5% and 5.6 + 3.2%, respectively; while that of ulcers treated with Probe Two were about three times higher, being 15.8 + 4.8% and 15.8 + 4.8%, respectively (Figs. 2 and 3).

Our overall small sample size—and the fact that two of our patients were deliberately treated with both probes as previously detailed—limited our ability to compare the above mean values statistically. Nonetheless, the superior effect of treatment with Probe Two seems clearly evident by the huge differences between means even when the standard deviations are taken into considerations. The following observations render further support for this assertion. Whereas none of the ulcers treated with Probe One healed completely, nine of the 12 (75%) treated with Probe Two healed fully [1]. Patients treated with Probe Two reported early pain relief—as early as one week of treatment—none of those treated with Probe One reported early relief from pain [2].

Table 1 Clinical characteristics of the patients

Patient	Age (Years)	Sex	Type	HbA1c (%)	Duration of Diabetes (Years)	Medication for Diabetes	Associated Medical Conditions	Duration of Ulcers (Months)
1	79	M	II	9.4	32	I	HBP	1
2	47	M	II	7.6	25	I; OM	HBP	9
3	50	M	II	10.4	12	I; OM	HBP	4
4	51	M	II	9.8	1.33	I; OM	CVI, HBP	10
5	63	M	II	12.1	1	OM	CVI	1
6	50	M	II	7.1	10	I	CVI, HBP	2
7	57	M	II	8.3	1.5		OM	4
8	77	F	II	7.4	15	OM	CVI, HBP	360
Mean	**59.0**			**9.1**	**12.22**			**48.87**
± SD	**12.61**			**1.73**	**11.50**			**125.76**

Key: I = Insulin; OM = Oral Medicine; HBP = High Blood Pressure; CVI = Chronic Venous Insufficiency.

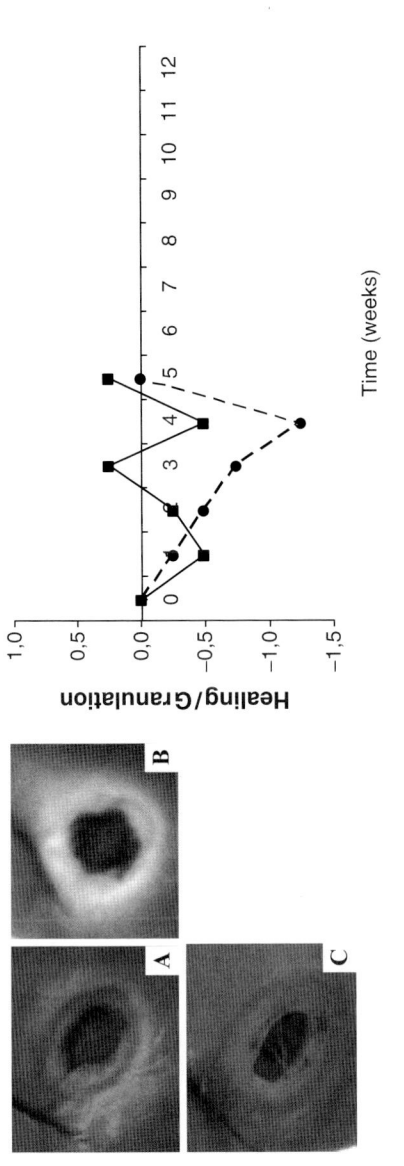

Fig. 2 Representative outcome of treatment with placebo Light Therapy Probe One (Case Number Two): This ulcer did not respond to treatment; indeed, the ulcer became larger and the patient had to be discontinued from the study. (A) shows the original size of the ulcer before treatment began; (B) is the size of the ulcer after two weeks, and (C) shows the size after four weeks. In the graph, the broken line with diamond blocks represents any increase or decrease in tissue healing (Area), while the line with square blocks is a measure of skin re-pigmentation, termed "granulation," in the graph

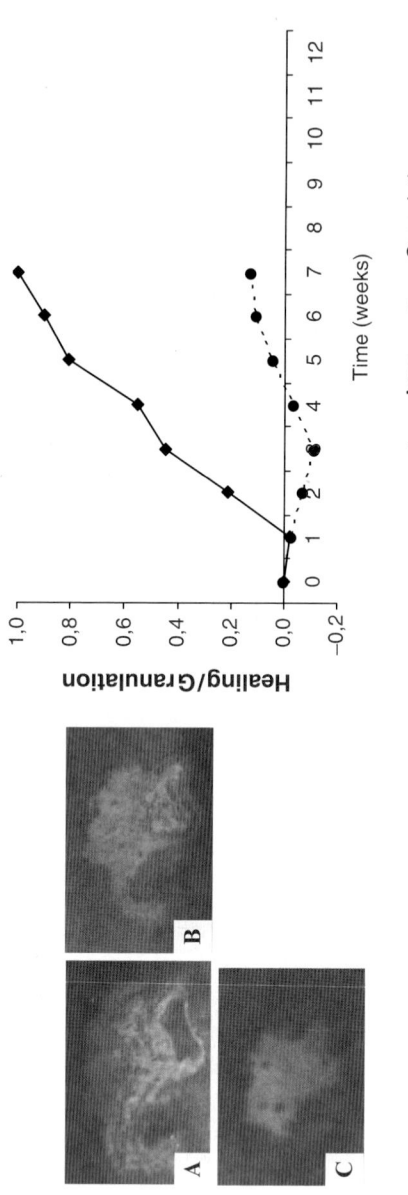

Fig. 3 Representative outcome of treatment with the real Light Therapy Probe Two (Case Number Four): This ulcer healed fully within seven weeks. (A) shows the original size of the ulcer before treatment began; (B) is the size of the ulcer after two weeks, and (C) shows the size after four weeks. In the graph, the broken line with diamond blocks represents any increase or decrease in tissue healing (Area), while the line with square blocks is a measure of skin re-pigmentation, termed "granulation" in the graph

Discussion

These findings provide initial evidence that combined 660 and 880 nm light promotes healing of diabetic ulcers that failed to respond to other forms of treatment. Absolute healing, absolute granulation and the rates of healing and granulation of ulcers treated with our real treatment probe, were superior to those treated with the placebo probe. These results support the notion that appropriate doses of light energy in the range of 600–1,000 nm promote tissue repair [10–16], particularly in cases of healing resistant ulcers [7–9, 34–37, 63–65].

Our finding that 75% of the ulcers that received the real treatment healed completely, is consistent with other reports [9, 63–65]. Previous studies indicate that 50–90% of diabetic ulcers respond positively to light therapy depending on treatment parameters [9, 63–65]. In their classic paper, Mester et al. [9] reported total healing or improvement in healing in 92.4% of the 1,120 cases treated with 4.0 J cm^{-2} 694 nm red light. Complete healing was achieved in 78.1% of the cases, and 80% of the 15 cases classified as "Diabetic lipodystrophy" healed completely. Similarly, Kleinman et al. [63] achieved full healing in 87.5% of recalcitrant leg ulcers using 785 nm light or a combination of 632.8 and 765 nm light. Lichtenstein and Morag [64], who treated 62 patients with healing resistant ulcers caused by chronic venous insufficiency, demonstrated full healing in 85.5% of the cases following irradiation with either 632.8 or 830 nm light [65].

The positive outcome of our study is encouraging. However, our small sample size limits our ability to generalize from our pilot data. Moreover, it was not possible to totally eliminate the effect of light from the placebo treatment group. As evidenced by the results, even the minimized amount of visible 660 nm light applied to the placebo group (0.03 J cm^{-2}) seemed to have some positive effect. However, the beneficial effect was clearly less than that of real treatment.

Conclusion

Our findings indicate that a combination of 660 and 880 nm SLD light promotes healing of diabetic ulcers that failed to respond to other forms of treatment. Further study with a larger sample size is on-going to confirm or refute these initial observations.

Acknowledgements This study was funded by grants from the following agencies and organizations: (1) FAPESP, São Paulo, Brazil [Grant No. 02/10723–5]; (2) CNPq, São Paulo, Brazil [Grant No. 8 141372/2004–4]; (3) FAEPA Foundation–Clinical Hospital of Faculty of Medicine, University of São Paulo, Ribeirão Preto, SP, Brazil; (4) Dynatronics Corporation, Salt Lake City, UT provided the Light Therapy Equipment and Conference Travel Grants; Sirius Alpha Corporation, Westbury, NY, provided additional travel support.

References

1. Phillips TJ. Chronic cutaneous ulcers: Etiology and epidemiology. J Invest Dermatol. 102: 38s–41s, 1994.
2. Nelzen O, Bergqvist D, Lindhagen A. High prevalence of diabetes in chronic leg ulcer patients: A cross-sectional population study. Diabetic Med. 10: 345–350, 1993.
3. Pecoraro RE, Reiber GE, Burgess EM. Pathways to diabetic limb amputation. Basis for preventation. Diabetes Care. 13: 513–521, 1990.
4. Harding KG, Morris HL, Patel GK. Science, medicine and the future: Healing chronic wounds. Brit Med J. 324: 160–163, 2002.
5. Sommer AP, Pinheiro ALB, Mester AR, Franke RP, Whelan, HT. Biostimulatory windows in low-intensity laser activation: Lasers, scanners, and NASA's light emitting diode array system. J Clin Laser Med Surg. 19: 29–33, 2001.
6. Enwemeka CS. Light is light. Photmed Laser Surg. 23: 159–160, 2005.
7. Powell MW, Carnegie DE, Burke TJ. Reversal of diabetic peripheral neuropathy and new wound incidence: The role of MIRE. Adv Skin Wound Care. 17: 143–147, 2004.
8. DeLellis SL, Carnegie DH, Burke TJ. Improved sensitivity in patients with peripheral neuropathy. Effects of monochromatic infrared photo energy. J Am Podiatr Med Assoc. 95: 143–147, 2005.
9. Mester E, Mester AF, Mester A. The biomedical effects of laser application. Laser Surg Med. 5: 31–39, 1985.
10. Conlan MJ, Rapley JW, Cobb CM. Biostimulation of wound healing by low-energy laser irradiation. A review. J Clin Periodont. 23: 492–496, 1996.
11. Yu W, Naim JO, Lanzafame RJ. Effects of photostimulation on wound healing in diabetic mice. Laser Surg Med. 20: 56–63, 1997.
12. Braverman B, McCarthy RJ, Ivankovich AD, Forde DE, Overfield M, Bapna MS. Effect of helium-neon and infrared laser irradiation on wound healing in rabbits. Laser Surg Med. 9: 50–58, 1989.
13. Longo L, Evangelista S, Tinacci G, Sesti AG. Effect of diodes-laser silver arsenide-aluminum (Gs-Al-As) 904 nm on healing of experimental wounds. Laser Surg Med. 7: 444–447, 1987.
14. Al-Watban FAH, Zhang XY. Stimulative and inhibitory effects of low incident levels of argon laser energy on wound healing. Laser Ther. 7: 11–18, 1995.
15. Lee P, Kim K, Kim K. Effects of low incident energy levels of infrared laser irradiation on healing of infected open skin wounds in rats. Laser Ther. 5: 59–64, 1993.
16. Ghamsari SM, Taguchi K, Abe N, Acorda JA, Sato M, Yamada H. Evaluation of low level laser therapy on primary healing of experimentally induced full thickness teat wounds in dairy cattle. Vet Surg. 26: 114–120, 1997.
17. Akai M, Usuba M, Maeshima T, Shirasaki Y, Yasuoka S. Laser's effect on bone and cartilage change induced by joint immobilization: An experiment with animal model. Laser Surg Med. 21: 480–484, 1997.
18. Bayat M, Delbari A, Almaseyeh MA, Sadeghi Y, Bayat M, Reziae F. Low-level Laser therapy improves early healing of medial collateral ligament injuries in rats. Photomed Laser Surg. 23 (6): 556–560, 2005.
19. Guzzaredla GA, Torriceli P, Rocca M, Tigani D, Brodano GB, Ferari D. Fini M, Giardino R. Assessment of low-power laser biostimulation on chondral lesions. An in vivo experimental study. Art Cells, Blood Subs & Immob Biotech. 28(5): 441–449, 2000.
20. Guzzaredla GA, Torriceli P, Fini M, Martini L, Morrone G, Giardino R. Low-power diode laser stimulation of surgical osteochondral defects. Results after 24 weeks. Art Cells, Blood Subs & Immob Biotech. 29(3): 235–244, 2001.
21. Enwemeka CS. Laser photostimulation. Clinical Manage. 10: 24–29, 1990.
22. Reddy GK, Stehno-Bittel L, Enwemeka CS. Laser photostimulation of collagen production in healing rabbit Achilles tendons. Laser Surg Med. 22: 281–287, 1998.

23. Enwemeka CS, Cohen E, Duswalt EP, Weber DM. The biomechanical effects of Ga-As laser photostimulation on tendon healing. Laser Ther. 6: 181–188, 1995.

24. Enwemeka CS. Ultrastructural morphometry of membrane-bound intracytoplasmic collagen fibrils in tendon fibroblasts exposed to He:Ne laser beam. Tissue & Cell 24: 511–523, 1992.

25. Ozawa Y, Shimizu N, Kariya G, Abiko Y. Low-energy laser irradiation stimulates bone nodule formation at early stages of cell culture in rat calvarial cells. Bone 22: 347–354, 1998.

26. Houghton PE, Brown JL. Effect of low level laser on healing in wounded fetal mouse limbs. Laser Ther. 11:54–69, 1999.

27. Luger EJ, Wollman Y, Kogan G, Dekel S. Effect of low-power laser irradiation on the mechanical properties of bone fracture healing in rats. Laser Surg Med. 22: 97–102, 1998.

28. Ozawa YSN, Kariya G, Abiko Y. Low-energy laser irradiation stimulates bone nodule formation at early stages of cell culture in rat calvarial cells. Bone. 22(4): 347–354, 1998.

29. Guzzarella GA, Porriceli P, Aldini NN, Giardino R. Laser technology in orthopedics: Preliminary study on low-power laser therapy to improve the bone-biomaterial interface. Int J Artif Organs. 24(12): 898–902, 2001.

30. Morrone GGG, Torriceli P, Rocca M, Tigani D, Brodano GB, Fini, M, Giardini R. Osteochondrial lesion repair of the knee in the rabbit after low-power diode Ga-Al-As laser biostimulation: An experimental study. Art Cells, Blood Subs & Immob Biotech. 28(4): 321–336, 2000.

31. Torriceli PGG, Fini, M, Guzzardella GA, Morrone G, Carpi A, Giardini R. Laser biostimulation of cartilage: In vitro evaluation. Biomed Pharmacother. 55: 117–120, 2001.

32. Morrone GGG, Guzzardella GA, Tigani D, Torriceli P, Fini M, Giardini R. Biostimulation of human chondrocytes with GA-Al-As diode laser: In vitro Research. Art Cells, Blood Subs & Immob Biotech. 28: 193–201, 2000.

33. Morrone GGG, Torriceli P, Fini M, Giardini R. In vitro experimental research of rabbit chondrocytes with diode laser GA-Al-As: A preliminary study. Art Cells, Blood Subs & Immob Biotech. 26(4): 437–439, 1998.

34. Schindl A, Schindl M, Schindl L. Successful treatment of a persistent radiation ulcer by low power laser therapy. J Am Acad Dermatol. 37: 646–648, 1997.

35. Schindl A, Schindl M, Schind L. Phototherapy with low intensity laser irradiation for a chronic radiation ulcer in a patient with lupus erythematosus and diabetes mellitus [letter]. Br J Dermatol. 137: 840–841, 1997.

36. Schindl A, Schindl M, Schon H, Knobler R, Havelec L, Schindl L. Low-intensity laser irradiation improves skin circulation in patients with diabetic microangiopathy. Diabetes Care. 21: 580–584, 1998.

37. Lundeberg T, Malm M. Low-power HeNe laser treatment of venous leg ulcers. Ann Plas Surg. 27: 537–539, 1991.

38. Al-Watban FAH, Zhang XY. Comparison of the effects of laser therapy on wound healing using different laser wavelengths. Laser Ther. 8: 127–135, 1996.

39. Al-Watban FAH, Zhang XY. Comparison of wound healing process using argon and krypton lasers. J Clin Laser Med Surg. 15: 209–215, 1997.

40. Braverman B, McCarthy RJ, Ivankovich AD, Forde DE, Overfield M, Bapna MS. Effect of helium-neon and infrared laser irradiation of wound healing in rabbits. Laser Surg Med. 9: 50–58, 1989.

41. Bouma MG, Buurman WA, van den Wildenberg FAJM. Low energy laser irradiation fails to modulate the inflammatory function of human monocytes and endothelial cells. Laser Surg Med. 19: 207–215, 1996.

42. Allendorf JDF, Bessler M, Huang J, Kayton ML, Laird D, Nowygrod R, Treat MR. Helium-neon laser irradiation at fluences of 1, 2 and 4 J/cm^2 failed to accelerated wound healing as assessed by both wound contracture rate and tensile strength. Laser Surg Med. 20: 340–345, 1997.

43. El Sayed SO, Dyson M. Effect of laser pulse repetition rate and pulse duration on mast cell number and degranulation. Laser Surg Med. 19: 433–437, 1996.

44. El Sayed SO, Dyson M. Comparison of the effect of multiwavelength light produced by a cluster of semiconductor diodes and of each individual diode on mast cell number and degranulation in intact and injured skin. Laser Surg Med. 10: 559–568, 1990.

45. Dyson M, Young S. Effect of laser therapy on wound contraction and cellularity in mice. Laser Med Sci. 1: 125–130, 1986.

46. Shiroto C, Sugawara K, Kumae T, Ono Y, Sasaki M, Ohshiro T. Effect of diode laser radiation in vitro on activity of human neutrophils. Original Articles. 135–140, 1990.

47. Young S, Bolton P, Dyson M, Harvey W, Diamantopoulos C. Macrophage responsiveness to light therapy. Laser Surg Med. 9: 497–505, 1989.

48. Haas AF, Isseroff R, Wheeland RG, Rood PA, Graves PJ. Low-energy helium-neon laser irradiation increases the motility of cultured human keratinocytes. J Invest Derm. 94: 822–826, 1990.

49. Abergel RP, Lyons RF, Castel JC, Dwyer RM, Uitto J. Biostimulation of wound healing by lasers: Experimental approaches in animal models and in fibroblast cultures. J Dermatol Surg Onc. 13: 127–133, 1987.

50. Graham DJ, Alexander JJ. The effects of argon laser on bovine aortic endothelial and smooth muscle cell proliferation and collagen production. Curr Surg. 47: 27–30, 1990.

51. Pogrel MA, Chen JW, Zhang K. Effects of low-energy gallium-aluminum-arsenide laser irradiation on cultured fibroblasts and keratinocytes. Laser Surg Med. 20: 426–432, 1997.

52. Steinlechner C, Dyson M. The effects of low level laser therapy on the proliferation of keratinocytes. Laser Ther. 5: 65–73, 1993.

53. Saperia D, Glassberg E, Lyons RF, Abergel RP, Baneux P, Castel JC, Dwyer RM, Uitto J. Demonstration of elevated type I and type III procollagen mRNA levels in cutaneous wounds treated with helium-neon laser. Proposed mechanism for enhanced wound healing. Biochem Biophys Res Commun. 138: 1123–1128, 1986.

54. Young S, Bolton P, Dyson M, Harvey W, Diamantopoulos C. Macrophage responsiveness to light therapy. Laser Surg Med. 9: 497–505, 1989.

55. Passarella S, Casamassima E, Molinari S, Pastore D, Quagliariello E, Catalano IM, Cingolani A. Increase of proton electrochemical potential and ATP synthesis in rat liver mitochondria irradiated in vitro by helium-neon laser. FEBS Lett. 175: 95–99, 1984.

56. Basford JR, Hallman HO, Sheffield CG, Mackey GL. Comparison of cold-quartz ultraviolet, low-energy laser, and occlusion in wound healing in a swine model. Arch Phys Med Rehabil. 67: 151–154, 1986.

57. Nussbaum EL, Biemann I, Mustard B. Comparison of ultrasound/ultraviolet-C and laser for treatment of pressure ulcers in patients with spinal cord injury. Phys Ther. 74: 812–823, 1994.

58. Basford JR. Laser therapy: Scientific basis and clinical role. Laser Ortho Surg. 16(5): 541–547, 1993.

59. McMeeken J, Stillman B. Perceptions of the efficacy of laser therapy. Australian J Physiother. 39: 101–106, 1993.

60. Basford JR. Low-energy laser therapy: Controversies and new research findings. Laser Surg Med. 9: 1–5, 1989.

61. Bouma MG, Buurman WA, van den Wildenberg FAJM. Low energy laser irradiation fails to modulate the inflammatory function of human monocytes and endothelial cells. Laser Surg Med. 19: 207–215, 1996.

62. Allendorf JDF, Bessler M, Huang J, Kayton ML, Laird D, Nowygrod R, Treat MR. Helium-neon laser irradiation at fluences of 1, 2 and 4 J/cm^2 failed to accelerated wound healing as assessed by both wound contracture rate and tensile strength. Laser Surg Med. 20: 340–345, 1997.

63. Kleinman Y, Simmer S, Braksma Y, Morag B, Lichtenstein D. Low level laser therapy in patients with venous ulcers: Early and long term outcomes. Laser Ther. 8: 205–208, 1996.

64. Lichtenstein D, Morag B. Low level laser therapy in ambulatory patients with venous stasis ulcers. Laser Ther. 11(2): 71–78, 1998.

65. Lee P, Kim K, Lim K. Effects of low incident energy levels of infrared laser irradiation on healing of infected open skin wounds in rats. Laser Ther. 5: 59–64, 1993.

Blue Light Photo-Destroys Methicillin Resistant Staphylococcus aureus (MRSA) In-Vitro

Chukuka S. Enwemeka, Deborah Williams, Steve Hollosi, and David Yens

Abstract Methicillin resistant Staphylococcus aureus (MRSA) is a virulent form of staphylococcus easily spread and to control. When contracted it can be exceedingly difficult to control especially where healthcare is not readily available. Blue light at wavelengths above 400 nm and doses near 60 J/cm^2 are shown to be effective at controlling infections.

Keywords: MRSA, blue light.

Introduction

Methicillin resistant strains of Staphylococcus aureus (MRSA) have emerged as community acquired pathogens capable of causing serious diseases in health care and non-health care settings [3, 6, 9, 11]. Community acquired MRSA infections, without prior exposure to the bacteria in medical facilities, have been described in both urban and rural settings. While most patients develop S. aureus infections from their own colonizing strains, reports have shown that MRSA may be acquired from other people and fomites. The cutaneous infections are often occasioned by trauma, disease and poor personal hygiene.

 Staphylococcus aureus (*S. aureus*) is the most virulent of staphylococcal species and is responsible for a variety of infections, ranging from superficial skin and soft tissue infections to life threatening system diseases. Treatment of *S. aureus* infections has become increasingly difficult because of widespread dissemination of plasmids containing the penicillin cleaving enzyme penicillinase. At present, less than 5% of staphylococci strains are susceptible to penicillin. In response to these increasing virulence factors, new pharmaceutical targets within the bacterial

C.S. Enwemeka
School of Health Professions, Behavioral and Life Sciences, New York Institute of Technology, Old Westbury, NY 11568-8000, USA, e-mail: Enwemeka@nyit.edu

R. Waynant and D.B. Tata (eds.), *Proceedings of Light-Activated Tissue Regeneration and Therapy Conference.*
© Springer Science + Business Media, LLC 2008

genome have been studied, resulting in the use of semisynthetic penicillinase-resistant penicillins—such as Methicillin—in the treatment of these penicillin-resistant isolates. However, about 40–50% of S. *aureus* isolates are resistant to Methicillin (Kasper et al., 2005); a problem that continues to worsen with uncontrolled use of antibiotics. Molecular typing studies have identified two clones of commonly acquired MRSA in the USA; designated as USA 300 and USA 400 [8, 10]. Outbreaks of MRSA skin and soft tissue infections among prisoners and athletes have been associated with the USA 300 [8]. While outbreaks associated with severe and fatal disease in children, as well as skin and soft tissue infections in native American populations have been associated with USA 400 [8, 9]. Both strains of MRSA—USA 300 and USA 400 clones—are resistant to β-lactams [8, 9]. Alternative measures to prevent the development of new or increased resistance are needed, and there is a need to focus on prevention of the spread of contamination and the eradication of nidi of infection [5]. Phototherapy seems a promising alternative treatment that could eradicate MRSA as evidenced by the works of [1, 2, 7, 13]. Studies have shown that gram positive bacteria such as S. *aureus* are susceptible to photodynamic inactivation [7]; and there are reports which suggest that blue light may produce a similar antibacterial effect [1, 2]. Therefore, we studied the effects of blue light ($\lambda > 400$ nm) on MRSA *in vitro*, and in particular, determined the dosages that can effectively eradicate the bacteria.

Methods

MRSA was streaked onto Tryptic Soy Agar (TSA) 35 mm plates and exposed to super blue light generated from superluminous diodes (SLDs). Each plate was then treated once at 0, 1, 3, 5, 7, 9, 11, 13, 15, 17, 19, 25, 30, 35, 40, 45, 50, 55, or 60 J cm^{-2}. Plates were then incubated at 35°C for 24 hours and colony counts determined using optical density analysis (Sigma Scan Pro5). The experiment was repeated three times to foster accuracy of results.

Results

As shown in the figure, the blue light produced a dose-dependent bactericidal effect on MRSA ($p < 0.001$), achieving a kill rate of 96.7%, 90.8% and 83.8% respectively during each of the three trials.

Discussion

At present, therapy for community-acquired MRSA presents many challenges. In the past the first lines of therapy for this organism were oral cephalosporin and penicillin, with some susceptibility to tetracycline and Bactrim. With the emergence of multi-antimicrobial resistance to MRSA consideration of alternative therapies seems appropriate. It is well accepted that certain wavelengths of light are germicidal. For example, ultraviolet light (UVL) is absorbed by a double bond in pyrimidine bases of DNA, opening the bond and allowing it to react with neighboring molecules. If the target bond lies next to a second pyrimidine base, the UVL modified base forms direct covalent bonds with it; forming either two new bonds between the neighboring bases—a tight four-membered ring, or a single bond between two carbon atoms on the rings—which results in a—six to four photoproduct [12]. Either way, the bacteria or cell is damaged. Perhaps this mechanism underlies recent reports by [13] who found significant improvement in patients with acne vulgaris, following treatment with combined blue (415 nm) and red (660 nm) light. They suggested that the blue and red light produced antibacterial and anti-inflammatory benefits respectively. More recently, [1] examined the effects of 405 and 470 nm light on two common aerobes, *Staphylococcus aureus* and *Pseudomonas aeruginosa*, and anaerobic Propionibacterium acnes, *in vitro*. Colony

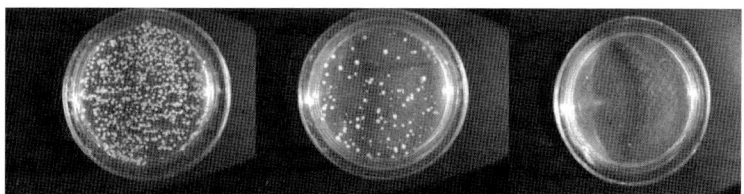

Colony Count for 19 Exposures as Proportion of Control, Three Trials

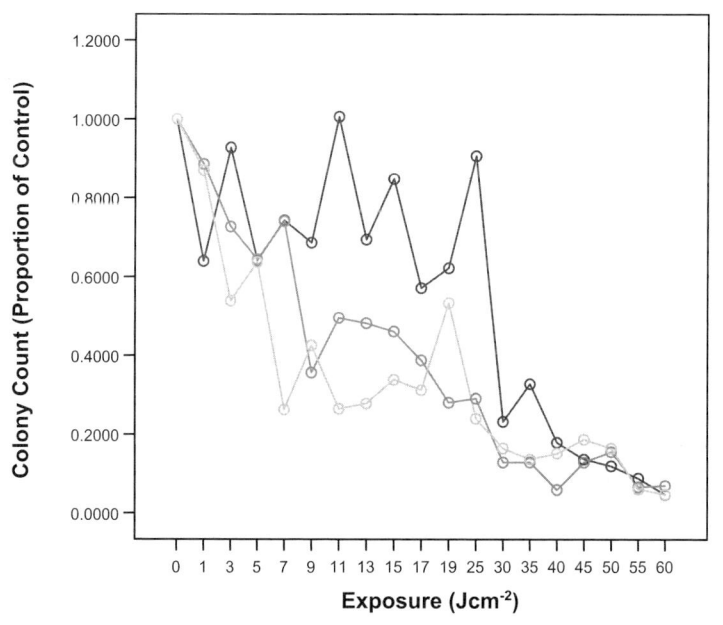

counts were then used to compare their abilities to kill bacteria. Bactericidal effects were demonstrated for both wavelengths but the kill rate was higher with the 405 nm light, reaching 90% for *S. aureus* and 95.1% for *P. aeruginosa*. Neither wavelength was effective for anaerobic *P. acne*. With the superior bactericidal effect demonstrated for 405 nm light, they then examined the *in vitro* effect of a combination of 405 nm blue light and 880 nm infrared light on *S. aureus* and *P. aeruginosa* in their second study [2]. Colony counts again showed strong bacteria kill rates—as much as 93.8% for *P. aeruginosa*, and as high as 72% for *S. aureus*. Our results are consistent with these recent reports [1, 2, 13], and demonstrate—for the first time—that blue light photo-destroys MRSA.

Conclusion

Our results warrant the conclusion that the blue light emitted by the SLD source used in this study, with predominant λ > 400 nm, photo-destroys MRSA *in vitro*; suggesting that a similar bactericidal effect may be achieved *in vivo*.

Acknowledgements This study was funded by Dynatronics Corporation, Salt Lake City, UT.

References

1. Guffey JS, Wilborn J (2006a): In vitro bactericidal effects of 405-nm and 470-nm blue light. Photomedicine and Laser Surgery 24: 684–688.
2. Guffey JS, Wilborn J (2006b): Effects of combined 405-nm and 880-nm light on *Staphylococcus aureus and Pseudomonas aeruginosa in vitro*. Photomedicine and Laser Surgery 24: 680–683.
3. Moellering RC (2006): The growing menace of community-acquired methicillin-resistant *Staphylococcus aureus*. Annals of Internal Medicine 144: 368–370.
4. Kasper DL, Braunwald E, Fauci AS, Hauser SL, Longo DL, Jameson JL, Isselbacher KJ (Eds.) (2005): *Harrison's Principles of Internal Medicine (16th Edition), Chapter 120*.
5. Thomas S (2004): *MRSA and the use of silver dressings: overcoming bacterial resistance*. World Wide Wounds.
6. Begier EM, Frenette K, Barrett NL, Mshar P, Petit S, Boxrud DJ, et al. (2004): A high morbidity outbreak of Methicillin–resistant *Staphylococcus aureus* among players on a college football team, facilitated by cosmetic body shaving and turf burns. Clinical Infectious Diseases 39: 1446–1453.
7. Gad F, Zahra T, Hasan T, Hamblin MR (2004): Effects of growth phase and extracellular slime on photodynamic inactivation of gram-positive pathogenic bacteria. Microbial Agents and Chemotherapy 48: 2173–2178.
8. McDougal LK, Steward CD, Killgore GE, Chaitram JM, McAllister SK, Tenover FC (2003): Pulsed field gel electrophoresis typing of oxacillin-resistant *Staphylococcus aureus* isolates from the United states: establishing a national database. Journal of Clinical Microbiology 41: 5113–5120.

9. Naimi TS, Le Dell KH, Como-Sabetti K, Bordchardt SM, Boxrud DJ, Etienne J (2003): Comparison of community and health-care associated Methicillin-resistant Staphylococcus aureus infection. JAMA 290: 2976–2984.

10. Okuma K, Iwakawa K, Turnidge JD, Grubb WB, Bell JM, O'Brien FG (2002): Dissemination of new Methicillin-resistant Staphylococcus aureus clones in the community. Journal of Clinical Microbiology 40: 4289–4294.

11. Morbidity and Mortality Weekly Report (MMWR) (2003): Methicillin-resistant Staphylococcus aureus among competitive sports–participants-Coloroda, Indiana, Pennsylvania and Los Angeles County 2000–2003 MMWR. Morbidity and Mortality Weekly Report 52: 793–795.

12. Goodsell DS (2001): The molecular perspective: ultraviolet light and pyrimidine dimmers. Oncologist 6: 298–299.

13. Papageorgiou P, Katsambas A, Chu A (2000): Phototherapy with blue (415 nm) and red (660 nm) light in the treatment of acne vulgaris. British Journal of Dermatology 142: 973–978.

Photobiomodulation for the Treatment of Retinal Injury and Retinal Degenerative Diseases

Janis T. Eells, Kristina D. DeSmet, Diana K. Kirk, Margaret Wong-Riley, Harry T. Whelan, James Ver Hoeve, T. Michael Nork, Jonathan Stone, and Krisztina Valter

Abstract Retinal injury and retinal degenerative diseases are a leading causes of visual impairment in the developed world. Mitochondrial dysfunction and oxidative stress play key roles in the pathogenesis of retinal injury and disease. The development and testing of strategies designed to improve mitochondrial function and attenuate oxidative stress are essential for combating retinal disease. One strategy involves the use of photobiomodulation. Photobiomodulation, low-energy photon irradiation by light in the far-red to near-infrared (NIR) range using low energy lasers or light-emitting diode (LED) arrays, has been applied clinically in the treatment soft tissue injuries and acceleration of wound healing for more than 30 years. The therapeutic effects of photobiomodulation have been hypothesized to be mediated by intracellular signaling mechanisms triggered by the interaction of far-red to NIR photons with the mitochondrial photoacceptor molecule cytochrome oxidase which culminate in improved mitochondrial energy metabolism, increased synthesis of cytoprotective factors and cell survival.

The therapeutic potential of 670 nm LED photobiomodulation administered once per day at a fluence of 4 J/cm^2 was investigated in established experimental models of retinal injury, retinal toxicity and retinal disease. Photobiomodulation stimulated retinal wound healing following high-intensity laser-induced retinal injury. Photobiomodulation not only enhanced the rate of wound healing, it also prevented the loss of retinal and cortical visual function induced by laser-induced retinal injury. In a rodent model of retinal mitochondrial toxicity, photobiomodulation preserved retinal function and prevented photoreceptor damage. Moreover, molecular studies revealed that photobiomodulation induced significant upregulation of gene expression pathways involved in mitochondrial energy production and

J.T. Eells

University of Wisconsin — Milwaukee, UMW College of Health Sciences, 2400 E. Hartford Avenue, Milwaukee, WI 53201, USA, e-mail: jeells@uwm.edu

R. Waynant and D.B. Tata (eds.), *Proceedings of Light-Activated Tissue Regeneration and Therapy Conference.*
© Springer Science + Business Media, LLC 2008

cytoprotection in the retina. Retinitis pigmentosa is the leading cause of vision loss due to retinal degeneration. Photobiomodulation administered during the critical period of photoreceptor development in a rat model of retinitis pigmentosa increased retinal mitochondrial cytochrome oxidase activity, upregulated the production of retinal antioxidants, increased the production of retinal neurotrophic factors and prevented photoreceptor cell death.

The molecular, biochemical and functional insights obtained from this research provide crucial information needed for a comprehensive FDA approval for the use of photobiomodulation in the treatment of retinal diseases. From a basic science perspective, they substantiate previous *in vitro* investigations and support the hypothesis that photobiomodulation augments mitochondrial function and stimulates cytoprotective pathways to prevent retinal damage. From a clinical perspective, they document the therapeutic potential of 670 nm photon therapy in experimental models of retinal injury, retinal toxicity and retinitis pigmentosa, thus setting the stage for clinical trials of photobiomodulation in human disease.

Keywords Photobiomodulation, mitochondrial dysfunction, oxidative stress, retinal injury, retinal degenerative disease.

Oxidative stress and mitochondrial dysfunction are central in retinal aging, retinal injury and in the pathogenesis of retinal degenerative diseases [2, 3, 24, 25]. Oxidative stress plays a key role in high intensity laser burn injury and retinal mitochondrial dysfunction is causative in methanol toxicity [12]. Retinal injury and retinal degenerative diseases are leading causes of visual impairment in the developed world [1, 9]. Although age-related macular degeneration is responsible for the vast majority of the cases of retinal degeneration, other important retinal dystrophies include, retinitis pigmentosa, cone-rod dystrophies and Stargardt disease. The pathogenesis of these diseases is incompletely understood and no efficient therapy or prevention exists to date. Therefore, the development of simple long-term strategies designed to improve mitochondrial function and attenuate oxidative stress are essential for combating neurodegenerative and retinal degenerative disease. One newly developed strategy involves the use of photobiomodulation.

Low energy photon irradiation by light in the far-red to near-infrared (NIR) range (630–1,000 nm) using low energy lasers or light-emitting diode (LED) arrays (collectively termed photobiomodulation) has been shown to accelerate wound healing, improve recovery from ischemic injury in the heart and attenuate degeneration in the injured retina and optic nerve [5, 12]. Moreover, red to near-infrared light therapy has been applied clinically in the treatment soft tissue injuries and to accelerate wound healing for more than 30 years [5, 12]. The therapeutic effects of red to near infrared light have been hypothesized to be mediated by intracellular signaling mechanisms triggered by the interaction of far-red to near-infrared light with the mitochondrial photoacceptor molecule cytochrome oxidase which culminate in improved cellular mitochondrial energy metabolism and antioxidant production (Fig. 1). In support of this hypothesis, we have demonstrated in primary neuronal cells that far-red to

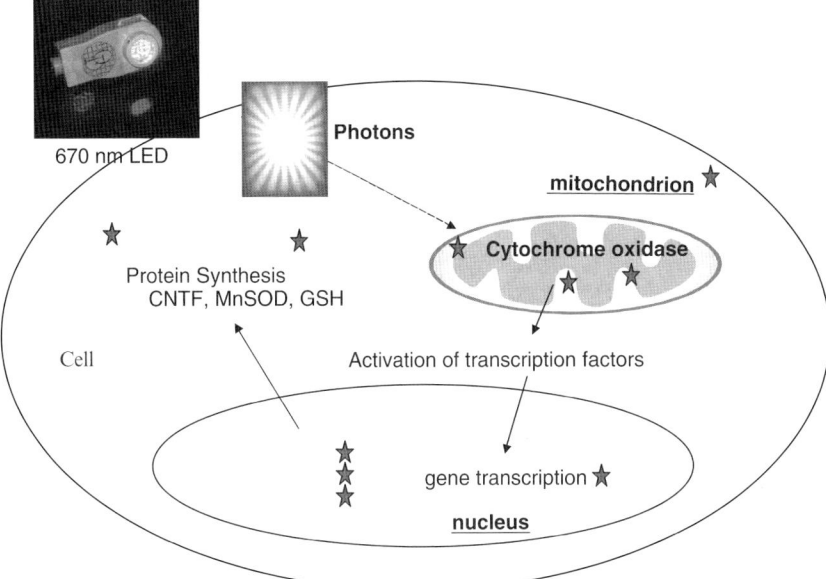

Fig. 1 Hypothesized signal transduction pathway for photobiomodulation

near-infrared LED photo-irradiation (4 J/cm^2) increases the production of cytochrome oxidase in cultured primary neurons, reverses the reduction of cytochrome oxidase activity produced by metabolic inhibitors and attenuates cyanide-induced apoptosis [20, 21]. We have also shown that the action spectrum of NIR light for stimulation of cytochrome oxidase activity parallels the near-infrared absorption spectrum of the oxidized form of cytochrome oxidase (Fig. 2) [21].

670 nm LED Treatment Promotes Retinal Wound Healing

Stimulation of wound healing by photobiomodulation is an attractive technology for instances where the lesion is located in tissue that might be compromised by more invasive approaches. Lesions in tissues such as the retina or brain thus might benefit from photobiomodulation as a primary or an adjunctive therapeutic approach. One such wound is a lesion in the retina that results from exposure to a laser beam. Lasers are increasingly used by the military and in industrial applications. At present, these types of retinal injuries are increasing, a trend that is expected to continue [30].

We have initiated studies of laser retinal injury in a nonhuman primate model. In each experiment one monkey was lased without LED treatment and one lased with LED treatment (670 nm, 4 J/cm^2). A laser grid (128 spots delivered to the macula and perimacula) was created in the central retina of right eye of each animal. This grid consisted of grade I and II burns, photocoagulating the photoreceptors and outer nuclear layer of the retina. Multifocal ERG was performed to assess the

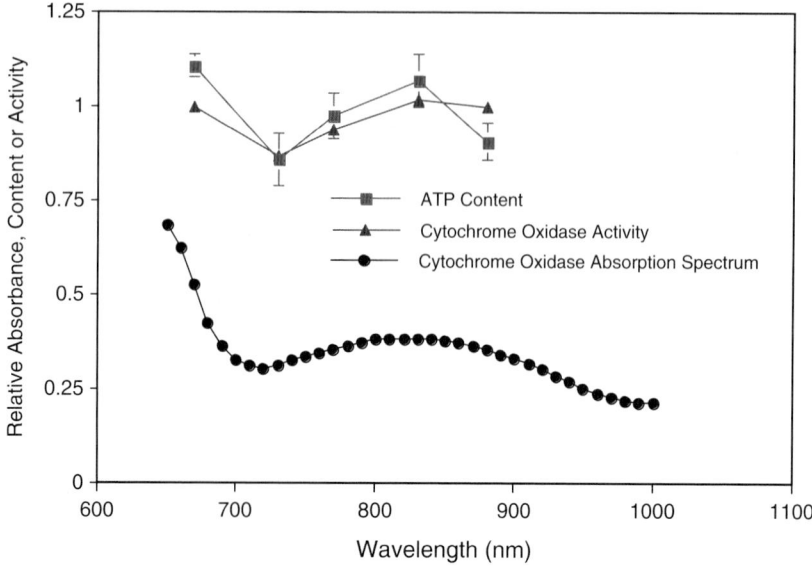

Fig. 2 Recovery of neuronal cytochrome oxidase activity and cellular ATP content correlates with cytochrome oxidase action spectrum

functional state of the retina. In the first experiment, the LED-treated monkey was treated at 1, 24, 72 and 96 hours post injury. ERG amplitude in both LED treated and untreated monkeys was temporarily increased shortly after laser injury and this increase was greater in the LED-treated monkey [31]. Assessment of the severity of the laser burn in LED treated and untreated animal demonstrated a greater that 50% improvement in the degree of retinal healing at 1 month post-laser in the LED-treated monkey (Fig. 3). In addition, the thickness of the retina measured at the fovea by optical coherence tomography did not differ from the pre-laser thickness in the LED-treated animal whereas it was 50% thinner in the untreated animal (Fig. 3). Importantly, LED treatment prevented the loss of cytochrome oxidase staining in the lateral geniculate nucleus [31] clearly showing that the brain was responding to visual input from the "healed" retina in the LED-treated animal much more effectively than in the untreated animal.

670 nm LED Treatment Attenuates Mitochondrial Injury Induced by Methanol Intoxication

Methanol intoxication produces toxic injury to the retina and optic nerve resulting in blindness. The toxic metabolite in methanol intoxication is formic acid, a mitochondrial toxin known to inhibit the essential mitochondrial enzyme, cytochrome oxidase. Studies were undertaken to test the hypothesis that exposure to

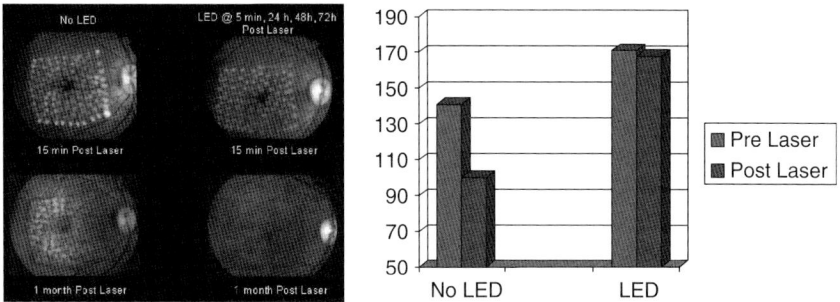

Fig. 3 Laser grid appearance and foveal thickness in control (*left*) and NIR-LED (*right*) treated monkey retina

monochromatic red radiation from 670 nm light-emitting diode (LED) arrays would protect the retina against the toxic actions of methanol-derived formic acid in a rodent model of methanol toxicity. Using the electroretinogram as a sensitive indicator of retinal function, we demonstrated that three brief (2 minutes 24 seconds) 670 nm LED treatments (4 J/cm^2), delivered at 5, 25 and 50 hours of methanol intoxication, significantly attenuated the retinotoxic effects of methanol-derived formate during intoxication (Fig. 2) and profoundly improved the recovery of retinal function following intoxication (Fig. 4). We further show that LED treatment protected the retina from the histopathologic changes induced by methanol-derived formate (Fig. 4). These findings provide a link between the actions of monochromatic red to near infrared light on mitochondrial oxidative metabolism *in vitro* and retinoprotection *in vivo*. They provide the basis for examining the efficacy of 670 nm LED treatment in retinal degenerative diseases.

The prolonged effect of 3 brief LED treatments in mediating the retinoprotective actions in methanol intoxication suggests that 670 nm LED photostimulation induces a cascade of signaling events initiated by the initial absorption of light by cytochrome oxidase. We have compared gene expression profiles in the neural retina of untreated rats with those from the neural retina of methanol-intoxicated rats and LED-treated methanol-intoxicated rats (ref). Results from these studies indicate that methanol intoxication and LED treatment altered the retinal expression of nearly 80 genes. At least 26 of these genes that were up-regulated in the retinas of methanol intoxicated rats were correspondingly down-regulated in the retinas of LED treated methanol intoxicated rats and the converse (Fig. 5). Several functional subcategories of genes regulated by 670 nm-LED treatment were identified in retinal samples including those encoding DNA repair proteins, antioxidant defense enzymes, molecular chaperones, protein biosynthesis enzymes, and trafficking and degradation proteins. It is important to note that a number of genes associated with protection against macular degeneration, including glutathione S-transferase (GST), are upregulated by 670 nm LED treatment and genes associated with retinal damage, including TIMP3, are down regulated. [31].

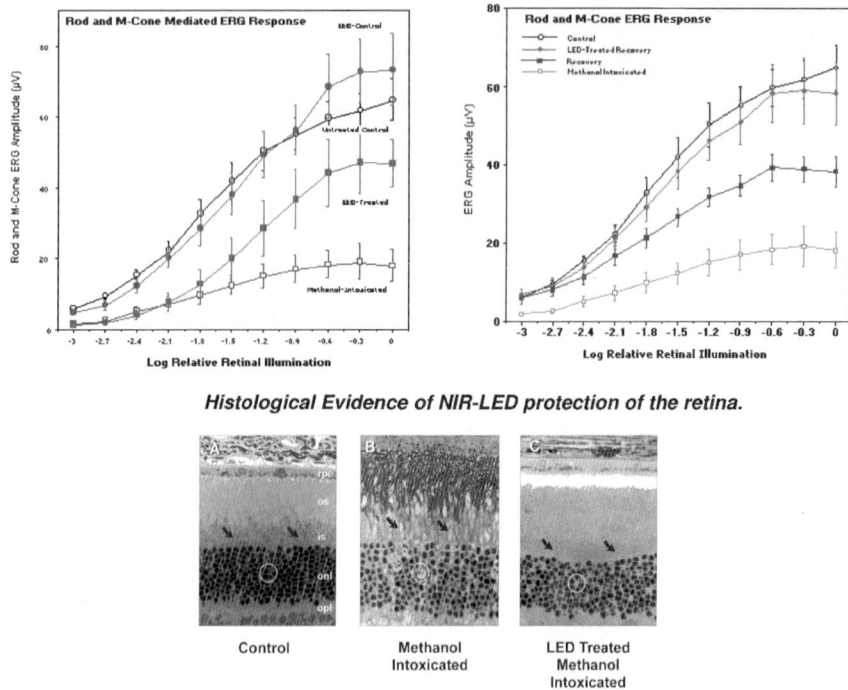

Fig. 4 670 nm LED treatment protects against methanol-induced retinal dysfunction and improves recovery

670 nm LED Therapy for the Treatment of Retinitis Pigmentosa

Retinitis pigmentosa (RP) is the leading cause of vision loss due to retinal degeneration. RP is a heterogenous group of retinal dystrophies characterized by photoreceptor death [8]. Autosomal dominant forms of RP, which cause a majority of RP cases, result from mutations in the rhodopsin gene. Rhodopsin is the visual pigment located within the outer segment of rod photoreceptor cells and is involved in phototransduction. The exact mechanism of photoreceptor degeneration is unclear but it is thought that the rhodopsin mutation alters the tertiary structure of the protein thus altering its function. Altered rhodopsin structure leads to misfolded opsin protein. Opsin is present in the photoreceptor inner segment where mitochondria are located. Opsin binds *cis*-retinal chromophore forming active rhodopsin which is subsequently translocated to the photoreceptor outer segment away from the mitochondria. It is thought that misfolded opsin does not sufficiently sequester *cis*-retinal chromophore which can initiate free radical reactions, increase oxidative stress, and leave the photoreceptor cells vulnerable to apoptosis [1].

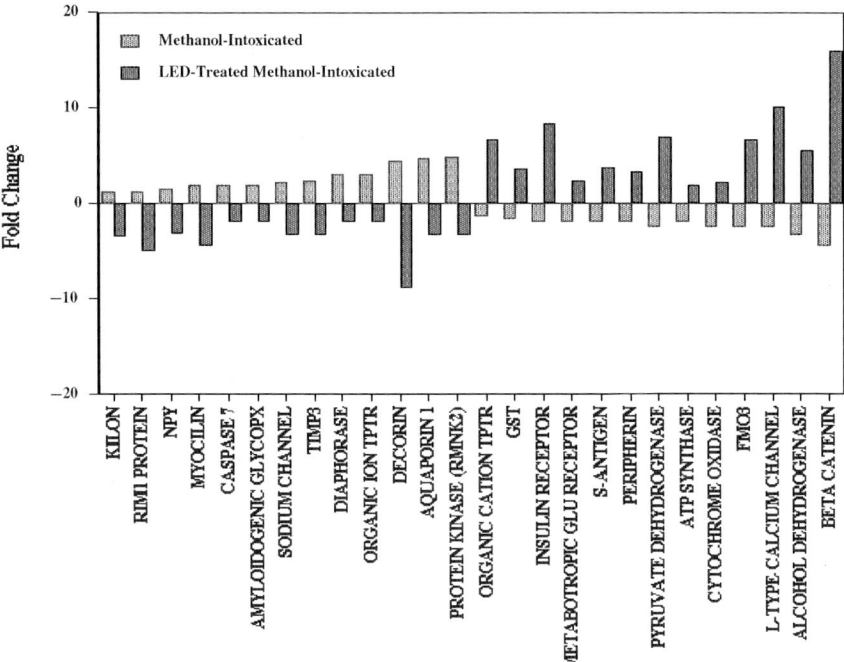

Fig. 5 Differential regulation of retinal genes by mitochondrial inhibition and 670 nm LED treatment

The P23H-3 rat is a well-established rodent model of human disease. The transgene in the P23H-3 rat mimics a rhodopsin mutation found in humans in North America, and photoreceptor death is excessive, most markedly during development (P15–30). [3, 17, 19]. Our studies describe the effect of 670 nm LED treatment administered during the critical period of photoreceptor cell loss in this animal model of RP.

P23H-3 rat pups were treated daily for 5 days during the critical period of photoreceptor development from postnatal day 16–20 with 670 nm LED array (GaAlAs LED arrays, 670 ± 20 nm at 50% power; Quantum Devices, Inc., Barneveld, WI). Rats pups were hand-held by the investigator and the 10 cm^2 670 nm LED array was positioned directly over the animal's head at a distance of 2 cm exposing both eyes. Treatment consisted of irradiation at 670 nm for 3 minutes resulting in a power intensity of 50 mW/cm^2 and an energy density of 4 J/cm^2. These stimulation parameters have been previously demonstrated to stimulate cell survival and cytochrome oxidase activity in cultured neurons [20–21] to promote wound healing clinically and in experimental animal models [4, 6] and to promote the survival and functional recovery of the retina and optic nerve *in vivo* after acute injury by a mitochondrial toxin [2]. Sham treatment involved holding the animals for 3 minutes with the LED array positioned over the eyes, but not illuminated. On postnatal day 21, animals were euthanized. One eye from each animal was prepared for immunohistochemical evaluation and the retina was dissected from the other eye and flash-frozen in liquid nitrogen for biochemical analysis.

670 nm LED-Induced Increases in Cytochrome Oxidase Concentrations in the P23H Retina

670 nm LED treatment significantly increased the concentration of cytochrome oxidase (CO), in the retina of the P23H-3 rat (N = 6, p < 0.001). Marked increases in CO concentrations were most apparent in inner segments of the photoreceptor cells consistent with the large numbers of mitochondria present in this region of the retina. These findings are consistent with our in vitro observations and support our hypothesis that upregulation of cytochrome oxidase activity is a critical component in the neuroprotective actions of photon therapy [5, 12] (Fig. 6). Moreover, recent studies have shown that cytochrome oxidase is the rate-limiting step in mitochondrial respiration [28, 29] and that increased cytochrome oxidase activity improves respiratory efficiency under conditions of oxidative stress, thus attenuating the production of reactive oxygen species [29].

Photobiomodulation-Induced Increases in Cytoprotective Enzymes and Cofactors in the P23H Retina

In addition to the increase in cytochrome oxidase activity and oxidative capacity, 670 nm LED treatment also significantly increased the concentrations of two important neuroprotective factors in the retina, the mitochondrial Mn-dependent form of superoxide dismutase, MnSOD, and ciliary neurotrophic factor (CNTF).

As shown in Fig. 7, 670 nm LED treatment significantly increased retinal concentrations of MnSOD (N = 6, p < 0.001). MnSOD is the mitochondrial

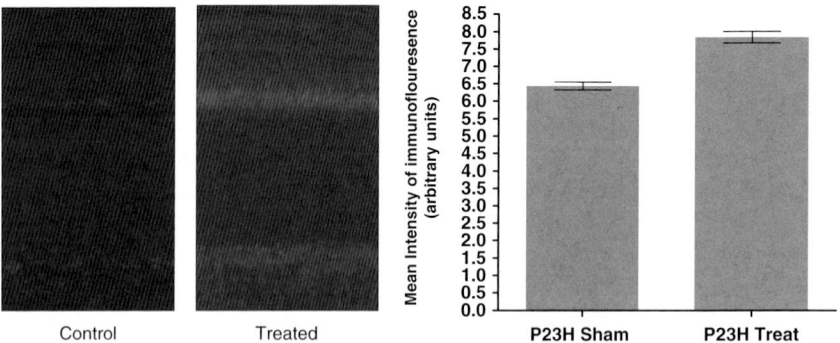

Fig. 6 670 nm LED treatment increases cytochrome oxidase activity in the P23H-3 retina

form of superoxide dismutase, an essential antioxidant enzyme responsible for the conversion of the highly reactive superoxide anion to the less reactive hydrogen peroxide. Mitochondria are the major source of superoxide production and are subjected to direct attack of reactive oxygen species (ROS). Mitochondria generate superoxide through a series of electron carriers arranged spatially according to their redox potentials. Mitochondrial dysfunction itself can lead to increased production of ROS, which can increase oxidative stress if the defense mechanisms of the cell are overwhelmed. ROS generated by mitochondria are considered to be integral factors in the retinal degenerative disease.

Figure 8 shows that 670 nm LED treatment significantly increases the concentration of CNTF in the developing P23H retina (N = 6, p < 0.001). We interpret these findings to indicate that 670 nm LED treatment upregulates the production of this neuroprotective factor by a mitochondrial mediated signaling mechanism. CNTF was first identified as a survival factor in studies involving ciliary ganglion neurons in the chick eye [9]. CNTF is a member of the IL-6 family of cytokines and

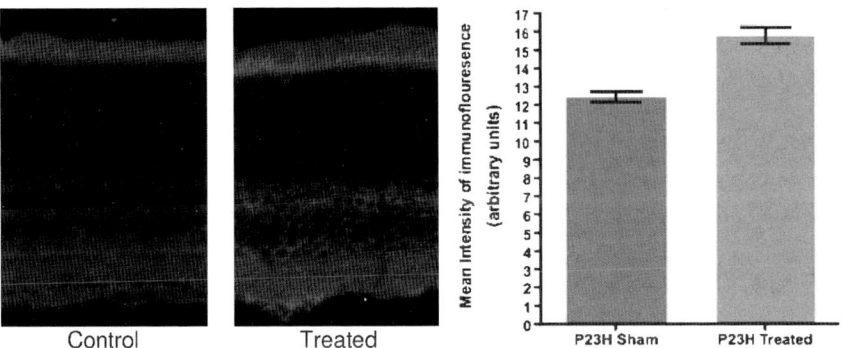

Fig. 7 670 nm LED treatment increases MnSOD activity in the P23H-3 retina

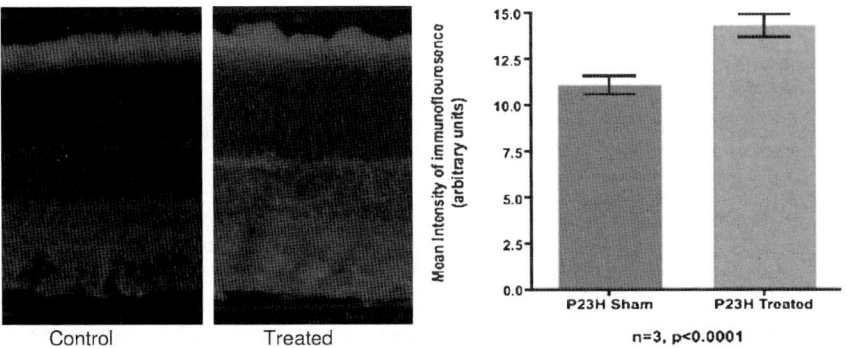

Fig. 8 670 nm LED treatment increases the protective factor CNTF in the P23H-3 retina

acts through a heterotrimeric receptor complex composed of CNTF receptor plus two signal-transducing transmembrane subunits. CNTF receptor is located on Müller glial membranes [9] and on rod and cone photoreceptors [9]. In animal studies, CNTF has been shown to be effective at retarding retinal degeneration in at least 13 RP models, including transgenic rats expressing the P23H mutation [9]. More recently, clinical studies have documented the potential of CNTF as therapeutic agent for RP [9].

Photobiomodulation Attenuates the Loss of Photoreceptors in the P23H Rat During the Critical Period

Our working hypothesis is that dysfunction of P23H-3 photoreceptors leads to a rise in oxygen tension in outer retina, which is toxic to photoreceptors [8], particularly during early development. We postulate that photobiomodulation acts to improve mitochondrial function and to increase antioxidant protection. Photobiomodulation-induced improvement of mitochondrial function would increase oxygen consumption in the P23H retina, reduce oxygen tension and thus reduce oxygen toxicity [8]. Secondly the photobiomodulation-induced increased production of cytoprotective factors and antioxidants would act in concert to protect the retina.

Figure 9 shows sections of P23H-3 retinas labeled for normal DNA (blue) and for the DNA fragmentation characteristic of apoptosis (red). The frequency of TUNEL + (dying) cells is an order of magnitude higher in the P23H-3 than in control (SD) animals (ref). 670 nm LED treatment reduced apoptotic photoreceptor cell death by more than 70% in the P23H-3 rat. These findings link the upregulation of antioxidants and cytoprotective factors to the prevention of photoreceptor apoptosis in the developing P23H-3 retina.

In summary, we have demonstrated that 670 nm LED treatment administered once per day for 5 days during the critical period of photoreceptor development in the P23H-3 rat increases retinal mitochondrial cytochrome oxidase activity,

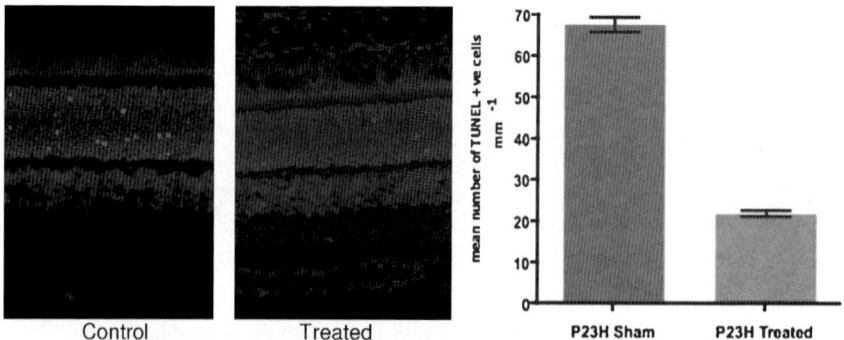

Fig. 9 670 nm LED treatment attenuates photoreceptor apoptosis in the P23H-3 retina

upregulates the production of antioxidant protective enzymes and cofactors in the retina, increases the production of retinal neuroprotective neurotrophic factors and prevents apoptotic photoreceptor cell death. These findings have profound implications for the use of photobiomodulation in the treatment of retinal degenerative diseases. They document the therapeutic potential of 670 nm photon therapy in a well-established rodent model of retinitis pigmentosa, thus setting the stage for clinical trials of photobiomodulation in human disease. The molecular, biochemical and functional insights obtained from this research will provide crucial information needed for a comprehensive FDA approval for the use of far-red to near-infrared LED devices in the treatment of retinal degenerative diseases. From a basic science perspective, they substantiate our previous findings *in vitro* and *in vivo* and strongly support our hypothesis that photobiomodulation augments mitochondrial function and stimulates antioxidant protective pathways in the neural retina to protect against retinal degeneration Finally, we believe that photobiomodulation has the potential to revolutionize the delivery of health care. In contrast to traditional medical approaches that rely on surgical or pharmacological intervention, photobiomodulation harnesses the cells' own potential for repair and provides an innovative and non-invasive therapeutic approach for the prevention and treatment of retinal dystrophies and degenerative disease.

Acknowledgments We gratefully acknowledge the technical support of Quantum Devices, Inc. for providing the light emitting diode arrays for this project and Anna Fekete for excellent technical assistance. This work was supported by Defense Advanced Research Projects Agency Grant DARPA N66001-01-1-8969 (H.T.W.), National Institute of Environmental Health Sciences Grant ES06648 (J.T.E), National Eye Institute Core Grant P30-EY01931 (J.T.E. and M.W. R., Core Investigators), National Eye Institute Grantand EY05439 (M.W.R.), the Bleser Foundation Endowed Professorship (H.T.W.), An Individual Investigator Grant TA-NE-0606-0348-UWI from the Foundation Fighting Blindness (J.T.E.) and the ARC Center for Excellence in Vision Research (K.V., J.S).

References

1. Hafezi, F. et al., *Molecular Ophthalmology: an update on animal models for retinal degenerations and dystrophies.* Br J Ophthalmol, 2000 84: 922–927.
2. Kanwar, M. et al., *Oxidative damage in the retinal mitochondria of diabetic mice: possible protection by superoxide dismutase.*
3. Ranchon, I. et al., *Free radical spin trap SPBN protects against light damage but does not rescue P23H and S334ter rhodopsin transgenic rats from inherited retinal degeneration.* J Neurosci, 2003 23: 6050–6057.
4. Liang, H.L. et al., *Photobiomodulation partially rescues visual cortical neurons from cyanide-induced apoptosis.* Neuroscience, 2006 139(2): 639–649.
5. Desmet, K.D. et al., *Clinical and experimental applications of NIR-LED photobiomodulation.* Photomed Laser Surg, 2006 24(2): 121–128.
6. Wong-Riley, M.T. et al., *Photobiomodulation directly benefits primary neurons functionally inactivated by toxins: role of cytochrome c oxidase.* J Biol Chem, 2005 280(6): 4761–4771.

7. Eells, J. et al., *Therapeutic photobiomodulation for methanol-induced retinal toxicity.* PNAS, 2003 100(6).
8. Stone, J. et al., *Mechanisms of photoreceptor death and survival in mammalian retina.* Prog Ret Eye Res, 1999 18: 689–735.
9. Sieving, P. et al., *Ciliary neurotropic factor (CNTF) for human retinal degeneration: phase 1 trial of CNTF delivered by encapsulated cell intraocular implants.* PNAS, 2006 103: 3896–3901.
10. Gargini, C. et al., *Electroretinogram changes associated with retinal upregulation of trophic factors: observations following optic nerve section.* Neuroscience, 2004 126(3): 775–783.
11. Valter, K. et al., *Time course of neurotrophic factor upregulation and retinal protection against light-induced damage after optic nerve section.* Invest Ophthalmol Vis Sci, 2005 46 (5): 1748–1754.
12. Eells, J.T. et al., *Mitochondrial signal transduction in accelerated wound and retinal healing by near-infrared light therapy.* Mitochondrion, 2004 4(5–6): 559–567.
13. Karu, T., *Primary and secondary mechanisms of action of visible to near-IR radiation on cells.* J Photochem Photobiol B, 1999 49: 1–17.
14. Eells, J. et al., *Near-infrared light therapy for retinitis pigmentosa.* Invest Ophthalmol Vis Sci, 2006: 47. ARVO Abstract 1022.
15. Geller, S. et al., *Toxicity of hyperoxia to the retina: evidence from the mouse,* in *Retinal Degenerative Disease,* J. Hollyfield, G. Anderson, and M. LaVail, Editors. 2005, Springer: Berlin/Heidelberg. pp. 425–438.
16. Algvere, P.V., J. Marshall, and S. Seregard, *Age-related maculopathy and the impact of blue light hazard.* Acta Ophthalmol Scand, 2006 84(1): 4–15.
17. Komeima, K. et al., *Antioxidants reduce cone cell death in a model of retinitis pigmentosa.* Proc Natl Acad Sci USA, 2006 103(30): 11300–11305.
18. Wellard, J. et al., *Photoreceptors in the rat retina are specifically vulnerable to both hypoxia and hyperoxia.* Vis Neurosci, 2005 22(4): 501–507.
19. Yu, D. et al., *Photoreceptor death, trophic factor expression, retinal oxygen status, and photoreceptor function in the P23H rat.* Invest Ophthalmol Vis Sci, 2004 45(6).
20. Valter, K. et al., *Photoreceptor dystrophy in the RCS rat: roles of oxygen, debris and bFGF.* Invest Ophthal Vis Sci, 1998 39: 2427–2442.
21. Nixon, P.J. et al., *The contribution of cone responses to rat electroretinograms.* Clin Exp Ophthalmol, 2001 29(3): 193–196.
22. Stone, J., et al., *Photoreceptor stability and degeneration in mammalian retina: lessons from the edge,* in *Macular Degeneration: Science and Medicine in Practice,* P. Penfold and J. Provis, Editors. 2005, Springer Verlag. pp. 149–165.
23. Bravo-Nuevo, A. et al., *Mitochondrial deletions in normal and degenerating rat retina,* in *Retinal Degenerations: Mechanisms and Experimental Therapy,* M. La Vail, J. Hollyfield, and G. Anderson, Editors. 2003, Kluwer/Plenum: New York. pp. 241–248.
24. Qi, X. et al., *Mitochondrial protein nitration primes neurodegeneration in experimental autoimmune encephalomyelitis.* J Biol Chem, 2006.
25. Shen, J. et al., *Oxidative damage is a potential cause of cone cell death in retinitis pigmentosa.* J Cell Physiol, 2005 203: 457–464.
26. Maslim, J. et al., *Tissue oxygen during a critical developmental period controls the death and survival of photoreceptors.* Invest Ophthal Vis Sci, 1997 38: 1667–1677.
27. Machida, S. et al., *P23H rhodopsin transgenic rat: correlation of retinal function with histopathology.* Invest Ophthalmol Vis Sci, 2000 41(10): 3200–3209.
28. Campian, J.L. et al., *Oxygen tolerance and coupling of mitochondrial electron transport.* J Biol Chem, 2004 279(45): 46580–46587.
29. Villani, G. et al., *Low reserve of cytochrome c oxidase capacity in vivo in the respiratory chain of a variety of human cell types.* J Biol Chem, 1998 273(48): 31829–31836.

30. Schuschereba ST, Cross MT, Pizarro JM, et al., *Pretreatment with hydroxyethyl starch-deferoxamine but not methylprednisolone reduces secondary injury to retina after laser irradiation.* Lasers Light, 1997 8: 1–14.
31. Whelan, HT, Wong-Riley, MTT, Eells, JT, VerHoeve, JN, Das, R, Jett, M., *DARPA soldier self care: rapid healing of laser eye injuries with light-emitting diode technology.* NATO RTO-MP-HFM, 2005 109: 1–18.

Irradiation with 780 nm Diode Laser Attenuates Inflammatory Cytokines While Upregulating Nitric Oxide in LPS-Stimulated Macrophages: Implications for the Prevention of Aneurysm Progression

Lilach Gavish, Louise S. Perez, Petachia Reissman, and S. David Gertz

Abstract Low level laser irradiation (LLLI) has been shown to reduce inflammation of tissue and we show that 780 nm radiation modifies certain processes fundamental to aneurysm progression. This study has been designed to determine the effect of LLLI on cytokine gene expression and secretion of inductible nitric oxide synthase (iNOS) and NO production in lipopolysaccharide (LPS) - stimulated macrophages.

Keywords: Cytokines, Nitric Oxide, 780 nm radiation, Aneurysm progression.

Introduction

Low level laser irradiation (LLLI) has been shown to reduce inflammation in a variety of clinical situations [2, 14, 15]. We have shown that LLLI (780 nm) modifies certain processes fundamental to aneurysm progression by increasing aortic smooth muscle cell proliferation and matrix protein secretion as well as modulating activity and expression of matrix metalloproteinases [6].

Inflammation is a major component of all arteriosclerotic diseases including aneurysm [17]. Macrophage recruitment and secretion of pro-inflammatory cytokines are central to most immune responses in the arterial wall. Activation of macrophages in vitro by Lipopolysaccharide (LPS), a component of the gram-negative bacterial wall, initiates de novo synthesis of a variety of factors including cytokines, chemokines, and inducible nitric oxide synthase (iNOS) through a well established signal transduction pathway [7, 10].

L. Gavish
Department of Anatomy and Cell Biology, The Hebrew University – Hadassah Medical School, Jerusalem, 91120, Israel, e-mail: lilachg@ekmd.huji.ac.il

R. Waynant and D.B. Tata (eds.), *Proceedings of Light-Activated Tissue Regeneration and Therapy Conference.*
© Springer Science + Business Media, LLC 2008

Nitric oxide (NO) serves as an effector in multiple physiological functions including vasodilatation, neurotransmission, and inflammation [12]. NO also has been shown to enhance production of collagen in several cellular systems including arterial smooth muscle cells [16] and to mediate growth factor-induced angiogenesis pathways thereby promoting neo-vascularisation [5, 11, 13].

The present study was designed to determine the effect of LLLI on cytokine gene expression and secretion as well as gene expression of inducible nitric oxide synthase (iNOS) and NO production in lipopolysaccharide (LPS)-stimulated macrophages.

Methods

This study was performed on the murine RAW 264.7 macrophage/monocyte cell line. The cells were seeded in subconfluent concentration 48 hours prior to irradiation. Macrophages were activated by addition of LPS (0, 0.1, and 1μg/ml) to the medium immediately before irradiation by 780 nm diode laser with 2 mW/cm^2 power density (total energy 2 J/cm^2). Following irradiation, cells were incubated for 2, 12 and 24 hours. Cytokine (IL-1-beta and MCP-1) and NO secretion were assessed from supernatant by ELISA and the Griess reaction, respectively. Gene expression of MCP-1, IL-1α, IL-1β, IL-6, IL-10, and TNFα was determined by total RNA extraction and reverse transcription polymerase chain reaction (RT-PCR) and normalized to the housekeeping gene G3PDH. Non-irradiated macrophages served as controls for all experiments.

Results

LLLI Inhibits LPS-Induced Cytokine-Chemokine Gene Expression

LLLI significantly reduced gene expression of the chemokine MCP-1, the pro-inflammatory cytokines IL1α, IL1β, IL-6 and TNFα, but also the anti-inflammatory cytokine IL10, when 1 μg/ml LPS was used as a stimulator. The extent of reduction of the LPS induced gene expression was 32 ± 7%, 25 ± 8%, and 23 ± 5% for MCP1, IL1α, and IL10 respectively (p < 0.01, by paired two-tailed t-test); and 16 ± 6%, 12 ± 5%, and 15 ± 8% for IL1β, IL6, and TNFα respectively (p < 0.05). However, LLLI did not influence cytokine-chemokine gene expression when cells were not stimulated with LPS or with those stimulated with low concentration of LPS (0.1 μg/ml).

We found that the inhibition of gene expression by LLLI was prominent only after a certain threshold of initial stimulation and that this effect increased with the level of initial stimulation. This effect of LLLI was found to deviate from linearity and to fit a parabolic inhibition behavior (r = 0.42, by non-linear regression analysis).

Low Level Laser Irradiation Inhibited LPS-Induced Secretion of MCP-1 and IL-1-β from Macrophages

Low Level Laser Irradiation reduced the LPS-induced secretion of macrophage MCP-1 over non-irradiated cells by $17 \pm 5\%$ ($p < 0.01$, by one-sample two-tailed t-test of the ratio, 0.1 µg/ml LPS) and $13 \pm 5\%$ ($p < 0.05$, 1 µg/ml LPS) at 12 hours. LLLI reduced the LPS-induced secretion of IL-1-beta by $22 \pm 5\%$ ($p < 0.01$, 0.1 µg/ml LPS) and $25 \pm 9\%$ ($p < 0.05$, 0.1 µg/ml LPS) at 24 hours.

LLLI Stimulates iNOS Gene Expression and NO Secretion

Low Level Laser Irradiation upregulated iNOS gene expression in cells not stimulated with LPS or those stimulated with low amounts of LPS by $108 \pm 31\%$ (pooled results, $p < 0.05$, by paired two-tailed t-test), 2 hours after treatment. However, LLLI did not affect cells stimulated with high levels of LPS. When subjecting data to baseline analysis we found that LLLI-induced stimulation of iNOS gene expression occurs only below a certain threshold of initial stimulation.

Low Level Laser Irradiation increased NO secretion compared to non-irradiated cells (LLLI vs Control: 1.72 ± 0.37 vs 0.95 ± 0.4 µM [without LPS, $p < 0.0005$ by paired, two-tailed t-test]; 7.46 ± 1.62 vs 4.44 ± 1.73 µM [0.1 µg/ml LPS, $p < 0.05$]; and 10.91 ± 3.53 vs 6.88 ± 1.52 µM [1 µg/ml LPS, $p < 0.02$]). The % increase of NO secretion following LLLI was found to be $100.3 \pm 59\%$, $90 \pm 86\%$ and $58 \pm 37\%$ for 0, 0.1, and 1 µg/ml LPS respectively.

Discussion

We have demonstrated that Low Level Laser Irradiation inhibits the gene expression of MCP-1, TNFα, IL-1α, IL-1-β, IL-6, and IL-10 in macrophages stimulated with high concentration of LPS. Likewise, LLLI reduced the secretion of MCP-1 and IL1β protein. On the other hand, LLLI increased the gene expression of iNOS and increased the secretion of NO. Several studies have emphasized the anti-inflammatory properties of LLLI [1, 3, 4]. We found that the effect of LLLI on the cytokines studied appeared to be non-specific—reducing gene expression of pro-inflammatory or anti-inflammatory cytokines alike. Therefore, further studies are necessary to test the possibility that LLLI might influence some upstream regulative factor that is common to these cytokines. Also, we have shown that the effect of LLLI is dependent on the extent of the inflammatory response. That these effects of laser may be elicited primarily on inflamed rather than normal tissues would appear to be of significant clinical importance.

Previous *in vitro* studies by Klebanov's group [8,9] showed that HeNe laser irradiation increased NO secretion from macrophages at a range compatible to that found in the current study. This suggests the possibility of a common maximum photoreactive potential for upregulation of the NO synthetic pathway.

We have shown previously that LLLI in 780 nm at 2 J/cm^2 enhances collagen synthesis in porcine aortic smooth muscle cells [6]. Oron and colleagues, using similar laser parameters, have shown that LLLI enhances vascular endothelial growth factor (VEGF) and iNOS expression as well as angiogenesis in a rat model of acute myocardial infarction [18]. It has also been shown that addition of exogenous NO to arterial smooth muscle cells enhances synthesis of extracellular collagen [16] and promotes angiogenesis through stimulation of VEGF production by rat vascular smooth muscle cells [5], endothelial cells [13] and cardiomyocytes [11]. Thus, when considered together, these studies suggest that LLLI stimulation of NO secretion from macrophages might result in enhanced collagen synthesis in smooth muscle cells and promote angiogenesis in the vascular wall.

These properties of LLLI, with its effects on smooth muscle cells reported previously, and the effects on macrophages reported here, may be of profound therapeutic relevance for aneurysm where inflammatory processes and weakening of the matrix structure of the arterial wall are major pathologic components. The finding that LLLI reduces cytokines only above a certain LPS stimulation threshold suggests that this suppressive effect may be minimal in non-inflammatory conditions.

Acknowledgments We thank Dr. Benjamin Gavish for help with the statistical analysis. Dr. Gertz is the Chutick Professor of Cardiac Studies, The Hebrew University, Faculty of Medicine, Jerusalem, Israel.

References

1. Aimbire F, Albertini R, Pacheco MT, Castro-Faria-Neto HC, Leonardo PS, Iversen VV, Lopes-Martins RA, Bjordal JM (2006) Low-level laser therapy induces dose-dependent reduction of TNFalpha levels in acute inflammation. Photomed Laser Surg 24:33–37
2. Bjordal JM, Couppe C, Chow RT, Tuner J, Ljunggren EA (2003) A systematic review of low level laser therapy with location-specific doses for pain from chronic joint disorders. Aust J Physiother 49:107–116
3. Bjordal JM, Lopes-Martins RA, Iversen VV (2006) A randomised, placebo controlled trial of low level laser therapy for activated Achilles tendinitis with microdialysis measurement of peritendinous prostaglandin E2 concentrations. Br J Sports Med 40:76–80; discussion 76–80
4. Byrnes KR, Wu X, Waynant RW, Ilev IK, Anders JJ (2005) Low power laser irradiation alters gene expression of olfactory ensheathing cells in vitro. Lasers Surg Med 37:161–171
5. Dulak J, Jozkowicz A, Dembinska-Kiec A, Guevara I, Zdzienicka A, Zmudzinska-Grochot D, Florek I, Wojtowicz A, Szuba A, Cooke JP (2000) Nitric oxide induces the synthesis of vascular endothelial growth factor by rat vascular smooth muscle cells. Arterioscler Thromb Vasc Biol 20:659–666

6. Gavish L, Perez L, Gertz SD (2006) Low-level laser irradiation modulates matrix metallo-proteinase activity and gene expression in porcine aortic smooth muscle cells. Lasers Surg Med 38:779–786

7. Guha M, Mackman N (2001) LPS induction of gene expression in human monocytes. Cell Signal 13:85–94

8. Klebanov GI, Ea P, Vladimirov Iu A (2003) Effect of low intensity laser light in the red range on macrophage superoxide dismutase activity. Biofizika 48:462–473

9. Klebanov GI, Poltanov EA, Chichuk TV, Osipov AN, Vladimirov YA (2005) Changes in superoxide dismutase activity and peroxynitrite content in rat peritoneal macrophages exposed to He-Ne laser radiation. Biochemistry (Mosc) 70:1335–1340

10. Kopydlowski KM, Salkowski CA, Cody MJ, van Rooijen N, Major J, Hamilton TA, Vogel SN (1999) Regulation of macrophage chemokine expression by lipopolysaccharide in vitro and in vivo. J Immunol 163:1537–1544

11. Kuwabara M, Kakinuma Y, Ando M, Katare RG, Yamasaki F, Doi Y, Sato T (2006) Nitric oxide stimulates vascular endothelial growth factor production in cardiomyocytes involved in angiogenesis. J Physiol Sci 56:95–101

12. Lowenstein CJ, Padalko E (2004) iNOS (NOS2) at a glance. J Cell Sci 117:2865–2867

13. Papapetropoulos A, Garcia-Cardena G, Madri JA, Sessa WC (1997) Nitric oxide production contributes to the angiogenic properties of vascular endothelial growth factor in human endothelial cells. J Clin Invest 100:3131–3139

14. Pinheiro AL, Cavalcanti ET, Pinheiro TI, Alves MJ, Miranda ER, De Quevedo AS, Manzi CT, Vieira AL, Rolim AB (1998) Low-level laser therapy is an important tool to treat disorders of the maxillofacial region. J Clin Laser Med Surg 16:223–226

15. Schindl A, Schindl M, Pernerstorfer-Schon H, Kerschan K, Knobler R, Schindl L (1999) Diabetic neuropathic foot ulcer: successful treatment by low-intensity laser therapy. Dermatology 198:314–316

16. Schmidt A, Geigenmueller S, Voelker W, Seiler P, Buddecke E (2003) Exogenous nitric oxide causes overexpression of TGF-beta1 and overproduction of extracellular matrix in human coronary smooth muscle cells. Cardiovasc Res 58:671–678

17. Thompson RW, Geraghty PJ, Lee JK (2002) Abdominal aortic aneurysms: basic mechanisms and clinical implications. Curr Probl Surg 39:110–230

18. Tuby H, Maltz L, Oron U (2006) Modulations of VEGF and iNOS in the rat heart by low level laser therapy are associated with cardioprotection and enhanced angiogenesis. Lasers Surg Med 38:682–688

New Aspects of Wound Healing

A. Lipovsky, Ankri R. Nitzan, Z.A. Landoy, J. Jacobi, and R. Lubart

Abstract Broadband light sources of light in the 400–800 nm region were found to be effective for treatment of *Staphylococcus aureus* if sufficient doses were applied.

Keywords: Broad-band light sources, *Staphylococcus aureus*, photobiomodulation.

Introduction

It has been repeatedly reported that visible and near infrared light induce various stimulatory effects in biological cells. This phenomenon is potentially useful as a non-harmful therapeutic mechanism, and has already been used for several therapeutic purposes.

Various devices have been implemented in light therapy especially in wound healing. The most prevalent to date are the low level lasers in the visible and near infrared region ($\sim$10 mW/cm^2), and LEDs, which typically produce low energy intensities (10–50 mW/cm^2) at a band width of around 10 nm.

Photobiomodulation should be distinguished from the effects of heating. Visible light is used to selectively heat tissue in order to treat conditions such as vascular lesions [1], but thermal effects are not considered as photobiomodulation. Typically, the light doses used for photobiomodulation are too small to cause significant heating. Effects of light that are mediated by cells with specialized photoreceptors, such as vision, circadian rhythms and photosynthesis, are also not considered photobiomodulation.

Photobiomodulation has been reported to induce many cell processes, such as proliferation [2–5], spermatozoa fertilization [6] and motility [7], action potentials [8] cell differentiation [9], protection of cells from damage [10] and recovery of damaged cells [11]. Other reported effects are the stimulation of collagen synthesis [12], release of cytokines [13], growth factors [14–16] and unidentified pro-mitotic factors [17].

R. Lubart

Bar Ilan University, Ramat Gan, Ramat Gan 52900 Israel, e-mail: Lubartr@mail.biu.ac.il

R. Waynant and D.B. Tata (eds.), *Proceedings of Light-Activated Tissue Regeneration and Therapy Conference.*

Therapeutic uses of photobiomodulation that have been tried include pain relief [18–21] and wound healing [22, 23], with mixed results. Clinical use of photobiomodulation is quite controversial, with charges of biased interpretation of results having been made on both sides [24, 25].

It is usually assumed that light must be absorbed in order to affect the cell. If, as is commonly assumed, light must be absorbed to affect cells, only molecules that absorb visible light can initiate photobiomodulation. Few of the many types of biological molecules absorb visible light. Furthermore, for most excited molecules, the excitation is dissipated quickly as heat. Because photobiomodulation, by definition, is not a phenomenon caused by heat, it must be initiated by molecules that initiate a chemical reaction before relaxation. The nature of the initiated chemical reaction is unknown, although various proposals have been put forward. The two clearest proposals are probably the creation of reactive oxygen species (ROS) and the acceleration of mitochondrial transport.

It is known that visible and near IR light can be absorbed by cellular photosensitizers such as cytochromes, flavins/riboflavins and NADP [35].

Absorption of light by these photosensitizers cause their excitation and relaxation by transferring electrons to O_2 thereby generating the ROS. ROS are probably best known in biology for their ability to cause oxidative stress. ROS can damage DNA, cell membranes and cellular proteins, leading to cell death or carcinogenesis [26]. This phenomenon is utilized by photodynamic therapy to kill cancer cells, and bacteria. However, moderate doses of ROS have been found to have numerous non-harmful effects in cells. It has been demonstrated that low ROS fluxes play an important role in activation and control of many cellular processes, such as transcription factors release, gene expression, muscle contraction and cell growth [31, 32].

In a very recent work of us [33], we tried to identify the endogenous photosensitizers responsible for ROS production by visible light. We used the electron paramagnetic resonance (EPR) coupled with the probe trapping technique, to monitor oxyradicals produced in various cell cultures as a function of illumination wavelength. We found that oxyradicals were created mainly by the flavins at 400–500 nm range of the visible light. Wavelengths above 500 nm probably stimulate the cell by accelerating the mitochondrial respiratory electron transfer as first suggested by Karu [34].

We claim that because the absorption bands of cytochrome oxidase and flavoproteins are quite wide [30], no expensive lasers are needed for their excitation. Until recently, broadband (400–800 nm), light emitting systems, were neglected. However, we have shown that broadband visible light, at the appropriate energy dose, mimic LLL [27, 28].

Realizing that a non-laser multiwave light source in the visible range is capable of ROS generation, and enhances fibroblasts proliferation as low energy lasers do, we decided to use this phototherapy device for wound healing. In this preliminary work we report our experience with broadband visible light device in the treatment of patients with chronic ulcers when conventional treatment failed. We also studied the sterilization effect of high intensity visible light since recently Feuerstein et al.

[29] have shown that intense blue visible light source exerts a phototoxic effect on *P. gingivalis* and *F. nucleatum*. We decided to examine the photo toxicity of our broadband (400–800 nm), visible light source on *Staphylococcus aureus* since it exists in many wounds. Two strains of the *Staphylococcus aureus* bacteria were chosen. We have found that white light at high intensity has a bactericidal effect while at low intensity it stimulates bacteria proliferation. This finding might be very important for wound healing.

Methods and Results

Wound Healing

Twenty seven patients with chronic diabetic or chronic venous ulcers, in which conventional treatment failed, were included in this study. Among them seven patients served as a placebo group. All patients underwent an evaluation that included a complete history, physical and neurological examinations and appropriate blood tests. Conventional treatment was administrated, including topical treatment and pressure bandages in case of venous ulcers, debridement, and oral antibiotics according to wound cultures and weigh off-loading. In the case of diabetic foot ulcers this was combined with local ulcer care (with various creams, ointments or saline). The light treatment was performed at home and the patients were seen every week or every other week.

Each patient was irradiated with 40 mW/cm^2 broadband visible light device 400–800 nm, 5 min, three times a day. During each treatment a dose of 12 J/cm^2 was irradiated to the surface of the ulcer. In the placebo group each patient was irradiated with a 5 mW/cm^2 broadband light device delivering a dose of 1.5 J/cm^2 to the ulcer surface. Conventional treatment was continued in all patients during the treatment period. Patients were monitored at least once weekly, and the ulcer appearance was recorded by the same observer. The primary end point was ulcer healing defined as complete closure of the ulcer.

The mean duration of the ulcers before the start of treatment was 3.1 ± 2 months. The mean duration of treatment was 40 ± 10 days. Complete closure of the ulcers was obtained in 14 (70%) of the patients. In the placebo group no complete closure was observed within 50 days of treatment.

Light-Induced Bacteria Destruction

Viability of *Staphylococcus aureus* 101 and 500 was assessed after bacteria in suspension were exposed to white light (400–800 nm), 380 mW/cm^2 for different time intervals. Bacteria viability is expressed by CFU/ml (Colony Forming Units).

We have found that for staphylococcus aureus 101 a reduction of survival begins when the time of irradiation exceeds 4 min. After 10 min, the reduction reached 2.9

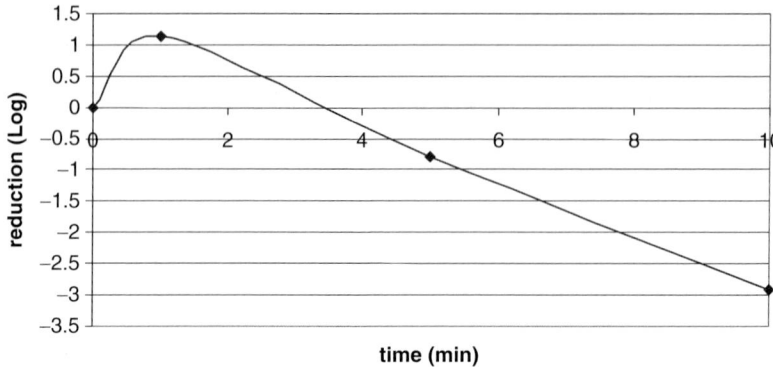

Fig. 1 Reduction of *Staphylococcus aureus* 101 as a function of the illumination time

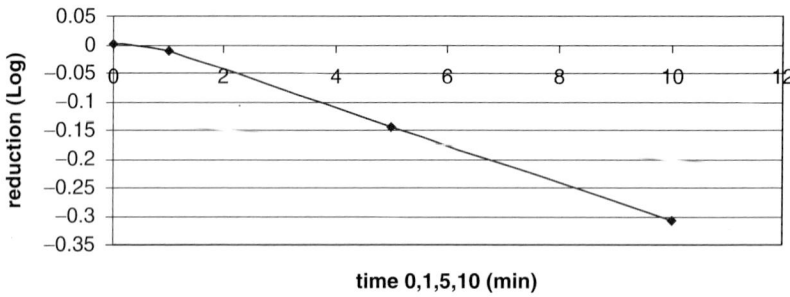

Fig. 2 Reduction of *Staphylococcus aureus* 500 as a function of the illumination time

log. Low energy doses of light increased the number of bacteria by 1.14 log (Fig. 1). On the other hand for *Staphylococcus aureus* 500, the maximal reduction was 0.3 log and no proliferation was observed at low energy doses (Fig. 2).

Discussion

Diabetic foot (DFU) and chronic venous ulcers (CVU) are cutaneous lesions that are difficult to treat and heal. Conventional therapies including local treatment/debridement and antibiotics are frequently ineffective. Even when successful, these treatments result in wound healing that occurs slowly and with a great deal of pain and disability. Our preliminary study showed a therapeutic effect of low energy white light. In 70% of our patients a complete ulcer closing was observed. Unlike lasers and LEDs in the visible range which stimulate cell activity at a single wavelength and at a very small area, a **broadband** of frequencies excites many cellular targets and also enables the coverage of **large surfaces**, which is a key to large surface applications such as wounds and burns. Note that there are LED arrays to answer this need, but the devices are usually cumbersome and produce fairly low intensity. The ability to irradiate large areas with relatively high energy, contributes to a significantly **shorter session time**, requiring only 2–5 minutes. Not less important, compared with lasers and LEDs systems, broadband visible light systems are **much less expensive**, opening the door for home application.

As to the sterilization effect of visible light, we have found that only very high light doses (over 90 J/cm^2) might have a bactericidal effect, depending on the bacteria strain. Low energy doses are dangerous for infected wounds since they might stimulate bacteria proliferation.

It comes out from this preliminary study that wounds can be sterilized with high doses of visible light and then their closure can be accelerated by low energy white light. Illumination of infected wounds with low power visible light is not recommended since it might increase bacteria viability.

References

1. Clark C, Cameron H, Moseley H, Ferguson J, Ibbotson SH. "Treatment of superficial cutaneous vascular lesions: experience with the KTP 532 nm laser" *Lasers Med Sci* 2004; 19:1–5.
2. Ben-Dov N, Shefer G, Irintchev A, Wernig A, Oron U, Halevy O. "Low-energy laser irradiation affects satellite cell proliferation and differentiation in vitro" *Biochim Biophys Acta* 1999; 1448:372–380.
3. Moore P, Ridgway TD, Higbee RG, Howard EW, Lucroy MD. "Effect of wavelength on low-intensity laser irradiation-stimulated cell proliferation in vitro" *Lasers Surg Med* 2005; 36:8–12.
4. Gulsoy M, Ozer GH, Bozkulak O, Tabakoglu HO, Aktas E, Deniz G, Ertan C. "The biological effects of 632.8-nm low energy He-Ne laser on peripheral blood mononuclear cells in vitro" *J Photochem Photobiol B* 2006; 82:199–202.
5. Grbavac RA, Veeck EB, Bernard JP, Ramalho LM, Pinheiro AL. "Effect of laser therapy in CO2 laser wounds in rats" *Photomed Laser Surg* 2006; 24:389–396.
6. Cohen N, Lubart R, Rubinstein S, Breitbart H. "Light irradiation of mouse spermatozoa: stimulation of in vitro fertilization and calcium signals" *Photochem Photobiol* 1998; 68:407–413.
7. Corral-Baques MI, Rigau T, Rivera M, Rodriguez JE, Rigau J. "Effect of 655-nm diode laser on dog sperm motility" *Lasers Med Sci* 2005; 20:28–34.
8. Walker JB, Akhanjee LK. "Laser-induced somatosensory evoked potentials: evidence of photosensitivity in peripheral nerves" *Brain Res* 1985; 344:281–285.
9. Stein A, Benayahu D, Maltz L, Oron U. "Low-level irradiation promotes proliferation and differentiation of human osteoblasts in vitro" *Photomed Laser Surg* 2005; 23:161–166.
10. Oron U. "Photoengineering of tissue repair in skeletal and cardiac muscles" *Photomed Laser Surg* 2006; 24:111–120.
11. Rochkind S, Ouaknine GE. "New trend in neuroscience: low-power laser effect on peripheral and central nervous system (basic science, preclinical and clinical studies) " *Neurol Res* 1992; 14:2–11.
12. Saperia D, Glassberg E, Lyons RF, Abergel RP, Baneux P, Castel JC, Dwyer RM, Uitto J. "Demonstration of elevated type I and type III procollagen mRNA levels in cutaneous wounds treated with helium-neon laser. Proposed mechanism for enhanced wound healing" *Biochem Biophys Res Commun* 1986; 138:1123–1128.

13. Funk JO, Kruse A, Kirchner H. "Cytokine production after helium-neon laser irradiation in cultures of human peripheral blood mononuclear cells" *J Photochem Photobiol B* 1992; 16:347–355.

14. Schwartz F, Brodie C, Appel E, Kazimirsky G, Shainberg A. "Effect of helium/neon laser irradiation on nerve growth factor synthesis and secretion in skeletal muscle cultures" *J Photochem Photobiol B* 2002; 66:195–2000.

15. Yu W, Naim JO, Lanzafame RJ. "The effect of irradiation on the release of bFGF from 3T3 fibroblasts" *Photochem Photobiol* 1994; 59:167–170.

16. Yu HS, Wu CS, Yu CL, Kao YH, Chiou MH. "Helium-neon laser irradiation stimulates migration and proliferation in melanocytes and induces repigmentation in segmental-type vitiligo" *J Invest Dermatol* 2003; 120:56–64.

17. Young S, Bolton P, Dyson M, Harvey W, Diamantopoulos C. "Macrophage responsiveness to light therapy" *Lasers Surg Med* 1989; 9:497–505.

18. Naeser MA. "Photobiomodulation of pain in carpal tunnel syndrome: review of seven laser therapy studies" *Photomed Laser Surg* 2006; 24:101–110.

19. Stasinopoulos DI, Johnson MI. "Effectiveness of low-level laser therapy for lateral elbow tendinopathy" *Photomed Laser Surg* 2005; 23:425–430.

20. Chow RT, Barnsley L. "Systematic review of the literature of low-level laser therapy (LLLT) in the management of neck pain" *Laser Surg Med* 2005; 37:46–52.

21. Chow RT, Heller GZ, Barnsley L. "The effect of 300 mW, 830 nm laser on chronic neck pain: a double-blind, randomized, placebo-controlled study" *Pain* 2006; 124:201–210.

22. Faria Amorim JC, Sousa GR, Silveira Lde B, Prates RA, Pinotti M, Ribeiro MS. "Clinical study of the gingiva healing after gingivectomy and low-laser therapy" *Photomed Laser Surg* 2006; 24:588–594.

23. Kopera D, Kokol R, Berger C, Haas J. "Does the use of low-level laser influence wound healing in chronic venous leg ulcers?" *J Wound Care* 2005; 14:391–394.

24. Bjordal JM, Bogen B, Lopes-Martins RA, Klovning A. "Can Cochrane reviews in controversial areas be biased? A sensitivity analysis based on the protocol of a Systematic Cochrane Review on low-level laser therapy in osteoarthritis" *Photomed Laser Surg* 2005; 23:453–458.

25. Lucas C, Criens-Poublon LJ, Cockrell CT, de Haan RJ. "Wound healing in cell studies and animal model experiments by low level laser therapy; were clinical studies justified? A systematic review" *Lasers Med Sci* 2002; 17:110–134.

26. Halliwell B, Aruoma OI. "DNA damage by oxygen-derived species. Its mechanism and measurement in mammalian systems" *FEBS Lett* 1991; 281:9–19.

27. Lavi R, Shainberg A, Friedmann H, Shneyvays V, Rickover O, Eichler M, Kaplan D, Lubart R. "Low energy visible light induced ROS generation and stimulates an increase of intracellular calcium concentration in cardac cells" *J Biol Chem* 2003; 278:40917–40922.

28. Lubart R, Landau Z, Jacobi J. "A new approach to ulcer treatment using broadband visible light" *Laser Therapy* 2007; 16(1):7–10.

29. Feuerstein O. "Mechanism of visible light phototoxicity on Porphyromonas gingivalis and Fusobacterium nucleatum" *Photochem Photobiol* 2004; 80:412–441.

30. Lubart R, Breitbart H. "Biostimulative effects of low-energy lasers and their implications for medicine" *Drug Develop Res* 2000; 50:471–475.

31. Suzuki YJ, Ford GD. "Redox regulation of signal transduction in cardiac and smooth muscle" *J Mol Cell Cardiol* 1999; 31:345–353.

32. Rhee SG. "Redox signaling: hydrogen peroxide as intracellular messenger" *Exp Mol Med* 1999; 31:53–59.

33. Eichler M, Lavi R, Shainberg A, Lubart R. "Flavins are source of visible-light-induced free radical formation in cells" *Lasers Surg Med* 2005; 37:314–9.

34. Karu TI. Photobiology of low-power laser therapy. Switzerland: Harwood Academic Publishers. 1989.

35. Lubart R, Friedman H, Lavie R. "Photobiostimulation as a function of different wavelengths" *Laser Therapy* 2000; 12:38–41.

Part III
Photodynamic Theraphy

An Introduction to Low-Level Light Therapy

Stuart K. Bisland

Abstract There have been numerous reports describing the phenomena of low-level light therapy (LLLT) within the clinic and its broad application to alleviate pain, enhance the rate of wound healing, including spinal cord injury, reduce inflammation, improve learning, bolster immunity and combat disease. Yet, despite the breadth of potential applications for which bio-stimulation may prove beneficial, there persists a dramatic ignorance in our understanding of the signal pathways that govern these effects. At the cellular level, there exist a variety of endogenous chromophores such as cytochrome c oxidase, NADPH, FAD, FMN and other factors intrinsic to the electron transport chain in mitochondria that absorb light of specific wavelength and will undoubtedly have their role in bio-stimulation, however the dose dependency of effect with regard to total light fluence and fluence rate, as well as the importance of specific subcellular targeting, remains elusive. Furthermore, the translation of cellular response(s) in vitro to in vivo needs to be expounded. Clearly, a rigorous examination of bio-stimulatory parameters as a function of cellular and tissue response is necessary if we are to attain optimized, reproducible protocols based on a true scientific rationale for using bio-stimulation as a therapeutic modality in clinic. This paper introduces a number of the challenges we now face for advancing the bio-stimulation phenomena into the scientific mainstream by highlighting our current knowledge in this field as well as some of the research that we are conducting using LLLT in combination with photodynamic therapy.

Keywords: Low-level light therapy, photobiomodulation, cytochrome c oxidase, apoptosis, mitochondria, laser; LED.

S.K. Bisland
Ontario Cancer Institute, University Health Network, University of Toronto, 610 University Avenue, Toronto, Ontario M5G 2M9, e-mail: sbisland@uhnres.utoronto.ca

R. Waynant and D.B. Tata (eds.), *Proceedings of Light-Activated Tissue Regeneration and Therapy Conference.*
© Springer Science + Business Media, LLC 2008

Low-Level Light Therapy

What Is Low Level Light Therapy (LLLT)?

LLLT refers to the application of low level light to biological systems with the intent to provide therapy. A simple enough concept but one clouded by discord and scepticism, for instance, the definition 'low level' remains ambiguous due to differing dosing regimens and may not necessarily translate to low intensity. Similarly, the use of the word 'therapy' could be misconstrued when a growing sector of LLLT is devoted to cosmetics. Finally, the underlying mechanisms of LLLT are a continuing source for conjecture. Regardless of this, one resounding fact is clear, that LLLT provides therapy to many thousands of people around the world for very different types of ailment. Indeed, it seems somewhat ironic that billions of dollars are spent each year researching new therapeutic strategies and devices, most of which never make it past preclinical investigation yet LLLT, despite the estimated 3,000 scientific papers published in the past 30 years, remains by all accounts a 'mystical medicine' with a bigger commercial market than perhaps any other medical therapy currently existing [1].

Current terminology confounds the field of LLLT. The application itself is often referred to as low power laser therapy (LPLT), bio-stimulation, low power red laser light therapy (LPRLLT) or light emitting diode therapy (LEDT). Similarly, the categories of lasers being used are also fraught with inconsistent nomenclature, from cold, soft and hard lasers to bio-regulatory lasers, therapeutic lasers and low intensity lasers, all basically describing the same group of lasers. This is compounded by the fact that a growing number of LLLT protocols use non-coherent sources of light such as light emitting diodes (LED) and halogen lamps, a distinction not usually clarified by the nomenclature. Ultimately, this lack of consensus only hampers the ability for LLLT to gain understanding and acceptance among the broader scientific and medical communities. Recently, in an attempt to consolidate thinking, the North American Space Agency (NASA) and the United States Military chose to adopt the unifying term, 'photobiomodulation' to describe the field and with similar insight the Laser Institute of America (LIA) and Food and Drug Administration (FDA) refer to LLLT lasers as low power medical lasers.

Tools of the Trade

Despite the principles defining the theory of light amplification by stimulated emission of radiation (LASER) first proposed in 1917 by Albert Einstein, it was not until the early 1960s that Theodore Maiman, based on the ideas of Schawlow and Townes at Bell Laboratories, was able to demonstrate the first working laser involving the use of a ruby crystal [2]. This was followed in consecutive years by other crystalline-based lasers including the yttrium-aluminum-garnet treated with 1–3% neodymium (Nd:YAG) laser and the helium/neon gas laser in 1961. Indeed, many of the lasers currently used in LLLT were developed in the early to mid 1960s involving the use of rare earth elements such as erbium (Er), gadolinium (Gd), holmium (Ho), praseodymium (Pr), thulium (Tm), uranium (Ur), and ytterbium (Yb). Within 3 years of this carbon dioxide lasers were being made that could cut through steel.

Legislation

According to federal legislative bodies most lasers used in LLLT fall into the category of low level medical lasers, class III (R; relaxed requirements) in accordance with guidelines prepared by the international commission on Non-ionizing radiation protection (ICNIRP) and presented to the world by the International Electrotechnical commission (IEC). These are lasers that can emit ≤5 mW of light, however, the majority of lasers currently being used for LLLT emit well beyond this level, up to 100 mW. IEC are recognised by the FDA as being comparable to the Centre for Devices and Radiological Health (CDRH) and all lasers for human and animal use in the United States and Canada must conform to Mandatory Performance standards as stipulated by CDRH office of Compliance and Surveillance. Readers should refer to the website: http://www.fda.gov/cdrh/comp/guidance for more information. Thus the FDA has introduced a new Pre-market Notification process [510(K)] that considers whether new laser devices meet the necessary standards to be sold commercially for pre-specified LLLT applications. Unfortunately, due to the overwhelming number of research and clinical protocols involving the use of lasers, it is very difficult for the FDA to maintain accurate lists of all laser manufacturers to ensure that their products meet these standards and as a result there are a growing number of lasers being used illegally. Similarly, physicians using lasers on humans for LLLT must be approved by an Institutional Review Board (IRB) or equivalent (Research Ethics Board in Canada) and investigators must seek approval under an Investigational Device Exemption (IDE) as proffered by the CDRH. However, comparable protocols for veterinary use do not require IDE or IRB approval.

Ultimately, much of the confusion as to whether a laser should be considered a low level laser or not stems not from the power output capacity of the laser but rather the total delivered light dose and fluence rate being used which may be substantial despite low output powers. Indeed, this realization together with the increasing use of incoherent broadband radiation such as LEDs and lamps prompted the inclusion of LEDs and certain lamps into the international laser safety standards (for Europe) in 1993 that define the maximum safe levels of exposure to skin and eye. The United States has yet to follow suit. LED arrays and lamps with near infra-red filters are becoming more and more common-place in LLLT protocols, offering relaxed legislation, reduced cost, flexibility in design and ease of use. This in turn has prompted a number of studies comparing the efficacy of lasers and LEDs for LLLT using either high or low frequency pulsed or continuous wave light and low

Table 1 Examples of lasers currently being used for LLLT

Clinical practice	Specific application	Laser(s) used	λ (nm)	References
Dermatology	Acne,[a] psoriasis, port wine stain,[b] rosacea, wrinkle[a,b] and hair removal,[b] wound repair,[c] infection[c] and burns[c]	KTP[a] Nd-YAG[b] Helium-neon[c]	532 1,064,595 632.8	[2–6]
Rheumatology	Musculoskeletal,[a] joint pain, inflammation[b–d] and infection[b,c]	GaAs[a] Helium-neon[b] GaAlAs[c]	904 632.8 780–870	[7–9, 12]
Immunology	Innate and adaptive immunity	Helium-neon	632.8	[1, 4, 7]
Psychology	Addictions, seasonal disorders, intelligence	Helium-neon	543,632.8	[11]
Reflexology	Pain,[a,b] migraine,[a] weight loss	Helium-neon[a] GaAs[b]	632.8 904	[9]
Dentistry	Tooth hypersensitivity,[a–c] TMJ disease, neuralgia,[d] inflammation[d]	GaAs[a] Helium-neon[b] Nd-YAG[c] GaAlAs[d]	904 632.8 1,064 790	[9, 10]
Orthopaedics	Bone repair,[a] neuralgia, infection	GaAlAs[a]	810	[13]
Chiropractology	Pain/stiffness associated with axial and appendicular skeleton	GaAlAs Helium-neon	810 632.8	[14]
Neurology	Nerve regeneration following spinal cord injury	GaAlAs	780–870	[15–17]

TMJ, Temporomandibular joint; KTP, Potassium titanyl phosphate; GaAlAs, Gallium aluminium arsenide.

or high power density, all of which has a definitive role in how well the LLLT works. A list of some lasers currently being used for LLLT is shown in Table 1.

Current Applications of LLLT

It is beyond the scope of this manuscript to attempt to describe all the applications for which LLLT is currently being used, however a number of defined clinical categories can be identified for which LLLT has proven to be particularly effective. Dermatologists use LLLT for cosmetic and medical conditions such as wrinkle and hair removal [3], treating wounds [4], acne vulgaris [5], port wine stain(s) and telangiectasia [6], while rheumatologists can alleviate pain in arthritic joints, muscle, nerves and bone as well as many types of inflammation such as tendonitis, brucitis, stomatitis and acute epididymitis [7]. Indeed the use of LLLT, by no means a panacea, has proven effective for many types of pain management associated with nerves (neuralgia), musculoskeletal conditions (such as fibromyalgia) [8] and migraine. Furthermore, there have been >100 double-blinded clinical trials now supporting the use of LLLT in alleviating pain [9, 10]. Some of the more controversial claims of LLLT include weight loss, alcohol [11] and smoke cessation and even the ability to increase your intelligence, which may or may derive from its ability to curb neuronal apoptosis (see below). Although it is doubtless that some of these claims prescribe to the placebo phenomenon, a growing number of 'believers' are testament to the credibility of LLLT, which this in turn stimulates renewed interest and research.

The Science of LLLT

Proposed Mechanisms

Currently, the literature provides no clear consensus for what is the optimal light source, wavelength and power dosimetry to use when treating a particular ailment using LLLT. In spite of this, there does appear to be some general agreement as to the underlying mechanisms responsible for the therapeutic effects achieved, albeit often in the absence of any concrete scientific evidence to support these hypotheses. The mediators of LLLT at the cellular level are suggested to include increased energy within cells in the form of adenosine triphosphate (ATP), increased deoxyribonucleic acid and ribonucleic acid synthesis, nitric oxide (NO˙) release, cytochrome c oxidase, reactive oxygen species, modifications to intracellular organelle membrane activity, calcium flux and stress proteins [18–22]. Any effect ultimately relies on the absorption of light by components (chromophores) within the cell of which there can be many including, aromatic amino acids, lipids, melanins, pyridinic co-factors, flavin coenzymes, porphyrins and water to name but a few [23]. At the tissue level the absorbed light can influence blood flow, following release of the vasodilator, NO˙ and/or 'warming' of local microenvironments [24]. Enhanced perfusion will in turn facilitate improved oxygenation and recruitment of macrophages, neutrophils and lymphocytes to areas undergoing repair and/or infection as well as further re-vascularization and proliferation of cells to aid healing. The opposite is also true; improved perfusion will facilitate clearance of inflammatory cells, fluids and debris (i.e. lymphatic drainage) more efficiently from arthritic knees or infected wounds. Altered nociception of somatic afferent (sensory) fibres via altered serotonin and bradykinin release have also been reported further supporting the neuralgesic properties attributed to LLLT [9].

The consequences of LLLT from cell to tissue will undoubtedly involve a myriad of effects with altered intracellular signalling and redox state (glutathione levels and transmembrane potentials), activation of redox-sensitive early/intermediate genes and related transcriptional factors including nuclear factor-$\kappa\beta$ [19].

These signalling pathways while no doubt implicit to the LLLT response remain somewhat obscure and unresolved. Dr. Tiina Karu at the Russian Academy of Sciences has been a strong advocate of LLLT and has done much to unravel the associated cellular mechanisms and emphasises the importance of exploiting new technologies to measure, analyse and quantify LLLT effect(s) [25].

The Role of Mitochondria

Mitochondria are perhaps the single most important entity within cells governing the LLLT response. They exist as 0.1–0.2 μm diameter organelles within most cells, occupying up to 25% of the cytoplasm and totalling 200–2,000 in number depending on the energy demands of the cell. They generate energy (ATP) for the cell via oxidative phosphorylation as nicotinamide adenine dinucleotide (NADH) is oxidized within the inner mitochondrial membrane providing a flow of electrons, called the electron transport chain (ETC) that sustains a proton gradient across the inner and outer mitochondrial membranes known as the mitochondrial membrane potential ($\Delta\Psi$) [26]. The flow of protons back into the mitochondrial matrix is coupled to the phosphorylation of adenine diphosphate and ATP synthesis. Interestingly, matured, mammalian erythrocytes, which are devoid of either a nucleus or mitochondria relying on glycolytic fermentation for energy instead, may provide valuable comparison for delineating alternate intracellular pathways affected by LLLT that are unrelated to mitochondria [27].

The role of calcium (Ca^{2+}) in regulating many if not all aspects of cellular function is well documented [26]. Its role in regulating mitochondrial function is of particular significance, necessitating tightly controlled transport across the mitochondrial membrane(s) into the cytosol via ion-exchange mechanisms and membrane pores. Excessive efflux of Ca^{2+} causes collapse of the $\Delta\Psi$ resulting in reduced ATP production and apoptosis. Alternatively, increased intramitochondrial Ca^{2+} can up-regulate nucleic acid synthesis and enhance metabolism by stimulating the ETC and faster ATP production [26, 28]. Indeed, this is implicated as a mechanism for LLLT although precise measurements of intra-mitochondrial Ca^{2+} following LLLT have not been reported and exactly how the ETC is stimulated remains unclear [29, 30]. However, the expression of anti-apoptotic proto-oncogene, B-cell lymphoma-2 (bcl-2), which prevents Ca^{2+} efflux from mitochondria has also been shown to increase following LLLT, further strengthening this idea of increased mitochondrial Ca^{2+} [31]. Another hypothesis involving bcl-2 also involves cytochrome c oxidase and it associated substrate, cytochrome c [32]. Cytochrome c provides electrons to cytochrome c oxidase which is the terminal ETC enzyme, reducing dioxygen to water. Bcl-2 inhibits the release of mitochondrial cytochrome c, thus preventing the activation of caspases implicit to apoptosis. In addition, the resulting increase in available mitochondrial cytochrome c may also act to increase the activity of cytochrome c oxidase hence the elevated ETC turnover. It is also important to consider that cytochrome c oxidase in its oxidized state maximally absorbs wavelengths of light commonly used for LLLT such as 670 and 830 nm which highlights the opportunity for direct stimulation of the enzyme to speed up the ETC.

Given the importance of mitochondria it will also be of interest to compare effects of LLLT in cells that rely predominantly on cytosolic glycolytic pathways for energy for instance tumour cells, as compared with those that use oxidative phosphorylation. Indeed, recent work by Naviaux et al., suggest that mitochondria in tumour cells are more chaotically distributed around the cytosol as compared to normal cells that exhibit highly organized spatial organization of mitochondria [33]. This provides distinguishing scattering profiles that may or may not reflect the heightened metabolic activity of tumour cells and raises the likelihood of differential LLLT effects due to differences in spatial distribution of mitochondria. A related hypothesis and certainly more controversial is the notion that mitochondria change their shape and surface area profiles in order to maximize light absorbance, much like a conformable mirror adjusts to collect light in a telescope. It has also been proposed that mitochondria are able to store light energy much like a capacitor converting it to chemical energy when necessary. Yet, despite the opportunities for using real-time microscopic imaging of LLLT-treated cells and isolated mitochondria models, there is little evidence to support either of these hypotheses.

Dosimetry

The complexities of light dosimetry into cells and tissues cannot be understated and as a result, dosimetry is often ignored or poorly implemented in the majority of LLLT protocols. Indeed, few if any standard protocols exist for LLLT; there are no

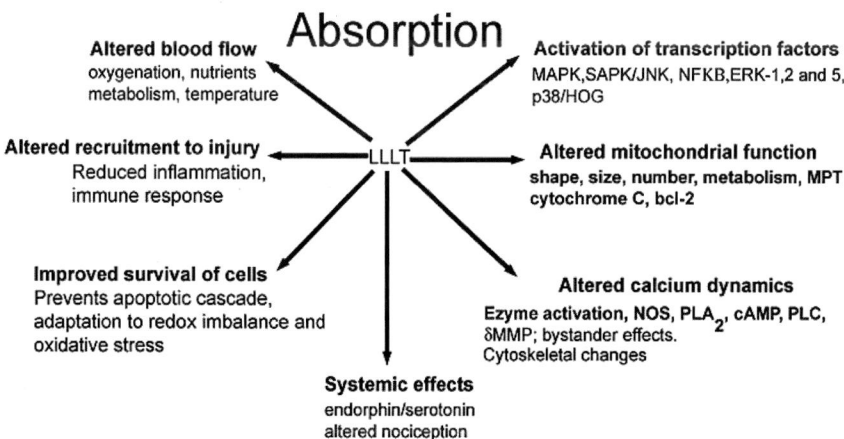

Fig. 1 Shows a schematic of some of the mechanisms thought to be involved in the LLLT response as light is absorbed into tissues and cells

reference tables that describe how to administer light into tissues to ensure optimal dosimetry for a particular ailment. In defence of this, it can certainly be argued as to whether stringent dosimetry is even necessary for LLLT given the many physiological parameters involved and differing light sources employed, it may be redundant to try and provide an algorithm that predicts response based on the delivered dose. The majority of LLLT protocols use wavelengths of light in the range of 620–870 nm to coincide with maximal absorbance profiles of endogenous chromophores [23]. Near infra-red wavelengths also provide improved penetration into tissues, overlapping with the so called optical window. The treatment parameters for wavelength, temporal mode of delivery whether pulsed or continuous, the radiant power (W), radiant exposure (W/cm^2), as well as the positioning of the light source relative to the tissue and the optical properties of the tissue all have a resounding influence on the final response at the cellular and tissue interface. The passage of photons through highly heterogeneous tissues can be accurately modelled according to Monte Carlo and/or diffusion approximations [34–36]. Both models require knowledge of the tissue's absorption and scattering coefficients as well as the scattering anisotropy factor g which can be obtained using diffuse reflectance measurements or direct transmittance measurements [35, 36]. For defined geometries, finite element methods are employed that allow calculation of the local light distribution for structures within the tissue for example, blood vessels which may be useful to LLLT for predicting changes in blood flow and temperature [34]. Still the complexity of conducting these measurements in practice and applying them to the models serves as a deterrent to most individuals using LLLT that do not possess a background in optical physics.

Although it is beyond the scope of this paper to examine the different methods for estimating light distribution for all the different types of tissue that LLLT is used, a brief consideration for light passage through skin is appropriate given the pertinence to LLLT. The portion of skin relevant to LLLT consists of the upper epidermis and lower dermis, approximately 100–150 μm and 1,500–3,000 μm thick, respectively and separated by a basement membrane. Wavelengths commonly employed for LLLT (620–870 nm) are typically attenuated before reaching the deeper fatty layer, the panniculus adiposus. The epidermis is devoid of blood vessels, consisting instead of keratinocytes with intermingled melanocytes, Langerhans cells and pressure-sensing Merkel cells. The major chromopores are melanin, which depending on whether they are compartmentalized into melanosomes or dispersed into melanin dust, can possess significant scattering and absorption [23, 37]. Absorption by melanins and most other endogenous chromophores within the epidermis including DNA, urocanic acid and aromatic amino acids is maximal at wavelengths <400 nm (i.e. in the ultra-violet range). However, exposure to ultra-violet light can promote the production of eumelanins that shift absorbance toward the red region of the spectra. Therefore non-Caucasians and individuals with highly pigmented (tanned) skin will absorb more red light during LLLT than those of fairer complexion [23]. The dermis, in contrast to the epidermis, is highly vascular with fibroblasts as the predominant cell type producing procollagen and elastin, together with mast cells, lymphatics and nerve endings.

Therefore, the effects of LLLT on blood flow and temperature will be largely orchestrated within the dermis, including collagen synthesis and wound repair. Temperature changes within the dermis can alter tissue scattering [37], although the temperature changes expected during LLLT ($<1–2°C$) are unlikely to have a significant effect on scattering.

Ultimately, it would appear that a compromise is necessary in terms of the extent of dosimetry that should be expected for LLLT. For unlike photodynamic therapy or other modalities that aim to kill target cells without disturbing other cells according to a well defined mechanism and where dosimetry should be explicit to ensure the best therapeutic outcome, the underlying mechanisms and physiology that govern LLLT are not so clearly discerned and may vary considerably from patient to patient. A database of effective regimens specifying the wavelength, mode of delivery, total delivered dose (J/cm^2) and fluence rate (mW/cm^2) would be a useful reference on which to base initial protocols. Consensus amongst clinicians and investigators for quantifiable methods of analysis and scoring charts (specific pain and mobility for example) on which to gauge response would also be important for subsequent optimization of the treatment.

Future Avenues for LLLT Research

New Technologies for Analysis

There are many new and evolving technologies that could provide quantitative measure of changes occurring at the cellular and tissue level following LLLT [25]. Microscopic techniques such as fluorescence correlation spectroscopy, fluorescence recovery after photobleaching or fluorescence resonance energy transfer can all provide information on intracellular protein dynamics and interactions that could help to understand LLLT response in real time in living cells. Two related techniques that exploit vibrational imaging are Raman spectroscopy and coherent anti-stokes Raman scattering (CARS) microscopy. Both can provide information on chemical composition and molecular structure non-invasively, although CARS has the distinct advantage of providing functional microscopy with the ability for time-resolved measurements of select chemical concentrations and for tracking their transport through cells and tissues [38]. Molecular analysis of gene and protein expression using real time reverse transcription polymerase chain reaction and microarrays will also prove invaluable in identifying key mediators at the cell and tissue level. Changing blood flow patterns within tissue are instrumental to LLLT. Current techniques for measuring blood flow include laser Doppler flow-metry and Doppler optical coherence tomography (DOCT) [39, 40]. DOCT can provide real-time topographical maps of micro-circulatory blood perfusion and direction in tissues although the depth of penetration is limited to <0.5 mm. However, recent studies from within our group confirm that DOCT can be adapted from a transcutaneous technique to an endoscopic one circumventing the depth issue [41]. This is an evolving technology with great promise which if combined with chemical composition capabilities or immunofluorescence could be used for in situ assessment of suspect tissues to complement tissue biopsy. In the meantime, it is certainly feasible that DOCT could be useful for non-invasively monitoring the onset of ischaemic insult following spinal cord injury and subsequently assessing the influence of LLLT. For more systemic, whole body assessment of blood flow and oxygenation functional magnetic resonance imaging can be used, or positron emission tomography to monitor metabolic behaviour using radio-labelled fluor-odeoxyglucose or DNA synthesis using radio-labelled thymine derivatives [42]. Being able to monitor apoptosis in vivo is a major achievement with many potential applications. Apoptotic cells express phosphatidylserine (PS) residues on their outer plasma membrane as part of the opsonization process and gamma camera imaging and MRI can be used to image these cells using labels conjugated to PS specific ligands, namely Technetium-99 m-labelled annexin V and superparamag-netic iron oxide-labelled synaptotagmin I, respectively [43]. It has also been demonstrated that apoptotic cells posses distinct ultrasound signals that can be recognised due to their condensed chromatin [44]. The question of whether LLLT can abate the apoptotic culling that manifests following spinal cord injury or with retinal degeneration (see below) could feasibly be investigated using these ad-vanced technologies.

A Potential Role for LLLT in Conditions Related to Apoptosis

Spinal cord injury occurs in an estimated 10,000–12,000 United States citizen every year most of whom are male and between the ages of 16 and 30 years old. The consequences can be fatal and/or lead to paralysis or which there is no cure. Furthermore, the full extent of damage often takes days to weeks to fully manifest as the initial injury sets off a cascade of physiological effects that leads to vast numbers of neurons dieing by apoptosis. The mechanism(s) for this delayed cell death is thought to be multi-factorial involving ischaemic injury as a result of damage to blood vessels, excitotoxicity, as excessive amounts of excitatory amino acid neurotransmitter are released leading to the formation of reactive radicals and increased intracellular Ca^{2+} [16, 17]. Tremendous amounts of research are currently focused on finding ways to circumvent this delayed apoptosis follow-ing spinal cord injury and a very recent study by Byrnes et al., in which rats with a spinal lesion were treated with transcutaneous 810 nm laser light over the course of 14 days suggest that LLLT may be useful in this regard [16]. The treatment resulted in greater axonal regeneration and improved functional recovery. Immunolabeling of specific cell types within treated tissue confirmed a reduction in pro-inflammatory/immune cells. Although apoptosis was not measured directly in this study, it is conceivable that by reducing the post-injury inflammation, blood vessels that would otherwise become compressed shut with increasing interstitial pressures, could remain open to sustain resident cells and prevent apoptosis. Further studies are needed in confirm the extent of apoptotic neuronal cell death following spinal cord injury and the response to differing LLLT treatment regimens.

N-retinyl-N-retinylidene ethanolamine (A2E) is an endogenous, lipophilic compound implicated in the dry form of age-related macular degeneration (AMD). A2E selectively binds to cytochrome c oxidase, in place of cytochrome c leading to apoptosis in mammalian retinal pigment epithelial cells [45]. Diphosphatidyl glycerol (cardiolipin) is concentrated in the inner mitochondrial membrane and competes with A2E for binding to cytochrome c oxidase. In this way cardiolipin may be able to counter the pro-apoptotic effects of A2E and prevent the progression of dry AMD. Despite the fact that cardiolipin appears to inhibit the action of A2E more effectively in the dark, little is known regards the underlying mechanisms of this light dependent inhibition. Given the ability for LLLT to promote anti-apoptotic factors, including those specific to cytochrome c oxidase, there may be rationale for testing the effects of LLLT on cardiolipin and A2E. There are no reports in the literature that describe the effects of LLLT on cardiolipin or A2E.

Another disease in which apoptosis is central to the pathology is amyotrophic lateral sclerosis (ALS). ALS is a neurodegenerative disease caused by progressive loss of cortical, spinal and brainstem motoneurons that presents clinically as muscle weakening and spasticity. Elevated intracellular (cytosolic) Ca^{2+} and irregular ETC activity within motoneuron mitochondria are thought to be key in ALS [46], although the question of whether altered Ca^{2+} homeostasis is a result of dysfunctional mitochondria or vice-versa remains to be answered. It is feasible however that LLLT could promote bcl-2 expression within motoneuron mitochondria to reduce Ca^{2+} efflux and prevent apoptotic cell death (see section, The Role of Mitochondria Above). An equally interesting and as yet relatively unexplored avenue of research relates to the light-mediated regulation of mitochondrial benzodiazepine receptors (MBR) and their associated inhibitory neurotransmitter γ-aminobutyric acid type A ($GABA_A$) receptors. Benzodiazepines are known to bind both MBRs and $GABA_A$ receptors as part of their anxiolytic, anti-convulsant action within the central nervous system. Unfortunately, benzodiazepine therapy can lead to addiction. An alternative mode of stimulating $GABA_A$ receptors using pulses of low level light was recently proposed by Leszkiewicz and Aizenman [47]. Moreover, in keeping with the effects of LLLT they also observed a change in redox status within the retinal cells being treated. Thus it would appear that there is certainly room for more work defining the dose and wavelength-dependency of this effect and in establishing a rationale for LLLT in treating conditions like depression or sleep disorders.

Conclusions

There is little doubt that for a growing number of people LLLT or photobiomodulation provides therapy to a broad range of condition, from arthritic pain and carpel tunnel syndrome to migraine and wound repair. A common denominator of mechanism appears to be that relating to improved microcirculatory blood flow, which can lead to increased cell proliferation and cell survival, improved lymphatic drainage and reduced interstitial pressures. Scepticism however persists as the scientific rationale for treatment regimen and intended response is lacking; and there often appears to be neither rhyme nor reason as to the choice of selected light source, wavelength or fluence dose being used. Guidelines of standard protocol could allow easier comparison between studies and minimize the likelihood that any discrepancies are due to methodology. In order that LLLT can evolve and advance, research with sound scientific basis must be encouraged, interfacing with new technologies that can provide quantifiable evidence of response and mechanisms implicit to the effect(s). This will in turn facilitate funding by granting agencies. Ultimately, there is still much we do not know about what happens inside a cell or tissue upon exposure to light and only through better scientific understanding can we ever hope to fully harness and optimize its application to medicine and therapeutics.

Acknowledgments The authors wish to acknowledge the technical assistance of Anoja Giles with culturing CNS-1 cells and Emily Pai for her work on biodynamic phototherapy. We also wish to thank Canadian Institute of Health Research for financial support.

References

1. Karu T. "Photobiology of low-power laser effects." Health Phys. **56**: 691–704 (1989).
2. Maiman TH. "Stimulated optical radiation in ruby." Nature, **187**: 493 (1960).
3. Lee MW. "Combination 532-nm and 1064-nm lasers for noninvasive skin rejuvenation and toning." Arch Dermatol. **139**: 1265–1276 (2003).
4. Maiya GA, Kumar P, Rao L. "Effect of low intensity helium-neon (He-Ne) laser irradiation on diabetic wound healing dynamics." Photomed Laser Surg. **23**: 187–190 (2005).
5. Baugh WP, Kucaba WD. "Nonablative phototherapy for acne vulgaris using the KTP 532 nm laser." Dermatol Surg. **31**: 1290–1296 (2005).
6. Woo WK, Jasim ZF, Handley JM. "Evaluating the efficacy of treatment of resistant port-wine stains with variable-pulse 595-nm pulsed dye and 532-nm Nd:YAG lasers." Dermatol Surg. **30**: 158–162 (2004).
7. Giuliani A, Fernandez M, Farinelli M, Baratto L, Capra R, Rovetta G, Monteforte P, Giardino L, Calza L. "Very low level laser therapy attenuates edema and pain in experimental models." Int J Tissue React. **26**: 29–37 (2004).
8. Gur A, Karakoc M, Nas K, Cevik R, Sarac J, Demir E. "Efficacy of low power laser therapy in fibromyalgia: a single-blind, placebo-controlled trial." Lasers Med Sci. **17**: 57–61 (2002).
9. Laakso EL, Cabot PJ. "Nociceptive scores and endorphin-containing cells reduced by low-level laser therapy (LLLT) in inflamed paws of Wistar rat." Photomed Laser Surg. **23**: 32–35 (2005).
10. Ferreira DM, Zangaro RA, Villaverde AB, Cury Y, Frigo L, Piccolo G, Longo I, Barbosa DG. "Analgesic effect of He-Ne (632.8 nm) low-level laser therapy on acute inflammatory pain." Photomed Laser Surg. **23**: 177–181 (2005).
11. Zalewska-Kaszubska J, Obzejta D. "Use of low-energy laser as adjunct treatment of alcohol addiction." Lasers Med Sci. **19**: 100–104 (2004).

12. Nussbaum EL, Lilge L, Mazzulli T. "Effects of 630-, 660-, 810-, and 905-nm laser irradiation delivering radiant exposure of 1–50 J/cm^2 on three species of bacteria in vitro." J Clin Laser Med Surg. **20**: 325–33 (2002).
13. Khadra M, Kasem N, Haanaes HR, Ellingsen JE, Lyngstadaas SP. "Enhancement of bone formation in rat calvarial bone defects using low-level laser therapy." Oral Surg Med Oral Pathol Oral Radiol Endod **97**: 693–700 (2004).
14. Simunovic Z. "Low level laser therapy with trigger points technique: a clinical study on 243 patients." J Clin Laser Med Surg. **14**: 163–167 (1996).
15. Chen YS, Hsu SF, Chiu CW, Lin JG, Chen CT, Yao CH. "Effect of low-power pulsed laser on peripheral nerve regeneration in rats." Microsurgery **25**: 83–89 (2005).
16. Byrnes KR, Waynant RW, Ilev IK, Wu X, Barna L, Smith K, Heckert R, Gerst H, Anders JJ. "Light promotes regeneration and functional recovery and alters the immune response after spinal cord injury." Lasers Surg Med. **36**: 171–185 (2005).
17. Rochkind S, Shahar A, Nevo Z. "An innovative approach to induce regeneration and the repair of spinal cord injury." Laser Ther. **9**: 151 (1997).
18. Karu T. "Primary and secondary mechanisms of action of visible to near-IR radiation on cells." J Photochem Photobiol B. **49**: 1–17 (1999).
19. Karu T, Pyatibrat LV, Afanasyeva NI. "Cellular effects of low power laser therapy can be mediated by nitric oxide." Laser Surg Med. **36**: 307–314 (2005).
20. Duan R, Liu TC, Li Y, Guo H, Yao LB. "Signal transduction pathways involved in low intensity He-Ne laser-induced respiratory burst in bovine neutrophils: a potential mechanism of low intensity laser biostimulation." Lasers Surg Med. **29**: 174–178 (2001).
21. Brown GC. "Regulation of mitochondrial respiration by nitric oxide inhibition of cytochrome c oxidase." Biochem Biophys Acta **1504**: 46–57 (2001).
22. Cooper CE. "Nitric oxide and cytochrome oxidase: substrate, inhibitor or effector?" Trends Biochem Sci. **27**: 33–39 (2002).
23. Young AR. "Chromopores in human skin." Phys Med Biol. **42**: 789–802 (1997).
24. Nunez SC, Nogueira GE, Ribeiro MS, Garcez AS, Lage-Marques JL. "He-Ne laser effects on blood microcirculation during wound healing: a method of in vivo study through laser Doppler flowmetry." Lasers Surg Med. **35**: 363–368 (2004).
25. Karu T. "High-Tech helps to estimate cellular Mechanisms of low power laser therapy." Laser Surg Med. **34**: 298–299 (2004).
26. Chakraborti T, Das S, Mondal M, Roychoudhury S, Chakraborti S. "Oxidant mitochondria and calcium: An overview." Cell Signal **11**: 77–85 (1999).
27. Kujawa J, Zavodnik L, Zavodnik I, Bryszewska M. "Low-intensity near-infrared laser radiation-induced changes of acetylcholinesterase activity of human erythrocytes." J Clin Laser Med Surg. **21**: 351–355 (2003).
28. Jouville LS, Pinton P, Bastianutto C, Rutter GA, Rizzuto R. "Regulation of mitochondrial ATP synthesis by calcium: Evidence for long-term metabolic priming." PNAS. **96**: 13807–13812 (1999).
29. Yu W, Naim JO, McGowan H, Ippolito K, Lanzafame RJ. "Photomodulation of oxidative metabolism and electron chain enzymes in rat liver mitochondria." Photochem Photobiol. **66**: 866–871 (1997).
30. Bortoletto R, Silva NS, Zangaro RA, Pacheco MT, Da Matta RA, Pacheco-Soares C. "Mitochondrial membrane potential after low-power laser irradiation." Lasers Med Sci. **18**: 204–206 (2004).
31. Shefer G, Partridge TA, Heslop L, Gross JG, Oron U, Halevy O. "Low-energy laser irradiation promotes the survival and cell cycle entry of skeletal muscle satellite cells." J Cell Sci. **115**: 1461–1469 (2002).
32. Wong-Riley MT, Liang HL, Eells JT, Chance B, Henry MM, Buchmann E, Kane M, Whelan HT. "Photobiomodulation directly benefits primary neurons functionally inactivated by toxins: role of cytochrome c oxidase." J Biol Chem. **280**: 4761–4771 (2005).

33. Gourley PL, Hendricks JK, McDonald AE, Copeland RG, Barrett KE, Gourley CR, Singh KK, Naviaux RK. "Mitochondrial correlation microscopy and nanolaser spectroscopy – new tools for biophotonic detection of cancer in single cells." Technol Cancer Res Treat. **4**: 585–592 (2005).
34. Zhang R, Verkruysse W, Aguilar G, Nelson JS. "Comparison of diffusion approximation and Monte Carlo based finite element models for simulating thermal responses to laser irradiation in discrete vessels." Phys Med Biol. **50**: 4075–4086 (2005).
35. Wilson BC, Jacques, SL. "Optical reflectance and transmission of tissues: principles and applications." IEEE J Quantum Elect. **26**: 2186–2199 (1990).
36. Kienle A, Lilge L, Patterson MS, Hibst R, Steiner R, Wilson BC. "Spatially resolved absolute diffuse reflectance measurements for non-invasive determination of the optical scattering and absorption coefficients of biological tissue." Appl Optics. **35**: 2304–2314 (1996).
37. Laufer J, simpson R, Kohl M, Essenpreis M, Cope M. "Effect of temperature on the optical properties of ex vivo human dermis and subdermis." Phys Med Biol. **43**: 2479–2489 (1998).
38. Evans CL, Potma EO, Puoris'haag M, Cote D, Lin CP, Xie XS. "Chemical imaging of tissue in vivo with video-rate coherent anti-Stokes Raman scattering microscopy." Proc Natl Acad Sci USA. **102**: 16807–16812 (2005).
39. Yang VXD, Gordon ML, Tang SJ, Marcon NE, Gardiner G, Qi B, Bisland S, Seng-Yue E, Lo S, Pekar J, Wilson BC, Vitkin IA. "High speed, wide velocity dynamic range Doppler optical coherence tomography (part III): in vivo endoscopic imaging of blood flow in the rat and human gastrointestinal tract." Opt Express. **11**: 2416–2424 (2003).
40. Larsson M, Nilsson H, Stromberg T. "In vivo determination of local skin optical properties and photon path length by use of spatially resolved diffuse reflectance with applications in laser Doppler flowmetry." Appl Optics. **42**: 124–134.
41. Yang VX, Mao YX, Munce N, Standish B, Kucharczyk W, Marcon NE, Wilson BC, Vitkin IA. "Interstitial Doppler optical coherence tomography." Opt Lett. **30**: 1791–1793 (2005).
42. Sun H, Mangner TJ, Collins JM, Muzik O, Douglas K, Shields AF. "Imaging DNA synthesis in vivo with 18F-FMAU and PET." J Nucl Med. **46**: 292–296 (2005).
43. Brauer M. "In vivo monitoring of apoptosis." Prog Neuro-Psychoph. **27**: 323–31 (2003).
44. Czarnota GJ, Kolios MC, Hunt JW, Sherar MD. "Ultrasound imaging of apoptosis. DNA-damage effects visualized." Methods Mol Biol. **203**: 257–277 (2002).
45. Shaban H, Borras C, Vina J, Richter C. "Phosphatidylglycerol potently protects human retinal pigment epithelial cells against apoptosis induced by A2E, a compound suspected to cause age-related macula degeneration." Exp Eye Res. **75**: 99–108 (2002).
46. von Lewinski F, Keller BU. "Ca^{2+}, mitochondria and selective motoneuron vulnerability: implications for ALS." Trends Neurosci. **28**: 494–500 (2005).
47. Leszkiewicz DN, Aizenman E. "Reversible modulation of GABA(A) receptor-mediated currents by light is dependent on the redox state of the receptor." Eur J Neurosci. **17**: 2077–2083 (2003).
48. Bisland SK, Pai E, Wilson BC, "Biodynamic phototherapy: *Priming* cells for 5-aminolevulinic acid-mediated photodynamic therapy." *In preparation*.

Enhancing Photodynamic Effect Using Low-Level Light Therapy

Stuart K. Bisland

Abstract Photodynamic therapy uses hirger doses of light and numerous photo-sensitizers to activate the killing of the sensitized tissue via singlet oxygen. The predominant target of PDT has been cancer, but it may also have application to other diseases. The combination of PDT with laser (light) therapy is very appropriate and valuable.

Keywords: Photodynamic therapy(PDT), singlet oxygen, mitochondria, ALA, PBR.

Healing Powers of Light

Low-level light therapy (LLLT) refers to the application of *low-level* light to biological systems with the intent to provide therapy, although, the parameters that represent *low-level light* remain ill defined. Nevertheless, LLLT is currently used, to treat a broad spectrum of clinical ailments. Dermatologists use LLLT for cosmetic and medical conditions such as wrinkle and hair removal [1], treating wounds [2], acne vulgaris [3], port wine stain(s) and telangiectasia [4], while rheumatologists can alleviate pain in arthritic joints, muscle, nerves and bone as well as many types of inflammation such as tendonitis, brucitis, stomatitis and acute epididymitis [5]. Photodynamic therapy (PDT) is another form of light therapy but unlike LLLT, involves the systemic or local administration of a photo-activating compound (photosensitizer) into the tissues prior to light exposure. The light doses used for PDT are typically on the order of 10–100-fold higher than that for LLLT, with the intent to cause cell death rather than biomodulation. As such, PDT has been developed successfully in the clinic to treat numerous types of cancers and non-cancer conditions [6, 7].

S.K. Bisland
Ontario Cancer Institute, University Health Network, University of Toronto, 610 University Avenue, Toronto, Ontario M5G 2M9, e-mail: sbisland@uhnres.utoronto.ca

R. Waynant and D.B. Tata (eds.), *Proceedings of Light-Activated Tissue Regeneration and Therapy Conference.*
© Springer Science + Business Media, LLC 2008

Light Interaction at the Cellular Level

Regarding mechanisms of action, PDT is fairly well understood, at least in terms of its photochemistry [8–13]. The photo-oxidative product most often implicated following type II photochemical interaction is singlet oxygen ($^1O_2$), which has been shown to be highly reactive with potent oxidizing capacity leading to damage of cellular membranes, intracellular organelles and proteins vital to cell survival. Other oxygen-derived species have been suggested including, hydrogen peroxide, superoxide anion and hydroxyl radical. Ultimately, local oxygen concentrations, the photochemical characteristics of the photosensitizer and light dose as well as the inherent properties of the targeted tissue will all influence the photodynamic action and biological response. The mechanisms underlying LLLT are less clear. Indeed, there is a growing list of potential *and confirmed* mediators of LLLT including, altered adenosine triphosphate levels, release of nitric oxide cytochrome c oxidase, reactive oxygen species, modifications to intracellular organelle membrane activity, calcium flux and stress proteins [14–18]. It is understood, however, that for there to be an effect following LLLT, light has be absorbed by components (chromophores) within the cell of which there can be many including, aromatic amino acids, lipids, melanins, pyridinic co-factors, flavin coenzymes, porphyrins and water to name but a few [19]. At the tissue level the absorbed light can influence blood flow, following release of the vasodilator, NO· and/or 'warming' of local microenvironments [20]. Enhanced perfusion will in turn facilitate improved oxygenation and recruitment of macrophages, neutrophils and lymphocytes to areas undergoing repair and/or infection as well as further re-vascularization and proliferation of cells to aid healing. The opposite is also true; improved perfusion will facilitate clearance of inflammatory cells, fluids and debris (i.e. lymphatic drainage) more efficiently from arthritic knees or infected wounds. Altered nocioception of somatic afferent (sensory) fibres via altered serotonin and bradykinin release have also been reported further supporting the neuralgesic properties attributed to LLLT [21].

The Importance of Mitochondria

The consequences of LLLT from cell to tissue will undoubtedly involve a myriad of effects with altered intracellular signalling and redox states (glutathione levels and transmembrane potentials), activation of redox-sensitive early/intermediate genes and related transcriptional factors including nuclear factor-$\kappa\beta$ [22]. Amongst all of these potential targets the mitochondrion remain perhaps the most important cellular organelle governing LLLT response, at least at the cellular level. The same can also be said for certain PDT photosensitizers that target mitochondria including photofrin, silicon phthalocyanine (Pc-4) and 5-aminolevulinic acid (ALA). ALA is of particular interest to us given that it is a naturally occurring substrate that is metabolized within mitochondria as part of the haem synthetic pathway. This pathway is common to most, if not all-eukaryotic cells; however, the potential to exploit ALA for PDT relates to the irregularity that some tumour cell types have that result in an accumulation of protoporphyrin IX (PpIX), which happens to be an efficient PDT photosensitizer. Much of the early work relating to ALA-PDT was pioneered by Kennedy and Pottier [23] and ALA-PDT is now used clinically for the treatment of a variety of cancers and non-cancer ailments. The use of ALA-PDT to treat brain cancers is a particularly exciting prospect and is currently in clinical trials [7]. Interestingly, work by Naviaux et al., suggests that mitochondria in tumour cells are more chaotically distributed around the cytosol as compared to normal cells that exhibit highly organized spatial organization of mitochondria [24]. This provides distinguishing scattering profiles that may or may not reflect the heightened metabolic activity of tumour cells and raises the likelihood of differential LLLT effects due to differences in spatial distribution of mitochondria. A related hypothesis and certainly more controversial is the notion that mitochondria change their shape and surface area profiles in order to maximize light absorbance, much like a conformable mirror adjusts to collect light in a telescope. Is it possible then that tumour cells are able to maximize the conversion of energies, be it heat, chemical or light and similarly, enhance the manufacture of vital proteins by increasing the number and organization of their mitochondria? Indeed, tumour cells with more mitochondria may be more resilient to certain treatment(s) [25].

Combining LLLT and PDT

Equipped with the knowledge that key chromophores within mitochondria are involved in the therapeutic response of low-level light and the potential for harnessing this effect toward PDT prompted our investigation of combined LLLT and ALA-PDT. The rationale for this approach and a clear role for mitochondria, acting as the common *link* between ALA-PDT and LLLT, related to the fact that mitochondrial peripheral-type benzodiazepine receptor (PBR), which are located in the outer mitochondrial membrane are responsible for transporting porphyrins, including PpIX in and out of the mitochondrial membrane [26–29]. PpIX is oxidized at the inner mitochondrial membrane prior to binding iron to produce haem. PBRs therefore have a governing role to play in the photodynamic action of ALA-PDT [30, 31]. Moreover, PBRs are thought to associate with a number of proteins previously implicated in LLLT including bcl-2, cytochrome C oxidase, voltage-dependent anion channels, adenine nucleotide translocase and cyclophilin D [22, 29, 32]. It is perhaps not surprising then to find that PBRs are involved with several cellular functions including haem synthesis, steroidogenesis, DNA synthesis, cell growth and differentiation, and apoptosis. Interestingly, PBRs are known to be highly expressed on some types of cancers found in the brain, predominantly those of glioma origin yet lacking from normal, brain neurons which express the distinct central-type of benzodiazepine receptors [32].

We found that by pre-exposing glioma-derived tumour cells (CNS-1) to a single round of LLLT (1 or 5 J/cm^2 at 10 mW/cm^2) using broad-spectrum red light (600–800 nm) or select monochromatic wavelengths of light, including 635 or 905 nm, that subsequent PDT treatment (with 15 J/cm^2 of 635 nm laser light) using ALA (1 mM; 4 hours) resulted in greater cell kill than PDT alone (Fig. 1). Interestingly, even without exogenous ALA, LLLT increased the relative cell kill compared to PDT alone (Fig. 1B, C) indicating that endogenous porphyrins may be involved. Measurement of porphyrin levels within the cells confirmed an increase in endogenous porphyrin production following LLLT (Fig. 2).

Thus LLLT-mediated increase in porphyrins manufactured within mitochondria is one mechanism that could account for the increased PDT response. The role of PBRs to facilitate this increased porphyrin production in cells was also confirmed. Western blot analysis of PBR protein expression from LLLT treated CNS-1 cells at 37°C (Fig. 3) demonstrated increased protein levels compared with untreated cells suggesting increased PBR expression following LLLT. Increased PBR expression was not evident if cells were incubated at 4°C during LLLT confirming the requirement for *de novo* protein synthesis.

It is certainly of interest to consider LLLT as adjuvant to PDT, in a sense *'priming'* the cells prior to photo-oxidation. It is important to realize, however, that not all cell types will behave the same way following LLLT. It will be important for future studies to identify distinctions between normal cells and malignant cells that will further increase the selectivity of response following LLLT and PDT, such as the case for PBRs. Another potential avenue for investigation would be to use LLLT to enhance endogenous porphyrin-related fluorescence to facilitate fluorescence-guided resection of tumours prior to PDT of malignancies. As far as we are aware there are no reports of using LLLT to facilitate fluorescence-guidance. The same principle may also apply to other non-malignant tissues which contain porphyrins, such as actinic keratosis or psoriasis. Ultimately, with the advent of new imaging technologies [33] and greater understanding of how to image intracellular events in real-time, together with the development of new light sources and detection probes, the ability to precisely define optimal parameters for using LLLT and PDT to treat specific conditions becomes a true science with clear guidelines for success.

This section will include some of the latest research in photodynamic therapy covering a broad-spectrum of topics from light dosimetry and light delivery strategies, optical imaging, cellular mechanisms, immunology and novel photosensitizers. Dr. Bin Chen of University of the Sciences in Philadelphia describes the use of fluorescence imaging to quantify vascular damage following PDT. Dr. Jarod Findlay of Pennsylvania University discusses the prospect of patient-specific light dosimetry based on light absorption by the photosensitizer rather than the tissues and the inherent problems therein. The intracellular targets of PDT are reported by Dr. Anna-Liisa Nieminen of the Department of Anatomy at Case Western Reserve University. In this report she dissects the sequence of events that result following PDT using silicon phthalocyanine 4 (Pc 4) with particular focus on mitochondria and lysosomes. Dr. Michael Hamblin at the Wellman Center for Photomedicine, Harvard Medical School, provides a comprehensive overview as to the role of the

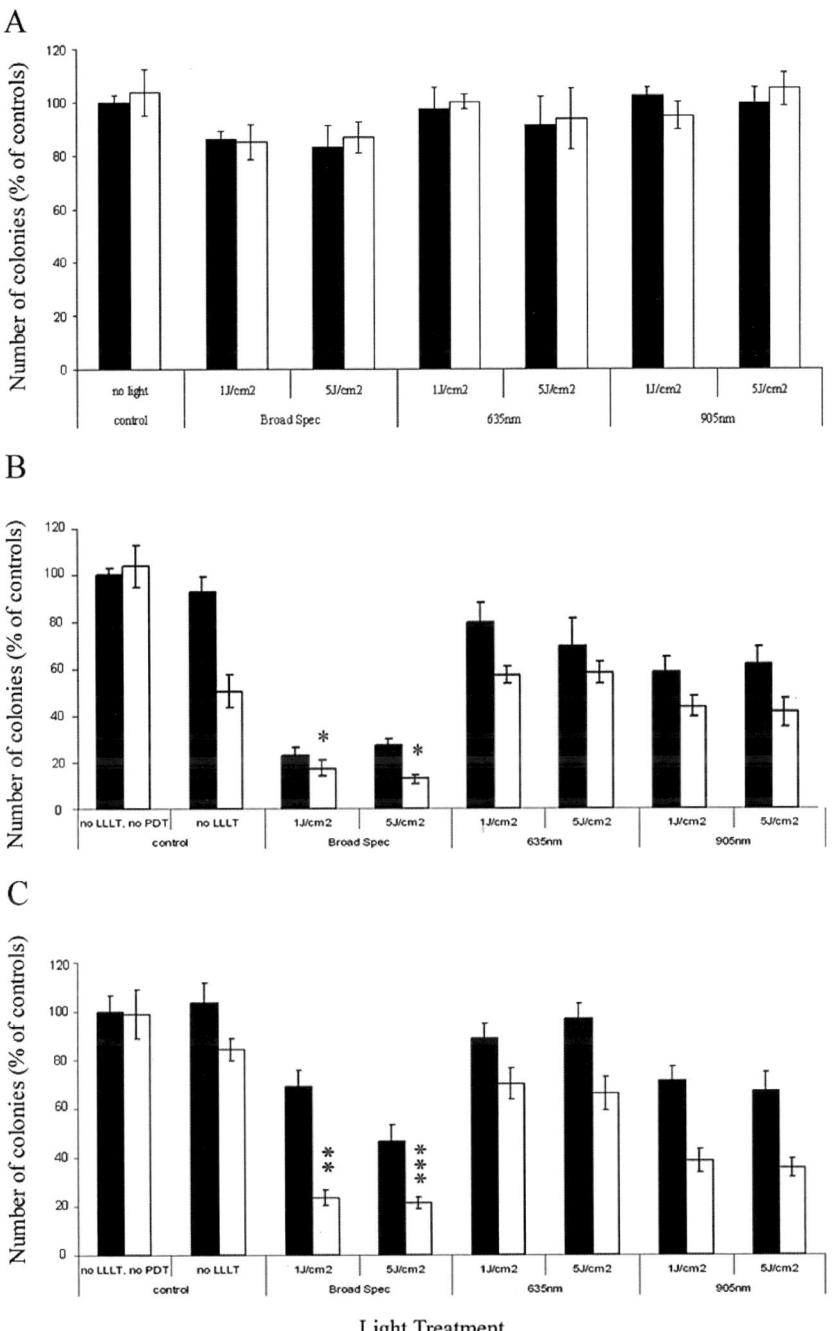

Fig. 1 Histograms showing the percentage of surviving cell colonies following different treatment regimens involving LLLT (A), LLLT + PDT at 37°C (B) or 4°C (C) in the presence (□) or absence (■) of exogenous ALA. Asterisks reveal to a significant difference between PDT treated groups and LLLT + PDT treated groups

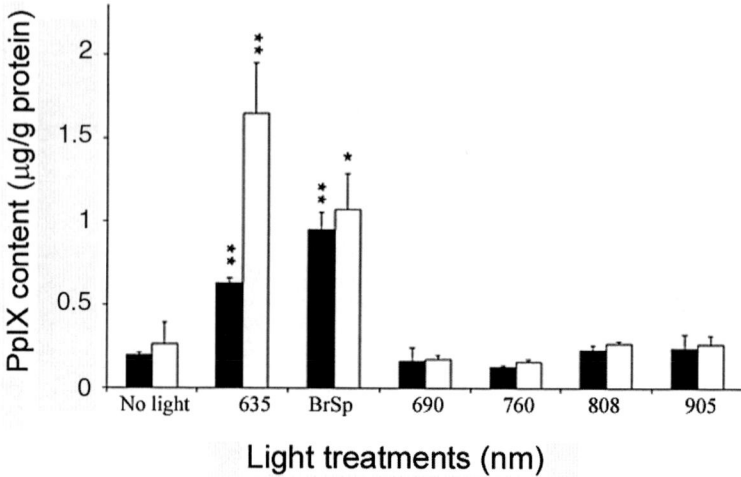

Fig. 2 Histograms showing PpIX (µg/g protein) content in CNS-1 following different light treatments. Significant increase in the amount of PpIX in cells was found comparing untreated cells (No light) and cells following LLLT at 635 nm and broad-spectrum light (BrSp) either with (□) or without (■) exogenous ALA

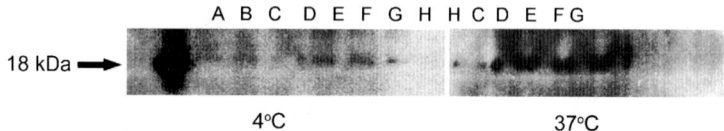

Fig. 3 Western blot showing the expression of 18 KDa PBR in CNS-1 following at 4°C or 37°C with no light (column H), 1 J/cm^2 (column B) or 5 J/cm^2, (column C) broad-spectrum light, 1 J/cm^2 (column D) or 5 J/cm^2, (column E) at 635 nm, 1 J/cm^2 (column F) or 5 J/cm^2, (column G) at 905 nm. Positive control lysate is also included (column A)

immune system in the PDT response and details experiments which describe the potential to exploit PDT-induced immunity for treating cancer(s). Dr. Dominic Robinson of The Centre for Optical Diagnostics and Therapy in Rotterdam has led some pioneering work using ALA-PDT and in this section describes some of his latest clinical findings involving alternate light delivery protocols for treating superficial basal cell carcinoma.

References

1. Lee MW. "Combination 532-nm and 1064-nm lasers for noninvasive skin rejuvenation and toning." Arch Dermatol. **139**: 1265–1276 (2003).
2. Maiya GA, Kumar P, Rao L. "Effect of low intensity helium-neon (He-Ne) laser irradiation on diabetic wound healing dynamics." Photomed. Laser Surg. **23**: 187–190 (2005).

3. Baugh WP, Kucaba WD. "Nonablative phototherapy for acne vulgaris using the KTP 532 nm laser." Dermatol. Surg. **31**: 1290–1296 (2005).

4. Woo WK, Jasim ZF, Handley JM. "Evaluating the efficacy of treatment of resistant port-wine stains with variable-pulse 595-nm pulsed dye and 532-nm Nd:YAG lasers." Dermatol. Surg. **30**: 158–162 (2004).

5. Giuliani A, Fernandez M, Farinelli M, Baratto L, Capra R, Rovetta G, Monteforte P, Giardino L, Calza L. "Very low level laser therapy attenuates edema and pain in experimental models." Int. J. Tissue React. **26**: 29–37 (2004).

6. Bisland SK. "Strategies for drug delivery: perspectives in photodynamic therapy." Recent Res. Devel. Bioconj. Chem. **2**: 1–43 (2005).

7. Bisland SK, Lilge L, Lin A, Rusnov R, Bogaards A, Wilson BC. "Metronomic PDT as a new paradigm for photodynamic therapy: rationale and pre-clinical evaluation of technical feasibility for treating malignant brain tumors." Photochem. Photobiol. **80**: 22–30 (2004).

8. Oschner M. "Photophysical and photobiological processes in the photodynamic therapy of tumors." J. Photoch. Photobiol. B. **39**: 1–18 (1997).

9. Moan J, Berg K. "Photochemotherapy of cancer: experimental research." Photochem. Photobiol. **55**: 931–948 (1992).

10. Henderson BW, Dougherty TJ. "How does photodynamic therapy work?" Photochem. Photobiol. **55**: 145–157 (1992).

11. Ochsner M. "Photophysical and photobiological processes in the photodynamic therapy of tumors." J. Photochem. Photobiol. B. **39**: 1–18 (1997).

12. Godar DE. "Light and death: photons and apoptosis." J. Investig. Dermatol. Symp. Proc. **4**: 17–23 (1999).

13. Kriska T, Korecz L, Nernes I, Gal D. "Physico-chemical modeling of the role of free radicals in photodynamic therapy III. Interactions of stable free radicals with excited photosensitizers studied by kinetic ESR spectroscopy." Biochem. Biophys. Res. Commun. **215**: 192–198 (1995).

14. Karu T. "Primary and secondary mechanisms of action of visible to near-IR radiation on cells." J. Photochem. Photobiol. B. **49**: 1–17 (1999).

15. Karu T, Pyatibrat LV, Afanasyeva NI. "Cellular effects of low power laser therapy can be mediated by nitric oxide." Laser. Surg. Med. **36**: 307–314 (2005).

16. Duan R, Liu TC, Li Y, Guo H, Yao LB. "Signal transduction pathways involved in low intensity He-Ne laser-induced respiratory burst in bovine neutrophils: a potential mechanism of low intensity laser biostimulation." Laser. Surg. Med. **29**: 174–178 (2001).

17. Brown GC. "Regulation of mitochondrial respiration by nitric oxide inhibition of cytochrome c oxidase." Biochem. Biophys. Acta. **1504**: 46–57 (2001).

18. Cooper CE. "Nitric oxide and cytochrome oxidase: substrate, inhibitor or effector?" Trends Biochem. Sci. **27**: 33–39 (2002).

19. Young AR. "Chromopores in human skin." Phys. Med. Biol. **42**: 789–802 (1997).

20. Nunez SC, Nogueira GE, Ribeiro MS, Garcez AS, Lage-Marques JL. "He-Ne laser effects on blood microcirculation during wound healing: a method of in vivo study through laser Doppler flowmetry." Laser Surg. Med. **35**: 363–368 (2004).

21. Laakso EL, Cabot PJ. "Nociceptive scores and endorphin-containing cells reduced by low-level laser therapy (LLLT) in inflamed paws of Wistar rat" Photomed. Laser Surg. **23**: 32–35 (2005).

22. Karu T. "High-Tech helps to estimate cellular mechanisms of low power laser therapy." Laser Surg. Med. **34**: 298–299 (2004).

23. Kennedy JC, Pottier RH. "Endogenous protoporphyrin IX, clinically useful photosensitizer for photodynamic therapy." J. Photochem. Photobiol. B. **14**: 275–292 (1991).

24. Gourley PL, Hendricks JK, McDonald AE, Copeland RG, Barrett KE, Gourley CR, Singh KK, Naviaux RK. "Mitochondrial correlation microscopy and nanolaser spectroscopy - new tools for biophotonic detection of cancer in single cells." Technol. Cancer Res. Treat. **4**: 585–592 (2005).

25. Sharkey SM, Wilson BC, Moorehead R, Singh G. "Mitochondrial alterations in photodynamic therapy-resistant cells." Cancer Res. **53**: 4994–4999 (1993).
26. Wendler G, Lindemann P, Lacapere J-J, Papadopoulos V. "Protoporphyrin IX binding and transport by recombinant mouse PBR." Biochem. Biophys. Res. Commun. **311**: 847–852 (2003).
27. Rebeiz N, Arkins S, Kelley KW, Rebeiz CA. "Enhancement of coproporphyrinogen III transport into isolated transformed leukocyte mitochondria by ATP." Arch. Biochem. Biophys. **2**: 475–481 (1996).
28. Verma A, Facchina SL, Hirsch DJ, Song S-Y, Dillahey LF, Williams JR, Snyder SH. "Photodynamic tumour therapy: mitochondrial benzodiazepine receptors as a therapeutic target." Mol. Med. **4**: 40–45 (1998).
29. Pastorino JG, Simbula G, Gilfor E, Hoak JB, Farber JL. "Protoporphyrin IX endogenous ligand of the peripheral benzodiazepine receptor, potentiates induction of the mitochondrial permeability transition and the killing of cultures hepatocytes by rotenone." J. Biol. Chem. **269**: 31041–31046 (1994).
30. Ratcliffe SL, Matthews EK. "Modification of the photodynamic action of d-aminolaevulinic acid (ALA) on rat pancreatoma cells by mitochondrial benzodiazepine receptor ligands." Br. J. Cancer. **71**: 300–305 (1995).
31. Mesenholler M, Matthews EK. "A key role for the mitochondrial benzodiazepine receptor in cellular photosensitization with delta-aminolaevulinic acid." Eur. J. Pharmacol. **406**: 171–180 (2000).
32. Krueger KE. "Molecular and functional properties of mitochondrial benzodiazepine receptors." Biochim. Biophys. Acta. **1241**: 453–470 (1995).
33. Bisland SK, Wilson BC. "To begin at the beginning: the science of bio-stimulation in cells and tissues." Proc. SPIE. **6140** (2006).

Light Fractionated ALA-PDT: From Pre-Clinical Models to Clinical Practice

D. Robinson, H.S. de Bruijn, E.R.M. de Haas, H.A.M. Neumann, and H.J.C.M. Sterenborg

Abstract Photodynamic therapy of superficial basal cell carcinoma using topical 5-aminolevulinic acid and a light fluence of 75–100 J cm^{-2} yields unsatisfactory long term clinical response rates. In a range of pre-clinical models illumination with two light fractions separated by 2 hours apart was considerably more effective than single illumination. Response is further enhanced if the fluence of the first light fraction is reduced while the cumulative fluence is maintained. We have demonstrated that these encouraging pre-clinical results are also evident for the clinical ALA-PDT of the treatment of superficial basal cell carcinoma. In a large scale randomised study including 505 primary sBCC we have shown that therapy using two light fractions of 20 and 80 J cm^{-2} performed 4 and 6 hours after the application of a single dose of 20% ALA results in a significant increase in complete response (P = 0.002, log-rank test). Twelve months after therapy, complete response rate following a two-fold illumination is 97% whereas the complete response to a single illumination is 89%. Numerous studies are underway investigating the mechanism underlying the increase in tissue response. Increased efficacy is not simply associated with an increasing PpIX content of the tissues during the treatment scheme and there is no direct relationship between the total amount of PpIX utilised and efficacy. We have shown that fractionated illumination does not enhance the efficacy of PDT using methyl-ester derivatives of ALA despite almost identical PpIX fluorescence kinetics during therapy. Our most recent data suggest that the in-vivo distribution of MAL and ALA and the exact site of PDT induced damage, is an important parameter in the mechanism underlying fractionated illumination for ALA-PDT. There is significant potential for the future use of light fractionation in other organs.

Keywords: Photodynamic therapy, protoporphyrin IX, optimisation, skin.

D. Robinson
Erasmus University Medical Center, P.O. Box 2040, Rotterdam, South Holland 3000 CA, The Netherlands, e-mail: d.robinson@erasmusmc.nl

R. Waynant and D.B. Tata (eds.), *Proceedings of Light-Activated Tissue Regeneration and Therapy Conference*.
© Springer Science + Business Media, LLC 2008

Introduction

Since its introduction in the 1990s photodynamic therapy (PDT) using porphyrin pre-cursors is now widely accepted as an attractive treatment option for a range of predominately cutaneous (pre-) malignancies. It has also been applied to a variety of non-malignant skin diseases and is under investigation for the treatment of other malignant conditions in various organs. Initial high complete response rates (CR), well above 90%, for the treatment of superficial basal cell carcinoma (BCC) using ALA were very encouraging [1–3]. However long term clinical results showed considerable variations, where CR has been found to be as low as 30%. This was a particular problem for nodular lesions [2, 5]. Clinically this led investigators to adopt approaches to enhance clinical response. The minimal side effects of PDT using topical ALA or its derivatives means that repeat treatments were a simple option [6, 7]. Indeed the recommended regime for the PDT using methyl-5-amino-levulinate (MAL) is two treatments 1 week apart. Further subsequent treatment sessions are often performed [8]. Experimentally, other approaches were taken to try to optimise the response of tissues to PDT in a single treatment. Penetration enhancers [9] and iron chelators [10] have been used in pre-clinical models. Also oral or systemic administration of ALA has been attempted in order to improve the biodistribution of PpIX [5, 11]. However, also after systemic ALA-PDT, only superficial necrosis was found in patients treated for dysplasia of the mouth [12] or the esophagus [13]. A number of animal studies have demonstrated that the response to PDT after systemic ALA administration can be improved by modifying the illumination scheme, for example, by reducing the fluence rate to improve oxygenation [14–16]. Another option is the use of light fractionation [17, 18]. Here the length of the dark interval is an important parameter. Short-term light fractionation (with one or more interruptions of seconds or minutes) may allow re-oxygenation during the dark interval that leads to more singlet oxygen deposition during therapy. We have also investigated the use of longer dark intervals in range of minutes to hours. This type of light fractionation is the subject of the present report. We have arbitrarily defined a long-term light fractionation scheme as an illumination scheme with two light fractions separated by an interval of 30 minutes or longer.

PpIX Re-synthesis and Light Fractionation

ALA-PDT relies on the capacity of cells to synthesise PpIX from exogenous ALA. This is obviously an important issue for light fractionated PDT. Another important consideration is photobleaching of the photosensitiser during PDT. This is a well known effect that is particularly important for ALA-induced PpIX [16]. Post treatment, cells within the illuminated volume can under some circumstances continue to synthesise PpIX. A number of studies have measured the continued

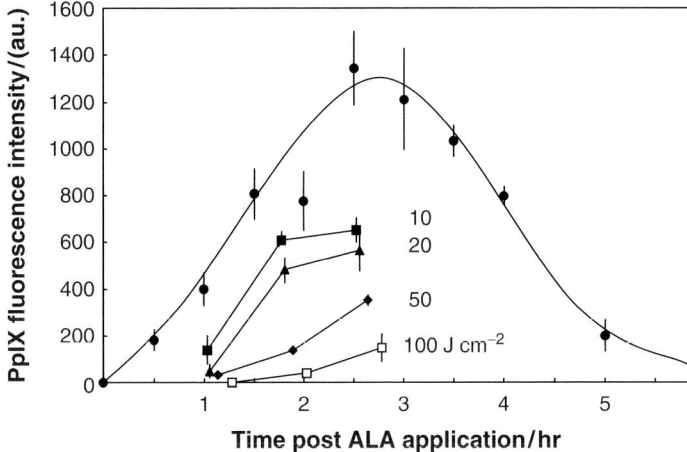

Fig. 1 (a) The pharmacokinetics of protoporphyrin IX (PpIX) following administration of 200 mg/kg 5-aminolevulinic acid (i.v.) administration in (●) rhabdomyosarcoma; and the return in PpIX fluorescence transdermal illumination (using 633 nm radiation at an irradiance of 100 mW cm^{-2}) to a light dose of (■) 10; (▲) 20; (◆) 50 and (□) 100 J cm^{-2}. Results are shown as mean ± SD and n = 5 (The figure is adapted from Fig. 1(b) in [31])

synthesis of PpIX after illumination in a range of tissues. The opportunity provided by this 'additional' photosensitiser was the primary reason underlying our choice of this type of illumination scheme. We have shown that the rate of PpIX re-synthesis after illumination is dependent on the PDT dose delivered [19]. There is a fluence dependence for a fixed fluence rate; the higher the fluence the less PpIX is synthesised and a fluence rate effect, where illumination at low fluence rate results in less re-synthesis for the same fluence. Figure 1 illustrates the effect of fluence on the re-synthesis of PpIX in a transplanted tumour model and is similar to that we have observed in other tissues.

The fact that the re-synthesis of PpIX after PDT is dependent on the PDT dose is a not surprising, since it is the cells themselves that synthesise PpIX. What is rather more complicated is determining an optimum treatment regimen that maximises response.

A number of issues are immediately apparent:

1. What is the most appropriate ALA-application time before therapy?
2. What is the optimum time interval between the two light fractions?
3. What is the optimum fluence and fluence rate of each fraction?

Our initial response data in animal models was achieved using equal light fractions of 75 J cm^{-2}, 1 and 2.5 hours after the systemic ALA [20]. These illumination parameters lead to a significant increase in tissue response as shown in Fig. 2: The vasculature, imaged in a rat skin fold observation chamber, is dramatically disrupted during the second of two light fractions. Further data from experiments using topical ALA in various normal and tumour tissues has shown

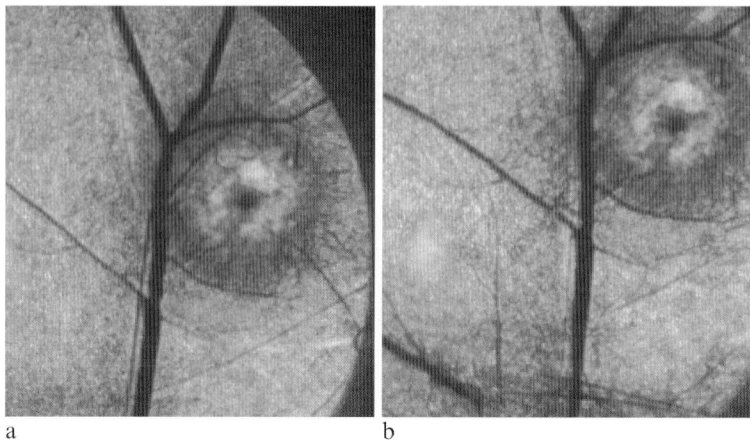

a   b

Fig. 2 The status of the normal and tumour microvasculature in the rat-skin fold observation chamber during ALA-PDT using (**a**) a single illumination and (**b**) during the second light fraction of a fractionated illumination scheme

that the increase in efficacy following fractionated ALA-PDT is a phenomenon that is independent of the animal model [21–24]. These experiments also showed that the ALA application period, the length of the dark interval and the fluence and fluence rate of each illumination strongly influence response. Initially we did not anticipate the critical role of these parameters but this is an indication of the complexity of the mechanism behind the response of tissues to light fractionation.

The length of the dark interval is an important parameter [25]. In skin it is necessary to have a significant dark interval before we observe an increase in response. Short dark intervals that would be associated with re-oxygenation of tissues do not lead to a significant increase in response [26]. Dark intervals longer than 30 minutes are necessary to show an effect. Efficacy increases with the increasing length of the dark interval up to the 2 hours that we have investigated. Longer dark intervals may be even more effective. We have not investigated these since the clinical suitability of such protocols becomes a significant factor.

In most of our pre-clinical work we have used a 4 hour ALA application period, similar to that used routinely used in the clinic. We have however shown that it is possible to reduce the overall treatment time, without compromising efficacy, by delivering the first light fraction earlier, 2 hours after ALA application with the second light fraction 2 hours later. This reduces the overall treatment time from 6 to approximately 4 hours. Here we were careful to choose an appropriate fluence for this early first light fraction. The concentration of PpIX available 2 hours after the application of ALA is approximately 50% of that at 4 hours. It is therefore necessary to compensate for the reduced concentration of PpIX by increasing the fluence of the first illumination from 5 to 10 J cm^{-2} [19].

One important result of these studies is that this increasing efficacy is not associated with an increasing PpIX content of the tissues during the treatment scheme [25]. There is no direct relationship between the amount of total amount

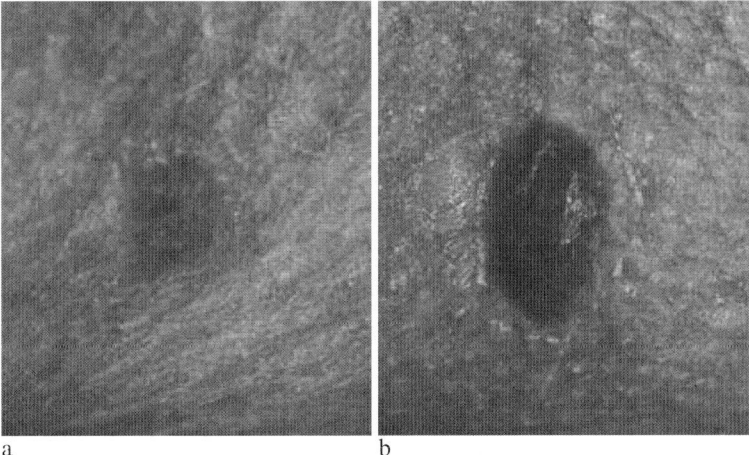

a   b

Fig. 3 The visual skin response 48 hours after ALA-PDT using (**a**) a single illumination scheme of 100 J cm^{-2} and (**b**) light fractionated ALA-PDT using 5 + 95 J cm^{-2} with a 2 hours dark interval

of PpIX utilised and efficacy. The complexity of the mechanism involved is also illustrated by the fact that there is an optimum PDT dose of the first and second light fractions [19]. The delivery of 5 J cm^{-2} at 50 mW cm^{-2}, 4 hours after the administration of ALA leads to a significant increase in response of normal mouse skin compared to 100 J cm^{-2} in a single fraction, as shown in Fig. 3. Delivering a higher dose negates the increase in effectiveness and 50 + 50 J cm^{-2} leads to significantly less response. We have also shown that it is necessary to have a large second light fraction to achieve an optimised response. Again in each of these studies we find that the increase in effectiveness is not associated with the utilisation of more PpIX [19].

Clinical PDT

While we investigate the mechanism behind the two-fold illumination scheme in pre-clinical models we have also been performing light fractionated ALA-PDT in the clinic to determine the clinical significance of our data. Our treatment protocols for clinical PDT have been strongly influenced by the results from our animal studies. We performed a clinical pilot-study treating superficial basal cell carcinoma using equal light fractions of 45 J cm^{-2} separated by a 2 hour dark interval [27]. We note that this protocol was designed before we had data on the importance of the PDT dose of the two light fractions. Treating 67 primary sBCC's we showed complete response of 84% with a mean follow-up of 59 months. While CR was greater than the CR of a single illumination scheme this difference is not statistically significant. Based on our more recent pre-clinical data we designed a randomized comparative study investigating the use of light fractionation using an optimized illumination scheme for ALA-PDT [28]. A treatment scheme involving the delivery of $20 + 80$ J cm^{-2} separated by a 2 hour dark interval, 4 and 6 hours after the application of ALA was devised. Note that a higher fluence of the first light fraction was chosen because of the increased thickness of superficial BCC. Five hundred and five primary sBCC were assigned into two treatment groups; 243 lesions were treated using a single illumination of 75 J cm^{-2} and 262 lesions received fractionated PDT after a single application of ALA. Both light fractions were delivered at a fluence rate of 50 mW cm^{-2}. An interim analysis of the clinical results, with a minimum follow-up of 12 months, showed CR after fractionated PDT was 97% compared to 89% following PDT in a single light fraction. This statistically significant increase in CR ($P = 0.002$, log-rank test) is to our knowledge the first study to show that optimization of clinical PDT is possible and adds weight to the results of our pre-clinical studies. Recently longer-term follow up data suggest that this CR is maintained 2 years after therapy. We have also shown an increase CR following fractionated ALA-PDT in squamous cell cancer in-situ, Bowen's disease [29]. Light fractionated ALA-PDT was slightly more painful the traditional approach to PDT but all patient completed therapy and cosmesis remained excellent. An example of the clinical response immediately after the end of the second light fraction is shown in Fig. 4.

It is important to note that the cumulative fluence in each of the illumination groups is not equal. Seventy-five joules per square centimeter was delivered in a single light fraction compared to 100 J cm^{-2} in the two-fold illumination scheme. This is a direct consequence of our intention to deliver a large fluence in the second light fraction. The influence of this additional cumulative fluence on the clinical response we observe is an important issue. A number of pre-clinical studies have shown that photobleaching of the PpIX limits the PDT dose that can be delivered in a single light fraction at fixed fluence rate. We have shown in normal mouse skin

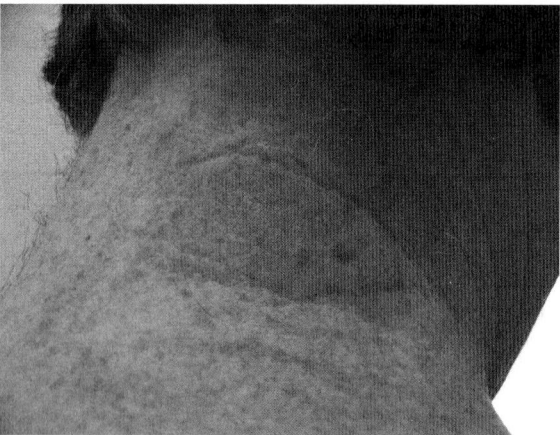

Fig. 4 The clinical response immediately after light fractionated ALA-PDT of superficial basal cell carcinoma

that 100 J cm^{-2} does not result in significantly more damage than 50 J cm^{-2}. The relationship between response to ALA-PDT and fluence has not been systematically investigated in the clinic. Only Oseroff has emphasised the importance of light fluence in the treatment of sBCC by topical ALA-PDT and this regards the delivered of very high light fluences at high fluence rate [30]. Most other investigators have applied both lower light fluence (rate) and cumulative fluences both below 75 J cm^{-2} and above 100 J cm^{-2}, with little evidence for a correlation between fluence and clinical response.

Mechanisms of Action

As we have described there is no clear relationship between the total amount of PpIX utilised and efficacy. This means that there must be a more subtle mechanism behind the increase in response. We have shown that there is a large increase in the number of circulating and tumour neutrophils immediately after ALA-PDT. We have investigated if theses phagocytes offer any therapeutic adjunct to the generation of single oxygen during PDT [31]. Administration of anti granulocyte serum treatment prevented the influx of neutrophils after ALA-PDT, but did not lead to a significant decrease in the efficacy of the PDT. These results indicate that the magnitude of damage inflicted on the tumour by ALA-PDT does not depend on the presence of neutrophils in the tumour or circulation and that they seem to play a bystander role in ALA-PDT.

There has been a considerable increase in the clinical use of ester derivatives of ALA such as methyl 5-aminolevulinate (MAL). We have recently investigated the role of light fractionation in pre-clinical models [32]. To our surprise light fractionation using an illumination scheme that is optimised for ALA does not lead to any increase in response over a single light fraction. Fractionated illumination does not enhance the efficacy of PDT using MAL as it does using ALA despite the fact that identical fluorescence kinetics during therapy. Only the initial rate of photobleaching was slightly greater during ALA-PDT although the difference was small. Previously we have hypothesized that cells surviving the first fraction are more susceptible to the second fraction. Since this is not true for MAL-PDT our data suggest that the in-vivo distribution of MAL and ALA and the exact site of PDT induced damage, is an important parameter in the mechanism underlying fractionated illumination for ALA-PDT. Indeed, preliminary in-vitro data in an adenocarcinoma cell line show that we are not able to increase the PDT response of cells incubated with either ALA or MAL using light fractionation.

Conclusions and Future Perspectives

We believe that there is a clear role for the use of light fractionation for the treatment of superficial basal cell carcinoma. If the ALA application time can be reduced from

4 to 2 hours, without any reduction in clinical efficacy this would be a further significant advance that reduces the overall treatment time to approximately 4 hours.

Non-light fractionated PDT has also been applied to thicker nodular lesions [33]. We believe that there is significant potential for the application of fractionation since this has been shown to be result in a thicker depth of necrosis compared to the traditional approach [19]. The maximum depth of response for the treatment of nodular lesions needs to be determined. It is also possible to envisage repeating an optimized fractionated illumination scheme for the treatment of nodular lesions in the skin.

There is also the potential to apply light fractionation for the treatment of other lesions beyond the skin, such as the oesophagus and brain, where ALA-PDT is an investigational treatment modality [34, 35]. We do note however that that there is a degree of complexity added by the use of 2 hour dark interval which could be difficult during a surgical procedure. It is also critical to perform efficient PDT dosimetry since the illumination parameters of the two light fractions are critical [36]. This can be a problem in hollow cavities where the tissue optical properties can significantly affect the distribution of light and the fluence (rate) [37].

References

1. Kennedy, J. C., Pottier, R. H. Endogenous protoporphyrin IX, a clinical useful photosensitizer for photodynamic therapy. J Photochem Photobiol B, 14: 275–292, 1992.
2. Calzavara-Pinton, P. G. Repetitive photodynamic therapy with topical δ-aminolaevulinic acid as an appropriate approach to the routine treatment of superficial non-melanoma skin tumours. J Photochem Photobiol B, 29: 53–57, 1995.
3. Meijnders, P. J. N., Star, W. M., de Bruijn, R. S., Treurniet-Donker, A. D., van Mierlo, M. J. M., Wijthoff, S. J. M., Naafs, B., Beerman, H., Levendag, P. C. Clinical results of photodynamic therapy for superficial skin malignancies or actinic keratosis using topical 5-aminolaevulinic acid. Lasers Med Sci, 11: 123–131, 1996.
4. Peng, Q., Warloe, T., Berg, K., Moan, J., Kongshaug, M., Giercksky, K. E., Nesland, J. M. 5-Aminolevulinic acid-based photodynamic therapy: clinical research and future challenges. Cancer (Phila), 79: 2282–2308, 1997.
5. Peng, Q., Warloe, T., Moan, J., Heyerdahl, H., Steen, H. B., Nesland, J. M., Giercksky, K. E. Distribution of 5-aminolevulinic acid-induced porphyrins in noduloulcerative basal cell carcinoma. Photochem Photobiol, 62: 906–913, 1995.
6. Morton, C. A., Whitehurst C., Moseley, H. McColl, J. H., oore, J. V., Mackie, R. M. Comparison of photodynamic therapy with cryotherapy in the treatment of Bowen's disease. Br J Dermatol 135: 766–771, 1996.
7. Haller, J. C., Cairnduff, F., Slack, G., Schofiled, J., Whitehurst, C., Turnstall, R., et al. Routine double treatments of superficial basal cell carcinomas using aminolaevulinic acid-based photodynamic therapy. Br J Dermatol 143: 1270–1274, 2000.
8. Rhodes, L. E., De Rie Enstrom, Y., Groves, R., Morken, T., Goulden, V., et al. Photodynamic therapy using topical methyl aminolevulinate vs surgery for nodular basal cell carcinoma. Arch Dermatol 140: 17–23, 2004.

9. Curnow, A., McIlroy, B. W., Postle-Hacon, M. J., Porter, J. B., MacRobert, A. J., Bown, S. G. Enhancement of 5-aminolaevulinic acid-induced photodynamic therapy in normal rat colon using hydroxypyridione iron chelating agents. Br J Cancer 78: 1278–1282, 1998.

10. Choudry, K., Brooke, R. C., Farrar, W., Rhodes, L. E. The effect of an iron chelating agent on protoporphyrin IX levels and phototoxicity in topical 5-aminolaevulinic acid photodynamic therapy. Br J Dermatol 149: 124–130, 2003.

11. Tope, W. D., Ross, E. V., Kollias, N., Martin, A., Gillies, R., Rox Anderson, R. Protoporphyrin IX fluorescence induced in basal cell carcinoma by oral δ-aminolevulinic acid. Photochem Photobiol, 67: 249–255, 1998.

12. Fan, K. F. M., Hopper, C., Speight, P. M., Buonaccorsi, G., MacRobert, A. J., Bown, S. G. Photodynamic therapy using 5-aminolevulinic acid for premalignant and malignant lesions of the oral cavity. Cancer (Phila), 78: 1374–1383, 1996.

13. Barr, H., Shepherd, N. A., Dix, A., Roberts, D. J. H., Tan, W. C., Krasner, N. Eradication of high-grade dysplasia in columnar-lined (Barrett's) oesophagus by photodynamic therapy with endogenously generated protoporphyrin IX. Lancet, 348: 584–585, 1996.

14. Hua, Z., Gibson, S. L., Foster, T. H., Hilf, R. Effectiveness of -aminolevulinic acid-induced protoporphyrin as a photosensitizer for photodynamic therapy *in vivo*. Cancer Res, 55: 1723–1731, 1995.

15. Pogue, B. W., Hasan, T. A theoretical study of light fractionation and dose-rate effects in photodynamic therapy. Radiat Res, 147: 551–559, 1997.

16. Robinson, D. J., de Bruijn, H. S., van der Veen, N., Stringer, M. R., Brown, S. B., Star, W. M. Fluorescence photobleaching of ALA-induced protoporphyrin IX during photodynamic therapy of normal hairless mouse skin: the effect of light dose and irradiance and the resulting biological effect. Photochem Photobiol, 67: 140–149, 1998.

17. Messmann, H., Mlkvy, P., Buonaccorsi, G., Davies, C. L., MacRobert, A. J., Bown, S. G. Enhancement of photodynamic therapy with 5-aminolaevulinic acid-induced porphyrin photosensitisation in normal rat colon by threshold and light fractionation studies. Br J Cancer, 72: 589–594, 1995.

18. Robinson, D. J., de Bruijn, H. S., van der Veen, N., Stringer, M. R., Brown, S. B., Star, W. M. Protoporphyrin IX fluorescence photobleaching during ALA-mediated photodynamic therapy of UVB-induced tumors in hairless mouse skin. Photochem Photobiol, 69: 61–70, 1999.

19. Robinson, D. J., de Bruijn, H. S., Star, W. M., Sterenborg, H. J. C. M. Dose and timing of the first light fraction in two fold illumination schemes for topical ALA-mediated photodynamic therapy of hairless mouse skin. Photochem Photobiol, 77: 319–323, 2003.

20. Van der Veen, N., van Leengoed, H. L. L. M., Star, W. M. In vivo fluorescence kinetics and photodynamic therapy using 5-aminolaevulinic acid-induced porphyrin: increased damage after multiple irradiations. Br J Cancer, 70: 867–872, 1994.

21. Van der Veen, N., Hebeda, K. M., de Bruijn, H. S., Star, W. M. Photodynamic effectiveness and vasoconstriction in hairless mouse skin after topical 5-aminolevulinic acid and single- or two-fold illumination. Photochem Photobiol, 70: 921–929, 1999.

22. De Bruijn, H. S., van der Veen, N., Robinson, D. J., Star, W. M. Improvement of systemic 5-aminolevulinic acid photodynamic therapy in-vivo using light fractionation with a 75-minute interval. Cancer Res, 59: 901–904, 1999.

23. Robinson, D. J., de Bruijn, H. S., de Wolf, J., Sterenborg, H. J. C. M., Star, W. M. Topical 5-aminolevulinic acid-photodynamic therapy of hairless mouse skin using two-fold illumination schemes: PpIX fluorescence kinetics, photobleaching and biological effect. Photochem Photobiol, 72: 794–802, 2000.

24. Thissen, M. R., de Blois, M. W., Robinson, D. J., de Bruijn, H. S., Dutrieux, R. P., Star, W. M., et al. PpIX fluorescence kinetics and increased skin damage after intracutaneous injection of 5-aminolevulinic acid and repeated illumination. J Invest Dermatol, 118: 239–245, 2002.

25. De Bruijn, H. S., van der Ploeg-van den Heuvel, A., Sterenborg, H. J. C. M., Robinson, D. J. Fractionated illumination after topical application of 5-aminolevulinic acid on normal skin of hairless mice; the influence of the dark interval. J Photochem Photobiol B, 85: 184–190, 2006.

26. Curnow, A., McIlroy, B. W., Postle-Hacon, M. J., MacRobert, A. J., Bown, S. G. Light dose fractionation to enhance photodynamic therapy using 5-aminolevulinic acid in the normal rat colon. Photochem Photobiol, 69: 71–76, 1999.

27. Star, W. M., van't Veen, A. J., Robinson, D. J., Munte, K., de Haas, E. R. M., Sterenborg, H. J. C. M. Topical 5-aminolevulinic acid mediated photodynamic therapy of superficial basal cell carcinoma using two light fractions with a two hour interval: long-term follow-up. Acta Derm Venereol, 86: 412–417, 2006.

28. De Haas, E. R., Kruijt, B., Sterenborg, H. J., Martino Neumann, H. A., Robinson, D. J. Fractionated illumination significantly improves the response of superficial basal cell carcinoma to aminolevulinic acid photodynamic therapy. J Invest Dermatol, 126: 2679–2686, 2006.

29. de Haas, E. R. M., Kruijt, B., Sterenborg, H. J. C. M., Neumann, H. A. M., Robinson, D. J. The response of Bowen's disease to ALA-PDT using a single and a two fold illumination scheme. Arch Dermatol (in press), 2007.

30. Oseroff, A. R. PDT for cutaneous malignancies: clinical applications and basic mechanisms. Photochem Photobiol, 67: 17S–18S, 1998.

31. de Bruijn, H. S., Sluiter, W., van der Ploeg-van den Heuvel, A., Sterenborg, H. J., Robinson, D. J. Evidence for a bystander role of neutrophils in the response to systemic 5-aminolevulinic acid-based photodynamic therapy. Photodermatol Photoimmunol Photomed, 22: 238–46, 2006.

32. de Bruijn, H. S., de Haas, E. R. M., Hebeda, K. M., van der Ploeg-van den Heuvel, A., Sterenborg, H. J. C. M., Neumann, H. A. M., Robinson, D. J. Light fractionation does not enhance the efficacy of methyl 5-aminolevulinate mediated photodynamic therapy in normal mouse skin. Photochem Photobiol (in press), 2007.

33. Rhodes, L. E., De Rie Enstrom, Y., Groves, R., Morken, T., Goulden, V., et al. Photodynamic therapy using topical methyl aminolevulinate vs surgery for nodular basal cell carcinoma. Arch Dermatol, 140: 17–23, 2004.

34. Bogaards, A., Varma, A., Zhang, K., Zach, D., Bisland, S. K., Moriyama, E. H., et al. Fluorescence image-guided brain tumour resection with adjuvant metronomic photodynamic therapy: pre-clinical model and technology development. Photochem Photobiol Sci, 4: 438–442, 2005.

35. Pech, O., Gossner, L., May, A., Rabenstein, T., Vieth, M., Stolte, M., et al. Long-term results of photodynamic therapy with 5-aminolevulinic acid for superficial Barrett's cancer and high-grade intraepithelial neoplasia. Gastrointest Endosc, 62: 24–30, 2005.

36. van Veen, R. L., Robinson, D. J., Siersema, P. D., Sterenborg, H. J. The importance of in situ dosimetry during photodynamic therapy of Barrett's esophagus. Gastrointest Endosc. 64: 786–788, 2006.

37. Boere, I., Robinson, D. J., de Bruijn, H. S., Kluin, J., Tilanus, H. W., Sterenborg, H. J., de Bruin, R. W. Protoporphyrin IX fluorescence photobleaching and the response of rat barret's esophagus following 5-aminolevulinic acid photodynamic therapy. Photochem Photobiol, 82, doi: 10.1562/2006-01-03-RA-763, 2006.

Combination Immunotherapy and Photodynamic Therapy for Cancer

Michael R. Hamblin, Ana P. Castano, and Pawel Mroz

Abstract Cancer is a leading cause of death among modern peoples largely due to metastatic disease. The ideal cancer treatment should target both the primary tumor and the metastases with the minimal toxicity. This is best accomplished by educating the body's immune system to recognize the tumor as foreign so that after the primary tumor is destroyed, distant metastases will also be eradicated. Photodynamic therapy (PDT) involves the IV administration of photosensitizers followed by illumination of the tumor with red light producing reactive oxygen species that eventually cause vascular shutdown and tumor cell apoptosis. Anti-tumor immunity is stimulated after PDT due to the acute inflammatory response, generation of tumor-specific antigens, and induction of heat-shock proteins, while the three commonest cancer therapies (surgery, chemotherapy and radiotherapy) all tend to suppress the immune system. Like many other immunotherapies, the extent of the immune response after PDT tends to depend on the antigenicity of the particular tumor. Combination regimens using PDT and immunostimulating treatments are likely to emerge in the future to even further enhance immunity. These are likely to include the so called biological response modifiers that generally consist of products obtained from pathogenic microorganisms against which mammals have evolved sophisticated defenses involving immune activation. A series of pattern recognition molecules including toll-like receptors have been identified that are activated by products derived from pathogens and lead to upregulation of transcription factors that induce expression of many cytokines and inflammatory mediators, which then cause activation of macrophages, dendritic and natural killer cells. There have been several reports of combinations of PDT with microbial derived products potentiating tumor response and leading to long-term anti-tumor immunity. In recent years the role of regulatory T-cells in suppressing anti-tumor immunity has been identified. Treatments such as low dose cyclophosphamide that selectively reduces T-regulatory cells can also be combined with PDT. Methods may be

M.R. Hamblin
Wellman Center for Photomedicine, Massachusetts General Hospital, 40 Blossom Street, BAR 414, Boston, MA 0211, USA, e-mail: Hamblin@Helix.Mgh.Harvard.Edu

R. Waynant and D.B. Tata (eds.), *Proceedings of Light-Activated Tissue Regeneration and Therapy Conference*.
© Springer Science + Business Media, LLC 2008

developed to increase the expression of particular tumor antigens before PDT. Although so far these combination therapies have only been used in animal models, their use in clinical trials should receive careful consideration.

Keywords: Photodynamic therapy, anti-tumor immunity, toll-like receptors, T regulatory cells, dendritic cells, cytotoxic T-cells, antigen presentation.

Introduction

Photodynamic therapy (PDT) is a promising new therapeutic procedure for the management of a variety of solid tumors and a number of non-malignant diseases. PDT is a two-step procedure that involves the administration of a photosensitizing dye or photosensitizer (PS), followed by activation of the drug with non-thermal light of a specific wavelength that is absorbed by the dye [1–3]. PDT generates singlet oxygen and other reactive oxygen species (ROS), which cause an oxidative stress and membrane damage in the treated cells leading to cell death. There are three main mechanisms that make PDT an effective anti-cancer procedure. Firstly, there is direct tumor killing by ROS and induction of apoptosis [4, 5]. Secondly, there is a tumor-associated vascular damage that deprives the tumor cells of oxygen and nutrients normally supplied by the microcirculation and resulting in tumor cell death [6, 7]. Thirdly and lastly there is an activation of anti-tumor immune response [8–10]. The relative contributions of these three separate mechanisms to the overall antitumor effects of PDT are difficult to unambiguously establish. The prevailing view is that all three mechanisms are necessary for the optimal tumor damage.

The earliest studies of antitumor immunity and PDT were carried out in mice by Canti and colleagues [11] who examined the effects of PDT with phthalocyanines on the antitumor immune response. Immunosuppressed and normal mice bearing the MS-2 fibrosarcoma treated with aluminum disulfonated phthalocyanine (AlS2Pc) were then exposed to laser light or treated with surgical excision of the tumor. All mice were cured and survived indefinitely, with no difference between the groups. The survivors were rechallenged with the parental MS-2. Some groups of surviving animals were immunosuppressed with cyclophosphamide before the re-challenge with the tumor. Resistance to rechallenge was observed only in normal animals cured with PDT, while the immunosuppressed animals and animals cured with surgery died of tumor. Finally, mice, cured of MS-2 by PDT and tumor-free were rechallenged with different syngenic L1210 and P388 murine leukemias and did not survive. These results suggested that a potent and specific 'antitumor immunity' is induced after PDT with photoactivated AlS2Pc.

Since then many reports have suggested the involvement of immune system in the PDT response, hence the aim of this review is to summarize the data concerning the immunotherapy in combination with PDT and to underline possible and existing targets for future development of combination therapy.

Pdt Combined with Immunoadjuvants

It has been established that PDT can initiate activation of the innate immune system. Studies in rodents proved that there is a significant infiltration of neutrophils, mast cells and monocytes/macrophages into the tumor area after PDT procedures [12, 13]. These cells are thought to enhance the PDT-mediated destruction of malignant tissue (Fig. 1 represents the possible immune events that follow PDT treatment). There are also an increasing number of studies showing that immunoadjuvants (usually injected intratumorally) can produce a somewhat similar infiltration of leukocytes into the tumor. Immunoadjuvants are frequently prepared from microbial cells and are thought to act via Toll-like receptors (TLRs) present on macrophages and dendritic cells [14] (Fig. 2). TLRs are a family of pattern recognition receptors and share significant homology with interleukin-1 receptor family members. So far 13 TLR family members have been identified. They are expressed by monocytes/macrophages, dendritic cells, mast cells and some epithelial cells [15–17]. Their expression is regulated by several cytokines like IL-2, IL-15, IFNγ and TNFα [18] and their principal role is thought to act as early warning or danger signals for imminent infection. The activation of TLR pathways can induce

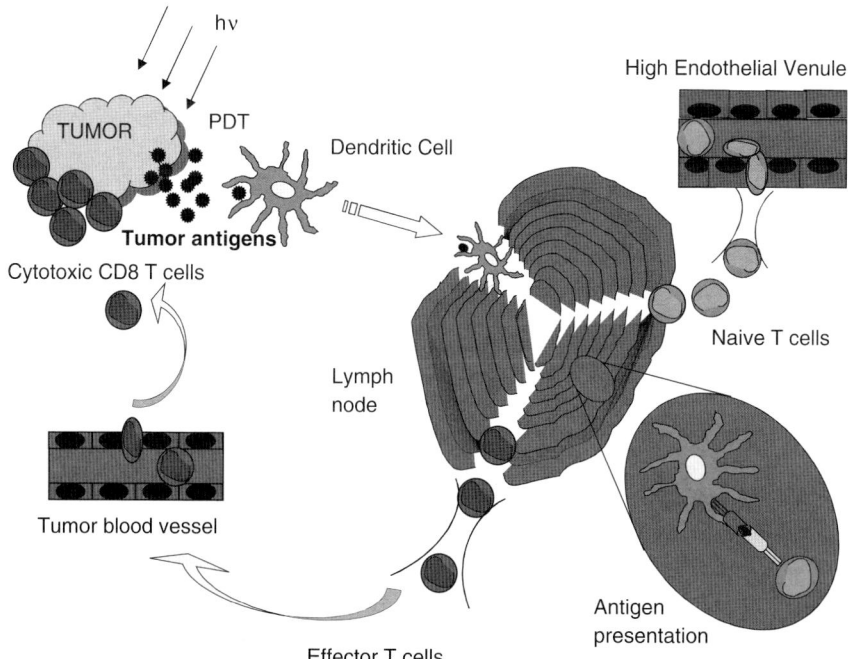

Fig. 1 Schematic figure illustrating the phagocytosis of tumor antigens by dendritic cells after PDT. DCs then migrate to tumor-draining lymph nodes and present antigens to naive T cells. These cytotoxic CD8-antigen specific T cells can then extravasate into tumor where they recognize specific antigens and kill tumor cells

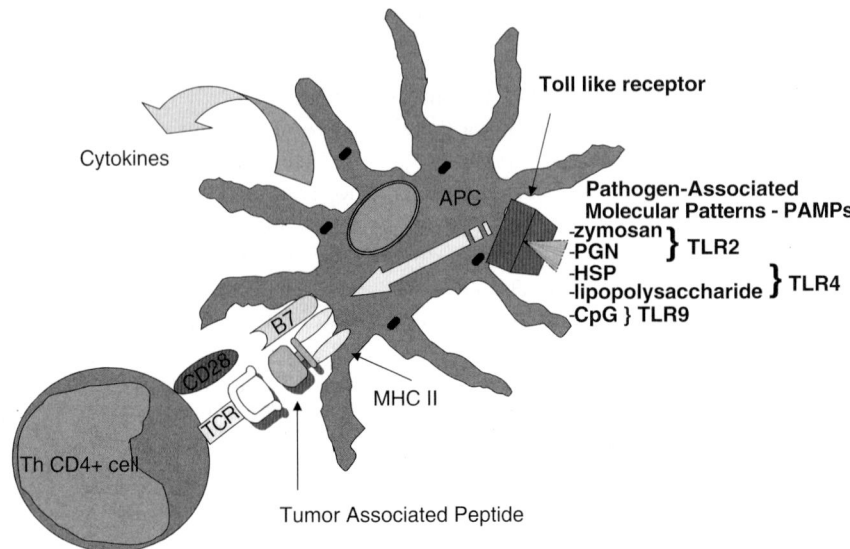

Fig. 2 Schematic figure illustrating the role of toll-like receptors (TLR) and their ligands in stimulating tumor-associated antigen presentation by dendritic cells in context of MHC class II to naive CD4+ T cells. PGN, peptidoglycan: HSP, heat shock protein, CpG, bacterial oligonucleotide, APC, antigen presenting cell.

transcription of several genes involved in immune system activation and which also are important for the antitumor immune response [19]. From these reports arose the idea of a combination therapy that involves the administration of immunoadjuvants, frequently potential TLR ligands, and different PDT regimens.

One of the first combinations of PDT with an immunostimulant was a study from 1989 [20] when Myers et al. combined HpD-PDT with a killed preparation of *Corynebacterium parvum* (CP, now *Propionibacterium acnes*) in a mouse model of subcutaneous bladder cancer (MBT2 in C3H mice). A low dose of CP was shown to enhance the effect of PDT while PDT reduced the benefit obtained with high dose of CP. However the administration of high dose of CP after PDT was shown to have a significantly greater effect than CP treatment before PDT.

The combination of PDT with Bacillus Calmette-Guerin (BCG) followed this successful example of adjuvant based therapy [21]. Subcutaneous mouse EMT6 tumors were treated with a single BCG administration in combination with six clinically tested photosensitizers: Photofrin, BPD, mTHPC, mono-L-aspartyl-chlorin e6, lutetium texaphyrin and ZnPC. Regardless the type of photosensitizer, the administration of BCG significantly increased the number of cured tumors. Moreover the combination therapy increased three-fold the number of memory T cells in tumor-draining lymph nodes in comparison to treatment with Photofrin-based-PDT alone. The therapeutic effects of BCG were observed even when the BCG administration was delayed to 7 days after PDT. A similar approach was applied by another group [22]. In a series of experiments they combined intra-tumor injection of Bacillus Calmette-Guerin 2 days before Photofrin-based-PDT in a C3H mice bearing MBT2 tumor. They observed a significant number of cures not seen with either treatment alone.

Korbelik and Cecic also reported the successful combination of Mycobacterium cell-wall extract (MCWE) with PDT [23]. MCWE is a potent non-specific immunostimulant, which elicits a local inflammatory response associated with an increase in the numbers of neutrophils, monocytes and macrophages. In a set of experiments several different photosensitizers–Photofrin, benzoporphyrin derivative monoacid (BPD), mTHPC and zinc (II)-phthalocyanine (ZnPc)–were combined with a single administration of MCWE immediately after light exposure. In the experiments combining MCWE with mTHPC-based-PDT an increased infiltration of neutrophils and other myeloid cells in the tumor area was observed. Despite different kinetics and intensity of damage of tumor microvasculature and other mechanism of antitumor activity between these photosensitizers, there appeared to be no obvious differences in the mechanism of MCWE mediated potentiation of PDT.

A related approach that takes advantage of the immune-stimulating effects of PDT is the preparation of cancer vaccines by applying in vitro PDT treatment to cell cultures. Gollnick et al. [24] compared the effectiveness of PDT-generated murine tumor cell vaccine (EMT6 and P815 tumor cells) to other methods of creating whole tumor vaccines, i.e., UV or ionizing irradiation. PDT vaccines were tumor specific, induced a cytotoxic T-cell response and the effectiveness did not require coadministration of any adjuvant, unlike in the other regiments. Moreover PDT-generated lysates were able to induce DC maturation and expression of IL-12.

Korbelik and Sun [25] produced a vaccine by incubating in vitro expanded mouse SCCVII cells with BPD, exposing to light, and finally to a lethal dose of X-ray. Treatment of established subcutaneous SCCVII tumors in syngenic C3H/HeN mice with PDT-vaccine produced a significant therapeutic effect, including growth retardation, tumor regression and cures. To demonstrate the tumor specificity of PDT-generated vaccine a mismatched tumor cell line was used and proved to be ineffective. Furthermore, vaccine cells retrieved from the treatment site at 1-h post injection were mixed with DCs, exhibiting heat shock protein 70 on their surface, and opsonized with complement C3 molecule. Tumor-draining lymph nodes from mice treated with the PDT-vaccine contained extensively increased numbers of DC as well as B and T lymphocytes. The high levels of surface-bound C3 molecules were detectable on DC and to a lesser extent on B cells.

Schizophyllan (SPG), together with lentinan, is an example of β-D-glucans, polysaccharides of fungal origin known to be potent inducers of humoral and cell-mediated immunity in humans and animals [26]. It has been proposed that beta-glucans can be involved in the activation of macrophages via the dectin-1 receptor [27], although neutrophils, lymphocytes and NK cells may also play a key role in beta-glucan action [28]. Krosl and Korbelik reported potentiation of the PDT effect by schizophyllan treatment in mice bearing squamous cell carcinoma (SCCVII). The tumor cure rate increased approximately three times when SPG was administered prior to PDT procedure, while the reverse sequence, the administration of SPG after PDT, had little significant effect.

The mouse SCCVII squamous cell carcinoma model was also used to examine the effectiveness of Photofrin-based-PDT in combination with serum vitamin D3-binding protein-derived macrophage-activating factor (DBPMAF) [29]. DBPMAF markedly enhanced the outcome of PDT, while showing no significant effect alone on the growth of SCCVII tumor. The most successful approach was to combine intraperitoneal and peritumoral injections of DBPMAF on days 0, 4, 8 and 12 after PDT. What is most interesting is that the addition of DBPMAF helped to overcome the PDT-induced immunosuppression, which was assessed by the evaluation of delayed-type contact hypersensitivity response in treated mice.

Korbelik's group has [30] observed activation of the complement system during PDT and proposed this as an additional mechanism of antitumor response. Tumor-localized treatment with zymosan, an alternative complement pathway activator and TLR 2 ligand, reduced the number of recurrent tumors after PDT. However, a similar treatment with heat-aggregated gamma globulin (a complement activator via the classical pathway) had no significant effect as a PDT adjuvant. Systemic complement activation with streptokinase treatment had no detectable effect on complement deposition at the tumor site without PDT, but it augmented the complement activity in PDT-treated tumors.

OK-432, a biological response modifier derived from killed preparation of streptococcal bacteria, has been shown to activate the immune system and to potentate the host defense mechanisms, especially against tumors [31]. The intra-tumoral injection of OK-432 3 h before HPD-PDT increased the tumor free time in mice with NRS1 tumors, while OK-432 injected immediately after PDT or OK-432 alone had little effect. The explanation for this observation was the recruitment of macrophages into the tumor area and synthesis of several cytokines including interferon gamma and TNFα [32]. A subsequent study by this group [33] used a high power laser that produced local hyperthermia as well as photoactivation of the HPD to carry out PDT in the same tumor model. OK-432 injected intratumorally 3 h prior to laser irradiation again gave a significant improvement in tumor control. The results from our laboratory (data not published) also confirm the beneficial effect of OK-432 on the outcome of PDT. We observed that combination of BPD-based-PDT with intra-tumor injection of OK-432 3 h after PDT cured all treated tumors and mice were resistant to rechallenge with the same number of tumor cells.

Castano and Hamblin have shown [34] that BPD-based-PDT of RIF1 tumors in wild-type C3H/HeN mice leads to initial tumor regression but not to permanent cures due to local recurrence. However, when the tumors were genetically engineered to express green fluorescent protein (GFP) from jellyfish, and then treated with PDT, not only 100% of cures but also a long-lasting resistance to rechallenge were observed. In contrast, the surgically removed RIF1-GFP tumors did not induced any significant immunity. It is very likely than that PDT induced significantly better immune recognition of the foreign GFP antigen and promoted long-lasting resistance.

Chen and colleagues developed the concept of "laser immunotherapy" as a novel approach that aims at the tumor-directed stimulation of the immune system [35]. It involves intratumoral administration of a laser-absorbing dye, indocyanine green (ICG), together with an immunoadjuvant, glycated chitosan (GC), followed by irradiation with a high power laser. The authors propose that the chief effect of the dye-laser combination is to produce an immediate photothermal effect that raises the temperature enough to kill the cancer cells by hyperthermia. They worked with DMBA-4 metastatic mammary tumors growing in female Wistar Furth rats and observed regression of untreated metastatic tumors at remote sites [36], resistance of cured rats to rechallenge [37] and induction of serum antibodies that bound to both living and fixed tumor cells.

The Role of Immune Cells in PDT Anti-Tumor Immunity

Korbelik and colleagues reported [38] that Photofrin-based-PDT of murine EMT6 mammary sarcoma cured all tumors in syngenic Balb/c mice but there was no long-term cures in either SCID or NUDE mice, the immunodeficient strains that share the same genetic background with Balb/c mice. However, the adoptive transfer of splenic T lymphocytes from Balb/c mice into SCID mice performed before PDT procedure postponed the recurrence of treated tumors. They also observed that neither adoptive transfer done immediately after PDT nor 7 days after PDT had any effect. This data suggests that the activity of host lymphoid populations, schematically illustrated on Fig. 3, was essential for preventing the recurrence of EMT6 tumors following PDT.

The same group [39] demonstrated full restoration of the therapeutic effect of PDT of EMT6 tumors in SCID mice that received splenocytes from BALB/c donors cured of EMT6 tumors with PDT 5 weeks before adoptive transfer. They also report that splenocytes obtained from donors cured of EMT6 tumors with X-rays irradiation were much less effective. Selective in vitro depletion of specific T-cell populations from engrafted splenocytes indicated that CD8 CTLs were the main immune effector cells conferring to the curative outcome to PDT. Figure 4. describes the possible mechanisms of interaction observed between antigen-specific CTLs and tumor cells. The immune specificity of these T-cell populations was demonstrated by the absence of cross-reactivity between the EMT6 and Meth-A tumor models (mismatch between tumors growing in splenocyte donors and recipients). These experiments indicate that PDT is highly effective at generating tumor-sensitized immune cells that can be recovered from lymphoid sites distant to the treated tumor area at protracted time intervals. PDT also creates the conditions necessary for converting the inactive adoptively transferred tumor-sensitized immune cells into fully functional antitumor effector cells.

A further study [40] examined the role of depletion of various leukocyte classes (neutrophils, macrophages, helper T cells, cytolytic T cells) in mice bearing EMT6 tumors after Photofrin-based-PDT. Immunodepletion of neutrophils and cytotoxic T cells performed immediately after PDT resulted in a distinct reduction in PDT-mediated tumor cures. Significant reduction in the EMT6 tumor cures was also

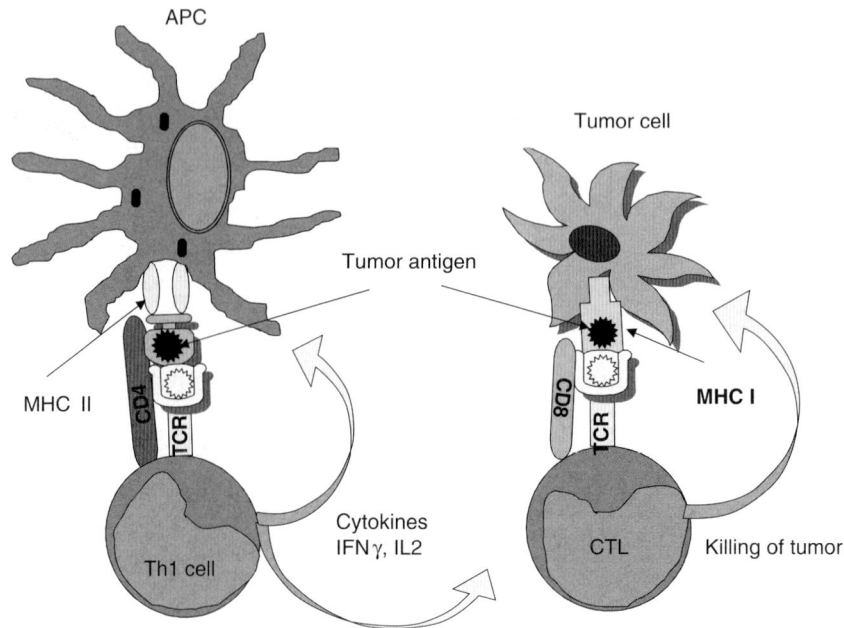

Fig. 3 Schematic figure illustrating antigen presentation to naive CD4 T-cells, leading to generation of cytotoxic CD8 cells capable of killing the tumor

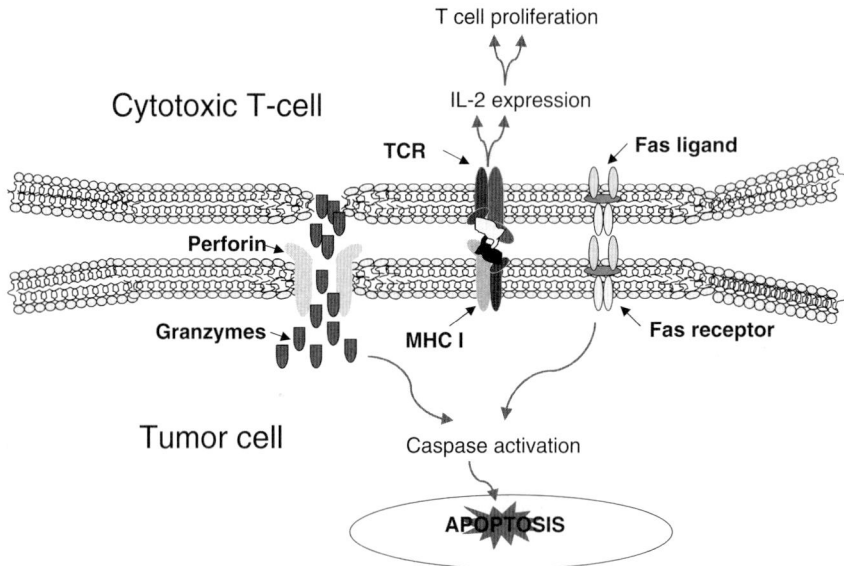

Fig. 4 Schematic figure illustrating mechanism for cytotoxic T cells to kill tumor cells via perforin/granzyme or Fas/Fas ligand

observed when T helper cells depletion or inactivation of macrophages by silica treatment was performed. The initial tumor ablation by PDT was not affected by any of the above depletion treatments.

Further insight into the mechanism of immune potentiation of PDT came from the studies of Hendrzak-Henion and coworkers [41]. By applying 2-iodo-5-ethylamino-9-diethylaminobenzo[a]-phenothiazinium chloride (2I-EtNBS) in PDT treatment of mice bearing EMT6 tumors they were able to cure between 75% to 100% of treated mice. However, PDT failed to inhibit tumor growth in T-cell-deficient NUDE mice. T-cell depletion studies with anti-CD8 antibody revealed that the CD8+ T-cell population was critical for an effective PDT response. Because anti-CD4 antibody inhibited tumor growth in the absence of PDT, the role of CD4+ T cells remains unclear. Depletion of natural killer (NK) cells in vivo with anti-asialo-GM1 antibody significantly reduced a suboptimal PDT effect relative to vehicle controls. However, splenic NK cells obtained from PDT-treated tumor-bearing mice were not cytotoxic in vitro against EMT-6 cells. These results suggest that NK cells contribute to the in vivo PDT effect by an indirect mechanism.

Korbelik and Sun devised another adjuvant procedure to potentiate PDT in immunosuppressed tumor bearing mice [42]. Adoptive transfer of human NK cell line genetically altered to produce interleukin-2 (NK92MI) was combined with PDT treatment of subcutaneous SiHa tumors (human cervical squamous cell carcinoma) growing in NOD/SCID mice. The most effective protocol for NK92MI cell transfer in conjunction with mTHPC-based-PDT was the injection of $5 \times 10(7)$ cells via peritumoral (p.t.) or intravenous routes immediately after PDT. This protocol produced a noticeable improvement in the therapeutic outcome of PDT in comparison to single treatment. The same positive result of the combination therapy was obtained with an HT-29 tumor model (human colorectal adenocarcinoma) engrafted in NOD/SCID mice, whereas such effect was not observed with the NK92 cell line that does not produce interleukin-2. Flow cytometry analysis revealed a higher percentage of p.t. injected NK92MI cells in PDT-treated than in non-treated HT-29 tumors. Further investigation revealed that the adoptive transfer of NK92MI is also a highly effective adjuvant modality for PDT treatment of BALB/c mice bearing EMT6 tumors. These experiments show also that adoptively transferred NK92MI cells are not rendered ineffective by an allogenic reaction of the host.

An interesting study by Golab et al. [43] examined the effectiveness of combination of Photofrin-based-PDT with administration of immature dendritic cells. Inoculation of immature DCs into the PDT-treated tumors resulted in effective homing to regional and peripheral lymph nodes and stimulation of cytotoxic T lymphocytes and natural killer cells. The combination treatment with PDT and DCs produced the best response against the primary tumor and also showed some effect against a second tumor challenge in the contralateral leg.

Cyclophosphamide (CY) is an alkylating agent widely used in cancer chemotherapy. It has a bimodal effect on the immune system, depending on the dose and schedule of administration. A single low dose of CY has an antimetastatic and immune stimulating effect [44], while higher doses cause immunosuppression.

It is proposed that low-dose CY selectively eliminates suppressor T cells or T-regulatory cells (Tregs) [45]. Tregs are a minor population of CD4$^+$ T cells ($\sim$10%) characterized by co-expression of CD25 and the transcription factor FoxP3. They are crucial for the control of autoreactive T cells *in vivo* [46]. It has been shown that the number of Tregs increases in certain tumors, which correlate with inhibition of the activity of CD4 Th 1 cells and CD8 cytotoxic T lymphocytes. This reaction may be mediated by cell contact mechanisms and induction of immunosuppressive cytokines (TGF-beta and IL-10) [47]. Our laboratory has studied [48] the role of Tregs in the response to J774 tumors (a Balb/c tumor model of metastatic sarcoma) in a vascular regiment of BPD-based-PDT. BPD-based-PDT alone cured the local tumors, but the mice died of metastases. When the CY was injected prior to PDT procedure the majority of the mice became long-term survivors and rejected a subsequent rechallenge with J774 cells. CY alone delayed tumor growth but gave no cures. Examination of splenocytes recovered from tumor bearing mice revealed that population of CD4+CD25+ T cells was significantly reduced and the splenocytes secreted significantly less TGF beta.

The Role of Cytokines in PDT-Induced Anti-Tumor Immunity

A study by Bellnier [49] found that Photofrin-based-PDT of subcutaneous SMT-F adenocarcinoma in DBA/2 mice was significantly enhanced by a single dose of intravenously administered recombinant human TNFα. This finding led this group to study the combination of PDT with 5,6-dimethylxanthenone-4-acetic acid (DMXAA), a agent that induce TNFα [50]. When this compound was administered at 1–3 h before Photofrin-based-PDT into mice bearing RIF-1 tumors the response to PDT was increased 2.8 fold. This observation was time specific, since the therapeutic effect was less at other time points examined within $+/-24$ h of PDT procedure. DMXAA can itself induce synthesis of TNFα in RIF-1 tumors whereas PDT did not and the neutralizing antibodies to TNFα reduced the tumor response to control levels. A subsequent study [51] reported similar results in Balb/c mice with CT26 tumors treated with HPPH-PDT at different fluences and fluence rates combined with DMXAA at two doses. Figure 5 summarizes all cytokine-based combination regimens.

Localized tumor treatment with granulocyte colony stimulating factor (G-CSF) in combination with Photofrin-based-PDT resulted in a significant reduction of tumor growth and prolongation of the survival time in BALB/c mice bearing two different tumors: colon-26 and Lewis lung carcinoma [52]. Moreover, 33% of C-26-bearing mice were completely cured from their tumors after combined therapy and developed a specific and long-lasting immunity. The tumors treated with both agents contained more infiltrating neutrophils and apoptotic cells then tumors treated with either G-CSF or PDT only. Importantly, simultaneous administration of Photofrin and G-CSF stimulated bone marrow and spleen myelopoiesis to

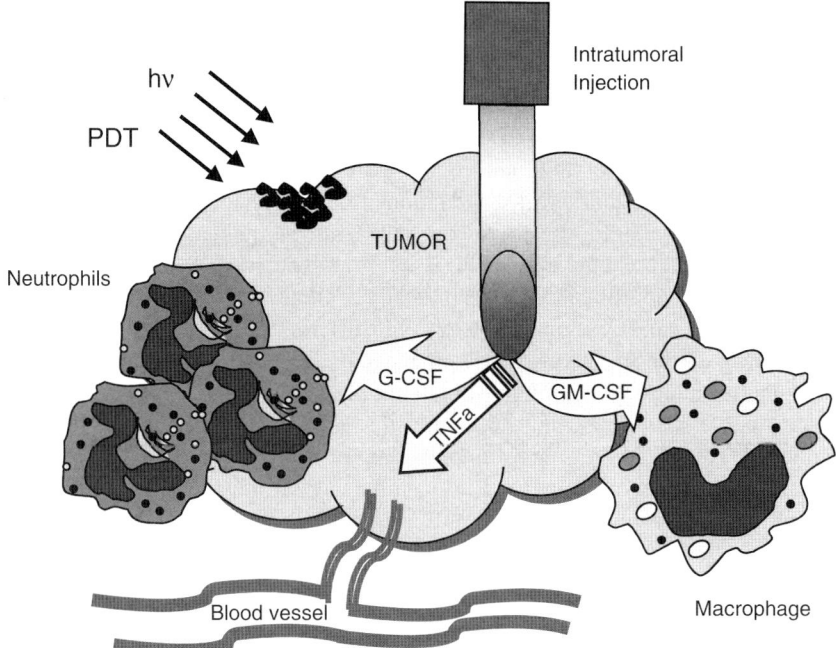

Fig. 5 Schematic figure illustrating intratumoral injection of cytokines that can act in combination with PDT

increase the number of neutrophils demonstrating functional characteristics of activation.

A comparable study was performed by Krosl et al. [53]. In a series of experiments they examined the influence of genetically engineered murine squamous cell carcinoma (SCCVII) producing murine granulocyte-macrophage colony-stimulating factor (GM-CSF) on the therapeutic outcome of PDT. The experiments revealed that repeated administration of lethally irradiated SCCVII tumor cells producing GM-CSF augmented the antitumor effectiveness of Photofrin-based-PDT and BPD-based-PDT. The treatment with GM-CSF resulted in higher cytotoxic activity of tumor-associated macrophages against SCCVII tumor cells.

Conclusions and Future Outlook

The mainstream cancer therapies such as surgery, chemotherapy and radiotherapy are all to some degree immunosuppressive, especially when delivered at the high doses necessary to produce a response in cases of advanced cancer. PDT however, has the potential capacity to destroy fairly large tumors, while at the same time may be able to potentiate the immune system to recognize attack and eliminate residual tumor whether at the site of the primary lesion, in draining lymph nodes or as distant metastases. PDT does produce a form of immunosuppression when tested in mice, characterized by the loss of the contact hypersensitivity (CHS) response to sensitization and rechallenge with a hapten such as dinitrofluorobenzene [54–57] However it is thought that this PDT-induced immunosuppression is mediated via MHC class I molecules, while anti-tumor immunity is mediated more by MHC class II molecules and that the PDT reduction of CHS may not be relevant to cancer therapy [58]. As the foregoing review has demonstrated all the experiments reported so far on combinations of PDT and immunotherapy have taken place in mouse and rat models of cancer. Despite many of thousands of patients worldwide having received various regimens of PDT for different cancers, almost no studies have looked at the effects of PDT on the patients' immune systems, and no combination studies with immunostimulants have been reported in patients. In years to come it is hoped that rationally designed clinical trials will establish whether PDT stimulation of anti-tumor immunity is a laboratory peculiarity confined to small rodents, or whether PDT combined with the appropriate immunostimulant could be applied to patients suffering from advanced cancer.

Acknowledgments This work was funded by the National Cancer Institute (grant number R01CA/AI838801 to MRH). APC was supported by Department of Defense CDMRP Breast Cancer Research Grant (W81XWH-04-1-0676).

References

1. A.P. Castano, T.N. Demidova and M.R. Hamblin, Mechanisms in photodynamic therapy: part one—photosensitizers, photochemistry and cellular localization, Photodiagn Photodyn Ther 1 (2004) 279–293.
2. A.P. Castano, T.N. Demidova and M.R. Hamblin, Mechanisms in photodynamic therapy: part three-photosensitizer pharmacokinetics, biodistribution, tumor localization and modes of tumor destruction, Photodiagn Photodyn Ther 2 (2005) 91–106.
3. A.P. Castano, T.N. Demidova and M.R. Hamblin, Mechanisms in photodynamic therapy: part two -cellular signalling, cell metabolism and modes of cell death, Photodiagn Photodyn Ther 2 (2005) 1–23.
4. P. Agostinis, E. Buytaert, H. Breyssens and N. Hendrickx, Regulatory pathways in photodynamic therapy induced apoptosis, Photochem Photobiol Sci 3 (2004) 721–729.
5. N.L. Oleinick, R.L. Morris and I. Belichenko, The role of apoptosis in response to photodynamic therapy: what, where, why, and how, Photochem Photobiol Sci 1 (2002) 1–21.
6. D.E. Dolmans, A. Kadambi, J.S. Hill, C.A. Waters, B.C. Robinson, J.P. Walker, D. Fukumura and R.K. Jain, Vascular accumulation of a novel photosensitizer, MV6401, causes selective thrombosis in tumor vessels after photodynamic therapy, Cancer Res 62 (2002) 2151–2156.
7. V.H. Fingar, Vascular effects of photodynamic therapy, J Clin Laser Med Surg 14 (1996) 323–328.
8. G. Canti, A. De Simone and M. Korbelik, Photodynamic therapy and the immune system in experimental oncology, Photochem Photobiol Sci 1 (2002) 79–80.
9. M. Korbelik, Induction of tumor immunity by photodynamic therapy., J Clin Laser Med Surg 14 (1996) 329–334.

10. F.H. van Duijnhoven, R.I. Aalbers, J.P. Rovers, O.T. Terpstra and P.J. Kuppen, The immunological consequences of photodynamic treatment of cancer, a literature review, Immunobiology 207 (2003) 105–113.

11. G. Canti, D. Lattuada, A. Nicolin, P. Taroni, G. Valentini and R. Cubeddu, Antitumor immunity induced by photodynamic therapy with aluminum disulfonated phthalocyanines and laser light, Anti-Cancer Drug 5 (1994) 443–447.

12. G. Krosl, M. Korbelik and G.J. Dougherty, Induction of immune cell infiltration into murine SCCVII tumour by photofrin-based photodynamic therapy, Br J Cancer 71 (1995) 549–555.

13. S.O. Gollnick, X. Liu, B. Owczarczak, D.A. Musser and B.W. Henderson, Altered expression of interleukin 6 and interleukin 10 as a result of photodynamic therapy in vivo, Cancer Res 57 (1997) 3904–3909.

14. K. Takeda, T. Kaisho and S. Akira, Toll-like receptors, Annu Rev Immunol 21 (2003) 335–376.

15. N. Kadowaki, S. Ho, S. Antonenko, R.W. Malefyt, R.A. Kastelein, F. Bazan and Y.J. Liu, Subsets of human dendritic cell precursors express different toll-like receptors and respond to different microbial antigens, J Exp Med 194 (2001) 863–869.

16. A. Krug, A. Towarowski, S. Britsch, S. Rothenfusser, V. Hornung, R. Bals, T. Giese, H. Engelmann, S. Endres, A.M. Krieg and G. Hartmann, Toll-like receptor expression reveals CpG DNA as a unique microbial stimulus for plasmacytoid dendritic cells which synergizes with CD40 ligand to induce high amounts of IL-12, Eur J Immunol 31 (2001) 3026–3037.

17. V. Supajatura, H. Ushio, A. Nakao, K. Okumura, C. Ra and H. Ogawa, Protective roles of mast cells against enterobacterial infection are mediated by Toll-like receptor 4, J Immunol 167 (2001) 2250–2256.

18. T. Matsuguchi, T. Musikacharoen, T. Ogawa and Y. Yoshikai, Gene expressions of Toll-like receptor 2, but not Toll-like receptor 4, is induced by LPS and inflammatory cytokines in mouse macrophages, J Immunol 165 (2000) 5767–5772.

19. T. Seya, T. Akazawa, J. Uehori, M. Matsumoto, I. Azuma and K. Toyoshima, Role of toll-like receptors and their adaptors in adjuvant immunotherapy for cancer, Anticancer Res 23 (2003) 4369–4376.

20. R.C. Myers, B.H. Lau, D.Y. Kunihira, R.R. Torrey, J.L. Woolley and J. Tosk, Modulation of hematoporphyrin derivative-sensitized phototherapy with corynebacterium parvum in murine transitional cell carcinoma, Urology 33 (1989) 230–235.

21. M. Korbelik, J. Sun and J.J. Posakony, Interaction between photodynamic therapy and BCG immunotherapy responsible for the reduced recurrence of treated mouse tumors, Photochem Photobiol 73 (2001) 403–409.

22. Y.H. Cho, R.C. Straight and J.A. Smith, Jr., Effects of photodynamic therapy in combination with intravesical drugs in a murine bladder tumor model, J Urol 147 (1992) 743–746.

23. M. Korbelik and I. Cecic, Enhancement of tumour response to photodynamic therapy by adjuvant mycobacterium cell-wall treatment, J Photochem Photobiol B 44 (1998) 151–158.

24. S.O. Gollnick, L. Vaughan and B.W. Henderson, Generation of effective antitumor vaccines using photodynamic therapy, Cancer Res 62 (2002) 1604–1608.

25. M. Korbelik and J. Sun, Photodynamic therapy-generated vaccine for cancer therapy, Cancer Immunol Immunother (2005) 1–10.

26. A.T. Borchers, J.S. Stern, R.M. Hackman, C.L. Keen and M.E. Gershwin, Mushrooms, tumors, and immunity, Proc Soc Exp Biol Med 221 (1999) 281–293.

27. P.R. Taylor, G.D. Brown, D.M. Reid, J.A. Willment, L. Martinez-Pomares, S. Gordon and S.Y. Wong, The beta-glucan receptor, dectin-1, is predominantly expressed on the surface of cells of the monocyte/macrophage and neutrophil lineages, J Immunol 169 (2002) 3876–3882.

28. L. Kubala, J. Ruzickova, K. Nickova, J. Sandula, M. Ciz and A. Lojek, The effect of (1–>3)-beta-D-glucans, carboxymethylglucan and schizophyllan on human leukocytes in vitro, Carbohydr Res 338 (2003) 2835–2840.

29. M. Korbelik, V.R. Naraparaju and N. Yamamoto, Macrophage-directed immunotherapy as adjuvant to photodynamic therapy of cancer, Br J Cancer 75 (1997) 202–207.

30. M. Korbelik, J. Sun, I. Cecic and K. Serrano, Adjuvant treatment for complement activation increases the effectiveness of photodynamic therapy of solid tumors, Photochem Photobiol Sci 3 (2004) 812–816.

31. G. Chihara, Y.Y. Maeda and J. Hamuro, Current status and perspectives of immunomodulators of microbial origin, Int J Tissue React 4 (1982) 207–225.

32. M. Uehara, K. Sano, Z.L. Wang, J. Sekine, H. Ikeda and T. Inokuchi, Enhancement of the photodynamic antitumor effect by streptococcal preparation OK-432 in the mouse carcinoma, Cancer Immunol Immunother 49 (2000) 401–409.

33. M. Uehara and T. Inokuchi, Hyperthermic photodynamic therapy combined with topical administration of OK-432 in the mouse carcinoma, Oral Oncol 39 (2003) 184–189.

34. A.P. Castano, Q. Liu and M.R. Hamblin, Green fluorescent protein expressing but not wild-type tumors in mice are cured by photodynamic therapy, Br J Cancer 33 (2006) 391–397.

35. W.R. Chen, R. Carubelli, H. Liu and R.E. Nordquist, Laser immunotherapy: a novel treatment modality for metastatic tumors, Mol Biotechnol 25 (2003) 37–44.

36. W.R. Chen, R.L. Adams, R. Carubelli and R.E. Nordquist, Laser-photosensitizer assisted immunotherapy: a novel modality for cancer treatment, Cancer Lett 115 (1997) 25–30.

37. W.R. Chen, W.G. Zhu, J.R. Dynlacht, H. Liu and R.E. Nordquist, Long-term tumor resistance induced by laser photo-immunotherapy, Int J Cancer 81 (1999) 808–812.

38. M. Korbelik, G. Krosl, J. Krosl and G.J. Dougherty, The role of host lymphoid populations in the response of mouse EMT6 tumor to photodynamic therapy, Cancer Res 56 (1996) 5647–5652.

39. M. Korbelik and G.J. Dougherty, Photodynamic therapy-mediated immune response against subcutaneous mouse tumors, Cancer Res 59 (1999) 1941–1946.

40. M. Korbelik and I. Cecic, Contribution of myeloid and lymphoid host cells to the curative outcome of mouse sarcoma treatment by photodynamic therapy, Cancer Lett 137 (1999) 91–98.

41. J.A. Hendrzak-Henion, T.L. Knisely, L. Cincotta, E. Cincotta and A.H. Cincotta, Role of the immune system in mediating the antitumor effect of benzophenothiazine photodynamic therapy, Photochem Photobiol 69 (1999) 575–581.

42. M. Korbelik and J. Sun, Cancer treatment by photodynamic therapy combined with adoptive immunotherapy using genetically altered natural killer cell line, Int J Cancer 93 (2001) 269–274.

43. A. Jalili, M. Makowski, T. Switaj, D. Nowis, G.M. Wilczynski, E. Wilczek, M. Chorazy-Massalska, A. Radzikowska, W. Maslinski, L. Bialy, J. Sienko, A. Sieron, M. Adamek, G. Basak, P. Mroz, I.W. Krasnodebski, M. Jakobisiak and J. Golab, Effective photoimmunotherapy of murine colon carcinoma induced by the combination of photodynamic therapy and dendritic cells, Clin Cancer Res 10 (2004) 4498–4508.

44. P. Matar, V.R. Rozados, S.I. Gervasoni and G.O. Scharovsky, Th2/Th1 switch induced by a single low dose of cyclophosphamide in a rat metastatic lymphoma model, Cancer Immunol Immunother 50 (2002) 588–596.

45. R.J. North, Cyclophosphamide-facilitated adoptive immunotherapy of an established tumor depends on elimination of tumor-induced suppressor T cells, J Exp Med 155 (1982) 1063–1074.

46. S. Sakaguchi, N. Sakaguchi, J. Shimizu, S. Yamazaki, T. Sakihama, M. Itoh, Y. Kuniyasu, T. Nomura, M. Toda and T. Takahashi, Immunologic tolerance maintained by CD25+ CD4+ regulatory T cells: their common role in controlling autoimmunity, tumor immunity, and transplantation tolerance, Immunol Rev 182 (2001) 18–32.

47. M. Stassen, S. Fondel, T. Bopp, C. Richter, C. Muller, J. Kubach, C. Becker, J. Knop, A.H. Enk, S. Schmitt, E. Schmitt and H. Jonuleit, Human CD25+ regulatory T cells: two subsets defined by the integrins alpha 4 beta 7 or alpha 4 beta 1 confer distinct suppressive properties upon CD4+ T helper cells, Eur J Immunol 34 (2004) 1303–1311.

48. A.P. Castano, Q. Liu and M.R. Hamblin, in Laser Interaction with Tissue and Cells XV (Jaques, S.L. and Roach, W.P., eds.), Vol. 5319, pp. 50–59, The International Society for Optical Engineering, San Jose, CA (2004).

49. D.A. Bellnier, Potentiation of photodynamic therapy in mice with recombinant human tumor necrosis factor-alpha, J Photochem Photobiol B 8 (1991) 203–210.

50. D.A. Bellnier, S.O. Gollnick, S.H. Camacho, W.R. Greco and R.T. Cheney, Treatment with the tumor necrosis factor-alpha-inducing drug 5,6-dimethylxanthenone-4-acetic acid enhances the antitumor activity of the photodynamic therapy of RIF-1 mouse tumors, Cancer Res 63 (2003) 7584–7590.

51. M. Seshadri, J.A. Spernyak, R. Mazurchuk, S.H. Camacho, A.R. Oseroff, R.T. Cheney and D. A. Bellnier, Tumor vascular response to photodynamic therapy and the antivascular agent 5,6-dimethylxanthenone-4-acetic acid: implications for combination therapy, Clin Cancer Res 11 (2005) 4241–4250.

52. J. Golab, G. Wilczynski, R. Zagozdzon, T. Stoklosa, A. Dabrowska, J. Rybczynska, M. Wasik, E. Machaj, T. Olda, K. Kozar, R. Kaminski, A. Giermasz, A. Czajka, W. Lasek, W. Feleszko and M. Jakobisiak, Potentiation of the anti-tumour effects of Photofrin-based photodynamic therapy by localized treatment with G-CSF, Br J Cancer 82 (2000) 1485–1491.

53. G. Krosl, M. Korbelik, J. Krosl and G.J. Dougherty, Potentiation of photodynamic therapy-elicited antitumor response by localized treatment with granulocyte-macrophage colony-stimulating factor, Cancer Res 56 (1996) 3281–3286.

54. G.O. Simkin, D.E. King, J.G. Levy, A.H. Chan and D.W. Hunt, Inhibition of contact hypersensitivity with different analogs of benzoporphyrin derivative, Immunopharmacology 37 (1997) 221–230.

55. G.O. Simkin, J.S. Tao, J.G. Levy and D.W. Hunt, IL-10 contributes to the inhibition of contact hypersensitivity in mice treated with photodynamic therapy, J Immunol 164 (2000) 2457–2462.

56. D.H. Lynch, S. Haddad, V.J. King, M.J. Ott, R.C. Straight and C.J. Jolles, Systemic immuno-suppression induced by photodynamic therapy (PDT) is adoptively transferred by macro-phages, Photochem Photobiol 49 (1989) 453–458.

57. S.O. Gollnick, D.A. Musser, A.R. Oseroff, L. Vaughan, B. Owczarczak and B.W. Henderson, IL-10 does not play a role in cutaneous Photofrin photodynamic therapy- induced suppression of the contact hypersensitivity response, Photochem Photobiol 74 (2001) 811–816.

58. D.A. Musser and A.R. Oseroff, Characteristics of the immunosuppression induced by cutane-ous photodynamic therapy: persistence, antigen specificity and cell type involved, Photochem Photobiol 73 (2001) 518–524.

Patient-Specific Dosimetry
for Photodynamic Therapy

Jarod C. Finlay, Jun Li, Xiaodong Zhou, and Timothy C. Zhu

Abstract Photodynamic therapy (PDT) has been in routine clinical use since the 1970s. During this period, there has been significant development in light source design, sensitizer chemistry, and clinical protocols. These advances have been paralleled by a continuous improvement in dosimetry, driven by increasing understanding of the underlying physical processes responsible for PDT-mediated tissue damage. A comprehensive dosimetry model must account for patient geometry, tissue optical properties, photosensitizer distribution, and tissue oxygenation. This paper summarizes the developments in PDT dosimetry designed to measure and account for individual variations among patients in these parameters. In each case, we present the results of measurements made during a Phase I clinical trial of PDT for recurrent prostate adenocarcinoma conducted at the University of Pennsylvania.

Keywords: Photodynamic therapy, dosimetry, spectroscopy, tissue optics.

Introduction

Photodynamic therapy (PDT) has been in routine clinical use since the 1970s. During this period, there has been significant development in light source design, sensitizer chemistry, and clinical protocols. These advances have been paralleled by a continuous improvement in dosimetry, driven by increasing understanding of the underlying physical processes responsible for PDT-mediated tissue damage. The simplest and most widely used method for prescribing PDT dose is the specification of an administered drug dose and the irradiance incident on the tissue surface. This definition has three significant drawbacks. First, the light fluence rate in the tissue may differ significantly from the incident irradiance. Second, the sensitizer concentration in tissue may differ significantly from the administered drug dose. These challenges have been appreciated for over two decades, [1] however solutions to

J.C. Finaly
University of Pennsylvania, 3400 Spruce Street, 2 Donner Building, Philadelphia, PA 19104, USA, e-mail: finlay@mail.med.upenn.edu

R. Waynant and D.B. Tata (eds.), *Proceedings of Light-Activated Tissue Regeneration and Therapy Conference.*
© Springer Science + Business Media, LLC 2008

them are only now reaching clinical implementation. Finally, the deposition of photodynamic dose depends critically on the local oxygenation of the tissue being treated, a factor not taken into account by the simple definition of PDT dose. This paper will summarize the developments in PDT dosimetry designed to address these shortcomings. In each case, we will present the results of measurements made during a Phase I clinical trial of PDT for recurrent prostate adenocarcinoma conducted at the University of Pennsylvania [2, 3].

Patient-Specific Light Dosimetry

Theoretical Background

The motivation for patient-specific light dosimetry comes from the fact that the light power delivered to the target tissue may differ significantly from the light power incident on the tissue surface due to multiple scattering of light by the tissue. In describing the amount of light delivered to a target, it is important to differentiate between the radiometric quantities of fluence rate (Φ) and irradiance (E). Both measure power per unit area, however the fluence rate includes contributions from all directions, while the irradiance considers only the component of the light incident in a particular direction. For practical purposes, commercially available power meters and flat-cut optical fibers are irradiance detectors. The absorption of light by photosensitizers, however, is isotropic, making the photodynamic effect dependent on fluence rate rather than irradiance.

There are two sources of scattering relevant to clinical PDT. The first arises from the geometry of the tissue. Light reflected from an external surface such as skin can be considered lost for the purposes of PDT dosimetry. In an enclosed cavity such as the esophagus or bladder, however, the light reflected from the surface will re-enter the tissue at another point. This 'integrating sphere effect' may increase the fluence rate by a factor of up to 5 compared with the same treatment on an external surface [4, 5]. This effect can be modeled theoretically using either radiative transport or integrating sphere theories with appropriate boundary conditions [4].

Second, even when there is no re-entry of light reflected from the tissue, there is considerable scattering of light *within* the tissue. This can lead to a fluence rate just under the tissue surface that is several times greater than the incident irradiance [6, 7]. The fluence rate distribution in tissue is highly dependent on the optical properties of the tissue, which can vary among patients and within a single patient. Accurate dosimetry therefore requires that measurements be made for each patient and each treatment site, as described below.

Measurement Techniques

Two strategies have been investigated to account for individual differences in optical properties and their effect on the *in vivo* light distribution. First, it is possible to place light detectors directly on the surface or within the tissue being treated. The importance of using isotropic detectors (which are sensitive to fluence rate) rather than irradiance detectors has been shown in clinical trials [8]. The most commonly used isotropic detectors for PDT dosimetry are optical fibers with scattering tips. These detectors can be made small enough to be placed interstitially and isotropic enough to serve as fluence rate detectors. The sensitivity of a scattering tip detector depends on the refractive index mismatch between the scattering tip and the medium in which it is immersed [9, 10]. The difference in calibration between air and water may be as high as a factor of two. In practice a detector is often placed inside an air-filled catheter or lumen which is then inserted into tissue. Our own experimental and theoretical results indicate, however, that the intervening medium (air in this case) has negligible effect, and that the detectors can be calibrated as if they were inserted directly into the tissue [11].

We have implemented real-time light dosimetry in a Phase I clinical trial for recurrent prostate adenocarcinoma at the University of Pennsylvania [2, 3]. This protocol uses the second-generation sensitizer Motexafin Lutetium (MLu) and a diode laser light source emitting at MLu's absorption peak of 732 nm. The entire prostate is illuminated *via* 12 to 16 linear diffusing optical fibers inserted into transparent catheters placed interstitially within the organ. Using a motorized positioning system, we have scanned isotropic detectors along catheters parallel to the treatment fibers during PDT, building up a profile of the *in vivo* fluence rate. A fluence rate profile obtained from a typical patient along one catheter in the organ is shown in Fig. 1.

The second method to determine the light fluence rate distribution *in vivo* is to measure the optical properties of the tissue being treated. In the red and infrared wavelengths, most tissue scatters much more strongly than it absorbs. As a result, the light field in tissue becomes diffuse at even short distances from the light source. The distribution of light in turbid media can be modeled using the radiative

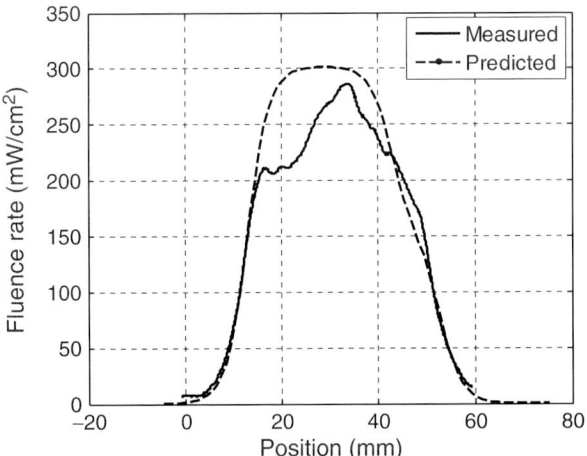

Fig. 1 Measured fluence rate profile along a catheter in a human prostate during PDT (*solid line*) and the profile predicted by a kernel model (*dashed line*)

transport equation [12] or its first-order approximation, the diffusion equation. In the regime where the diffusion equation is valid, it can be shown that the light distribution can be completely characterized by its absorption coefficient (μ_a) and its reduced scattering coefficient (μ_s'). In our prostate protocol, we measure the fluence rate profile as a function of distance from an isotropic point source to reconstruct the optical properties of the prostate. One challenge in this measurement is the separation of μ_a and μ_s'. In the diffusion approximation in infinite media, the fluence rate is given by:

$$\phi = \frac{S \cdot 3\mu_s'}{4\pi r} \cdot e^{-\mu_{eff} \cdot r} . \tag{1}$$

where S is the power of the point source and the effective attenuation coefficient is given by $\mu_{eff} = \sqrt{3 \cdot \mu_a \cdot \mu_s'}$ The shape of the fluence rate profile is sufficient only to constrain μ_{eff}, not μ_a and μ_s' separately. To separate the two, an absolute measurement of fluence rate is needed. This requires careful absolute calibration of the isotropic detectors and sources used for the measurement. By combining optical properties measurements obtained at a large number of source and detector positions, we have created a 3-D map of the μ_a and μ_s' of tissue, as shown in Fig. 2.

Recently, we have developed a kernel-based method to reconstruct the fluence rate at all points in the tissue of interest based on the known source distribution and the 3-D absorption and scattering map of the individual prostate. The kernel model is based on an analytic solution of the diffusion equation. The dashed line in Fig. 1 shows the predictions of this model for a typical prostate patient, which is in approximate agreement with the measured profile (solid line).

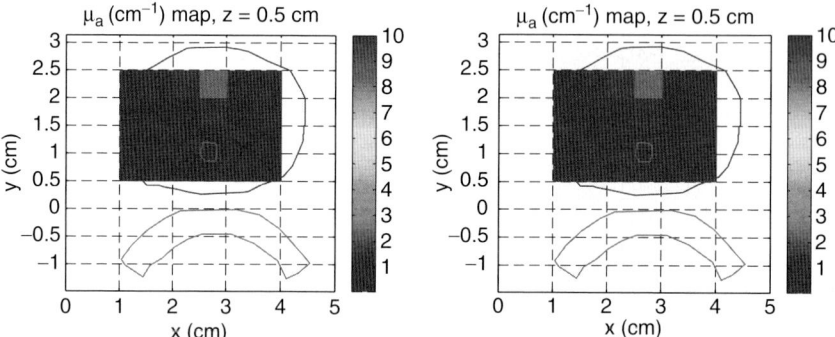

Fig. 2 Optical properties map of a human prostate before PDT. The map was constructed from measurements of the diffuse fluence rate at various distances from an isotropic point source placed at each location in the tissue

Sensitizer-Based Dosimetry

Background

Even if it were possible to eliminate the variability in light dose across the target tissue, the PDT response within the tissue would still be subject to variation due to the inhomogeneity in sensitizer distribution. It has long been appreciated that different tissue types and tumors take up differing amounts of photosensitizer [13, 14]. Recently, however, it has been shown that a significant amount of the variation in PDT response from one tumor to another can be reduced if the prescribed dose takes into account the sensitizer concentration measured *via* fluorescence spectroscopy [15]. This result confirms the basic assumption underlying Patterson, et al.'s definition of PDT dose [16]. Under this definition, the PDT dose is given by

$$D = \int_{0}^{t} \varepsilon c \cdot \frac{\phi(t')}{h\nu} \cdot \frac{1}{\rho} dt', \qquad (2)$$

where ρ is the density of tissue, $h\nu$ is the energy of a photon, c is the drug concentration in tissue, and ε is the extinction coefficient of the photosensitizer. This measure of PDT dose counts the number of photons that give up energy to sensitizer molecules, thereby accounting for differences in tissue optical properties (*via* the fluence rate) and sensitizer concentration.

Strategies for Measuring Sensitizer Distribution

Measurement of sensitizer concentration and its spatial heterogeneity presents several challenges. If *ex vivo* samples are available, it is relatively straightforward to measure their sensitizer concentrations by fluorescence extraction [17, 18] or other laboratory methods. However, because these methods require invasive sampling and significant time for analysis, they are not suitable for informing real-time treatment planning. To incorporate sensitizer concentration into a PDT treatment planning system requires a rapid, non-invasive method for measuring it with spatial resolution compatible with its expected variability and sensitivity sufficient to differentiate the variations in concentration seen *in vivo*. These requirements limit the potential technologies. The most successfully applied of these have both been optical methods, namely absorption and fluorescence spectroscopes.

A fundamental requirement of a good sensitizer is the absorption of light. Therefore, any sensitizer will necessarily change the absorption spectrum of the tissue in which it is located, allowing it to be detected *via* absorption spectroscopy. Absorption spectroscopy has the added advantage of providing information about other absorbers present in the tissue, including oxy- and deoxyhemoglobin. This advantage brings with it a complication: Sensitizers with absorption peaks at shorter wavelengths than approximately 650 nm sufferer from a significant overlap with the visible absorption bands of hemoglobin. This interference renders Photofrin, for example, very difficult to detect *via* absorption spectroscopy in bloody tumors. Sensitizers with longer-wavelength emission peaks are easier to detect *via* their absorption.

In our prostate clinical trial, we have measured the absorption spectra of human prostate tissue *in vivo* by means of multiple diffuse transmission measurements at varying source-detector separations. To improve the robustness of the fitting, we have corrected errors in probe positioning by aligning the measured diffuse transmission spectra to the corresponding single-wavelength diffuse transmission scans. We then use the diffusion approximation to solve for the absorption spectrum at each source-detector separation, assuming a standard shape for the reduced scattering spectrum [19]. A typical series of diffuse transmission spectra is shown in Fig. 3a. The minima corresponding to the absorption maxima of hemoglobin (~580 nm) and the sensitizer (~730 nm) are clearly visible. The resulting absorption spectrum from one of these spectra is shown in Fig. 3b, along with the contributions from oxy- and deoxyhemoglobin, water, and the photosensitizer (MLu) as determined by linear fitting.

By combining several of these scans at various source positions, we can build up a linear profile of hemoglobin concentration and profiles of hemoglobin saturation and photosensitizer concentration, as shown in Fig. 4a, b, respectively. In these plots, each data point represents a separate detector position, and each line or points was taken at a different source position.

The second optical method of sensitizer detection, fluorescence spectroscopy, has several advantages over absorption spectroscopy. First, the sensitivity of fluorescence is potentially much greater than that of absorption spectroscopy. Fluorescence spectroscopy is typically performed with very short source-detector separations or with a single fiber serving as both source and detector [20, 21].

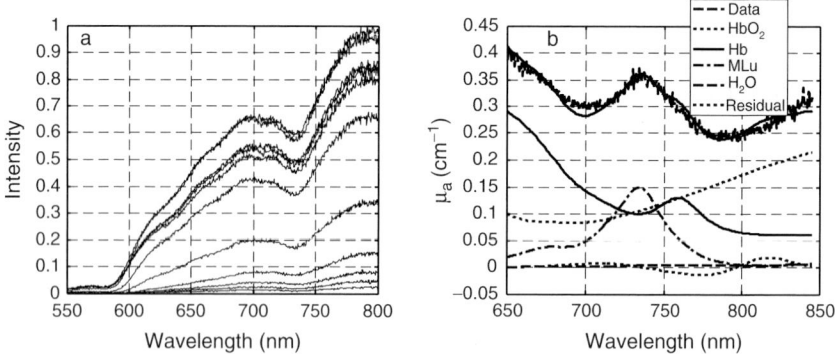

Fig. 3 (a) *In vivo* diffuse transmission spectra obtained at source-detector separations and (b) the corresponding linear fit used to determine the concentrations of the absorbers present in the tissue

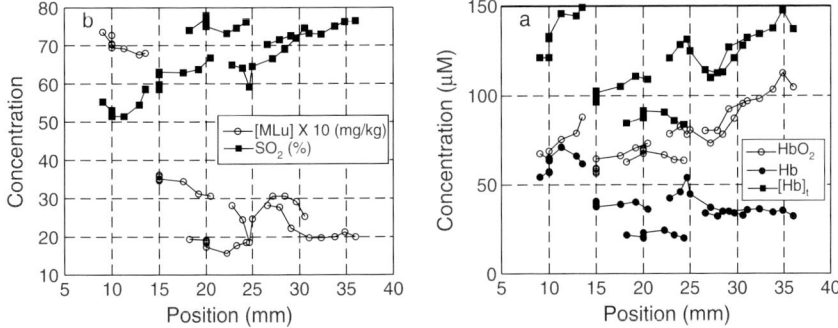

Fig. 4 *In vivo* profiles measured in a human prostate of (**a**) concentrations of hemoglobin in its two forms and total hemoglobin and (**b**) hemoglobin oxygen saturation and sensitizer concentration

The resulting small probes are compatible with minimally invasive endoscopic or interstitial measurement. Finally, fluorescence does not suffer from the same overlap problems with hemoglobin, as absorption spectroscopy does. Hemoglobin is non-fluorescent, and the endogenous fluorophores typically present in tissue have emission maxima at wavelengths shorter than the emission of most sensitizers [22, 23], making the identification of the fluorescence signal arising from the sensitizer relatively unambiguous. Fluorescence can also be used to identify fluorescence photoproducts that result from PDT-induced photobleaching of the sensitizer [24–27]. One significant challenge in quantitative fluorescence spectroscopy is the ambiguity caused by variations in the absorption and scattering coefficients of tissue at the excitation and emission wavelengths, which may be mistaken for changes in fluorescence emission. Several methods have been developed to address this problem, ranging from specially designed probes which minimize the effects of optical property variations [20] to corrections based on concurrent measurements of reflectance [26, 28, 29], optical properties, [21] or absorber concentration [30].

In our prostate PDT protocol, we have measured used fluorescence spectroscopy to measure the distribution of sensitizer along each catheter in the human prostate, using a single side-firing fiber to deliver light and collect the emitted fluorescence [21]. The fluorescence is corrected for optical properties using an empirical algorithm informed by the measured optical properties at 732 nm. Figure 5 illustrates the agreement in MLu concentration between the fluorescence measurement (solid circles), the absorption spectroscopy results shown in Fig. 4 (solid triangles) and the concentration determined by assuming that it varies linearly with the value of μ_a at 732 nm [31] (solid squares).

Model-Based Explicit Dosimetry

The final component of PDT outcome, oxygen, is not accounted for by the dosimetry strategies discussed thus far. To incorporate the local oxygen concentration into patient-specific dosimetry requires a model that relates the local concentration of oxygen to the efficiency of conversion of light energy to singlet oxygen. Foster et al. have demonstrated the applicability of a model constructed from relatively simple differential equations describing the possible interactions among sensitizer and oxygen molecules in their various energy states to measurements in multicell tumor spheroids [32–35]. The supply and consumption of oxygen are modeled as chemical diffusion with consumption. The corresponding *in vivo* model is more complicated in that it takes into account the supply of oxygen from the vasculature [36]. The adaptation of such a model to clinical PDT dosimetry is challenging because the microscopic information used to inform the spheroid model is not available *in vivo*. An approximate model based on volume-averaged optical properties can be constructed, however [37]. Rather than predicting the oxygen distribution from first principles, the *in vivo* model relies on measurements of tissue oxygenation *via* optical measurement of the tissue hemoglobin saturation. The hemoglobin concentration and saturation profiles shown in Fig. 4 demonstrate the feasibility of mapping the tissue oxygenation at least in a volume-averaged sense. This, the sensitizer distribution shown in Fig. 5 and the modeling of *in vivo* fluence rate illustrated in Fig. 1 yield the 3-dimensional distributions of sensitizer, light, and oxygen, the three components needed to determine a 3-dimensional map of PDT dose for an individual patient.

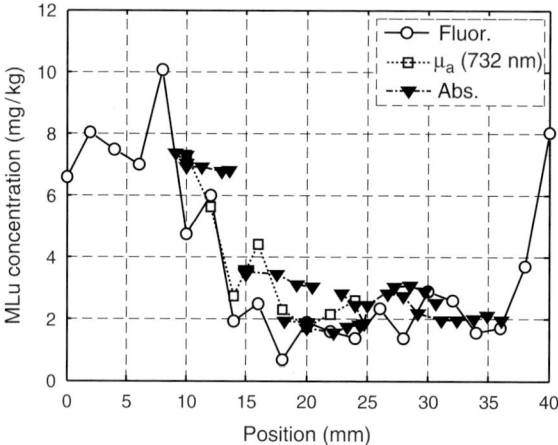

Fig. 5 Distribution of sensitizer along a line in a human prostate after PDT as determined by fluorescence measurement (*open circles*), absorption spectroscopy (*solid triangles*) and μ_a at 732 nm (*open squares*)

Conclusions

One of the limiting factors in the clinical acceptance of PDT has been the variability and unpredictability in outcome. A large portion of this variability can be attributed to the use of dosimetry methods that do not take into account the complex details of the underlying physics of the treatment. As the more sophisticated dosimetry methods outlined in here make their way into clinical use, we expect the consistency and predictability of PDT outcome to improve. The next generation of PDT clinical trials under development in our center and others will involve not only real-time light dosimetry, but also real-time treatment planning [38].

Despite the complexity of model-based PDT dosimetry, it represents a significant simplification of the true *in vivo* situation. PDT treatment has been shown to induce vascular changes in preclinical and clinical trials, [39, 40] and ongoing research in various labs is beginning to elucidate the role of immune response to PDT [41–43]. As these processes are better understood, it is likely that the current dosimetric models will have to be further refined to better predict PDT outcome.

Acknowledgments This work was supported in part by Department of Defense grant DAMD17-03-1-0132 and by National Institutes of Health P01 grant CA87971-01 and R01 grant CA109456.

References

1. B. C. Wilson and M. S. Patterson, "The physics of photodynamic therapy," Phys Med Biol 31, 327–360 (1986).
2. K. Verigos, D. C. Stripp, R. Mick, T. C. Zhu, R. Whittington, D. Smith, A. Dimofte, J. Finlay, T. M. Busch, Z. A. Tochner, S. Malkowicz, E. Glatstein, and S. M. Hahn, "Updated results of a phase I trial of motexafin lutetium-mediated interstitial photodynamic therapy in patients with locally recurrent prostate cancer," J Environ Pathol Toxicol Oncol 25, 373–388 (2006).
3. D. Stripp, R. Mick, T. C. Zhu, R. Whittington, D. Smith, A. Dimofte, J. C. Finlay, J. Miles, T. M. Busch, D. Shin, A. Kachur, Z. Tochner, S. B. Malkowicz, E. Glatstein, and S. M. Hahn, "Phase I trial of Motexfin Lutetium-mediated interstitial photodynamic therapy in patients with locally recurrent prostate cancer," Proc SPIE 5315, 88–99 (2004).
4. W. Star, "The relationship between integrating sphere and diffusion theory calculations of fluence rate at the wall of a spherical cavity," Phys Med Biol 40, 1–8 (1995).
5. H. J. van Staveren, M. Keijzer, T. Keesmaat, H. Jansen, W. J. Kirkel, J. F. Beek, and W. M. Star, "Integrating sphere effect in whole-bladder-wall photodynamic therapy: III. Fluence multiplication, optical penetration and light distribution with an eccentric source for human bladder optical properties," Phys Med Biol 41, 579–590 (1996).
6. S. L. Jacques, "Simple optical theory for light dosimetry during PDT," Proc SPIE 1645, 155–165 (1992).
7. W. M. Star, "Comparing the P_3-approximation with diffusion theory and with Monte Carlo calculations of light propagation in a slab geometry," SPIE Institute Series IS5, 46–54 (1989).
8. T. G. Vulcan, T. C. Zhu, C. E. Rodriguez, A. Hsi, D. L. Fraker, P. Baas, L. H. Murrer, W. M. Star, E. Glatstein, A. G. Yodh, and S. M. Hahn, "Comparison between isotropic and

nonisotropic dosimetry systems during intraperitoneal photodynamic therapy," Laser Surg Med 26, 292–301 (2000).

9. J. P. Marijnissen and W. M. Star, "Performance of isotropic light dosimetry probes based on scattering bulbs in turbid media," Phys Med Biol 47, 2049–2058 (2002).

10. J. P. Marijnissen and W. M. Star, "Calibration of isotropic light dosimetry probes based on scattering bulbs in clear media," Phys Med Biol 41, 1191–1208 (1996).

11. T. C. Zhu, A. Dimofte, J. C. Finlay, E. Glatstein, and S. M. Hahn, "Detector calibration factor for interstitial *in vivo* light dosimetry using isotropic detectors with scattering tip," Proc SPIE 5689, 174–185 (2005).

12. A. Ishimaru, *Wave propagation and scattering in random media.* (IEEE Press, New York, 1997).

13. M. Solonenko, R. Cheung, T. M. Busch, A. Kachur, G. M. Griffin, T. Vulcan, T. C. Zhu, H.-W. Wang, S. M. Hahn, and A. G. Yodh, "*In vivo* reflectance measurements of optical properties, blood oxygenation and motexafin lutetium uptake in canine large bowels, kidneys and prostates," Phys Med Biol 47, 857–873 (2002).

14. H. W. Wang, T. C. Zhu, M. E. Putt, M. Solonenko, J. Metz, A. Dimofte, J. Miles, D. L. Fraker, E. Glatstein, S. M. Hahn, and A. G. Yodh, "Broadband reflectance measurements of light penetration, blood oxygenation, hemoglobin concentration, and drug concentration in human intraperitoneal tissues before and after photodynamic therapy," J Biomed Opt 10, 14004 (2005).

15. X. Zhou, B. W. Pogue, B. Chen, E. Demidenko, R. Joshi, J. Hoopes, and T. Hasan, "Pre-treatment photosensitizer dosimetry reduces variation in treatment response," Int. J. Radiation Oncology Biol. Phys. 64, 1211–1220 (2006).

16. M. S. Patterson, B. C. Wilson, and R. Graff, "In vivo tests of the concept of photodynamic threshold dose in normal rat liver photosensitized by aluminum chlorosulphonated phthalocyanine," Photochem Photobiol 51, 343–349 (1990).

17. K. W. Woodburn, Q. Fan, D. Kessel, Y. Lou, and S. W. Young, "Photodynamic therapy of B16F10 murine melanoma with lutetium texaphyrin," J Invest Dermatol 110, 746–751 (1998).

18. T. C. Zhu, A. Dimofte, J. C. Finlay, D. Stripp, T. Busch, J. Miles, R. Whittington, S. B. Malkowicz, Z. Tochner, E. Glatstein, and S. M. Hahn, "Optical properties of human prostate at 732 nm measured *in vivo* during Motexafin Lutetium–mediated photodynamic therapy," Photochem Photobiol 81, 96–105 (2005).

19. J. C. Finlay, T. C. Zhu, A. Dimofte, D. Stripp, S. B. Malkowicz, R. Whittington, J. Miles, E. Glatstein, and S. M. Hahn, "*In vivo* determination of the absorption and scattering spectra of the human prostate during photodynamic therapy," Proc SPIE 5315, 132–142 (2004).

20. K. R. Diamond, M. S. Patterson, and T. J. Farrell, "Quantification of fluorophore concentration in tissue-simulating media by fluorescence measurements with a single optical fiber," Appl Optics 42, 2436–2442 (2003).

21. J. C. Finlay, T. C. Zhu, A. Dimofte, D. Stripp, S. B. Malkowicz, T. M. Busch, and S. M. Hahn, "Interstitial fluorescence spectroscopy in the human prostate during motexafin lutetium-mediated photodynamic therapy," Photochem Photobiol 82, 1270–1278 (2006).

22. S. Andersson-Engels, J. Johansson, and K. Svanberg, "Fluorescence imaging and point measurements of tissue: Applications to the demarcation of malignanat tumors and atherosclerotic lesions from normal tissue," Photochem Photobiol 53, 807–814 (1991).

23. R. Richards-Kortum and E. Sevick-Muraca, "Quantitative optical spectroscopy for tissue diagnosis," Annu Rev Phys Chem 47, 555–606 (1996).

24. D. J. Robinson, H. S. de Bruijn, N. van der Veen, M. R. Stringer, S. B. Brown, and W. M. Star, "Protoporphyrin IX fluorescence photobleaching during ALA-mediated photodynamic therapy of UVB-induced tumors in hairless mouse skin," Photochem Photobiol 69, 61–70 (1999).

25. D. J. Robinson, H. S. de Bruijn, N. van der Veen, M. R. Stringer, S. B. Brown, and W. M. Star, "Fluorescence photobleaching of ALA-induced protoporphyrin IX during photodynamic

therapy of normal hairless mouse skin: The effect of light dose and irradiance and the resulting biological effect," Photochem Photobiol 67, 140–149 (1998).

26. J. C. Finlay, D. L. Conover, E. L. Hull, and T. H. Foster, "Porphyrin bleaching and PDT-induced spectral changes are irradiance dependent in ALA-sensitized normal rat skin *in vivo*," Photochem Photobiol 73, 54–63 (2001).

27. J. C. Finlay, S. Mitra, and T. H. Foster, "Photobleaching kinetics of Photofrin *in vivo* and in multicell tumor spheroids indicate multiple simultaneous bleaching mechanisms," Phys Med Biol 49, 4837–4860 (2004).

28. R. W. Weersink, M. S. Patterson, K. Diamond, S. Silver, and N. Padgett, "Noninvasive measurement of fluorophore concentration in turbid media with a simple fluorescence/reflectance ratio technique," Appl Optics 40, 6389–6395 (2001).

29. J. Wu, M. S. Feld, and R. P. Rava, "Analytical model for extracting intrinsic fluorescence in turbid media," Appl Optics 32, 3583–3595 (1993).

30. J. C. Finlay and T. H. Foster, "Recovery of hemoglobin oxygen saturation and intrinsic fluorescence using a forward adjoint model of fluorescence," Appl Optics 44, 1917–1933 (2005).

31. T. C. Zhu, J. C. Finlay, and S. M. Hahn, "Determination of the distribution of light, optical properties, drug concentration, and tissue oxygenation in-vivo in human prostate during motexafin lutetium-mediated photodynamic therapy," J Photochem Photobiol B 79, 231–241 (2005).

32. T. H. Foster, D. F. Hartley, M. G. Nichols, and R. Hilf, "Fluence rate effects in photodynamic therapy of multicell tumor spheroids," Cancer Res 53, 1249–1254 (1993).

33. M. G. Nichols and T. H. Foster, "Oxygen diffusion and reaction kinetics in the photodynamic therapy of multicell tumour spheroids," Phys Med Biol 39, 2161–2181 (1994).

34. I. Georgakoudi and T. H. Foster, "Singlet oxygen- *versus* nonsinglet oxygen-mediated mechanisms of sensitizer photobleaching and their effects on photodynamic dosimetry," Photochem Photobiol 67, 612–625 (1998).

35. I. Georgakoudi, M. G. Nichols, and T. H. Foster, "The mechanism of Photofrin photobleaching and its consequences for photodynamic dosimetry," Photochem Photobiol 65, 135–144 (1997).

36. K. K. Wang, S. Mitra, and T. H. Foster, "A comprehensive mathematical model of microscopic dose deposition in photodynamic therapy," Med Phys 34, 282–293 (2007).

37. T. C. Zhu, J. C. Finlay, X. Zhou, and J. Li, "Macroscopic Modeling of the singlet oxygen production during PDT," Proc SPIE 6427, 642–708 (2007).

38. M. D. Altschuler, T. C. Zhu, J. Li, and S. M. Hahn, "Optimized interstitial PDT prostate treatment planning with the Cimmino feasibility algorithm," Med Phys 32, 3524–3536 (2005).

39. G. Yu, T. Durduran, C. Zhou, H. W. Wang, M. E. Putt, H. M. Saunders, C. M. Sehgal, E. Glatstein, A. G. Yodh, and T. M. Busch, "Noninvasive monitoring of murine tumor blood flow during and after photodynamic therapy provides early assessment of therapeutic efficacy," Clin Cancer Res 11, 3543–3552 (2005).

40. G. Yu, T. Durduran, C. Zhou, T. C. Zhu, J. C. Finlay, T. M. Busch, S. B. Malkowicz, S. M. Hahn, and A. G. Yodh, "Real-time in situ monitoring of human prostate photodynamic therapy with diffuse light," Photochem Photobiol 82, 1279–1284 (2006).

41. M. Korbelik and J. Sun, "Photodynamic therapy-generated vaccine for cancer therapy," Cancer Immunol Immunother 55, 900–909 (2006).

42. M. Korbelik, "PDT-associated host response and its role in the therapy outcome," Laser Surg Med 38, 500–508 (2006).

43. A. Oseroff, "PDT as a cytotoxic agent and biological response modifier: Implications for cancer prevention and treatment in immunosuppressed and immunocompetent patients," J Invest Dermatol 126, 542–544 (2006).

Novel Targeting and Activation Strategies for Photodynamic Therapy

Juan Chen, Ian R. Corbin, and Gang Zheng

Abstract Photodynamic treatments can be enhanced by (1) selective, precise delivery, (2) specific delivery of the sensitizer to the cancer tissue, and (3) precise control of the sensitizer's ability to produce singlet oxygen. Fiber optic delivery can assist with light control. This paper focuses on loading sensitizer into low and high density lipoproteins and using nanoparticles into specific receptors for delivery. In addition, special glucose transporting molecules have also been developed to assist in cell targeting.

Keywords: Nanoparticles, low density lipoproteins, high density lipoproteins, receptors, molecular beacons, lysine residues.

Photodynamic therapy (PDT) is an emerging cancer treatment modality involving the combination of light, a photosensitizer (PS) and molecular oxygen. It offers unique control in the PS's action because the key cytotoxic agent, singlet oxygen ($^1O_2$) is only produced *in situ* upon irradiation [1]. Therefore the PDT selectivity can be controlled by three approaches: (1) controlling the selectivity of light irradiation to the disease tissue, (2) enhancing the specificity of delivery of the PS to the tumor tissue, and (3) exerting precise control of the PS's ability to produce $^1O_2$ by responding to specific cancer-associated biomarkers. The first approach is the easiest to implement as light can be readily manipulated and positioned, particularly through the judicious use of advanced fiber optics (e.g. prostate interstitial fibers) [2]. Thus a certain level of PDT selectivity can be achieved by irradiating only the disease tissues. However, this approach can not achieve a

G. Zheng
Ontario Cancer Institute and University of Toronto, Canada; Department of Radiology, University of Pennsylvania, Philadelphia, PA, USA; Joey and Toby Tanenbaum/Brazilian Ball Chair in Prostate Cancer Research, University Health Network, Canada, e-mail: gang.zheng@uhnres.utoronto.ca

R. Waynant and D.B. Tata (eds.), *Proceedings of Light-Activated Tissue Regeneration and Therapy Conference*.
© Springer Science + Business Media, LLC 2008

high level of selectivity due to the limited tumor localization of existing PDT agents, which in turn causes treatment-related toxicity to surround normal tissues as well as systemic sunlight-induced skin toxicity. In regards to the latter two approaches, we have recently developed some novel targeting and activation strategies for achieving improved PDT selectivity. Here are some examples:

Novel PDT Targeting Strategy Based on Low-Density Lipoprotein (LDL) Nanoparticles

The potential use of LDL as nanocarriers for targeted delivery of cancer diagnostics and therapeutics has long been recognized because numerous tumors overexpress LDL receptors (LDLR) to provide the substrates (cholesterol and fatty acids) needed for active membrane synthesis [3–7]. These naturally occurring nanoparticles have a high payload carrying capacity and are biocompatible, biodegradable and nonimmunogenic. In addition, the size of LDL particle is precisely controlled ($\sim$22 nm) by its apoB-100 component through a network of amphipathic α-helix protein-lipid interactions, setting it apart from liposomes and other lipid emulsions. In recent years, we have developed a series PDT agent loaded LDL particles and have shown that they enhance PDT selectivity toward cancer cells overexpressing LDLR [8, 9]. More recently, we have introduced a novel concept of rerouting LDL nanoparticles by conjugating tumor homing ligands to the LDL apoB-100 component [10]. This modification targets the LDL particles to alternate surface receptors and epitopes while abrogating its recognition and binding to LDLR. Thus, this new approach overcomes the narrow purview of LDLR-positive cells and drastically broadens the utility of the LDL nanocarrier to a wider range of tumor types.

Targeting PS-Reconstituted LDL to LDLR

Enhanced PDT selectivity and efficacy for treating LDLR-positive cancer cells can be achieved by loading high quantities of PDT agents inside the core of LDL. To accomplish this, selected PDT agents are modified to contain a lipid moiety which serves to anchor the agent within the lipoprotein hydrophobic core. The first of such anchors is cholesterol oleate. The oleate moiety is used to facilitate reconstitution of the PDT agent into LDL and the cholesterol moiety is utilized to anchor the PS into the lipoprotein phospholipid monolayer, thereby preventing its leakage from the LDL core. Based on this strategy (see Scheme 1), pyropheophorbide cholesteryl oleate (Pyro-CE) and BChl cholesteryl oleate (BChl-CE) were synthesized [8]. These cholesterol oleate-based PS were then successfully reconstituted into LDL to form r-(Pyro-CE)-LDL or r-(BChl-CE)-LDL respectively with PS payload (expressed as molar ratios to apoB-100) in the range of 50:1. Further

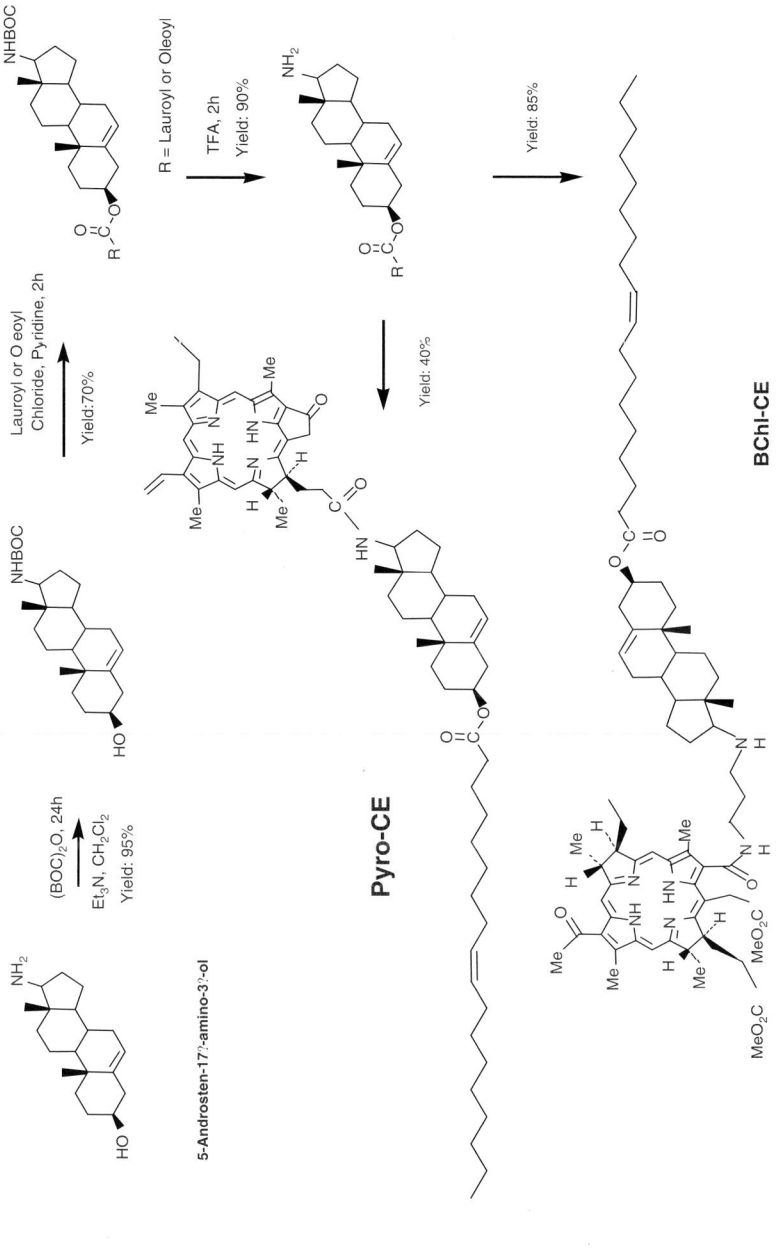

Scheme 1 Synthesis of cholesterol oleate anchored Pyro and BChl conjugates

increases in the PS payload could not be achieved due to the strong tendency of these planar porphyrin macrocycles to aggregate. To address this issue, we designed a novel lipid-anchoring strategy to improve the PS's core-payload based on bisoleate conjugates of silicon (Si) phthalocyanine_(Pc) and naphthalocyanines (Nc). The structures of tetra-*t*-butyl silicon phthalocyanine bisoleate (SiPcBOA) [9] and tert-*t*-butyl silicon naphthalocyanine bisoleate (SiNcBOA) are shown in Scheme 2. The rationale for this strategy is described as follows: (1) Si coordination allows the binding of two axial oleate ligands at the top and bottom of the planar macrocycle to prevent Pc aggregation and provide strong lipid anchors to the LDL phospholipid monolayer; (2) phthalocyanine (Pc) and naphthalocyanine (Nc) dyes are neutral, porphyrin-like compounds but are much more stable photochemically and photophysically than corresponding porphyrin analogs; (3) Pc and Nc have photophysical properties (strong absorption in the NIR region and good singlet oxygen quantum yield) consistent with its being an effective photosensitizer for PDT; (4) four lipophilic *t*-butyl groups are incorporated at the symmetrical peripheral positions to improve the lipophilicity of Pc or Nc and to further prevent their aggregation. Using this approach, we were able to increase the LDL core-payload for Pc and Nc by six fold and two fold respectively. The resulting SiPcBOA-reconstituted LDL (r-Pc-LDL) and SiNcBOA-reconstituted LDL (r-Nc-LDL) each contain 300 Pc or 100 Nc molecules per LDL nanoparticle. More importantly, LDL nanoparticles with such high PS payloads maintain both the structure (e.g.,

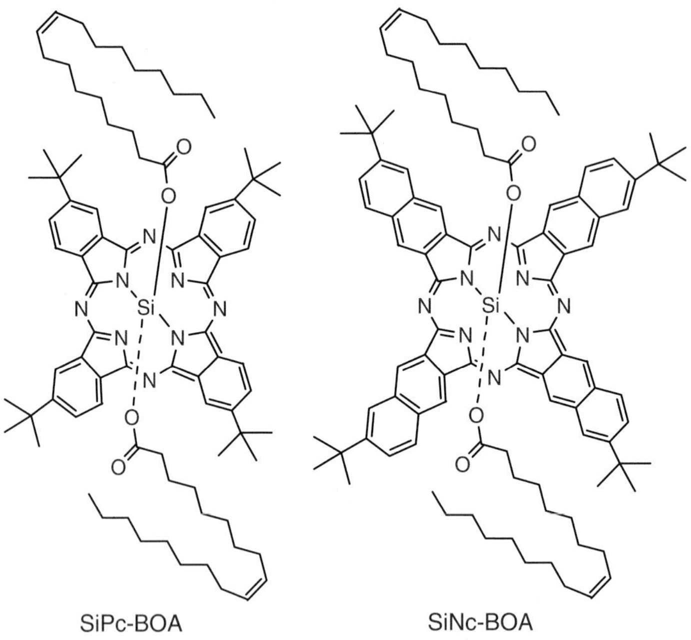

SiPc-BOA   SiNc-BOA

Scheme 2 The structures of SiPc-BOA and SiNc-BOA

size and shape) and functional integrity (ability to be recognized by LDLR, bind to LDLR and to be internalized by LDLR).

Using various imaging techniques, we have validated that these LDL-transported PDT agents do indeed specifically target cancer cells overexpressing LDLR and equally important their PDT efficacy has been shown to be LDLR dependent. Some of the representative studies are demonstrated in Figs. 1–3.

Targeting PS-Reconstituted LDL to a Receptor of Choice

While LDL has proven to be a useful vehicle for delivery of lipophilic PDT agents to tumors, its application is largely limited to LDLR-related diseases. As such it has limited utility for cancer therapy because many tumors do not overexpress the LDLR, whereas several normal tissues do. Moreover, a number of other receptors, so called "cancer signatures, " have proven to be more tumor-specific than LDLR, such as Her2/neu [11], △ EGF [12], somatostatin [13], folate (FA) [14], $\alpha_v\beta_3$ integrins [15], etc. Our lab recently developed and validated a general strategy for rerouting LDL particles to desired receptors which not only provides a method for targeting cancer cells with greater specificity, but also facilitates selective delivery of PDT agents to cancer cells not expressing LDLR.

The rerouting strategy takes advantage of a highly basic domain that contains lysine (Lys) residues with an anomalously low pK_a in the receptor-targeting moiety of apoB-100. Of the 357 Lys residues in apoB-100, 225 are exposed on the protein surface, and the remaining 132 Lys sidechain nitrogens are shielded by either lipid-protein or protein-protein interactions [16]. Of the 225 exposed Lys residues, 53 Lys ε-amino groups are "active" with a low pK_a value of 8.9, and 172 Lys are

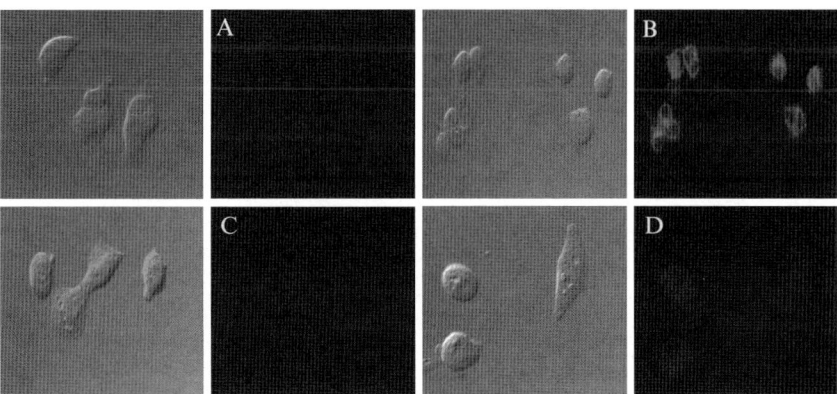

Fig. 1 Fluorescence confocal images of HepG$_2$ tumor cells incubated with (A) unlabeled LDL, (B) r-(Pyro-CE)-LDL, (C) r-(Pyro-CE)-LDL+ 25-fold of native LDL and (D) non-LDL-reconstituted Pyro-CE. The letters refer to each pair of images, with the bright field images on the left and the fluorescence images on the right

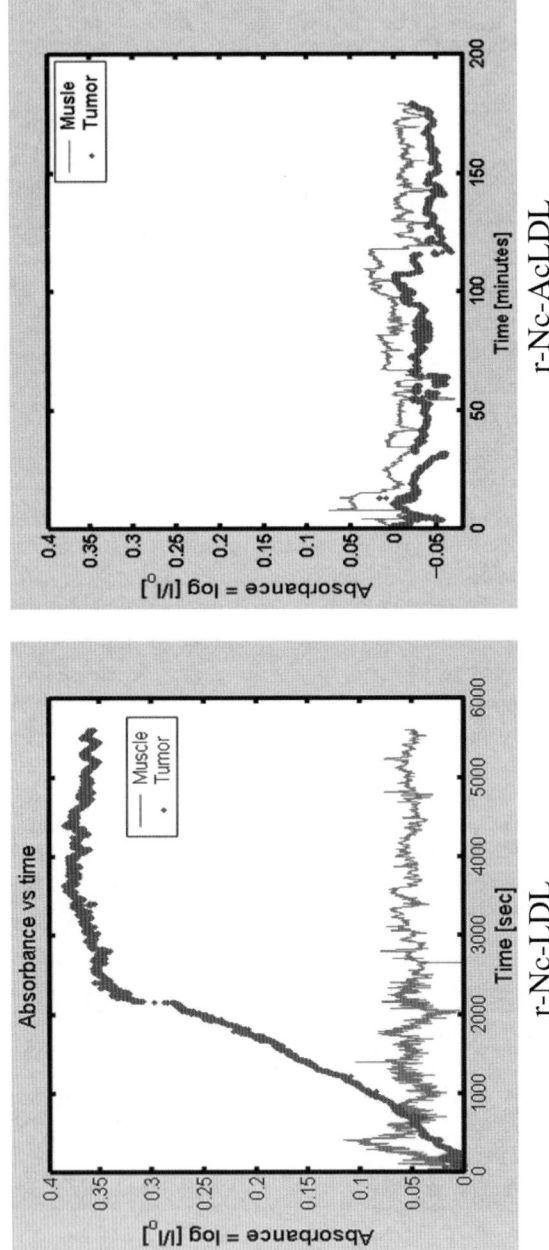

Fig. 2 *In vivo* I&Q spectrum of HepG$_2$ tumor and normal muscle after r-Nc-LDL (*left*) or r-Nc-AcLDL (*right*) intravenous injection. Significant absorption enhancement in tumor tissue compared to the surrounding muscle tissue was observed in r-Nc-LDL injected mouse (tumor/normal muscle = 8:1). Conversely, injection of r-Nc-AcLDL that lack of LDLR binding affinity did not cause any absorption increase either in tumor or in normal muscle. This independently validated the LDLR mediated *in vivo* tumor uptake of r-Nc-LDL

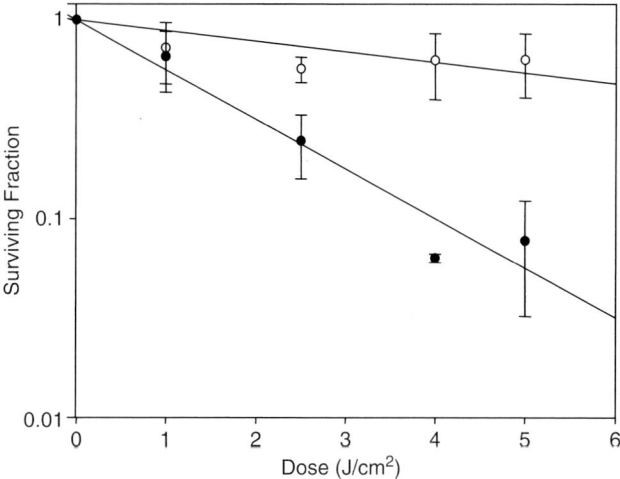

Fig. 3 *In vitro* PDT response of HepG$_2$ cells to r-Pc-LDL (*closed circles*) and SiPcBOA (*open circles*) using a clonogenic assay. Compared to free SiBcBOA, r-Pc-LDL particles kill significantly more cell at the same drug and light doses and its photocytotoxicity is drug and light dose dependent. The slopes of the linear regression fit to the logarithmic data for these two plots are significantly different from each other indicating greatly enhanced efficacy of LDLR-targeted PDT in an LDLR-overexpressing cell line. SEMs were generated for the arithmetic mean

"normal" with a pK$_a$ value of 10.5 [16]. The Lys residues in the receptor binding region have active ε-amino groups (pK$_a$ = 8.9). Thus, if one titrates apoB-100 with agents that alkylate the Lys ε-amino groups, the active lysines are alkylated first, and when about 20% of lysines are modified, the binding capability of this protein to the LDLR is essentially abolished [16]. By using alkylating groups that target non-LDL receptors (e.g., FA [14], peptides or peptidomimetics that bind to somatostatin [13], Her2/neu [17], ΔEGF [12], α$_v$β$_3$ integrins [15], etc.), one could both abolish the LDLR targeting capability of LDL and simultaneously reroute the particle to these alternate receptors.

Proof of this rerouting strategy was first demonstrated *in vitro* with fluorescent labeled or PS-reconstituted folic acid (FA)-conjugated LDL. These FA conjugated particles were avidly taken up and accumulated in folate receptor (FR)-overexpressing KB cells while minimal uptake was seen in cells lacking FR (CHO and HT1080) or in LDLR overexpressing cells (HepG$_2$) [10] (Fig. 4). Later, this strategy was further validated *in vivo* using a folate-directed NIR fluorophore-labeled LDL (DiR-LDL-FA). Real time NIR fluorescent imaging revealed avid uptake and retention of DiR-LDL-FA within the KB tumor, conversely little fluorescence was detected in HT1080 tumors. This result was consistent with *ex vivo* biodistribution studies of excised tumors and host tissues. Further validation of the FR-mediated uptake of DiR-LDL-FA was demonstrated *in vivo* with a co-injection of free FA (30 folds excess) in tumor bearing mice. In this experiment the fluorescence signal within KB

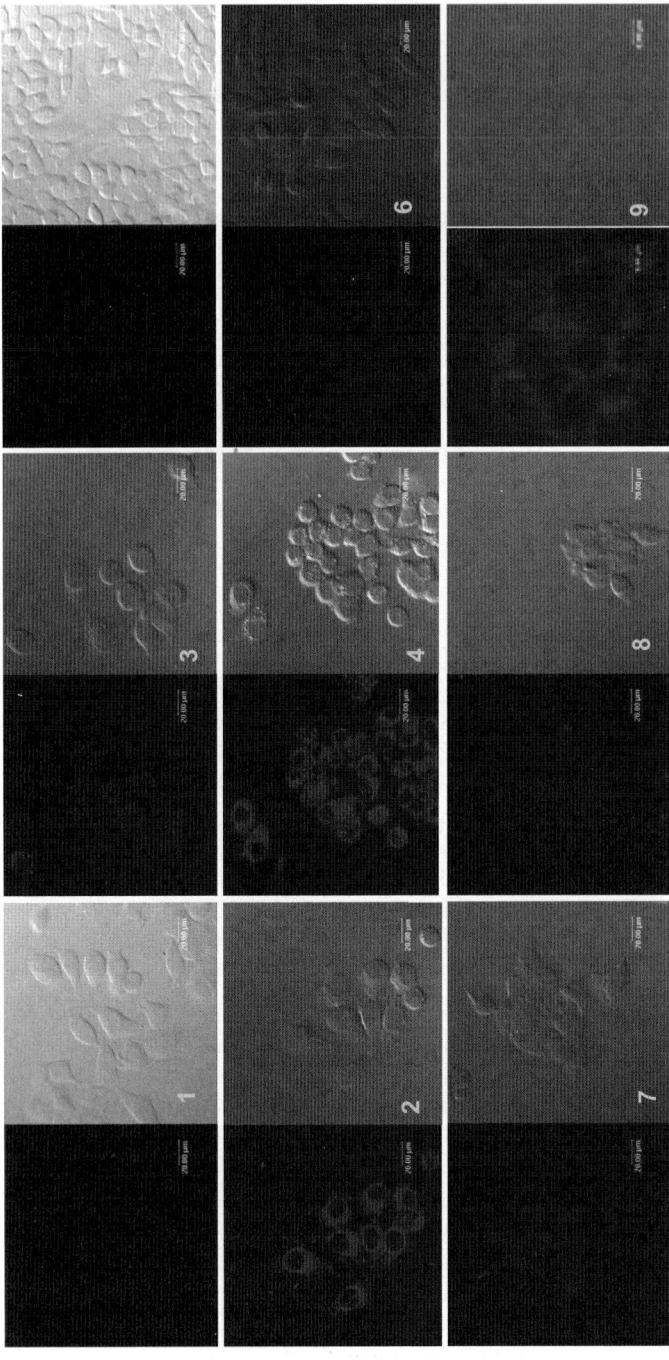

Fig. 4 Confocal study of r-Pc-LDL-FA uptake in KB (FR positive), CHO (FR negative) and HepG$_2$ cells (LDLR positive) [1] KB cells alone [1] KB cells + 0.58 μM of r-Pc-LDL-FA; [3] KB cells + 0.58 μM of r-Pc-LDL-FA + 250-fold of free FA; [4] KB cells + 0.58 μM of r-Pc-LDL-FA + 25-fold of native LDL; [5] CHO cells alone; [6] CHO cells + 0.58 μM of r-Pc-LDL-FA; [7] HepG$_2$ cells alone; [8] HepG$_2$ cells + 0.58 μM of r-Pc-LDL-FA; [9] HepG$_2$ cells + 0.29 μM of r-Pc-LDL. These images confirm the redirection of r-Pc-LDL-FA from LDLR to Folate receptor. The numbers refer to each pair of images, with the fluorescence images on the left and the bright field images on the right

tumor was greatly diminished indicating that excess FA abolished the uptake of DiR-LDL-FA in the FR positive tumor.

Novel PDT Targeting Strategy Based on High-Density Lipoprotein (HDL) Nanoparticles

The new LDL rerouting strategy also opens the door for utilizing other lipoproteins as cancer-targeted nanocarriers for PDT agent. For example, HDL is recognized by HDL receptor, commonly known as the scavenger receptor type I (SR-BI) [18]. Unlike LDLR, SR-BI is not widely overexpressed in cancer cells. As the result, the utility of native HDL for cancer application is extremely limited. Interestingly, HDL has a number of attractive features which make it an ideal candidate to serve as a nanocarrier. First, HDL is the smallest lipoprotein (7–12 nm) and its apoA-1 component offers exquisite size control similar to that of LDL [19]. Secondly, the small size of HDL permits it to easily transit across normal endothelial cells granting it access to most tissues in the body. Moreover, the endogenous nature of HDL along with its small size allows it to mitigate reticular endothelial system (RES) surveillance thus enabling long circulation kinetics for this particle. Finally, reconstituted HDL (rHDL) is self assembled from individual components (apoA-I and commercial lipids) [20] this makes it suitable for large scale production and provides opportunities to introduce alternate components to the nanoparticle formulation.

Given these favorable attributes we explored whether the general rerouting strategy could also apply to HDL particles. To demonstrate this HDL was reconstituted with a lipophilic NIR optical dye (DiR-BOA), this served as a marker to trace the activity and distribution of the nanoparticle. In a similar manner as with LDL, folic acid moieties were conjugated with the lysine residues of HDL's apoprotein component (apoA-1). After successful preparation of this FA conjugated HDL(DiR-BOA) characterization studies revealed that formulation of HDL with exogenous agents (including PS as SiPc) does not compromise the structural or functional integrity of the HDL particle. Exquisite size control and monodispersity are maintained similar to that of the native particle. Functional studies went on to show that the conjugation of folic acid allowed us to effectively reroute the HDL nanoparticle to cancer cells expressing the folate receptors (Fig. 5). Rerouting of HDL was demonstrated both with *in vitro* and *in vivo* experiments. The *in vivo* data showed particularly impressive results as exceptional *in vivo* performance was displayed by the FA-conjugated HDL nanoparticle, high levels of nanoparticle accumulation were detected within the target FR expressing tumor while minimal uptake was evident in tissues of the RES. Ongoing experiments are being conducted in our lab to further develop the rerouted HDL nanoplatform as this nanoparticle shows great promise for clinical applications.

In summary, an ever increasing selection of cancer targeting ligands such as small peptides, antibody fragments or aptamers can be conjugated to lipoproteins to actively reroute these nanoparticles to selected tumors epitopes. The versatility and

| KB Cells | KB Cells | HT-1080 cells | ldl(mSR-BI) cells |

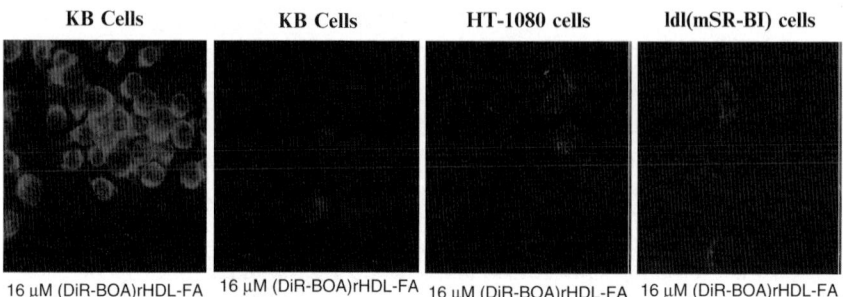

16 μM (DiR-BOA)rHDL-FA   16 μM (DiR-BOA)rHDL-FA   16 μM (DiR-BOA)rHDL-FA   16 μM (DiR-BOA)rHDL-FA
+130 fold Free folic acid

Fig. 5 Folate-conjugated high density lipoprotein, (DiR-BOA)rHDL-FA, a versatile and biocompatible nanocarrier for cancer imaging and treatment. Confocal experiments demonstrate selective uptake of (DiR-BOA)rHDL-FA in KB cells (FR positive) but not in HT-1080 (FR negative) or in ldl(mSR-BI) cells (HDL receptor positive). The accumulation of (DiR-BOA)rHDL-FA in KB cells can be fully inhibited by an excess of free folic acid

modularity of the lipoprotein nanoplatform offers the ability to 'tailor' the nanoparticle in terms of payload (PDT and other diagnostic/therapeutic agents) and targeting (homing molecule to specific tumor markers) for personalized treatment and diagnosis of cancer.

Novel Targeting Strategies for PDT

Glucose Transporters-Targeted PDT Agents

The high glycolytic rate of cancers is an obvious characteristic of tumors, which can be used in distinguishing the insatiable cancer from normal-growing tissues. For example, most of tumor cells overexpress glucose transporters (GLUTs) and have an increased activity of mitochondria-bound hexokinase [21]. Thus, as a cancer signature GLUTs has been widely used as a molecular imaging target for the detection of a wide range of human cancers [22]. To increase tumor selective accumulation of PS, we have synthesized a series of PDT agents that target glucose transporters including pyropheophorbide 2-deoxyglucosamide (Pyro-2DG) and bacteriochlorin 2-deoxyglucosamide (BChl-2DG) [23, 24] (Scheme 3). We have demonstrated that these 2DG-based NIRF/PDT agents can be selectively delivered to and trapped in tumor cells via the GLUT/hexokinase pathway [24] (Fig. 6). Furthermore photoactivation of these compounds efficiently causes selective damage to the mitochondrial within treated tumors, while leaving the adjacent unirradiated tissues unharmed (Fig. 7).

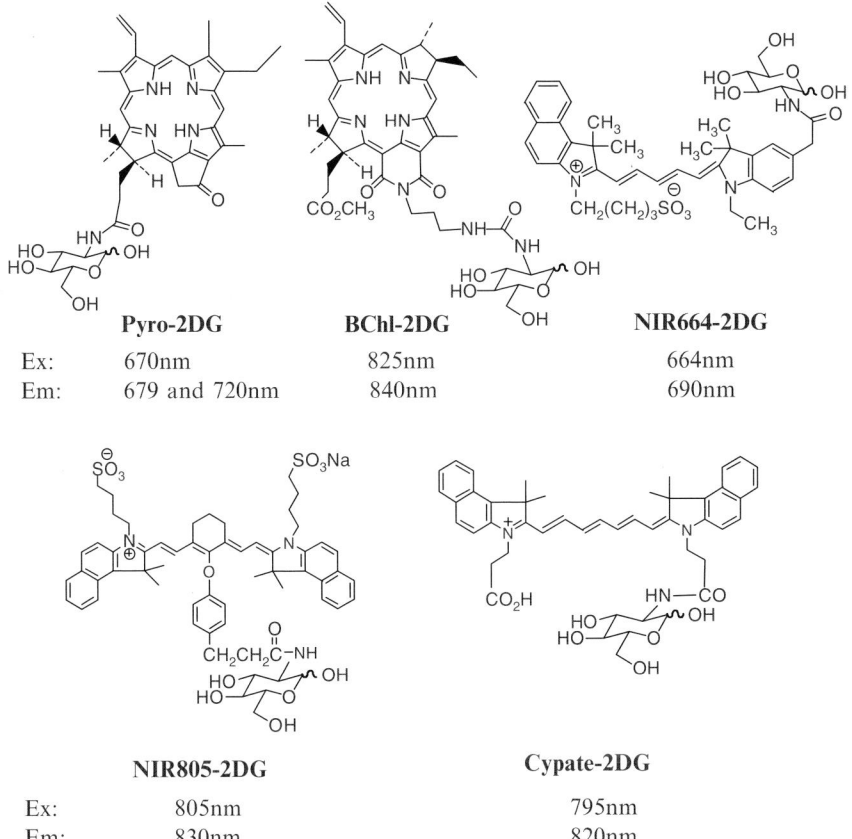

	Pyro-2DG	BChl-2DG	NIR664-2DG
Ex:	670nm	825nm	664nm
Em:	679 and 720nm	840nm	690nm

	NIR805-2DG	Cypate-2DG
Ex:	805nm	795nm
Em:	830nm	820nm

Scheme 3 The structures of 2-deoxyglucose-based contrast agents

Peptide-Modulated PDT Agents Targeting Folate Receptors

Folate receptor (FR) is also considered as a "cancer signature" since a number of epithelial cancers (breast, ovary and colon) overexpress the folate receptor and normal tissues have low expression levels [14]. Thus attaching a folate molecule to a PS could enhance its PDT selectivity toward cancer cells. We recently developed a FR-targeted, peptide-modulated PDT agent that selectively detects and destroys the targeted cancer cells while sparing normal tissue [25] (Fig. 8). This was demonstrated as this agent displayed minimal normal tissue uptake (e.g., liver and spleen) and was discriminating between tumors with different levels of FR expression. This construct (Pyro-peptide-Folate, **PPF**) is comprised of three components: (1) Pyropheophorbide *a* (Pyro) as an imaging and therapeutic agent, (2) peptide sequence as a stable linker and modulator improving the delivery efficien-

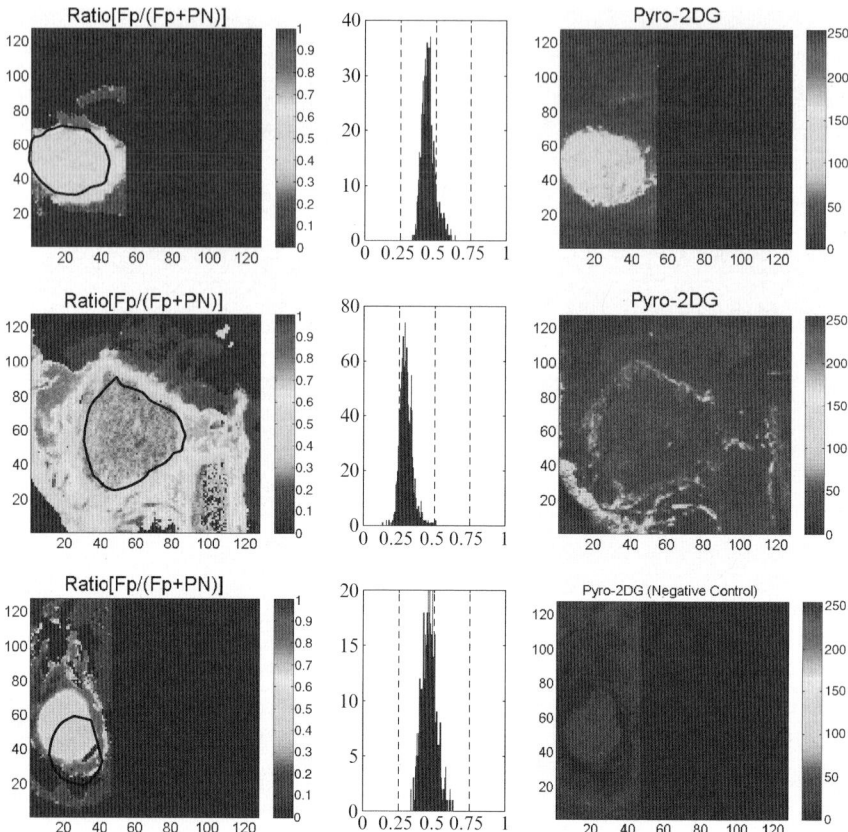

Fig. 6 Fluorescence images of (1) 9L glioma tumor (*top row*), (2) normal tissue (*middle row*) following intravenous injection of Pyro-2DG, and (3) control tumor in mouse without Pyro-2DG (*bottom row*). This Figure shown the redox ratio images (regions for analysis are outlined) in the first column, the histograms of redox ratios from corresponding marked regions in the second column and fluorescent images of Pyro-2DG in the third column. This data confirms that Pyro-2DG selectively accumulates inside tumor compared to surrounding normal tissue. In addition, Pyro-2DG has no effect on mitochondrial activity in the absence of light

cy, and (3) Folate as a homing molecule targeting FR-expressing cancer cells. We observed an enhanced accumulation of **PPF** in KB cancer cells (FR positive) compared to HT 1080 cancer cells (FR negative), resulting in a more effective post-PDT killing of KB cells over HT 1080 or normal CHO cells (FR negative). The accumulation of **PPF** in KB cells can be inhibited (up to 70%) by an excess of free folic acid. The FR targeting effect of the folate component in the Pyro-peptide construct was also confirmed *in vivo* by the preferential accumulation of **PPF** in KB tumors (KB vs. HT 1080 tumors 2.5:1) (Fig. 8). In contrast, no significant difference between the KB and HT 1080 tumor was observed with the untargeted probe

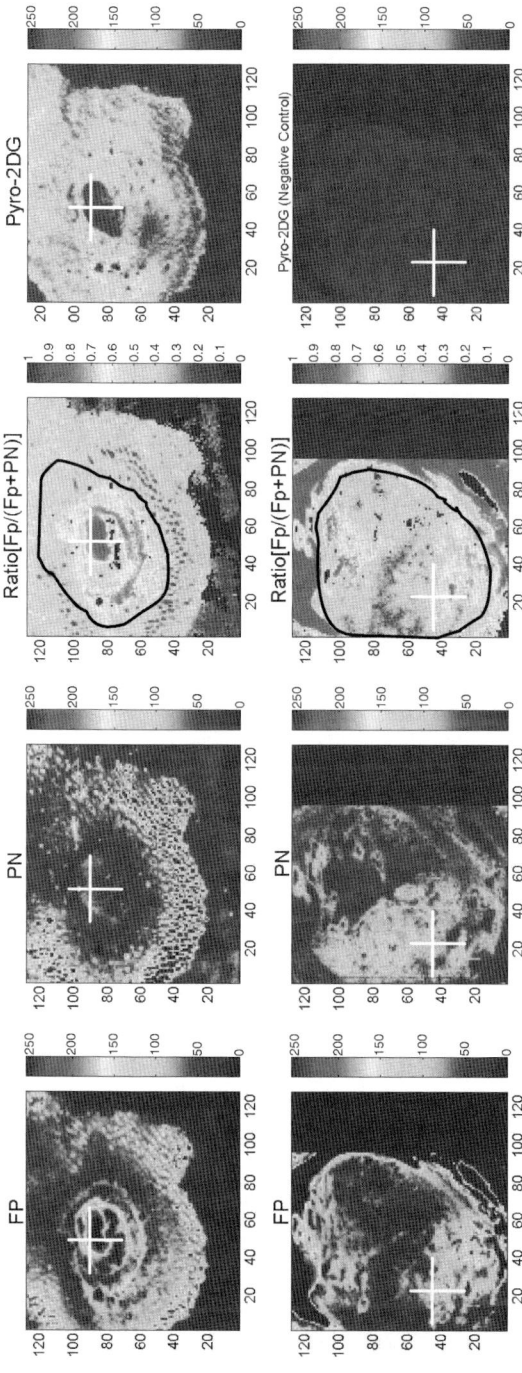

Fig. 7 Selective destruction of tumors by PDT corresponds to a marked change in the intrinsic fluorescence (decreased NADH, increased FP). The selective photobleaching of Pyro-2DG is also displayed. (*Top row*: PDT of tumor in Pyro-2DG-treated mouse; *bottom row*: light control, PDT of tumor in mouse without Pyro-2DG). Note: The irradiated region is marked by a white cross, and the tumor margin is outlined with a black circle. This data clearly demonstrates that the PDT response is the result of Pyro-2DG' photosensitization

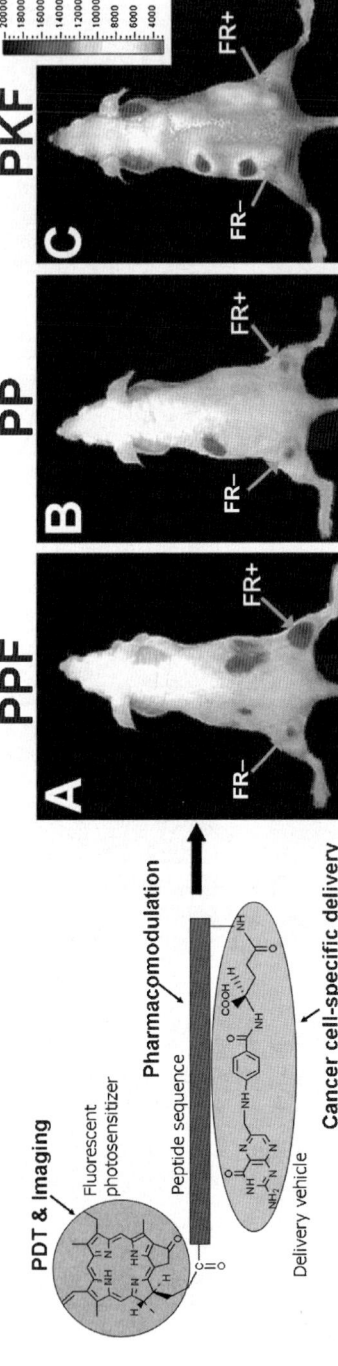

Fig. 8 The structure of PPF *in vivo* fluorescence imaging of dual tumor mice (KB tumor (FR$^+$) on the right and HT1080 tumor (FR$^-$) on the left) following administration of either PPF, PP or PKF. These data demonstrate the selective accumulation of PPF in KB (FR+) tumor

(Pyro-peptide, **PP**). In addition, incorporating a short peptide sequence significantly improved the delivery efficiency of the probe to the target tumor. This 'peptide-based pharmacomodulation' was demonstrated by a 50-fold reduction in **PPF** accumulation in liver and spleen when compared to the peptide-lacking probe (Pyro-K-Folate, **PKF**).

Although PPF has shown promising results further investigation is required on this construct. First of all the PPF targeting to FR serves merely as a proof-of-concept model, the same principal could be applied to alternate cell surface receptors and epitope, such as more tumor specific receptor ΔEGF or $\alpha v \beta_3$ integrins. In addition, extensive study is also needed to identify the molecular mechanism underlying this peptide-induced pharmacomodulation. Many drugs suffer from adverse toxicity and many imaging probes have insufficient signal-to-noise at their target site due to their high normal tissue uptake and low accumulation within targeted tumor. Thus, this peptide-based pharmacomodulation provides a glimpse into future strategies that may address these challenging issues of delivery efficiency.

Novel PDT Activation Strategy Based on Photodynamic Molecular Beacons

Molecular beacons are target-activatable optical probes based on fluorescence resonance energy transfer (FRET) [26–29]. They feature a fluorophore-quencher pair held together with a cleavable or separable linker that can interact specifically with a biomarker. Therefore, the beacons offer exquisite control of fluorescence emission in response to specific cancer targets and thus are useful tools for *in vivo* cancer imaging. On the other hand, PDT is a cell killing process by light activation of a photosensitizer in the presence of oxygen. The key cytotoxic agent is singlet oxygen ($^1O_2$) [30]. By combining these two well known principles (FRET and PDT), we introduced the novel concept of photodynamic molecular beacons (PMB) to achieve the selective PDT-induced cell killing by exerting the selective control over the photosensitizer' ability to produce $^1O_2$ [31]. The PMB comprises of a disease-specific linker, a photosensitizer and a $^1O_2$ quencher. With these novel constructs no photosensitization ($^1O_2$ production) occurs until the linker interacts with a specific biomarker, such as a tumor-specific enzyme (e.g., MMP7) or mRNA (e.g., c-Raf-1, Kras). Once activated the PMB emits fluorescence and destroys the cancer cells, while leaving normal cells undetectable and unharmed. The core principle of the PMB concept is that the selective PDT-induced cell killing can be achieved by exerting precise control of the PS's ability to produce $^1O_2$ by responding to specific cancer-associated biomarkers. Thus, the PDT selectivity will no longer depend solely on how selectively the PS can be delivered to cancer cells. Rather, it will depend on how selective a biomarker is to cancer cells and how selective the interaction of PMB is to this biomarker.

Protease Activated PMB

The first PMB utilizing $^1O_2$ quenching and activation was designed to comprise of a carotenoid molecule (CAR) conjugated to a pyropheophorbide molecule (Pyro) through a caspase-3 protease cleavable peptide linker (GDEVDGSGK) (Scheme 4). Using this model PMB, Pyro-peptide-CAR (PPC), we have validated that the $^1O_2$ generation of Pyro is effectively inhibited by the CAR quencher and is further specifically activated by caspase-3-induced peptide cleavage [31] (Fig. 9). We also demonstrated that CAR-mediated $^1O_2$ quenching in PPC fully protects viable cancer cells (no caspase-3 expression) from the harmful toxic effects of $^1O_2$. Conversely, Pyro-peptide (PP) without the CAR moiety readily kills these cells even at a 30-fold lower concentration. These data clearly show the photoprotective role of CAR when conjugated to Pyro through this flexible peptide linker and suggest that the PDT selectivity is protease sequence-dependent.

It should be noted that caspase-3, used in this PPC design, initially served only as a model protease since this enzyme is traditionally viewed as a key participant in the apoptotic cascade. Only recently, it was found that caspase-3 is overexpressed in hepatocellular carcinomas, thus providing a unique opportunity to actually use caspase-3 as the trigger to achieve selective photodynamic cytotoxicity for this specific cancer type [32].

To further evaluate the PDT selectivity of PMB in living tumor cells, we have synthesized a tumor associated protease MMP7-activated PMB, $PP_{MMP7}B$, comprising of a black hole quencher 3 (BHQ3) conjugated to Pyro through a MMP7 substrate linker. As shown in Fig. 10 [33], upon PDT treatment, $PP_{MMP7}B$ only induced cell killing in KB cells (overexpressing MMP7). We further demonstrated that its efficacy is drug and light doses dependent, confirming that $PP_{MMP7}B$ is specifically photoactivated by MMP7 and its photodynamic cytotoxicity is MMP7 sequence specific. In addition, a preliminary *in vivo* PDT study of $PP_{MMP7}B$ provides initial evidence of the MMP7-triggered PDT efficacy in KB tumor-bearing mice.

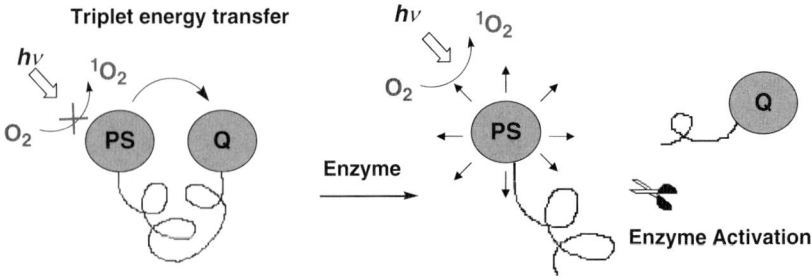

Scheme 4 The concept of protease-activated photodynamic molecule beacon (PMB)

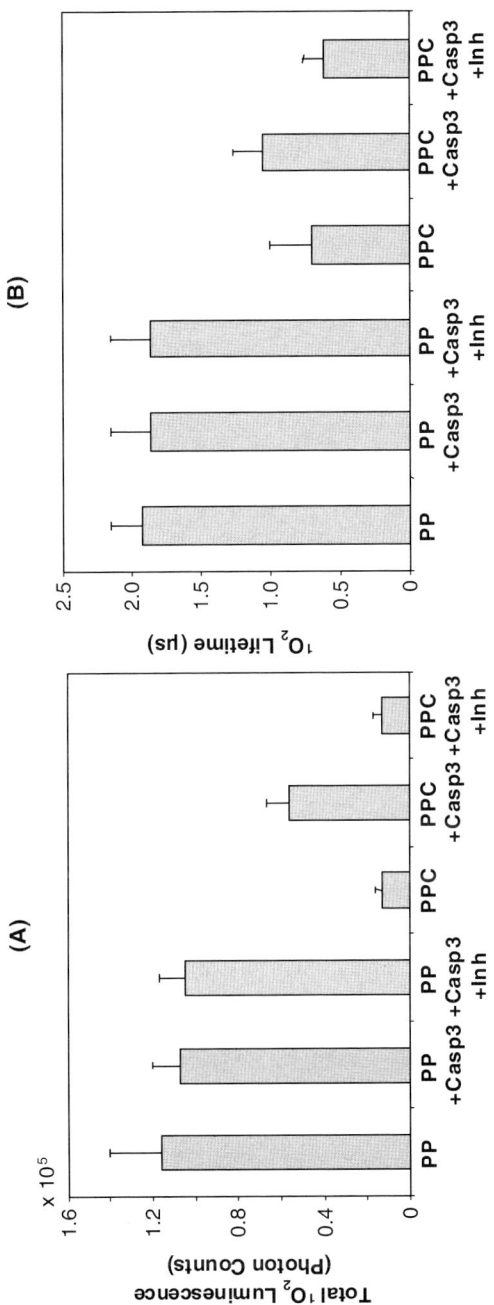

Fig. 9 Total $^1O_2$ luminescence counts for PP, PP + caspase-3, PP + caspase-3 + inhibitor, PPC, PPC + caspase-3, PPC + caspase-3 + inhibitor. (B) Corresponding $^1O_2$ lifetime. Pyro-peptide (PP) without quencher moiety serves as positive control

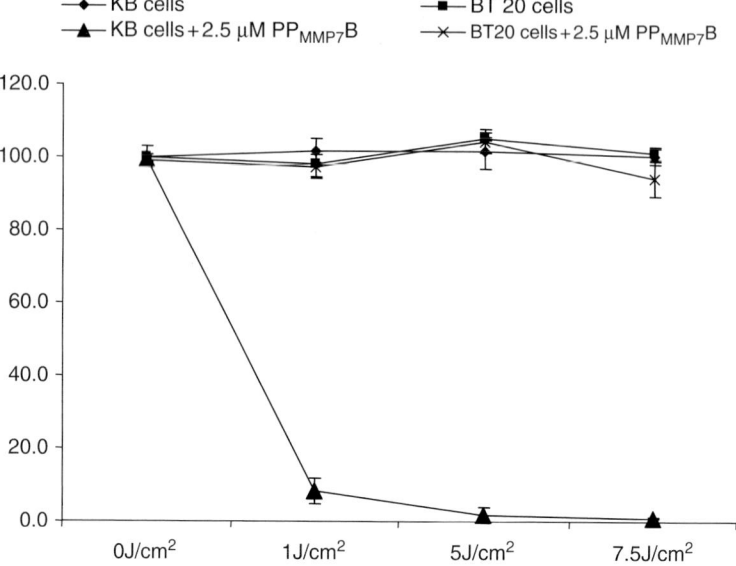

Fig. 10 The viability of KB (overexpressing MMP7) and BT20 (lack of MMP7) cells after incubation with 2.5 μM PP$_{MMP7}$B for 16 hours and treatment with different PDT light doses (1, 5, 7.5 J/cm^2) as determined by the MTT assay. Data shown are based on three different experiments and the results were expressed as mean ± standard error

mRNA Activated PMB

Tumor specific mRNAs are recognized as some of the most specific markers for cancer. Given that the hybridization of mRNA to its complementary oligonucleotide sequence by Watson-Crick base pairing is a highly selective and efficient process, antisence oligonucleotides (AS-ONs) have emerged as very attractive therapeutic agents. By taking advantage of the AS-ON technique, we have designed a novel of PMB to precisely control the $^1O_2$ generation of PS by tumor specific mRNA. As depicted in Scheme 5, this so called mRNA-activated PMB (mRNA-PMB) consists of a PS and a quencher (Q) that are attached to at either end of a single-stranded oligonucleotide, which forms a stem-loop structure. The loop is an AS-ON sequence complementary to an mRNA which is specifically expressed in the target cancer cell. Two complementary arm sequences on either side of the loop sequence hybridize with each other, thus forming the stem-loop structure. We first developed a c-raf-1 mRNA activatable PMB and our initial experiments revealed that: (1) the stem-loop structure of mRNA-activated PMB was able to keep PS and Q locked in very close proximity, thus resulting in effective fluorescence and $^1O_2$ quenching of the PS by energy transfer; (2) in the presence of the tumor specific mRNA, the loop sequence hybridizes with the mRNA and disrupts the hydrogen bonds of the stem. This removes the Q from the immediate vicinity of the PS, and

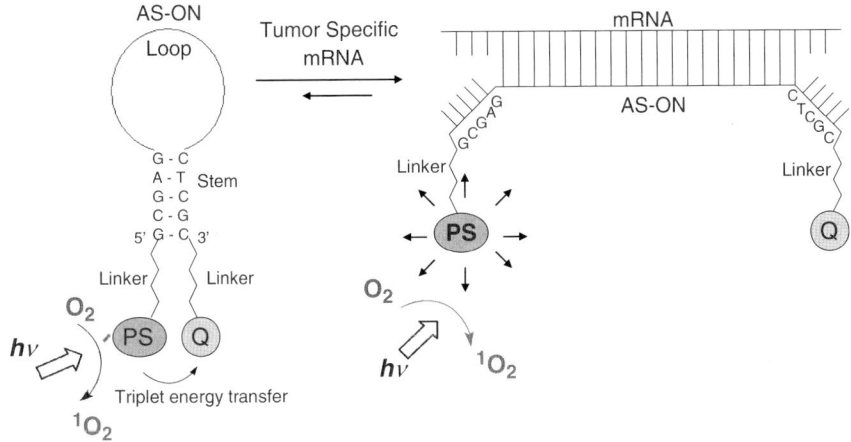

Scheme 5 The concept of mRNA-activated photodynamic molecule beacon (PMB)

upon irradiation with light, results in fluorescence and $^1O_2$ reproduction of the PS [34].

Although we have proved the PMB concept for the precise control of the PS's ability to produce $^1O_2$ by responding to tumor-associated protease or mRNA, the same principal could be applied to other activation schemes that separate the PS from its quencher. For example, this concept could be applied to DNA sequence-controlled on-and-off switchable singlet oxygen sensitizers [35], in which a PS and a $^1O_2$ quencher are kept in close contact in the "off-state" by DNA-programmed assembly and the PS will be activated after releasing from the quencher though a process of competitive DNA hybridization. It also could be applied to a phospholipase-activated PMB, in which a tumor-associated phospholipase is used as the triggering device [36]. Hence, this PMB approach has potential to generate a wide range of clinical useful strategies to enhance the specificity of PDT to treat cancer. As well as, it could significantly minimize the PDT complications by protecting adjacent non-targeted tissues from photodamage, making PDT a more safe and selective clinical strategy.

Summary

We have developed various PDT targeting strategies in last few years. Since a variety of molecules are unique to or overexpressed in cancer, these strategies are inherently flexible and tunable and offer excellent control of PDT agent localization or activation in cancer cells. Moreover, target-specific delivery of high quantities of PDT agents has been accomplished by using the lipoprotein nanoparticles, which are naturally biocompatible, biodegradable and non-immunogenic; small with precisely controlled size; versatile in receptor of choice and RES mitigating, thus are very promising in future clinical application. We have also introduced a novel concept of photodynamic molecular beacons (PMB) that shift the control of PDT selectivity away from how PDT agents are delivered to how PDT agents are activated by the specific cancer biomarkers (e.g., tumor specific protease and mRNAs). The PMB strategy is particularly attractive for future PDT applications, especially considering our increasing knowledge about the human genome in health and disease. Thus, it is anticipated that an unprecedented level of the PDT selectivity can be achieved through one or a combination of: (1) selective delivery light to tumors, (2) selective localization of PS in tumor, and (3) selective $^1O_2$ production in responding to cancer-associated biomarker.

Acknowledgements This work was supported by grants from National Institute of Health, Department of Defense and Oncologic Foundation of Buffalo.

References

1. Oleinick, N. L. & Evans, H. H. (1998) *Radiation Research* **150**, S146–156.
2. Pinthus, J. H., Bogaards, A., Weersink, R., Wilson, B. C., & Trachtenberg, J. (2006) *The Journal of Urology* **175**, 1201–1207.
3. Versluis, A. J., van Geel, P. J., Oppelaar, H., van Berkel, T. J., & Bijsterbosch, M. K. (1996) *British Journal of Cancer* **74**, 525–532.
4. Samadi-Baboli, M., Favre, G., Canal, P., & Soula, G. (1993) *British Journal of Cancer* **68**, 319–326.
5. Rensen, P. C., de Vrueh, R. L., Kuiper, J., Bijsterbosch, M. K., Biessen, E. A., & van Berkel, T. J. (2001) *Advanced Drug Delivery Reviews* **47**, 251–276.
6. Masquelier, M., Vitols, S., & Peterson, C. (1986) *Cancer Research* **46**, 3842–3847.
7. Firestone, R. A., Pisano, J. M., Falck, J. R., McPhaul, M. M., & Krieger, M. (1984) *Journal of Medicinal Chemistry* **27**, 1037–1043.
8. Zheng, G., Li, H., Zhang, M., Lund-Katz, S., Chance, B., & Glickson, J. D. (2002) *Bioconjugate Chemistry* **13**, 392–396.
9. Li, H., Marotta, D. E., Kim, S., Busch, T. M., Wileyto, E. P., & Zheng, G. (2005) *Journal of Biomedical Optics* **10**, 41203.
10. Zheng, G., Chen, J., Li, H., & Glickson, J. D. (2005) *Proceedings of the National Academy of Sciences of the United States of America* **102**, 17757–17762.
11. Slamon, D. J., Godolphin, W., Jones, L. A., Holt, J. A., Wong, S. G., Keith, D. E., Levin, W. J., Stuart, S. G., Udove, J., Ullrich, A., et al. (1989) *Science* **244**, 707–712.
12. Wong, A. J., Ruppert, J. M., Bigner, S. H., Grzeschik, C. H., Humphrey, P. A., Bigner, D. S., & Vogelstein, B. (1992) *Proceedings of the National Academy of Sciences of the United States of America* **89**, 2965–2969.
13. Becker, A., Hessenius, C., Licha, K., Ebert, B., Sukowski, U., Semmler, W., Wiedenmann, B., & Grotzinger, C. (2001) *Nature Biotechnology* **19**, 327–331.
14. Leamon, C. P. & Low, P. S. (1991) *Proceedings of the National Academy of Sciences of the United States of America* **88**, 5572–5576.
15. Ruoslahti, E. (2002) *Nature Reviews Cancer* **2**, 83–90.
16. Lund-Katz, S., Ibdah, J. A., Letizia, J. Y., Thomas, M. T., & Phillips, M. C. (1988) *Journal of Biological Chemistry* **263**, 13831–13838.

17. Park, J. W., Kirpotin, D. B., Hong, K., Shalaby, R., Shao, Y., Nielsen, U. B., Marks, J. D., Papahadjopoulos, D., & Benz, C. C. (2001) *Journal of Controlled Release* **74**, 95–113.
18. Acton, S., Rigotti, A., Landschulz, K. T., Xu, S., Hobbs, H. H., & Krieger, M. (1996) *Science* **271**, 518–520.
19. Saito, H., Dhanasekaran, P., Nguyen, D., Holvoet, P., Lund-Katz, S., & Phillips, M. C. (2003) *The Journal of Biological Chemistry* **278**, 23227–23232.
20. Pittman, R. C., Glass, C. K., Atkinson, D., & Small, D. M. (1987) *The Journal of Biological Chemistry* **262**, 2435–2442.
21. Medina, R. A. & Owen, G. I. (2002) *Biological Research* **35**, 9–26.
22. Czernin, J. & Phelps, M. E. (2002) *Annual Review of Medicine* **53**, 89–112.
23. Chen, Y., Zheng, G., Zhang, Z. H., Blessington, D., Zhang, M., Li, H., Liu, Q., Zhou, L., Intes, X., Achilefu, S., et al. (2003) *Optics Letters* **28**, 2070–2072.
24. Zhang, M., Zhang, Z., Blessington, D., Li, H., Busch, T. M., Madrak, V., Miles, J., Chance, B., Glickson, J. D., & Zheng, G. (2003) *Bioconjugate Chemistry* **14**, 709–714.
25. Stefflova, K., Li, H., Chen, J., & Zheng, G. (2007) *Bioconjugate Chemistry* **18**(2), 379–388.
26. Zhang, J., Campbell, R. E., Ting, A. Y., & Tsien, R. Y. (2002) *Nature Reviews* **3**, 906–918.
27. Tyagi, S. & Kramer, F. R. (1996) *Nature Biotechnology* **14**, 303–308.
28. Weissleder, R., Tung, C. H., Mahmood, U., & Bogdanov, A., Jr. (1999) *Nature Biotechnology* **17**, 375–378.
29. Matayoshi, E. D., Wang, G. T., Krafft, G. A., & Erickson, J. (1990) *Science* **247**, 954–958.
30. Dougherty, T. J., Gomer, C. J., Henderson, B. W., Jori, G., Kessel, D., Korbelik, M., Moan, J., & Peng, Q. (1998) *Journal of the National Cancer Institute* **90**, 889–905.
31. Chen, J., Stefflova, K., Niedre, M. J., Wilson, B. C., Chance, B., Glickson, J. D., & Zheng, G. (2004) *Journal of the American Chemical Society* **126**, 11450–11451.
32. Persad, R., Liu, C., Wu, T. T., Houlihan, P. S., Hamilton, S. R., Diehl, A. M., & Rashid, A. (2004) *Modern Pathology* **17**, 861–867.
33. Chen, J., Stefflova, K., Warren, M., Bu, J., Wilson, B. C., & Zheng, G. (2007) *SPIE* **6449**.
34. Chen, J., Stefflova, K., Niedra, M. J., Wilson, B. C. & Zheng, G. (2004) *Molecular Imaging* **3**, 277.
35. Clo, E., Snyder, J. W., Voigt, N. V., Ogilby, P. R., & Gothelf, K. V. (2006) *Journal of the American Chemical Society* **128**, 4200–4201.
36. Mawn, T. M., Popov, A. V., Milkevitch, M., Kim, S., Zheng, G. & Delikatny, E. J. (2006) *Molecular Imaging* **5**, 315.

Part IV
Cardiovascular

Light Therapy for the Cardiovascular System

Hana Tuby, Lydia Maltz, and Uri Oron

Abstract Coronary artery disease, which may lead to myocardial infarction (MI), is the primary cause of mortality world-wide. In a series of studies in which the rat and canine models were used, the effect of LLLT on infarct size after chronic myocardial infarction in rats was investigated. LLLT caused a profound (50–70%) reduction in infarct size and ventricular dilatation in the rat heart after chronic MI. This phenomenon was achieved by the cardioprotective effect of LLLT on mitochondria, elevation of cytoprotective heat shock proteins and enhanced angiogenesis in the myocardium following laser irradiation. The effect of LLLT on the expression of vascular endothelial growth factor (VEGF) and inducible nitric oxide synthase (iNOS) in the infarcted heart was also investigated. It was found that VGEF and iNOS expression in the infarcted rat heart is markedly upregulated by LLLT and is associated with enhanced angiogenesis and cardioprotection. The possible beneficial effects on implantation of autologeous mesenchymal stem cells (MSCs) that had been laser irradiated prior to their implantation into the infarcted rat heart was also investigated. These findings provided the first evidence that LLLT can significantly increase survival and/or proliferation of MSCs post implanation into the ischemic/infarcted heart, followed by a marked reduction of scarring, and enhanced angiogenesis. The results of the animal studies may also have clinical relevance. It can be postulated that the use of laser following MI is most probably safe and our observations indicate that delivery of laser energy to the heart may have an important beneficial effect on patients after acute MI or ischemic heart conditions.

Keywords: Low level laser irradiation, Heart muscle, Angiogenesis, Cardioprotection.

Coronary artery disease, which may lead to myocardial infarction (MI), is the primary cause of mortality world-wide. Cardiac repair after infarct is a complex process involving diverse inflammatory components, extracellular matrix remodeling, and responses of the cardiomyocytes to the ischemia. Many studies have been

U. Oron

Department of Zoology, The George S. Wise Faculty of Life Sciences, Tel-Aviv University, Tel-Aviv, 69978, Israel, e-mail: oronu@post.tau.ac.il

R. Waynant and D.B. Tata (eds.), *Proceedings of Light-Activated Tissue Regeneration and Therapy Conference.*

directed towards the use of drugs, growth factors and various interventional technologies in reducing myocardial infarct size and improving heart function post MI in experimental animals and humans. Novel approaches to enhancing angiogenesis in the ischemic myocardium by introducing growth factors [mainly of the vascular endothelial growth factors (VEGF) family] were adopted and found to have a beneficial effect on patients with severe angina [1]. We have been using LLLT in order to attenuate the complex process that occurs in the heart post MI in experimental models of induction of MI in mice, rats, dogs and pigs [2–6].

In a series of studies in which the rat model was used [3–5] the effect of LLLT on infarct size after chronic myocardial infarction (MI) and ischemia-reperfusion injury in rats was investigated. The left anterior descending (LAD) coronary artery was ligated in rats to create MI or ischemia-reperfusion injury. The hearts of the laser-irradiated (LI) rats received irradiation after LAD coronary artery occlusion and 3 days post-MI in order to enhance formation of new blood vessels (angiogenesis). Twenty-six laser irradiation was delivered at 5 mW/cm^2 power density for 2 minutes. At 14, 21, and 45 days post-LAD coronary artery permanent occlusion, infarct sizes (percentage of left ventricular volume) in the non laser-irradiated (NLI) rats were 52 ± 12 (SD), 47 ± 11, and $34 \pm 7\%$, respectively, whereas in the LI rats they were significantly lower, being 20 ± 8, 15 ± 6, and $10 \pm 4\%$, respectively. Left ventricular dilatation (LVD) in the chronic infarcted rats was significantly reduced (50–60%) in LI compared with NLI rats. LVD in the ischemia-reperfusion-injured LI rats was significantly reduced to a value that did not differ from intact normal non-infarcted rats. Laser irradiation caused a significant 2.2-fold elevation in the content of inducible heat shock proteins (specifically HSP70i) and a 3.1-fold elevation in newly-formed blood vessels in the heart compared with NLI rats. It is concluded from this study26 that LLLT caused a profound reduction in infarct size and LVD in the rat heart after chronic MI and caused complete reduction of LVD in the ischemic reperfused heart. This phenomenon may be partially explained by the cardioprotective effect of the HSP70i and enhanced angiogenesis in the myocardium following laser irradiation.

The aim of another study was to investigate the possibility that LLLT attenuates infarct size formation after induction of MI in rats and dogs [3–6]. Laser irradiation was applied to the infarcted area of rats and dogs at various power densities (2.5–20 mW/cm^2) after occlusion of the coronary artery. In infarcted laser-irradiated rats that received laser irradiation immediately and 3 days after MI at power densities of 2.5, 5, and 20 mW/cm^2), there was a 14%, 62% (significant; $P < 0.05$), and 2.8% reduction of infarct size (14 days after MI) relative to non-laser-irradiated rats, respectively. The results of this study indicate that delivery of low-energy laser irradiation to the infarcted myocardium in rats and dogs has a marked effect on the infarct size after MI but also that the laser irradiation effect depends on many factors, such as power density, frequency, and timing of laser delivery to biological tissues.

Another study further explored the effect of LLLT on the infarcted heart of rats and dogs with special attention to the mechanisms of the laser therapy [4]. Myocardial infarction (MI) was induced in 50 dogs and 26 rats by ligation of the left anterior descending coronary artery. Following induction of MI the laser-irradiated

(LI) group received the laser irradiation (infrared laser, 803 nm wavelength) epicardially. Control MI-induced non laser-irradiated (NLI) dogs were sham operated and laser was not applied. All dogs were sacrificed at 5–6 weeks post MI. Infarct size was determined by TTC staining and histology. The laser treatment significantly ($p < 0.05$) lowered mortality from 30% to 6.5% following induction of MI. The infarct size in the LI dogs was significantly ($p < 0.0001$) reduced (52%) as compared to NLI dogs. Histological observation of the infarct revealed a typical scar tissue in NLI dogs and cellularity in most of the LI dogs. Only $14 \pm 3\%$ of the mitochondria in the cardiomyocytes in the ischemic zone (4 hours post MI) of LI MI-induced rats were severely damaged as compared to $36 \pm 1\%$ in NLI rats. Accordingly, ATP content in that zone was 7.6-fold significantly higher in LI vs. NLI rats.

The results of all the above studies indicate that LLLT given at the proper power density and timing attenuates the sequential complex processes that take place after induction of MI, as reflected in rats and dogs. The rationale to use both rats and dogs in this study was two-fold. First, it was of interest to explore whether the power density to achieve an optimal biological effect on the ischemic heart is similar in small and large animals. Second, ischemia in the heart (and infarct size) is different in rats and dogs, because rats have practically no collateral coronary arteries, whereas, in dogs, the collateral coronary arteries are well developed. The results achieved in the studies indicate that the power density necessary to achieve reduction of infarct size is similar in small and large animals. These results are of importance for determination of the optimal laser power density for human application. The results also indicate that a possible cardioprotective effect of the laser irradiation may take place even under acute ischemic conditions, as in the infarcted rat heart. The mechanisms associated with the cardioprotective phenomenon by the laser irradiation are not as yet clearly understood. However, several explanations can be posited. Irradiation given within a short time interval after occlusion of the coronary arteries may attenuate the very rapid decrease in ATP, and the consequent irreversible adverse effects that take place in the cardiomyocyte mitochondria and myofibrillar structures in the ischemic zone. Indeed, laser irradiation has been found to increase mitochondrial respiration and ATP synthesis in ischemic tissues [4]. Thus, the injured laser-irradiated cells may have a much slower rate of degeneration due to ATP elevation, as found in the laser-irradiated dogs compared with non-irradiated dogs, lending further credence to the cardioprotective effect of the cardiomyocytes in the ischemic zone induced by the laser irradiation during the initial phase after MI. Myocardial ischemic injury is also mediated, at least in part, by the generation and accumulation of reactive oxygen species in the cells under ischemic conditions, which also take part in their degenerative process.

The results of the animal studies reviewed in this article may also have clinical relevance. It can be postulated that the use of laser following MI is most probably safe on the basis of previous results showing that direct low-energy laser irradiation on myoblasts in culture does not affect their differentiation in vitro [7–8] and that laser applications to the rat brain at various laser doses [9] has no pathological effects. Also the use of proper levels of laser in humans has no known deleterious effects [10]. Our observations indicate that delivery of laser energy to the heart may have an important beneficial effect on patients after acute MI or on ischemic heart conditions that are not accessible to current revascularization procedures. A laser treatment can also be used prior to a planned surgical procedure (i.e. bypass, valve replacement in the heart or surgical invasive procedure in skeletal muscles) in order to reduce adverse effects of transient ischemia/anoxia on the tissue during the procedure. Indeed, we have recently shown [11] that application of LLLT to the intact heart prior to experimental induction of myocardial infarction (MI) in rats caused a marked reduction in scar formation post MI, in comparison to rats that were not treated with the laser prior to MI.

The aim of another recent study [11] was to investigate the effect of LLLT on the expression of vascular endothelial growth factor (VEGF) and inducible nitric oxide synthase (iNOS). VEGF is known as a key mediator of ischemia-driven angiogenesis. It induces sprouting and capillary growth toward the ischemic tissue and matches the vascular density according to development and physiological increase in oxygen consumption [12]. iNOS is also known as one of the major cardioprotective proteins [13] and has been used as such successfully in gene therapy [14]. Myocardial infarction was induced by occlusion of the left descending artery in 87 rats. LLLT was applied to intact and post infarction. VEGF, iNOS and angiogenesis were determined. Both the laser-irradiated rat hearts post infarction and intact hearts demonstrated a significant increase in VEGF and iNOS expression compared to non laser-irradiated hearts. LLLT also caused a significant elevation in angiogenesis. It was concluded that VEGF and iNOS expression in the infarcted rat heart is markedly upregulated by LLLT and is associated with enhanced angiogenesis and cardioprotection.

The aim of a recent (unpublished results, this laboratory) study was to evaluate the possible beneficial effects of implantation of autologous mesenchymal stem cells (MSCs) that had been laser irradiated prior to their implantation into the infarcted rat heart. The cardiomyocytes in the adult mammalian heart are considered to be in a terminally differentiated stage, with no capacity to proliferate under normal conditions or following insult (i.e. ischemia, injury) [15, 16]. This poses a major medical problem, as inadequate regeneration contributes to myocardial scarring, heart failure, arrhythmia and death [17]. In recent years it has been demonstrated that despite the destructive progression that takes place, mitotic divisions and DNA synthesis occur and cardiac stem cells are to be found among the mammalian heart cells [18, 19]. Despite these encouraging findings, the ability of the heart to regenerate itself is very limited, and can not compensate for the loss of cardiomyocytes post MI.

One approach to overcoming the limited capacity of the cardiomyocytes to proliferate after ischemia is to attempt to implant various cell types (skeletal muscle satellite cells, mesenchymal stem cells (MSCs), etc.) into the myocardium post MI. The results of these studies have indicated that, overall, such implantation of cells is feasible and results in improvement of heart function as compared to saline-injected (control) infarcted hearts [20, 21]. Despite the above promising reports regarding experimental animal models, the studies in humans have shown mixed results. In one study intracoronary administration of bone marrow cells was associated with improved recovery of left ventricular contractile function in patients with acute MI [22]. In another study MSCs injected into the myocardium during coronary artery bypass grafting (CABG) in humans demonstrated a significant improvement in heart function over CABG alone [23]. Recently in three double-blind clinical trials on a total of 426 patients with ischemic heart failure, there was a minimal or no improvement of heart function up to 6 months post implantation in patients that received autologous MSCs as compared to saline [24].

In our study mesenchymal stem cells (MSCs) have been isolated from rat bone marrow and grown in culture. The cells were laser irradiated or with a Ga-As laser (810 nm wavelength), labeled with 5-Bromo-2′deoxyuridine (BrdU), and then implanted (control or laser-treated) into infarcted rat hearts. Hearts were excised 3 weeks later and cells were stained for BrdU and c-kit immunoreactivity (a marker of MSCs). Hearts that were implanted with laser-treated cells showed a significant ($p < 0.006$) reduction of 53% in infarct size compared to hearts that were implanted with non laser-treated cells. The hearts that were implanted with laser-treated cells prior to their implantation demonstrated a 5 and 6.3-fold significant ($p < 0.02$ and $p = 0.04$ respectively) increase, in cell density that positively reacted to BrdU and c-kit respectively as compared to control. A significantly ($P < 0.001$ and $p = 0.01$) 2.5 and 2-fold higher level of angiogenesis and vascular endothelial growth factor was seen in infarcted hearts that were implanted with laser treated cells as compared to non-treated implanted cells. The findings of this recent study provided the first evidence that LLLT can significantly increase survival and/or proliferation of MSCs post implantation into the ischemic/infarcted heart, followed by a marked reduction of scarring, and enhanced angiogenesis. The mechanisms associated with this phenomenon remain to be elucidated in further studies.

In conclusion we have reviewed here the various approaches that can be taken by application of light therapy to the ischemic/infarcted heart. In all of them when light was used in proper timing and parameters one could observe beneficial effects over non treated hearts. The observation has clinical applications and will have to further explore in order to elucidate the mechanism associated with these interesting phenomena.

References

1. Losordo DW, Vale PR, Symes JF, Dunnington CH, Esakof DD, Maysky M, Ashare AB, Lathi K, Isner JM. Gene therapy for myocardial angiogenesis: initial clinical results with direct myocardial injection of phVEGF165 as sole therapy for myocardial ischemia. Circulation 1998; 98:2800–2804.
2. Yaakov N, Bdolah A, Wolberg Z, Ben-Haim S, Oron U. Recovery from sarafotoxin-b induced cardiopathological effects in mice following low energy laser irradiation. Basic Res Cardiol 2000; 95:385–389.
3. Yaakobi T, Shoshani Y, Levkovitz S, Rubin O, Ben-Haim SA, Oron U. Long term effect of low energy laser irradiation on infarction and reperfusion injury in the rat heart. J Appl Physiol 2001; 90:2411–2419.
4. Oron U, Yaakobi T, Oron A, Mordechovitz D, Shofti R, Hayam G, Dror U, Gepstein L, Wolf T, Haudenschild C, Ben Haim SA. Low energy laser irradiation reduces formation of scar tissue following myocardial infarction in dogs. Circulation 2001; 103:296–301.

5. Oron U, Yaakobi T, Oron A, Hayam G, Gepstein L, Wolf T, Rubin O, and Ben Haim SA. Attenuation of the formation of scar tissue in rats and dogs post myocardial infarction by low energy laser irradiation. Lasers Surg Med 2001; 28:204–211.
6. Oron U. Photoengineering of tissue repair in skeletal and cardiac muscles. Photomed Laser Surg 2006; 24:111–120. Review.
7. Ben-Dov N, Shefer G, Irinitchev A, Wernig A, Oron U, Halevy O. Low-energy laser irradiation affects satellite cell proliferation and differentiation in vitro. Biochim Biophys Acta 1999; 1448:372–381.
8. Shefer G, Oron U, Irintchev A, Wernig A, Halevy O. Skeletal muscle cell activation by low energy laser irradiation: a role for the MAP/ERK pathway. J Cell Physiol 2001; 187:73–80.
9. Ilic S, Leichliter S, Streeter J, Oron A, DeTaboada L, Oron U. Effects of power densities, continuous and pulse frequencies, and number of sessions of low-level laser therapy on intact rat brain. Photomed Laser Surg 2006; 24(4):458–466.
10. Tuner J, Hode L. Low level laser therapy – clinical practice and scientific background. Grangesberg, Sweden: Prima Book in Sweden AB. 1999.
11. Tuby H, Maltz L, Oron U. Modulations of VEGF and iNOS in the rat heart by low level laser therapy are associated with cardioprotection and enhanced angiogenesis. Laser Surg Med 2006; 38:682–688.
12. Zachary I, Gliki G. Signaling transduction mechanisms mediating biological action of the vascular endothelial growth factor family. Cardiovascular Res 2001; 49:568–581.
13. Dawn B, Bolli R. Role of nitric oxide in myocardial preconditioning. Ann NY Sci 2002; 962:18–41.
14. Li Q, Guo Y, Tan W, Stein AB, Dawn B, Wu WJ, Zhu X, Lu X, Xu X, Siddiqui T, Tiwari S, Bolli R. Gene therapy with inducible nitric oxide synthase provides long-term protection against myocardial provides long term protection against myocardial infarction without adverse functional consequences. Am J Physiol 2006; 290:584–589.
15. Rumyantsev PP. Interrelations of the proliferation and differentiation processes during cardiac myogenesis and regeneration. Int Rev Cytol 1977; 51:186–273. Review.
16. Harsdorf R.Von, P Poole-Wilson, R Dietz. Regenerative capacity of the myocardium: implications for treatment of heart failure. Lancet 2004; 363:1306–1313.
17. Thom T, Haase N, Rosamond W, Howard VJ, Rumsfeld J, Manolio T, Zheng ZJ, Flegal K, O'Donnell C, Kittner S, Lloyd-Jones D, Goff DC, Jr. Hong Y. Heart disease and stroke statistics – 2006 update: a report from the American Heart Association Statistics Committee and Stroke Statistics Subcommittee. Circulation 2006; 113:85–151.
18. Anversa P, Fitzpatrick D, Argani S, Capasso JM. Myocyte mitotic division in the aging mammalian rat heart. Circ Res 1991; 69(4):1159–1164.
19. Beltrami AP, Barlucchi L, Torella D, Baker M, Limana F, Chimenti S, Kasahara H, Rota M, Musso E, Urbanek K, Leri A, Kajstura J, Nadal-Ginard B, Anversa P. Adult cardiac stem cells are multipotent and support myocardial regeneration. Cell 2003; 114(6):763–776.
20. Wang QD, Sjoquist PO. Myocardial regeneration with stem cells: pharmacological possibilities for efficacy enhancement. Pharmacol Res 2006; 53:331–340.
21. Dawn B. and R Bolli. Adult bone marrow-derived cells: regenerative potential, plasticity, and tissue commitment. Basic Res Cardiol 2005; 100:494–503.
22. Schachinger V, Erbs S, Elsasser A, Haberbosch W, Hambrecht R, Hölschermann H, Yu J, Corti R, Mathey DG, Hamm CW, Süselbeck T, Assmus B, Tonn T, Dimmeler S, Zeiher AM. REPAIR-AMI Investigators. Intracoronary bone marrow-derived progenitor cells in acute myocardial infarction. N Engl J Med 2006; 355(12):1210–1221.
23. Stamm C, Kleine HD, Choi YH, Dunkelmann S, Lauffs JA, Lorenzen B, David A, Liebold A, Nienaber C, Zurakowski D, Freund M, Steinhoff G. Intramyocardial delivery of CD133+ marrow cells and coronary artery bypass grafting for chronic ischemic heart disease: safety and efficacy studies. J Thora Cardiovas Surg 2007; 133(3):717–725.
24. Rosenzweig A. Cardiac cell therapy – mixed results from mixed cells. N Eng J Med 2006; 355:1274–1277.

Part V
Dentistry

Introduction: Overview

Donald E. Patthoff

Abstract This session focuses on optical techniques such as OCT for assessing the effectiveness of LLLT for periodontal repair, bone growth stimulation and numerous other dental techniques. Details of periodontal repair and bone growth as well as a summary of all LLLT uses and details of some details of the mechanisms are also given.

Keywords: Optical Coherence Tomography (OCT), implant failure, photobiomodulation, TGF-β1.

History

The first lasers specifically designed for use in dentistry became a reality in the mid nineteen-eighties. The Food and Drug Administration gave marketing approved to a CO_2 laser for oral surgery procedures in 1987, a Nd:YAG laser for soft tissue dental procedures in the early 1990s, and an Er:YAG laser for tooth cutting in 1997. The history of accepted laser use in dentistry in the United States, then, barely covers 20 years. Even considering the few earlier years of their use in other parts of the world, lasers and various other therapeutic light devices are a relatively new phenomenon in dentistry and have not yet attracted the amount and kind of clinical research needed to develop their huge potential.

In the first years, most dental uses of lasers in the United States focused on properties of these devices that would allow them to be adapted as cutting instruments, blood management aids, bacterialcidal agents, and anesthetics. Since these early units were often seen as tools that could be used to make reasonable modifications of the usual and customary procedures of dentistry, their safety and effectiveness tended to be evaluated against normal dental procedures and practices.

Focus, then, was placed on practical dental applications and ongoing clinical evaluation. Evaluations were both practice-based and evidence-based and is still

D.E. Patthoff
300 Fox Goft Ave., Martinsburg, West Virginia

R. Waynant and D.B. Tata (eds.), *Proceedings of Light-Activated Tissue Regeneration and Therapy Conference*.
© Springer Science + Business Media, LLC 2008

evolving. All of this, in terms of FDA marketing clearance, is under the section of devices; the approach to evidence for evaluating safety and effectiveness is not the same as its section on drugs. The use of lasers and their evaluation is, and must take place, then, within a complex constructed system that eventually relates dental diagnoses to common treatments.

Clinical research variables, teaching outcomes, and learning competencies about lasers and laser use in dentistry, then, were often defined in terms of these commonly accepted treatments and their codes such as the procedural codes of the American Dental Association in the United States. These codes are still tied to insurance reimbursements and the usual and customary fees set by each dentist. This, in turn, helps generate the background for evaluating usual and customary treatment procedures and outcomes. As a result, these same procedural codes are often used to generate clinical research data, insurance actuarial tables, and liability evidence.

Developers and users of laser dental devices, then, tended to stay within the limits of this semi- integrated system framed by these comparative procedures. This allowed for a reasonable and practical return on investments and gradual observation of their own practices concerning lasers and laser use.

In the mean time, though, the development and evaluation of all types of systems was growing within a field known as systems thinking. Along with this field, systems-ethics structures were also being imagined and explored within the context of, for example, access to care. The Journal of the American Dental Education Association published such a review as a special issue in November 2006 on professional promises and access to care. Systems ethics and systems thinking, though, were not only influencing access to care, but also the development of lasers, and other dental, medical, and social needs. A key notion of this systems ethics thinking draws attention to the significance about what is the end good that is given attention in systems and how they are accepted. Systems can be and are built, for example, around diseases, procedures, and/or people. And, whatever the primary base of a system, that base tends to influence, if not dominate, the focus of every action, whether or not people pay attention to it. When working within a-particular-based-system, then, neutrality within the system can be considered a bias towards the preservation of the system if and when that base is no longer the focus that is desired.

One example of how this systems thinking plays out in dentistry, is in the area of periodontal disease and treatment. Underlying the treatment of all periodontal disease is basic home care, hygiene, and proper educational information at the right time for a specific person and a desire to keep their teeth healthy for life. Despite home care, though, professional interventions are often indicated. Some of those treatment interventions involved surgical procedures. At one time, it was questioned if full thickness surgical flaps were better than partial thickness flaps. Attempts to investigate and control for common underlying disease patterns and practitioner skills were at first minimal, and then became more complex. This evolved into various research comparisons of surgical procedures to procedures involving scaling and root planning, and curettages. Within this genuine concern about the benefits and best approaches to these procedures, were other issues concerning access to care and qualifications of care providers. These comparisons of surgical procedures and results, for example, were mixed with broader access to care questions; other formats of research involving patient education, and turf/boarder questions about who was doing the treatment were also being considered. Furthermore, focus was turning away from the specific procedures and more towards the bacterial causes and the immune responses involved in periodontal diseases. At the same time, additional turf and boarder questions arose about who should be qualified to do what in terms of specific procedures and what qualified as a surgical procedure within the context of licenses.

Within this complex set of questions, a minimum surgical procedure, ENAP (Excisional New Attachment Procedure), evolved. It required less surgical exposure, fewer risks and demonstrated equivalency of results to other surgical procedures and approached to treating periodontal disease. Some thought of it more as a modified scaling and curettage, others thought it to be a more prudent surgical flap procedure. In the years it took to develop the new procedure codes that evolved with these defined treatments, the old ones were adapted and partially accepted in day-to-day office policy protocols. When lasers came into use, then, they too were applied to and used in all of these surgical and bacterial procedures. One application eventually resulted in the FDA approval of an LNAP (Laser New Attachment Procedure) with a specific laser instrument and patented teaching method based on a demonstrated equivalency to ENAP.

In the background of all this laser and non-laser developments, but never totally out of view, a few education and research pioneers were also focusing on other properties of lasers. These properties were not easily communicated within this system, though, and did not seem as relevant as the usual practical clinical information that was growing behind the mainstream clinical marketing of these early dental laser devices. Research on these properties of lasers and light, then, was not aimed at influencing the most obvious desires for, and common adaptations of, this evolving technology and its techniques. The language of this more basic research was expressed in terms of complex light refractory formulas and their developing theories. These, in turn, were used to build the devices that measure the interactions of light with molecules, enzymes, intracellular structures like mitochondria, and other such things as the relationship of how cells communicate with each other. As that information grew, the need for newer physics formula became necessary to quantify these new observations. This, then, also became mixed with the languages of biochemistry, physiology, pathology, and a host of other areas of relevance.

One of these laser properties that was less commonly discussed among dental clinicians, though, held potential for imaging biological tissues in ways that went beyond the capabilities of X-rays, magnetic imaging, and ultra-sound. It also offered advantages of possible cellular and sub cellular imaging in real time and in three dimensions. If possible, this could help integrate and better relate, then, the information from some of these diverse and overlapping fields of study. The reality of this mechanism came around the year 2000 with the introduction of the first Optical Coherence Tomography Units (OCT).

Other characteristics of lasers were also in the background in the United States and were often considered associated observations. In other countries, like Brazil, these characteristics, however, were becoming the most explored and used benefits of dental lasers. Light was being used as medications and/or for biological stimulations/inhibitions (also known as modulation). These uses, though, could not be easily packaged or sold within the usual U.S. markets of systemic and topical drugs. That is, these kinds of effects fell, not under the claims of devices in terms of FDA marketing limits for manufacturers, but under the FDA's drug division and its rules and regulations.

For marketing purposes, these claims needed to be tied to the pharmacological and biochemical measures of medications and other physiological effectors. Still the significant therapeutic effects of light on multiple classes of diseases, injuries and medical disorders was becoming well known among clinicians using lasers. In particular, lights effectiveness on wound healing, as well as for reduction of inflammation and swelling, and for pain control was more than common conference talk. Some devices were able to demonstrate enough evidence that they gradually became cleared FDA cleared for marketing therapeutic claims for specific conditions like tennis elbow.

One approach, for practitioners, then, was to use the device in "off label" ways. This meant that practitioners could use lasers for these other effects if there were other forms of practice based evidence for their beneficial use; manufacturers and their representatives, however, could not, mention them in their promotions or distribution information. Dentists who represented or were sponsored by laser manufacturers, then, were often on questionable grounds when talking or teaching about laser use for off label effects that were observed when used within specific practice based techniques. Describing those techniques and limits, though, were difficult; not only were there a complex set of variables regarding the use of lasers, but also because changing any of these variables seemed to make some specific techniques ineffective.

The organizers of a recent SPIE conference call for papers summarized the complexity of rationally choosing amongst a large number of illumination parameters and included such variables as wavelength, fluence, power density, pulse structure, and treatment timing which has led to the publication of a number of negative as well as many positive studies. Another complex set of variables they did not list, but still needing consideration, fell under the areas of diagnoses, body environment, treatment setting, practitioner styles, and a host of other clinical and biological options.

The limits about discussing the effects of off label uses, combined, with trade secrets with in different laser manufacturers, also made collaborative efforts to clarify these variables more complex. These kinds of evaluations also, to be done well and gain any sense of worthy information, needed two forms of informed consent/refusal, both research and the clinical. And, each of these different kinds of consent/refusal needs had different intents and were often contradictory.

Jan Tuner, a dentist in Scandanavia, and Lars Hode began, then, to collate all the research of these bio-stimulation and bio-inhibition characteristics no matter where they were used; many of the applications fit closer to the fields of physical therapy and complementary medicine, like acu-puncture, than to clinical allopathic medicine and dentistry. It was being called laser therapy. The term was first coined by, a researcher in Russian Tina Karu. She began detail measurements of the phenomenon in the 1970s, after a Hungarian researcher, Endre Mester, first reported his observations in 1966.

The key common desires for all involved, however, was to design and produce the essential basic and clinical research, and eventually connect them in an acceptable way. This was needed to demonstrate any realistic, safe, and effective benefits of laser therapy. These therapies, however, were still not easy to compare to other standard treatment protocols of medicine and surgery that aimed at similar benefits. There were, also, those many additional factors that needed consideration because of how light was being, for example, applied and pulsed. It was such, then, that the research methodologies of the different interventions that needed to be compared could never be accomplished with the same approaches or same understandings of acceptable scientific rigor. The mix of hard science data with market survey data added further complexity. Still, their was strong desire to at least separate market and practitioner claims from placebo effects for both therapeutic and competition market reasons. This lead to increased emphasis on disclosures about conflicts of interest but did not remove the importance of designing basic research that could be grounded in the best physics theories, and at the same time, the best ethics theories and all the biological and biochemical observations in between.

Because lasers and lights can be treated as photons and equated to electrons, this allowed the development of a research strategy that would use light as both an independent and a dependent variable. That is, light could be used to gather input and output measurements. Long standing spectrophotometric devises continued to evolve for the collection of sub-cellular and molecular data, and, in 2006, the OCT was demonstrating that it was ready for real time collection of clinical data that would build on and improve the long standing standard of dental clinical measurements - x-rays and the newer forms of ultrasound based devices. Hidden within these developments, though, was an opportunity to gain understanding about the underlying mechanisms that were not light-based. That is, it could supplement the methods already in place for evaluating various other approaches that were claiming similar effects and that fell under the label of complementary medicine rather than allopathic medicine. Examples were uses of hydrogen peroxide, acu-puncture, and colostrums.

Researchers specializing in the workings of mitochondria in animal cells, which were similar to the light absorption organelles in plant cells, began at about the same time lasers were introduced to dentistry in early 1990, to collect their research within such publications as the Journal of Mitochondria. It was believed that the primary cellular chromophores, that absorbs low levels of red and near-infrared light in mitochondria, was cytochrome c oxidase. This absorption of energy was evidently leading to increase in ATP synthesis and release of reactive oxygen species from the electron transport chain. This, subsequently, was shown to activate transcription factors and lead to cell proliferation and migration. Other photoreceptors were being identified in the mitochondria and cell walls. Still this did not fully explain how cells communicated with each other to decide what they would be or how they would relate. Much, however, was becoming known in immunology and oncology concerning hundreds of forms of G proteins that was beginning to explain that mechanism. Within one group of proteins, for example a rogue protein, P2Y, was found to be a photoreceptor.

In summary, then the topic of laser therapy, now better known as Light-Activated Tissue Regeneration and Therapy, then, has existed for nearly 40 years. It was discovered soon after the first laser was invented. Laser therapy also has a number of other names: low level light therapy or (LLLT), cold laser, soft laser, and photobio-modulation. Exploring light-activated tissue regeneration and applying it clinically to humans, though, involves multidisciplinary approaches; it also involves clinical patients, cell and molecular biology, laser biophysics and many other forms of scholarship. All of these approach questions of cause and effect differently and many of their methods cannot be easily integrated with all the forms of reasoning commonly used in our complex world of science, clinical practice, law, and marketing.

Last Step

Every researcher involved in this early work, then, began to face resistance from the mainstream biology research journals and highly regarded health care journals. This was partly because the mechanisms were new and often did not easily fit the normal publishing standards of a particular field of study. It was also because the field was so broad and it was difficult to sometimes find qualified reviewers. For most studies, a broad spectrum of views was required. The traditional approaches to systematic reviews and meta-analyses were also a concern. It was difficult to separate out all the key variables in a way that would not include those studies with the obvious research flaws that would produce non-significant findings or even negative findings. If traditional approaches to scientific reviews were not able to control for these variables. Identity of many of the variables were not simple word searches, or on collection of papers around a particular biochemical reaction, but rather complex evaluation of methodology and comparison of related conditions with unrecognized common causality or different clinical labels.

Many different kinds of knowledge and skills, then, needed to be brought together in a way that would allow the depth of their research to be expressed and shared with others who were willing to listen and learn and, at the same time, allow them to share their expertise in a similar manner. The first Gordon style conference on light-activated tissue regeneration, then, was first held in Hawaii in 2004. This international gathering concluded that the effects of laser therapy were real, that a better understanding of its mechanism was critical, and that attention should be focused on optimal dosing.

Dentistry was part of that conference. Dr. Paul Bradley presented information on his lab at ANOVA and described his early research on TMJ and Head and Neck Pain. He also gave observations about improved healing and less pain in many oral surgery procedures and proposed his theories. Arun Dubar presented a few case studies involving laser uses in his office. Don Patthoff, as editor of the Journal of the Academy of Laser Dentistry, after conducting a workshop on the more common ways ethics is approached, described some ethical challenges regarding: (1) the ability to do adequate research, (2) establish a community and a journal to help focus the research within the specific skills, knowledge, and general practices of dentistry, and (3) stay open and contribute to other fields of health care and other needs of society.

Next Steps

In late 2006, Ron Waynant announced that the results of recent research indicated a mechanism that would explain all of the diverse effects being listed under light activated tissue regeneration and laser therapy. This lead to this second ECI Gordon like conference; Don Patthoff again agreed to chair the dental session.

Waynant's Announcement

At this conference the main organizers plan to introduce and discuss a dominant mechanism by which light therapy works. Dr. Waynants group at FDA, Dr. Anders's group at USUHS, and Dr. Mitra at Florida Institute of Technology were in the process of proving that light interacts with the mitochondria of cells and generates hydrogen peroxide, as implied by Karu and Lubart. Neither realized, however, that the smallest amount of H_2O_2, about $3-15$ mol/10^7 cells, was enough to optimize the stimulation of cells leading to the benefit to nearly one hundred medical problems. The benefits of hydrogen peroxide had been known for hundreds of years, but they had largely been abandoned by modern medicine in favor of more costly (and profitable) drugs. Dr. Waynant and the organizers believed that this H_2O_2 mechanism explained, to a large extent, the mystery of laser therapy's success in the treatment of every one of the approximately 100 diseases and conditions evaluated. It therefore suggests an extremely safe way of pinpointing the treatment on the surface of the body. It also called renewed attention to a cheap, effective drug. A drug naturally produced and used by the body, and that could potentially play a much larger role in health care. It also implied that the use of light (and lasers) is not necessary to utilize this cure, but could play an important role in evaluating and better understanding of this cure. Colostrums, the first milk produced after all mammalian mothers give birth to their child was well known for treating many of the same diseases that were being expected through laser therapy, hydrogen peroxide therapy, and acu-puncture. In many cases colostrums would be topically applied with a cotton ball applicator and generate rather than using lasers costing hundreds or thousands of dollars and would produce similar effects. The realization that the new measurement and imaging tools from light devices, as well as the demonstrated effects of light as a biomodulator, could be used to further refine the evaluation of claims from these other areas of complementary medicine, and, perhaps, use their data to further explain the underlying mechanisms. Using light to generate the primary underlying drug near the surface of tissue, however, was also promising and had some advantages. Light can pinpoint it effects to the cells that need it. It is safe, non-messy, and the curing drug, or initiating factors can penetrate deeply through cell layers to seek and destroy diseased cells, stimulate nerve growth, and generate other benefits. Higher concentrations can lead to inhibition of the stimulation, inhibition of cell proliferation or to cell death. Quantification and optimization of light generation of hydrogen peroxide is in progress and was hope to be completed before the conference. Updates on a specific Proline-rich Polypeptides (PRPs), that are within naturally produced and preserved colostrums and currently bottled for use as an oral spray, were also hoped to be unveiled.

Ron's announcement was possible because of years of research by hundreds of individuals, institutions, and funding agents around the world. Ron's summary references multiple effects that occur with these common products and lasers. It mixes biochemistry, biology, microbiology, physics, and many other forms of basic research. It impacts clinical research, clinical medicine, marketing research, and a host of other disciplines.

Don's Duty

In my role as editor of what has now become the Journal of Laser Dentistry, I was responsible for communicating the uses and effects of lasers in dentistry through a professional dental organization dedicated to laser education, the Academy of Laser Dentistry. My experiences in communicating the proceedings of the Hawaii ECI conference to the dental profession led to some interesting observations. I first began to notice the issues when I decided to use the Hawaii mitochondria material as my first area of focus. That material lead to more people doing more of their homework and gradually helped transform the previous discussions from ones of disbelief and doubt to curiosity and excitement, and in some cases, is rapidly achieving a general level of acceptance in the medical and biomedical communities. These explorations and applications, though, also raised some interesting ethical issues about clinical trials and the normal development and growth of clinical practices that were similar to those ethical issues beginning to be raised about the differences between medical research and surgical research. Besides the similarities, though there were differences that needed consideration. I therefore started by collaborating with members of the American Society of Dental Ethics (ASDE) who began asking some constructive rather than empirical research questions.

The nature of laser therapy combines several forms of clinical research and creates new issues. How and who decides, for example, when a clinical act is research or simply a reasonable clinical judgment? With lasers, because they simultaneously produce both surgical and medical effects and can be used for cosmetic purposes, this old ethical question, now takes on new twists.

The origin of research informed consent is different than clinical informed consent and raises other ethical issues when the lines of research and clinical application cross; researchers, for example, focus on the documentation, while clinicians require an ongoing uncertain conversational process. The intent, purpose, and end -good of these legal/ethical acts are also different. Without a commonality about the ultimate end good, reasonable or integrated processes, that help generate meaning about data and information from and within science and clinical practice, are impossible.

The matter of trust also takes on new challenges. Trust between a patient and a clinician is built on a fiduciary relationship. Clinicians promise to prioritize the well being of the individual person or group. Each clinician brings a particular understanding of science, clinical based experiences, general knowledge of people and society, art, technology, techniques, skills, and ethical/moral values to a particular moment for a particular patient/group in a particular circumstance. It is more than evidenced based. How do simple therapies that have broad systemic effect impact licensure issues and limitations of expertise?

A person meeting a researcher may think he will receive help for a particular concern, but that is not the case. Differences need to be made clear between a researcher's desire for a relationship and a clinician's. For one, the interventions offered may not help, intentionally do nothing, or may actually harm the subject. The end is directed primarily towards science, health care, and humanity, yet must still address patient needs.

The desire for the relationship and the need for it are also not the same. The difference between need and desire even adds another consideration; that is, how a subject defines need is different than how researchers define need.

There are many other differences that add complexity; these are only five examples:

How are research and clinical findings and progress published?

How does the FDA separate research in cosmetics, devices, and drugs?

How does education address complementary and alternative medicines when the notion of integrative medicine combines them, and when there is some evidence for their effectiveness and safety?

How should placebo and usual and customary treatment be controlled? In surgery, for example, practitioners must be intimately involved in the act and cannot be blinded in the same manner as medicine?

Technology produces both surgical and medical effects as well as microbiological effects. What methodologies give sufficient data when same subject or half-subject studies are negated because the laser therapy effect is not just local?

Intellectual Property Rights (IP) and business practice rights are also adding complexity to the business, legal, and technological world. These developments cross business ethics with research, professional, and bioethics and are increasingly becoming the product for organizations involved in IP licensing.

Ethical/research differences also arise in legal, social, clinical, animal, cellular and biochemical studies regarding the very nature and meaning of causality. Reasoning protocols, research and clinical methodologies, as well as formats for informed consent/refusal considerations regarding all the reasonable benefits and harms, then, are all basic to developing any clinical research regarding the mechanisms and applications of tissue regeneration protocols.

The purpose of this introductory ethics reflection is not to answer every question or solve all the problems, rather it is to raise awareness of some unspoken critical issues that undermine all of our efforts and give them voice so everyone can fairly work with them and make more next steps. I propose that dentist work closely with ASDE to help articulate these ethical concerns and promote the continuing collaboration necessary for their practical resolution. Since ASDE is part of the American Society of Bioethics and Humanities, it will further complement the research developments and clinical applications of this field to all areas of health care.

Craig's Way of Seeing

Craig Gimbel presented some exciting developments in Optical Coherence Tomography (OCT). Awareness of OCT, and gaining a clearer sense of the ethical issues uncovered in the opening session, suggested some new ways for doing better clinical research in dentistry regarding tissue regeneration and healing. OCT is a noninvasive, non-ionizing method for imaging dental microstructure that has the potential of evaluating the health of periodontal tissue. OCT provides an optical biopsy" of tissue 2–3 mm in depth. This sixth modality of imaging was pioneered at Lawrence Livermore National Laboratory. OCT is based on the optical scattering signatures within tissue structure. With the use of a broad spectrum bandwidth light source, high resolution images, 10 times the resolution of x-ray, can detect important tissue interfaces within the periodontal sulcus and its' relationship to the attachment apparatus of the tooth. Multiple cross-sectional tomograms can be stacked to create two and three-dimensional images.

Optical Coherence Tomography has the potential to follow the progression of plaque-induced periodontal disorders that result in the degeneration of connective tissue attachment. Clinicians and researchers can use it to evaluate the effects of photo bio modulation on soft tissue periodontal regeneration. X-rays cannot image soft tissue and therefore cannot identify early active disease or the reversal to health of the periodontal unit; OCT records soft tissue contour and microstructure, quantifying the soft tissue changes that occur in real time. Soft tissue health is a major area in dentistry where OCT diagnostic imaging will make a significant impact. Monitoring health, rather than disease not only adds a new tool to prevention, but also introduces another level of ethical consideration regarding the very subtle and extremely important use and/or misuse of similar information for diagnosing, preventing, and/or marketing. Information that confirms the diagnosis of a known disease is not the same as information that may partially correlate with a level of health, or information that may influence a target customer to purchase a product.

Akira's Periodontal Strategies

Akira Aoki approached the use of photo-bio-modulation for periodontal disease in a very creative way that crossed many areas of research and challenged many of our basic clinical understandings about the nature of the problem and its treatment. The most common forms of periodontal diseases are classified as chronic, inflammatory and infectious diseases, leading to progressive loss of periodontal tissues. Lasers have been used extensively in periodontal therapy, and they are considered by their users to be basically effective because of their physical properties, namely ablation, hemostasis, bacterial killing and cell stimulation. High-power lasers are conventionally used for soft tissue treatments such as gingivoplasty and frenulotomies. An Er:YAG laser, that can be used on both soft and hard dental tissues, was recently developed, and is now available for treating tooth roots and bone surfaces; it has, consequently, become one of the more promising laser units for periodontal treatment.

The laser application strategy now mainly used for the treatment of periodontal diseases is high-power laser ablation (HLLT). Laser ablation of diseased tissues is widely performed and a simultaneous photo-bio-modulation effect (LLLT) may be partly expected in the surrounding tissues. In non-surgical or closed periodontal pocket therapy, lasers can not only ablate the diseased tissues but also stimulate or activate the surrounding gingival and bone tissues, and that can result in improved pocket healing and tissue regeneration.

In previous studies, the group Akira works with observed increased new bone formation in dogs after surgical periodontal and peri-implant treatment using an Er:YAG laser. The improved bone regeneration was shown to be partly due to the photo-bio-modulation (LLLT) during ablation with a high energy level laser (HLLT). Regarding osteogenesis, it had been reported that some low energy level lasers could increase the ALP activity and m-RNA expression of osteogenic markers in osteoblasts and promote bone module formation in vitro. Using a low energy level of Er:YAG laser on gingival fibroblasts, Akira's group demonstrated that there was a stimulatory action on cell proliferation through the production of PGE_2 via the expression of COX-2. This could be considered as one of the important regulatory pathways that enhance cell proliferation. Regarding the bactericidal property of photodynamic therapy (PDT), they found that both argon laser and halogen light produced a sufficient bacterial killing effect without use of a photo-sensitizer. By elucidating the photo-bio-modulation effect in detail, this effect could be used more effectively; laser therapy would then offer additional advantages to the non-surgical and surgical therapies of periodontitis either as an adjunct, or as an alternative to our current mechanical, medical, and pharmaceutical treatment.

Furthermore, as a future strategy of laser therapy, phototherapy can be developed for the prevention and control of periodontal diseases. In the initial stages of periodontal disease, elimination of periodontopathic bacteria and reduction of the inflammatory condition might possibly be controlled by using periodic external laser/light irradiation aimed only at the gingival surface. Also, in the case of moderate and advanced stages of periodontal disease, internal irradiation via the periodontal pocket, as well as external irradiation, would be useful for the enhancement of soft and hard tissue regeneration in combination with, or without, current regeneration procedures.

They conclude that photo-bio-modulation/activation and photo-dynamic effects can be applied more aggressively for the treatment of periodontal diseases.

Bouquot's Bone Building

Jerry Bouquot' paper described the clinical research that he conducted with Peter Brawn on the effects of photo-bio-modulation on dental bone. Low bone density (LBD) and ischemically damaged, desiccated bone both have a poor ability to remodel and are contraindications for dental implants. Unfortunately, readily available diagnostic imaging devices lack the ability to adequately identify such bone. The new technology of through-transmission or quantitative ultrasound (QUS), is specifically cleared by the FDA, however, to localize jawbone regions of LBD, desiccated bone, and medullary cavitations (a sign of chronic ischemia) in a non-invasive manner. While the specificity (proportion of false negatives) of this technology is not well established, the sensitivity (proportion of false positives) is higher than almost any other imaging technique on the market today, with fewer than 3% false positive scans in jawbone. QUS evaluation prior to implant placement should be effective in warning the clinician of a high potential for failure. Identifying the problem is only half of the equation, however. Ideally, regions of LBD or desiccation can be treated to bring them to a level of healthy bone capable of remodeling around and holding an implant. Low-level laser therapy had been shown in cell culture and animal studies to stimulate bone healing, enhance endothelial cell proliferation, and angiogenesis, stimulate T-cell production of angiogenetic factors, and improve both microcirculation flow and lymphatic drainage. NIR-LED photobiomodulation appears to have the same influences.

Their objective was to use QUS to determine the efficacy of LED phototherapy to improve LBD or hydration of jawbone prior to implant placement.

They used an FDA cleared dental QUS device, the Cavitat 4000, to evaluate edentulous alveolar sites in 68 patients (48 females; mean age: 57 years; range: 37–100 years). When LBD or desiccated bone was found a 3 month course of 15 minutes/day, five times weekly LED therapy ensued using a LED 840 nm @ 20 mW/cm^2 and 660 nm @ 15 mW/cm^2).

Of 1,148 QUS jawbone scans, each representing the area of one tooth, 294 were positive for damaged bone. Using an established 5-point scale (0 = normal; 4 = most severe), 79 positive sites were grade I, 69 were grade II, 86 were grade III and 61 were grade IV. The average grade for positive sites was 2.43. After LED photomodulation the average grade was 1.33 and 120 sites (41.5%) had returned to completely normal bone. Another 54 (18.4%) sites were grade I after therapy, a mild level of pathology probably acceptable for implant placement. The mean difference, i.e. improvement of bone quality, of 1.11 was very statistically significant (matched pair analysis: Std error 0.06914; t-ratio -15.9896; DF 293; prob [t] less than 0.0001; 95% confidence interval 0.558–1.242).

They concluded that the NIR-LED photobiomodulation appears to enhance bone health *in vivo* by increasing bone density and/or hydration, which accords with published *in vitro* experimentations. Whether improvements are long-lasting or actually equate with improved implant health and longevity must be addressed by future research. Their preliminary results, however, are encouraging and suggest that this combination of new technologies has the potentially to extraordinarily impact dental implant success rates.

Praveen's Pleiotropic Effects

Praveen R. Arany then gave some molecular insights into photobiomodulation, the activation of Latent TGF-β1 by low power laser. He started by saying the term 'Photobiomodulation' was coined by Tina Kuru and her colleagues to encompass the pleiotropic effects of low power lasers on biological processes. Many studies have demonstrated the efficacy and range of clinical applications of laser treatment but a major stumbling stone for its wider acceptance into mainstream medicine has been its elusive biological effector pathways. Various studies have demonstrated changes in redox levels, mitochondrial function as well as direct effects on proliferation or apoptosis of cells.

Praveen's group began to focus on a multifaceted growth factor, Transforming Growth Factor Beta 1 (TGF-β1) as they observed many similarities in the literature on biological responses when TGF-β1 or Laser radiation was used in similar clinical contexts specifically wound healing. TGF-β1 is secreted as a latent complex and requires physicochemical activation to be biologically potent. They used a cell-free in vitro activation system with a far infrared (904 nm) laser unit, ELISAs and a functional reporter-based system to assay for Latent TGF-β activation by low power laser radiation. They demonstrate that laser irradiation is also capable of 'priming' these latent complexes and making them more amenable to physiological modes of activation. They then confirmed these findings in vivo using an oral healing model and observe an increased TGF-β1, but not β3, expression by immunohistochemistry immediately following laser irradiation compared to non-irradiated control site in the same patient. Praveen presented data suggesting that activation of the latent TGF-β1 complex might be a central mediator in the effects of low power laser photobiomodulation.

The Redox angle as a causal factor has been long implicated but it is truly great to finally put the speculative nature of these theories to rest with hard data. Praveen believes this is a pivotal event in laser biology. His work involved human clinical patients and they found that TGF-beta is directly involved in the laser biostimulatory process in vivo. Interestingly, hydrogen peroxide has also been implicated directly by Dr. Mary Helen Barcellos-Hoff's lab at Berkley. It was involved in the activation process of TGF beta which Praveen's group found is directly affected by low power laser irradiation; they therefore believe it is the key downstream "biological effector. " As this work involved multidisciplinary approaches involving clinical patients, cell and molecular biology, laser biophysics among other aspects, Praveen faced stiff challenges, because of the natural boundaries of expertise, to getting his work reviewed and published in the mainstream biology research journals.

Optical Coherence Tomography Imaging for Evaluating the Photobiomodulation Effects on Tissue Regeneration in Periodontal Tissue

Craig B. Gimbel

Abstract Optical Coherence Tomography (OCT) is a noninvasive method for imaging dental microstructure which has the potential of evaluating the health of periodontal tissue. OCT provides an "optical biopsy" of tissue 2–3 mm in depth. This sixth modality of imaging was pioneered at Lawrence Livermore National Laboratory. OCT is based on the optical scattering signatures within tissue structure. With the use of a broad spectrum bandwidth light source, high resolution images, up to 10 times the resolution of x-ray, can detect important tissue interfaces within the periodontal sulcus and its' relationship to the attachment apparatus of the tooth. Multiple cross-sectional tomograms can be stacked to create two and three dimensional images.

Optical Coherence Tomography has the potential to follow the progression of plaque-induced periodontal disorders that result in the degeneration of connective tissue attachment. This noninvasive imaging can be used by the clinician and researcher to evaluate the effects of photo bio modulation on soft tissue periodontal regeneration. X-rays cannot image soft tissue and therefore cannot identify early active disease or the reversal to health of the periodontal unit. Photo bio modulation effects on the periodontal tissue can be monitored because OCT records soft tissue contour and microstructure, quantifying the soft tissue changes that occur in real time. Soft tissue health is a major area in dentistry where OCT diagnostic imaging will make a significant impact.

Keywords: Optical Coherence Tomography (OCT), dentistry, periodontal tissue regeneration, real time diagnostic imaging, non invasive, cross sectional tomograms, optical biopsy, microstructure, high resolution.

C.B. Gimbel
Academy of Laser Dentistry, Coral Springs, Florida

R. Waynant and D.B. Tata (eds.), *Proceedings of Light-Activated Tissue Regeneration and Therapy Conference.*
© Springer Science + Business Media, LLC 2008

The history of medical imaging has spanned 121 years since the late 1800s with the development of the x-ray. This was followed by ultrasound in 1965 which emits high frequency sound waves to image internal biological structures. With the development of Computed Tomography (CT Scan) in 1973, images were created by combining x-ray and computer technology to capture thin slices of tissue. Magnetic Resonance Imaging (MRI) in 1983 used a magnetic field and pulses of radio wave energy to make images of primarily soft tissue. In 1999, Positron Emission Tomography (PET Scan) was developed to measure metabolic changes in soft tissue cells. X-ray has become the mainstay imaging in dental medicine over the years. But, there are severe limitations of this imaging modality and it is invasive, emitting ionizing radiation. It cannot image soft tissue and does not have the ability to image microstructure details. Optical Coherence Tomography has the potential to image periodontal tissue and oral mucosa to follow the tissue changes of health and disease, both quantitatively and qualitatively. This concept of using light and optics to perform imaging in biological tissues was first proposed by M. Duguay in 1971 [1, 9].

Optical Coherence Tomography (OCT) is a noninvasive method for imaging dental microstructure which has the potential of evaluating the health of periodontal tissue and mucosa. OCT provides an "optical biopsy" of tissue 2–3 mm in depth. It was first reported as an in vivo method for cross sectional imaging of the tissues of the eye by Huang et al. in 1991 [3–6]. The dental applications for this sixth modality of imaging was subsequently pioneered in 1991 by Dr. Linda Otis and colleagues at the University of California, San Francisco and at Lawrence Livermore National Laboratory [6–9]. OCT is based on the optical scattering signatures within tissue structure. With the use of a broad spectrum bandwidth light source, high resolution images, 10 times the resolution of x-ray, can detect important tissue interfaces within the periodontal sulcus and its' relationship to the attachment apparatus of the tooth, as well as images of the oral mucosal tissue. Image resolution of 1–15 um can be achieved. This is one to two orders of magnitude higher than conventional ultrasound. High–speed, real time imaging can be performed in situ and nondestructively. Multiple cross-sectional tomograms can be stacked to create two and three dimensional images. Optical Coherence tomography enables an optical biopsy, thru imaging of the tissue structure and its' pathology, on a resolution scale approaching that of histopathology. Being a noninvasive imaging modality, there is absolutely no need to excise specimens as in conventional biopsy and tissue sample processing.

Optical Coherence Tomography has the potential to follow the progression of plaque-induced periodontal disorders that result in the degeneration of the connective tissue attachment. This noninvasive imaging can be used by the clinician and researcher to evaluate the effects of photo bio modulation on soft tissue periodontal regeneration. X-rays cannot image soft tissue and therefore cannot identify early active disease or the reversal to health of the periodontal unit. Photo bio modulation effects on the periodontal tissue can be monitored because OCT records soft tissue contour and microstructure, quantifying the soft tissue changes that occur in real time. Soft and hard tissue boundaries, as well as visualization and measurement of soft tissue thickness, sulcular depth and length of the periodontal attachment around teeth are possible through the use of OCT [10]. Doppler OCT has the ability to obtain an in situ three-dimensional tomographic image and velocity profiles of blood profusion in tissue at discrete spatial locations in superficial and deeper layers of soft tissue. Vascular perfusion is related to health of the tissue. Soft tissue health is a major area in dentistry where OCT diagnostic imaging will make a significant impact.

OCT is a fiber-optically based technology which can easily be integrated with a wide range of delivery systems and imaging probes. Image information is generated in an electronic form that can be processed and analyzed by computer software for direct imaging as well as transmission, storage and retrieval. The OCT system can be engineered to be compact in form with a small footprint and low cost for clinical and research utilization. OCT is analogous to ultrasound B mode imaging The only exception is that it uses light instead of sound. The images are generated by performing axial measurements of the echo time delay and magnitude of reflected light at different transverse positions. OCT images are a two or three dimensional data set representing differences in optical back reflection in a cross sectional volume of tissue. A comparison of ultrasound and OCT is in order. In ultrasound, high frequency sound waves are launched into the tissue using an ultrasonic transducer. These sound waves pass into the tissue and are reflected from the internal structures which have various acoustic properties. The time behavior developed by the reflected sound waves or echoes are detected by the probe which determines the dimensions of the internal tissue structure. OCT differs in that measurements of distance and microstructure are performed by directing a beam of light via an optical fiber probe onto the tissue in order to measure the back reflecting light from the internal tissue microstructures. This light is a continuous wave, short coherence length light in the near infrared region of the photonic electromagnetic spectrum. The velocity of light is approximately a million times faster than the velocity of sound. Distances within the tissue are determined by measuring the "echo" time delay of back reflected light waves. Therefore distance measurement requires ultra fast time resolution. A direct electronic detection is not possible on this time scale and therefore measurement of this "echo" time delay is based on a correlation technique that compares the back reflected light signal (Tissue Arm) to a reference light traveling a known path length (Reference Arm). A big advantage of OCT is that direct contact of the probe with tissue, as in ultrasound sound waves, is not required for transmission. Light requires no physical contact to the imaged tissue. The OCT image can be obtained in a field of water, saliva or blood contamination.

Optical coherence tomography uses Michelson type interferometric detection and correlation to measure the back scattered light time delay. Sir Isaac Newton first described this classical optical measurement known as low coherence interferometry. This concept was first used to measure eye length [2]. Michelson low coherence interferometry measures the field of the optical beam rather than the intensity. An incident optical light wave from a superluminescent diode (SLED) light source enters the interferometer which contains a partially reflecting mirror or beamsplitter, which yields two beams, one of which acts as the reference beam or arm and the other as a measurement or signal beam known as the sample arm. The two beams travel the same given distance in both arms of the interferometer. The reference arm light beam is reflected from the reference mirror and the sample arm beam is reflected from the tissue sample being imaged. The two beams then recombine back and interfere at the beamsplitter. The interferometer output is the sum of the two fields, the reference mirror and the signal beam reflected from the tissue being imaged. A detector measures the intensity of the optical beam output that is the square of the electromagnetic field. The position of the reference mirror is varied so that the path lengths of the optical beam traveling in the reference arm changes. An interference effect will be observed in the intensity detected by the detector if the path length is changed by scanning the reference mirror. Back reflected light from the tissue sample is correlated with light that travels a known reference path delay. Because the reference mirror is moved in known increments, the exact position of the reflected light within the tissue can be determined. The optical scattering properties of the tissues determine the magnitude of the reflected signal. The sample arm optical fiber is scanned across the tissue surface to obtain the two or three dimensional image.

Cross sectional imaging in OCT is achieved by performing successive axial measurements of back reflected light at different transverse positions [4]. High speed successive axial measurements of the back reflection creates a two dimensional data set which represents the intensity of reflection of the optical light beam as a function of depth in the tissue. The image data is acquired by computer software and displayed as a two dimensional gray scale or false color on the monitor for viewing. The vertical direction corresponds to the direction of the incident beam and the axial data sets. In a highly scattering tissue, light is rapidly attenuated with propagation depth resulting in a gradation of signal in the image. The imaging depth is not strongly limited when OCT is performed in very weakly scattering media. Transparent tissues have very low back scattering. Image contrast is a result of differences in the back scattering properties of tissues. Most biological tissues are highly scattering in nature.

The modern OCT system consists of a fiber optic Michelson type interferometer with a low coherence light source coupled into it and the interference detected by a photodiode at the output end (Figs. 1 and 2). The sample arm of the interferometer emits a light beam that is directed on the sample tissue being imaged and the other reference arm has a scanning delay.

Optical Coherence Tomography can resolve changes in architecture morphology in tissue that is associated with health and pathologic changes. Studies have been performed *in vitro* to correlate OCT imaging with histology for pathologies of the gastrointestinal, biliary, female reproductive, pulmonary and urinary tracts [11–21].

Feldchtein, F.I., Gelikonov,V.M. et al. [22] have imaged and demonstrated *in vivo* and *in vitro* structural changes of oral mucosal tissue. Masticatory (gingival and hard palate), lining (alveolar, soft palate, labial, buccal, floor of mouth and ventral surface of tongue) and specialized mucosa (lips, dorsum of tongue) have been imaged and compared to histological excisional biopsy tissue sections. Images of masticatory mucosa demonstrated orthokeratinized squamous epithelium, projecting into the overlying epithelium, lamina propria, and bundles of collagen fibers

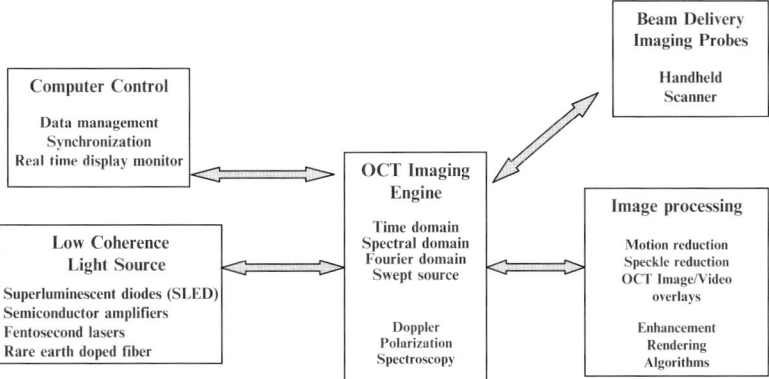

Fig. 1 Typical OCT system

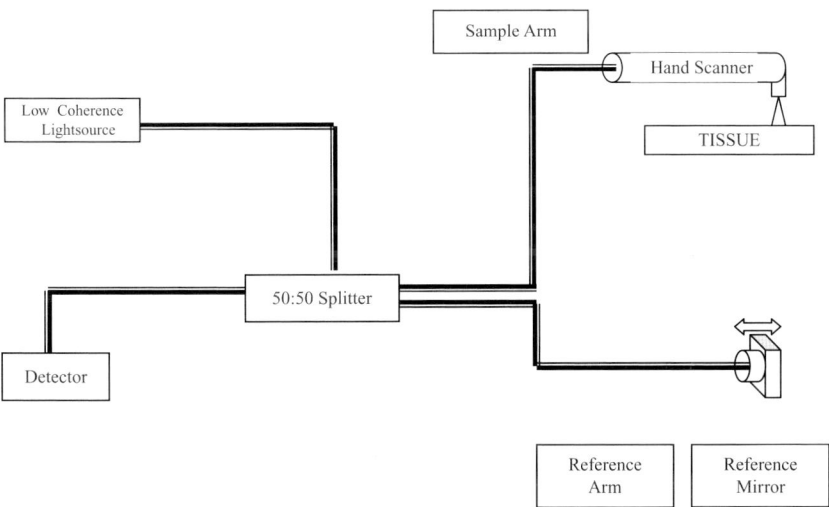

Fig. 2 Michelson Interferometer

interlaced into the periosteum. Gingival mucosa images display keratinized strati-
fied squamous epithelium. Lining mucosa of the soft palate had an absence of a
corneous layer and the presence of a submucosa along with fatty elements, salivary
ducts and muscular fibers. Buccal mucosa displayed the various zones: zona
maxilaris, zona mandibularis and zona intermedia. Images of specialized mucosa
of the lingual dorsum of the tongue had no submucosa, being directly attached to
the muscles of the tongue. The various types of lingual papillae were evident in the
images.

Optical Coherence Tomography can differentiate soft and hard tissue boundaries
of the periodontal unit. It is possible to visualize and measure soft tissue thickness,

sulcular depth and length of the collagen fiber attachment apparatus that surrounds teeth and the shape and contour of the alveolar crest [23]. *In vivo* studies have correlated clinical periodontal probing depths as measured with OCT to those of a Michigan 0 Probe. The OCT images included important anatomical features including soft tissue surface contour and thickness, gingival crest, sulcus and DEJ. Similarly, those sulcular depths probed *in vitro* on porcine jaws and then photomicrographed, had reference points corresponding to OCT images taken of the same areas. The dentinenamel junction (DEJ), cementoenamel junction (CEJ), and sulcular depths strongly correlated in both OCT and photomicrographs [23].

Implant health can be clinically determined by OCT because it provides high-resolution, cross sectional images of the soft tissue that surrounds the implant. Animal histological studies demonstrate a scar-like fibrous connective tissue formed by the gingival tissue adjacent to titanium implant interfaces. A disorganized connective tissue surface containing more vascular tissue characterizes a disease state of the implant integration called peri-implantitis. OCT imaging of healthy implant integration sites demonstrates an organized alignment of the collagen fibers, producing a high OCT intensity signal due to its' birefringent nature. Failing implants imaging are characterized by linear signal deficits, low intensity collagen signals and increased vasculature [23].

Optical Coherence Tomography imaging information can further be analyzed and utilized in various different ways when using it with other detection schemes. Doppler Flowmetry, when combined with OCT, produces images that detect and measure blood flow rates in superficial labial vessels [24–28]. In situ three dimensional tomographic image and velocity profiles of blood perfusion in tissue at discrete spatial locations can be obtained. This can be utilized to examine tissue viability, regeneration and health. Monitoring of photo biomodulation therapy over time may be possible through the use of this technique. Polarization-sensitive OCT (PS-OCT) uses the information carried by the transverse polarization character of light to extract extra information from the tissue sample. PS-OCT provides high resolution spatial information based on the polarization state of light reflected from tissue. PS-OCT can be used to examine the macroscopic orientation of the collagen fibers in gingival connective tissue and attachment apparatus of the tooth due to its' birefringence. Two types of birefringence exist in tissue. A form birefringence results from ordered linear structures which are exhibited by both soft and hard tissue. Intrinsic birefringence is due to the molecular form arrangement within the tissue. These tissue types include tendons, muscle, nerve, bone, cartilage, fibrous tissue and teeth. Enamel and dentin exhibit this structural anisotropy because of the orientation of enamel rods and dentinal tubules that are linearly oriented. Greater detail of the structural features can be obtained by combining the polarization sensitive system vertical and horizontal polarized signals. Potential biological applications of PS-OCT are just beginning to be explored. It provides additional contrast that can be used to image and identify structural components of tissue. Certain functional information within tissue is associated with transient changes in birefringence. Monitoring of photo biomodulation tissue effects are possible through the use of PS-OCT.

Ultrahigh resolution cellular level OCT imaging is possible through the use of short pulse, fentosecond solid state laser light sources. The ability to image sub cellular structure is an important tool for studying mitotic activity and cell migration during development as well as an aid in diagnosis of early neoplasia. Standard OCT image axial resolutions of 10–15 um enable imaging of early architectural morphological changes. The quantitative and qualitative information is unparalleled and the high pixel density image characterization extends diagnostic capabilities in real-time.

Optical Coherence Tomography has the ability to image, on a micron scale, the internal microstructure of tissue *in situ* and in real time. It can function as a noninvasive optical biopsy. The impact on the diagnosis and monitoring of soft tissue disease and health will be immense. This will allow for noninvasive monitoring of the photo biomodulation effects on regeneration of periodontal tissue. Because OCT may be capable of resolving subtle scattering differences and increased blood flow caused by inflammation of soft tissue, periodontal disease will be monitored back to health. OCT will impact both clinical treatment and research.

References

1. Duguay, MA. Light photographed in flight. *Am. Sci.*, 1971; 59, 551.
2. Duguay, MA and Mattick, AT. Ultrahigh speed photography of picosecond light pulses and echoes. *Appl. Optics*, 1971; 10, 2162.
3. Huang, D, Swanson, EA, Lin, CP, Schuman, JS, Stinson, WG, Chang, W, Hee, MR, Flotto, T, Gregory, K, Puliafito, CA, Fujimoto, JG. Optical coherence tomography. *Science*, 1991; 25495035, 1178–1181.
4. Huang, D, Swanson, EA, Lin, CP, Fujimoto, JG, et al. Optical coherence tomography. *Science*, 1991; 254, 1178.
5. Brezinski, ME and Fujimoto, JG. Optical coherence tomography: high resolution imaging in non transparent tissue. *IEEE J. Sel. Top. Quantum Electron*, 1999; 5, 1185.
6. Fujimoto, JG, Pritis, C, Boppart, SA, and Brejinski, ME. Optical coherence tomography: an emerging technology for biomedical imaging and optical biopsy. *Neoplasia*, 2000; 2, 9.
7. Coleston, BW Jr, Sathyam, US, DaSilva, LB, Everett, MJ, Stroeve, P, and Otis, LL. Dental OCT. *Opt. Express*, 1998; 3(6), 230–238.
8. Otis, LL, Colston, BW, Armitage, G, and Everett, M. Optical imaging of periodontal tissues. *J. Dent. Res.*, 1997; 76(Special issue), 383, abstract 2956.
9. Colston, BW, Everett, MJ, DaSilva, LB, Otis, LL, Stroeve, P, and Nathel, H. Imaging of hard and soft tissue structure in the oral cavity by optical coherence tomography. *Appl. Optics*, 1998; 37(16), 3582–3585.
10. Otis, LL. Dental optical coherence tomography past and future. *L. Laser Dent.*, 2007, 15(1), 14–19.
11. Park, H, Chodorow, M, and Kompfner, R. High resolution optical ranging system. *Appl. Optics*, 1981; 20, 2389.
12. Fujimoto, JG, Desilvestri, S, Ippen, EP, Puliafito, CA, Margolis, R, and Oseroff, A. Fentosecond optical ranging in biological systems. *Opt. Lett.*, 1986; 11, 150.
13. Youngquist, RC, Carr, S, and Davies, DEN. Optical coherence-domain reflectometry: a new optical evaluation technique. *Opt. Lett.*, 1987; 12, 158.

14. Takada, K, Yokohama, I, Chida, K, and Noda, J. New measurement system for fault location in optical waveguide devices based on an interferometric technique. *Appl. Optics*, 1987; 26, 1603.
15. Giligen, HH, Novak, RP, Salathe, RP, Hodel, W, and Beaud, P. Submillimeter optical reflectometry. *IEEE J. Lightwave Technol.*, 1989; 7, 1225.
16. Fercher, AF, Mengedoht, K, and Werner. W. Eye-length measurement by interferometry with partially coherent light. *Opt. Lett.*, 1988; 13, 1867.
17. Clivaz, X, Marquis-Weible, F, Salathe, RP, Novak, RP, and Gillgen, HH. High-resolution reflectometry in biological tissues. *Opt. Lett.*,1992; 17, 4.
18. Schmitt, JM, Knuttel, A, and Bonner, RF. Measurement of optical-properties of biological tissues by low-coherence by reflectometry. *Appl. Optics*, 1993; 32, 6032.
19. Huang, D, Wang, J, Lin, CP, Puliafito, CA, and Fulimoto, JG. Micron- ranging of cornea and anterior chamber by optical reflectometry. *Laser Surg. Med.*, 1991; 11, 419.
20. Swanson, EA, Izatt, JA, Hee, MR, Huang, D, Lin, CP, Schuman, JS, Puliafito, CA, and Fujimoto, JG. In vivo retinal imaging by optical coherence tomography. *Opt. Lett.*, 1993; 18, 1864.
21. Hee, MR, Izatt, JA, Swanson, EA, Huang, D, Lin, CP, Schuman, JS, Puliafito, CA, and Fujimoto, JG. Optical coherence tomography of the human retina. *Arch. Opthalmol.*, 1995; 113, 325.
22. Feldchtein, FI, Gelikonov, GV, Gelikonov, VM, Iksanov, RR, Kuranov, RV, and Sergeev, AM. In vivo imaging of hard and soft tissue of the oral cavity. *Opt. Express*, 1998; 3(6), 239–250.
23. Otis, LL. Dental optical coherence tomography past and future. *J. Laser Dent.*, 2007; 15(1), 14–19.
24. Otis, LL, Piao, D, Gibson, CW, and Zhu, Q. Quantifying labial blood flow using optical Doppler tomography. *Oral Surg. Oral Med. Oral Pathol. Oral Radiol. Endod.*, 2004; 98(2), 189–194.
25. Piao, D, Otis, LL, and Zhu, Q. Doppler angle and flow velocity mapping by Combined Doppler shift and Doppler bandwidth measurements in optical Doppler tomography. *Opt. Lett.*, 2003; 28(13), 1120–1122.
26. Otis, LL and Zhu Q. Optical Doppler tomography (ODT) and blood flow. *J. Dent. Res.*, 2002; 81(Special issue A) abstract 09789 (www.dentalresearch.org)
27. Piao, D, Otis, LL, Dutta, NK, and Zhu, Q. Quantitative assessment of flow velocity-estimation algorithms for optical Doppler tomography imaging. *Appl. Optics*, 2002; 41(29), 6118–6127.
28. Chen, Y, Otis, LL, Piao, D, and Zhu, Q. Characterization of dentin, enamel, and carious lesions by a polarization-sensitive optical coherence tomography system. *Appl. Optics*, 2005; 44(11), 2041–2048.

Photobiomodulation Laser Strategies in Periodontal Therapy

Akira Aoki, Aristeo Atsushi Takasaki, Amir Pourzarandian, Koji Mizutani, Senarath M.P.M. Ruwanpura, Kengo Iwasaki, Kazuyuki Noguchi, Shigeru Oda, Hisashi Watanabe, Isao Ishikawa, and Yuichi Izumi

Abstract Laser is considered basically effective for treating periodontal diseases because of its excellent physical properties namely ablation, hemostasis, bacterial killing and cell stimulation. The current laser application mainly used for the treatment of periodontitis is high-power laser ablation (HLLT). Laser ablation of diseased periodontal tissues using the HLLT is widely performed, partly expecting a simultaneous photo-bio-modulation effect (LLLT) in the surrounding tissues. In periodontal pocket therapy, laser can not only ablate the diseased tissues but also stimulate or activate the surrounding gingival and bone tissues, which would result in improved pocket healing and tissue regeneration. By elucidating the photo-bio-modulation effect in detail, this effect could be used more effectively and laser therapy would be more advantageous in non-surgical and surgical therapies of periodontitis as an adjunctive or alternative means to current mechanical treatment. As a future strategy of periodontal therapy, the photo-therapy using photo-bio-modulation/activation and photo-dynamic effects could be developed increasingly for prevention and control of periodontal diseases.

Keywords: Photo-bio-modulation, periodontal diseases, periodontitis, periodontal pocket, bone, Er:YAG laser.

Y. Izumi
Section of Periodontology, Department of Hard Tissue Engineering, Graduate School, Tokyo Medical and Dental University (TMDU), 1-5-45 Yushima Bunkyo-ku, Tokyo 113-8549, Japan, e-mail: aoperi@tmd.ac.jp

R. Waynant and D.B. Tata (eds.), *Proceedings of Light-Activated Tissue Regeneration and Therapy Conference.*
© Springer Science + Business Media, LLC 2008

Introduction

The most common form of periodontal disease is a chronic, inflammatory and infectious disease, leading to episodic and progressive loss of periodontal tissues including gingival tissue, periodontal ligament and bone tissue. Recently, a variety of lasers have been used for a broad range of oral-facial conditions including periodontal therapy [1, 2]. The use of lasers is considered safe and effective for treating such inflammatory and infectious diseases as periodontitis. Lasers have numerous physical properties that can effect a broad range of biological responses that are suitable for treating a variety of periodontal conditions, such as ablation, hemostasis, microbial inhibition and destruction, cell stimulation, as well as modulation of metabolic activity.

High-power lasers were first used successfully as a variation of conventional approaches for soft tissue treatment such as gingivectomy and gingivoplasty in the clinic. Recently, for example, an Er:YAG laser was developed which can be used on both dental soft and hard tissues due to its low thermal side-effects. Consequently, the Er:YAG laser has been used to treat gingiva, tooth roots, and bone tissue, thus becoming one of the more promising laser units for periodontal treatment [1, 3].

Current Laser Strategy

The main laser application strategy currently used for the treatment of periodontal diseases is high-power laser ablation or high-level laser treatment (HLLT). Laser ablation of diseased tissues is widely performed, partly expecting a simultaneous photo-bio-modulation (PBM) effect (LLLT) in the surrounding tissues [7].

In periodontal pocket therapy, laser devices can not only ablate the diseased tissues and decontaminate and detoxify the pockets and root surfaces but also stimulate or activate the surrounding gingival and bone tissues. If properly used, this would result in improved pocket healing with soft and bone tissues regeneration by reduction of inflammatory condition and promotion of cell proliferation and differentiation (Fig. 1) [1].

Such additional PBM effects during HLLT would be not so strong but the effects are also another advantageous property of laser pocket treatment and would provide a great therapeutic benefit producing improved clinical outcomes. Some researchers and clinicians have recently recognized and realized those PBM effects in the laser pocket treatment to some extent and have been using lasers intentionally expecting those effects. Interestingly, they have experienced improved pocket healing and increased bone regeneration following laser treatment. However, clinical studies concerning the PBM effects in laser pocket treatment have not been clearly proposed and demonstrated so far. Although laser pocket treatment has been increasingly reported in the non-surgical or closed periodontal pocket therapy, most researchers have not sufficiently noticed and understood the PBM effects during laser pocket treatment and scientific publications showing positive results of PBM are delayed and still insufficient.

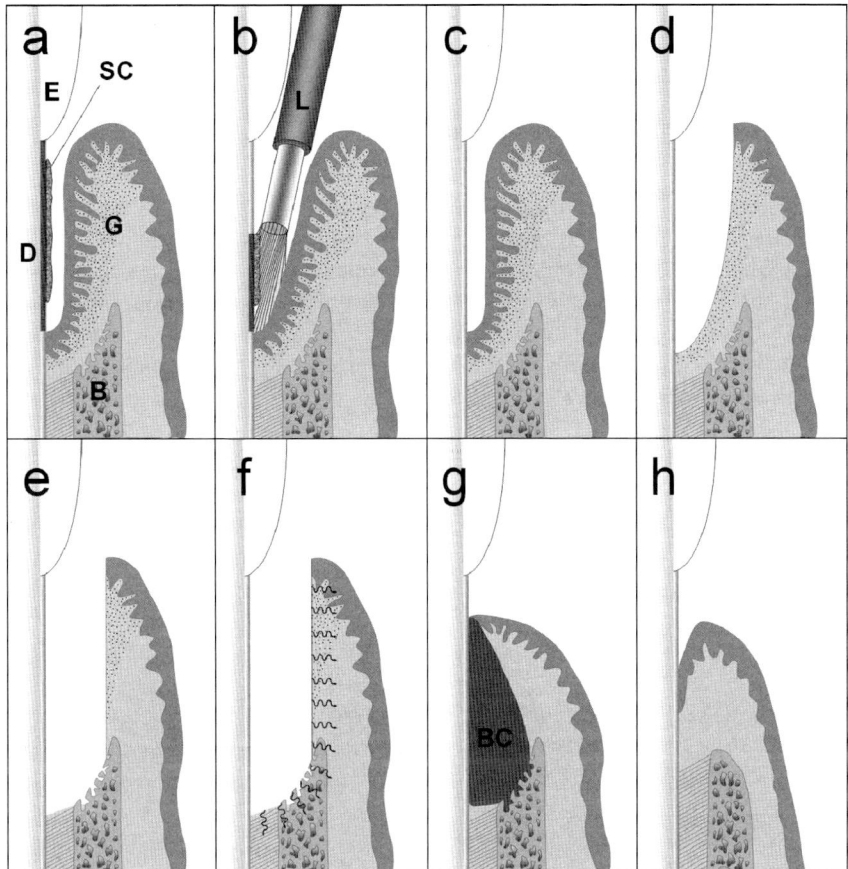

Fig. 1 Schematic illustration of the procedures of laser pocket therapy. Laser can not only ablate the diseased tissue by HLLT but also exerts photo-bio-modulation (PBM) effect (LLLT) to the surrounding tissues. (**a**) Advanced periodontal pocket showing gingival tissue detachment, sub-gingival calculus deposition and contamination of the tooth root surface, epithelial down-growth and lining of the inner surface of gingival connective tissue with inflammation, vertical bone resorption and diseased connective tissue formation in the defect. (**b**) Laser ablation of the deposited calculus and decontamination and detoxification of the root surface. (**c, d**) Ablation of lining epithelium. (**e**) Ablation of diseased connective tissue on the inner surface of the gingival tissue and in the vertical bone defect. (**f**) Simultaneous PBM effects stimulating or activating the surrounding gingival and bone tissues, which would result in improved pocket healing with soft and bone tissue regeneration by reduction of inflammatory condition and promotion of cell proliferation and differentiation. (**g**) Blood clot (BC) formation in the pocket and defect. (**h**) Favorable pocket healing with gingival connective tissue attachment and bone tissue regeneration. E: enamel of tooth crown, D: dentin of tooth root, SC: subgingival calculus, B: alveolar bone, G: gingival tissue, L: laser tip

Potential Applications of PBM for Wound Healing and Tissue Regeneration in Periodontal Therapy

Promotion of New Bone Formation

In our previous studies, we observed increased new bone formation in dogs after surgical periodontal and peri-implant treatment using a high energy-level Er:YAG laser. Mizutani et al. [5] compared the periodontal tissue healing following periodontal flap surgery using an Er:YAG laser with that of conventional mechanical curette surgery. In six dogs, bilateral premolars with experimentally-induced periodontitis were treated by periodontal surgical procedure. Degranulation and root debridement were effectively performed with the Er:YAG laser irradiation without major thermal damage. At 3 months post surgery, interestingly, the amount of newly-formed bone was significantly greater in the laser group than in the curette group in the histological analysis (Fig. 2). This study showed that the Er:YAG laser irradiation has the potential to promote new bone formation.

Also, Takasaki et al. [13] evaluated the utility of the application of an Er:YAG laser for the surgical treatment of peri-implant infection. In four dogs, the peri-implant surgery was performed using an Er:YAG laser or a plastic curette for

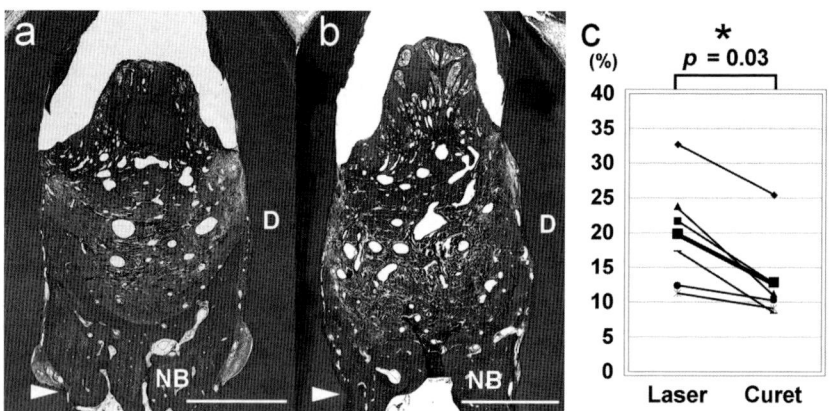

Fig. 2 Histological photomicrographs of mesio-distal sections of furcation at 12 weeks after periodontal flap surgery using the Er:YAG laser (**a**) and curette (**b**), and the histometric analysis of the ratio of newly-formed bone (NB) area (**c**). In both the laser and curette sites, periodontal soft tissue attachment with some degree of bone regeneration was noted in the furcation area. The NB was coronally extended along the dental root surface (D) in the defect above the notch (arrow heads). Note the greater new bone formation in the laser-treated site (**a**) than the curette-treated site (**b**) (bar = 800 μm, original magnification 27×). In the histometric analysis (**c**), all measurement data obtained in square millimeters were converted to a percentage relative to the area of each original defect. The thick line shows the mean in the graph. (*$P < 0.05$; Wilcoxon signed-rank test, n = 6) (Photographs and figure from [5]. With kind permission. Lasers in Surgery and Medicine © copyright (2006) Wiley)

degranulation and implant surface debridement. After 6 months, histologically, the laser-treated implant surface did not inhibit new bone formation but rather the laser group showed a tendency to produce greater new bone-to-implant contact than the curette group (Fig. 3). The results indicated that the Er:YAG laser therapy has a potential to induce favorable bone healing in the surgical treatment of peri-implantitis.

Thus, both studies demonstrated the increased or favorable new bone formation after laser treatment compared to mechanical treatment. There would be several reasons for the increased bone formation, but the improved bone regeneration may be partly due to the PBM of low-level laser which was scattered or penetrated during HLLT.

Regarding osteogenesis, previous several in vitro studies have suggested that low-level laser irradiation could promote new bone formation by inducing proliferation and differentiation of osteoblasts [8, 12]. It has been reported that low-level lasers increased the ALP activity [8] and mRNA expression of osteoblastic differentiation markers such as osteopontin [12], osteocalcin [8] and bone sialoprotein [12] in osteoblasts and promote bone nodule formation [8]. These PBM effects would be useful for periodontal regenerative therapy. Therefore, further basic and clinical studies are required for the establishment of a clinically effective and reliable procedure.

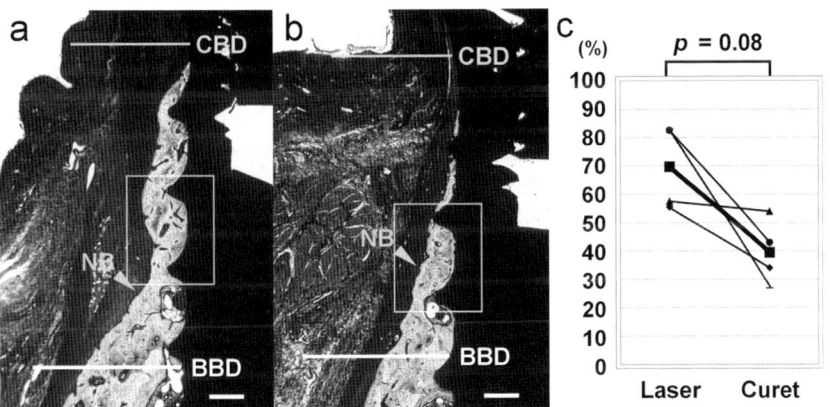

Fig. 3 Histological photomicrographs of buccolingual sections parallel to the long axis of the implant in the center of the dehiscence defect at 24 weeks following surgical therapy of peri-implant infection using an Er:YAG laser (**a**) or a plastic curette (**b**), and the histometric analysis of the ratio of new bone-to-implant contact (NBIC) (**c**). Both histological sections show the highest new bone (NB) formation in each group. In the laser group (**a**), the NB was more coronally-extended along and in direct contact to the implant surface from the bottom of the bone defect (BBD) than the curette group (**b**). CBD: Coronal level of original bone defect (bar = 500 μm; original magnification 30x). In the histometric analysis (**c**), all measurement data obtained in millimeters were converted to a percentage relative to the length of implant surface of the bone defect following debridement. The thick line shows the mean in the graph. (Student paired t-test, n = 4) (Photographs and figure from [13]. With kind permission of Springer Science and Business Media. Lasers in Medical Science © copyright (2007) Springer)

Promotion of Cell Proliferation

Low-level laser irradiation has been reported to enhance wound healing. Activation of gingival fibroblasts has a potential for early wound healing in periodontal treatment. Our group demonstrated that a low energy-level of Er:YAG laser exerts stimulatory effects on cell proliferation of human gingival fibroblasts (HGFs).

First, Pourzarandian et al. [9] investigated the effect of low-level Er:YAG laser irradiation on HGF proliferation. Cultured HGFs were exposed to low-level, pulsed Er:YAG laser irradiation with various energy densities ranging from 1.68 to 5.0 J/cm^2 and 20 Hz. The cultured fibroblasts after irradiation were analyzed by means of light microscopy and transmission electron microscopy (TEM). As a result, light microscopy revealed that the number of cells increased and the shape of the cells was irregular and more mature than that of control (Fig. 4a, b). In TEM observation, the fibroblasts seemed to be metabolically active. A comparison of cell growth at day 1 and day 3 after treatment showed a significant increase in the number of cells in the Er:YAG laser irradiation groups of 1.68, 2.35 and 3.37 J/cm^2 (Fig. 4c). The comparison of cell death determined by the level of lactate dehydrogenase (LDH) between laser-treated and untreated control cultures showed no significant differences after laser irradiation at 1.68–3.37 J/cm^2. However, an energy density at 5.0 J/cm^2 showed a significant increase of the LDH level and decreased the cell number after 3 days. The results showed that the low-level Er:YAG laser irradiation stimulates the proliferation of cultured HGFs and suggests that the low-level Er:YAG laser irradiation may be of therapeutic benefit for wound healing.

Secondly, regarding the mechanism of increased cell proliferation, we focused on prostaglandin E$_2$ (PGE$_2$) production after laser irradiation. PGE$_2$ is one of the important early mediators in the natural healing process and it regulates cell proliferation through interaction with its specific receptors and modification of the levels of second messengers such as calcium and cAMP [11]. PGE$_2$ production is induced in HGFs via de novo synthesis of cyclooxygenase-2 (COX-2), which is an inducible PG synthase, in response to proinflammatory stimuli [6].

Pourzarandian et al. [10] investigated the change of PGE$_2$ production and COX-2 gene expression in HGF after Er:YAG laser irradiation in vitro. Cultured HGFs were exposed to low-level Er:YAG laser irradiation with an energy density of 1.68–3.37 J/cm^2. The levels of PGE$_2$ production were measured by enzyme-linked immunosorbent assay. Total RNA was extracted and COX-2 mRNA expression was analyzed by reverse transcriptase—polymerase chain reaction (RT-PCR). The Er:YAG laser significantly increased PGE$_2$ production in a laser energy-dependent manner (Fig. 5a). COX-2 mRNA expression was also enhanced with an increase in energy level (Fig. 5b). Indomethacin, a non-specific COX-1/COX-2 inhibitor, and NS398, a specific COX-2 inhibitor, completely inhibited the PGE$_2$ synthesis stimulated by Er:YAG laser irradiation. These results showed that the Er:YAG laser irradiation appears to exert its stimulative action on HGF proliferation through the production of PGE$_2$ via the expression of COX-2. This could be considered as one of the important regulatory pathways that enhance cell proliferation for tissue regeneration.

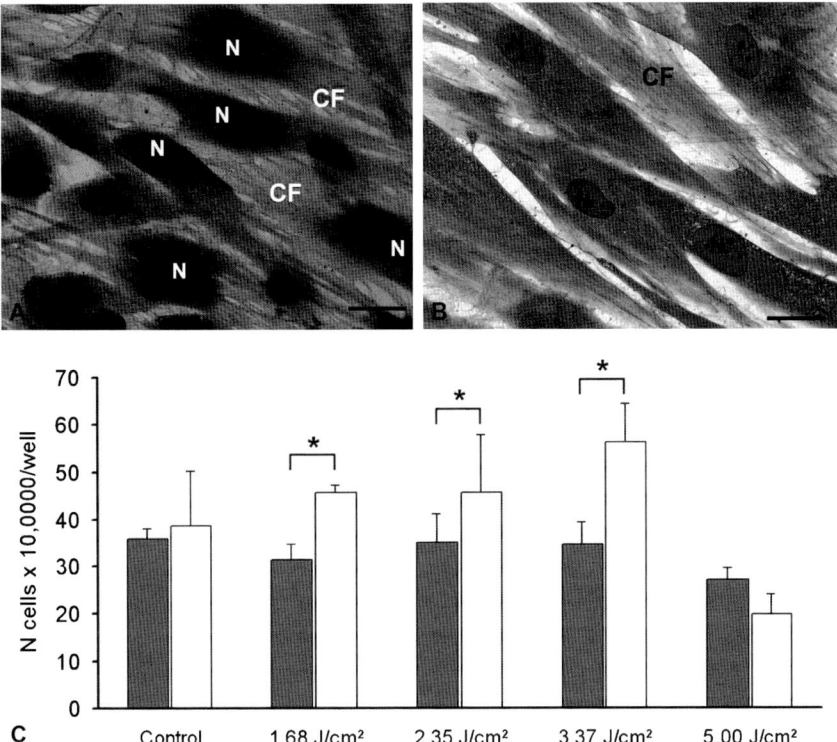

Fig. 4 Light micrograph of cultured human gingival fibroblasts (HGF) 3 days after Er:YAG laser treatment with 3.37 J/cm^2 energy density (**a**) and cultured non-lased control HGFs 3 days after (**b**), and a graph showing the HGF cell numbers after laser treatment using various energy levels (**c**). In the HGFs treated by Er:YAG laser (**a**), the number of cells seems to have increased and the cells were surrounded by a network of collagen fiber (CF). In the control HGF (**b**), cells are elongated and parallel to each other, surrounded by collagen fibers (CF). N: nucleus (original magnification 1,000x; bar = 10 μm). In the graph (**c**), each bar represents the mean ± SD of cell numbers. Black column: average of cell numbers 24 hours after treatment. White column: average of cell numbers 3 days after treatment. *Significant difference ($P < 0.001$, Fisher's test) (Photographs and figure from [9]. With kind permission. Journal of Periodontology © copyright (2004) the American Academy of Periodontology)

Photo-Dynamic Therapy to Induce Bacterial Death

Recently, application of the bactericidal property of photo-dynamic therapy (PDT) has been considered as a novel treatment modality to control the bacterial infection in the field of periodontal therapy. Several studies demonstrated the high bactericidal effect of PDT and suggested that it may be a valuable alternative [4]. PDT is based on the principle that a photoactivatable substance, the photosensitizer, binds to the target cell and can be activated by light of a suitable wavelength. During this process, free radicals are formed, which then produce an effect that is toxic to the bacteria.

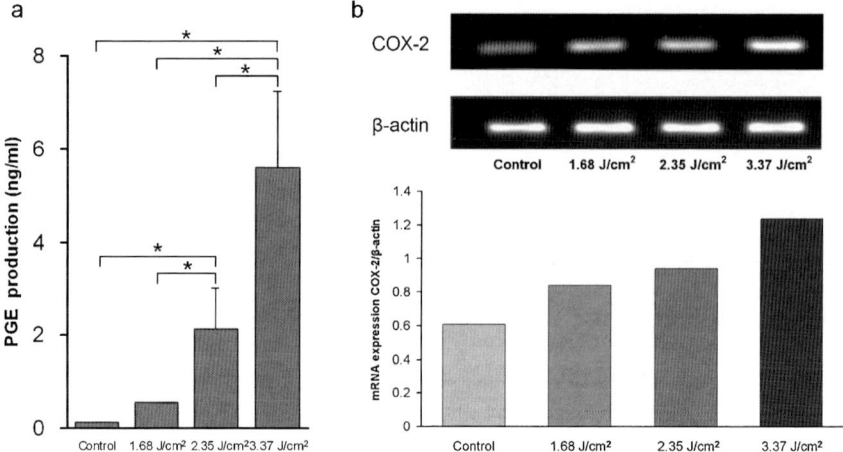

Fig. 5 Prostaglandin E$_2$ (PGE$_2$) release from HGFs irradiated with low-level Er:YAG laser **(a)** and cyclooxygenase-2 (COX-2) mRNA expression in HGFs **(b)**. In graph **(a)**, each bar represents the mean PGE$_2$ ± SD. Laser irradiation stimulated PGE$_2$ production in a laser-energy dependent manner. *Significant difference ($P < 0.05$) (Fisher's test). In graph **(b)**, total RNA was extracted and COX-2 mRNA expression was analyzed by RT-PCR. The COX-2 mRNA was highly induced in HGF cells irradiated by Er:YAG laser in a laser-energy dependent manner (Figures from [10]. With kind permission. Journal of Periodontal Research © copyright (2005) Blackwell Munksgaard)

Regarding the bactericidal effect of low-level laser or light without use of photo-sensitizer, we have already found that both argon laser (457–502 nm: peak 486 nm, 150 and 250 mW) and halogen light (400–500 nm: peak 492 nm, 450 mW) had a sufficient killing effect on periodontopathic bacteria, *Porphyromonas gingivalis* (unpublished data). In this study, the 250 mW argon laser showed the highest bactericidal effect with approximately 30%, 77%, 90%, and 99% of the bacterial death after 1, 3, 5 and 10 minutes of exposure, respectively. Although there was a moderate temperature elevation according to the increase of the irradiation time and therefore the bactericidal effect would be partly due to the heat effect, it appeared that PDT or photo-therapy might have promise as a novel method of eliminating bacterial infection.

PBM Laser Strategies in Periodontal Therapy

It is considered that lasers would help the phase of the tissues and cells in the diseased site change rapidly from the inflammatory and destructive state into that of healing and regeneration by modulating or activating cell metabolism. By elucidating the various effects of PBM in detail, these effects could be used more effectively and laser therapy would be more advantageous in non-surgical and surgical therapies of periodontitis as an adjunctive or alternative means to current mechanical and chemical regeneration procedures.

Furthermore, as a future strategy of laser therapy, the photo-therapy can be developed increasingly for prevention and control of periodontal diseases. In the initial stages of periodontal disease, elimination of periodontopathic bacteria and reduction of the inflammatory condition might be possibly controlled more effectively using the periodical external laser/light irradiation from the gingival surface than the conventional mechanical means alone. In order to maintain periodontal hygiene and to prevent infection and reinfection, constant effort to remove bacteria is indispensable but teeth and periodontal tissues form a very complicated structure and thus conventional mechanical means cannot achieve a complete bacterial elimination. For example, as a patient's home care, a photo-toothbrush, if necessary in combination with a tooth paste including a photo-sensitizer, might be developed for the control of infection and inflammation of the periodontal tissues. Also, in case of moderate and advanced stages of periodontal disease, internal irradiation via the periodontal pocket as well as external irradiation would be useful for enhancement of soft and hard tissue regeneration in combination with or without current regeneration procedures.

In conclusion, photo-therapy using photo-bio-modulation/activation and photodynamic effects should be studied more extensively for the treatment of periodontal diseases.

References

1. Aoki A, Sasaki K, Watanabe H, Ishikawa I (2004) Lasers in non-surgical periodontal therapy. Periodontology 2000 36: 59–97
2. Cobb CM (2006) Lasers in periodontics: a review of the literature. J Periodontol 77: 545–564
3. Ishikawa I, Aoki A, Takasaki A (2004) Potential applications of erbium: YAG laser in periodontics. J Periodont Res 39: 275–285
4. Meisel P, Kocher T (2005) Photodynamic therapy for periodontal diseases: state of the art. J Photochem Photobiol B 79: 159–170
5. Mizutani K, Aoki A, Takasaki A, Kinoshita A, Hayashi C, Oda S, Ishikawa I (2006) Periodontal tissue healing following flap surgery using an Er:YAG laser in dogs. Laser Surg Med 38: 314–324
6. Noguchi K, Iwasaki K, Shitashige M, Endo H, Kondo H, Ishikawa I (2000) Cyclooxygenase-2-dependent prostaglandin E_2 down-regulates intercellular adhesion molecule-1 expression via EP_2/EP_4 receptors in interleukin- 1ß-stimulated human gingival fibroblasts. J Dent Res 79: 1955–1961
7. Ohshiro T, Calderhead RG (1991) Development of low reactive-level laser therapy and its present status. J Clin Laser Med Surg 9: 267–275
8. Ozawa Y, Shimizu N, Kariya G, Abiko Y (1998) Low-energy laser irradiation stimulates bone nodule formation at early stages of cell culture in rat calvarial cells. Bone 22: 347–354
9. Pourzarandian A, Watanabe H, Ruwanpura SM, Aoki A, Ishikawa I (2005) Effect of low-level Er:YAG laser irradiation on cultured human gingival fibroblasts. J Periodontol 76: 187–193

10. Pourzarandian A, Watanabe H, Ruwanpura SM, Aoki A, Noguchi K, Ishikawa I (2005) Er:YAG laser irradiation increases prostaglandin E_2 production via the induction of cyclooxygenase-2 mRNA in human gingival fibroblasts. J Periodontal Res 40(2): 182–186

11. Sanchez T, Moreno JJ (2002) Role of EP_1 and EP_4 PGE_2 subtype receptors in serum-induced 3T6 fibroblast cycle progression and proliferation. Am J Physiol Cell Physiol 282: C280–288

12. Stein A, Benayahu D, Maltz L, Oron U (2005) Low-level laser irradiation promotes proliferation and differentiation of human osteoblasts in vitro. Photomed Laser Surg 23: 161–166

13. Takasaki A, Aoki A, Mizutani K, Kikuchi S, Oda S, Ishikawa I (2007) Er:YAG laser therapy for peri-implant infection: a histological study. Laser Med Sci 22: 143–157

Combined New Technologies to Improve Dental Implant Success and Quantitative Ultrasound Evaluation of NIR-LED Photobiomodulation

Jerry E. Bouquot, Peter R. Brawn, and John C. Cline

Abstract The importance of LLLT (photobiomodulation) to improve and maintain the health of dental implants is described.

Keywords: Dental implants, ultrasonic evaluation, photobiomodulation, LED.

Dental implants have become extremely popular for tooth replacement [1]. After several failed designs, the current and relatively successful implant is a metal cylinder or tapered cylinder with surface screw-like grooves and with various surface coatings. Each is designed to replace a single tooth and it is common for numerous implants to be placed in one patient (Fig. 1). There is an obvious need to place the implant into healthy tissues [1–6]. The bone must reshape around the implant (osteointegration) to provide stability, without which the constant biting forces on the device will lead to increasing mobility and eventual failure. Likewise, since the implant extends into the bacteria-laden environment of the mouth, the surface mucosa must tightly abut the implant surface to provide an acceptable barrier and reduce the potential for infection (Aimplantitis@). The effectiveness of this barrier is considerably reduced when the implant is loose because of poor integration with the underlying bone.

Prognosis is dependent on several factors, of course. The design of the implant is very important, as are continued cleanliness of the implant site, control of occlusal forces (especially lateral or excursive forces) and the lack of systemic diseases, such as diabetes, which are inherently associated with poor healing. Low bone density (LBD) and ischemically damaged, desiccated bone both have a poor ability to remodel and are also major contraindications for dental implants [7–11]. Moreover, each of the latter problems increases in frequency with age and implants are typically placed in older individuals [9–11]. Unfortunately, readily available diagnostic imaging devices lack the ability to adequately identify such bone.

J.E. Bouquot
Department of Diagnostic Sciences, University of Texas Dental Branch at Houston, Texas, U.S.

R. Waynant and D.B. Tata (eds.), *Proceedings of Light-Activated Tissue Regeneration and Therapy Conference.*
© Springer Science + Business Media, LLC 2008

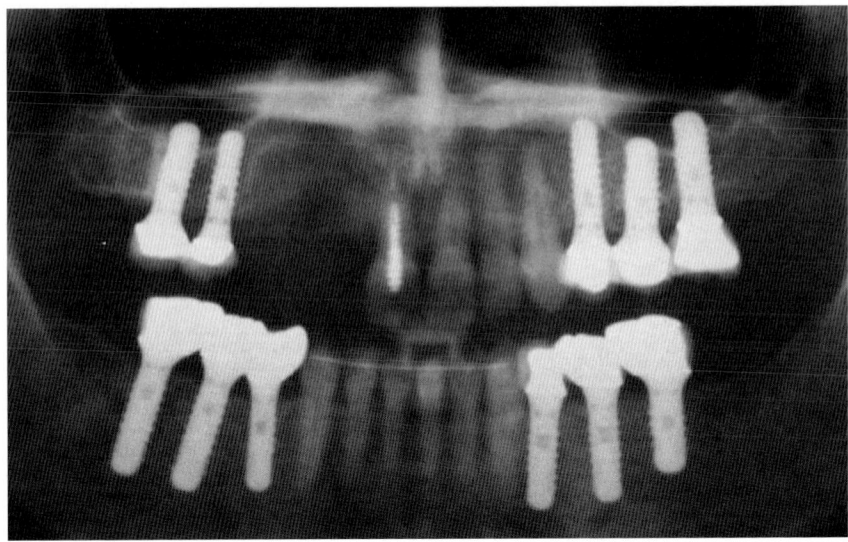

Fig. 1 Multiple implants in a patient with several edentulous alveolar sites

The radiographs used so extensively by dentists are particularly dismal in this regard and it has been demonstrated repeatedly that the jaws can lose more than half of their internal structure without an obvious alteration of the radiographic appearance [7, 8]. This state of affairs leaves the implant dentist with little or no ability to properly assess a major parameter of bone health prior to placing an implant. Moreover, without this knowledge there can be no thought of improving the bone quality prior to the placement of the implant.

A Potential Imaging Solution

The new technology of through-transmission or quantitative ultrasound (QUS) may provide a partial solution to this dilemma [12–16]. QUS uses sound at the low end of the ultrasound spectrum. While sound typically reflects off bone, sound waves in this region travel through bone as well as or better than they travel through water. A few specific changes, however, will greatly reduce the quantity of sound which is able to traverse from one cortex to the other. QUS is specifically cleared by the FDA to safely and non-invasively identify such changes: (1) regions of LBD; (2) regions of desiccated or dehydrated bone (usually the result of poor blood flow); and (3) intramedullary voids or cavitations (a sign of chronic ischemia).

while the specificity (proportion of false negatives) of this technology is not well established, the sensitivity (proportion of false positives) is higher than almost any other imaging technique, with fewer than 3% false positive scans in jawbone [15]. Perhaps more significantly, the generated ultrasound scan (Fig. 2) is much more easily interpreted and affected bone sites are much more readily identified than

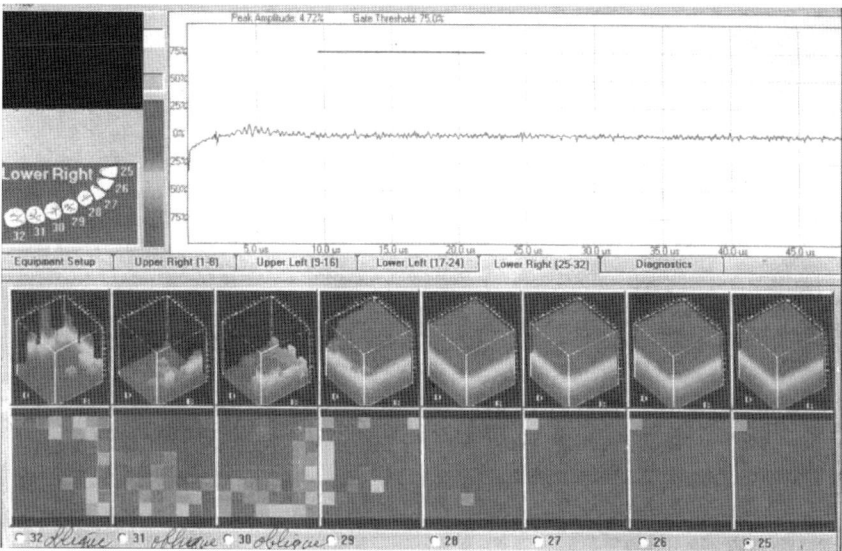

Fig. 2 Screen save of the Cavitat 4000 showing severe involvement of the right mandibular molar region, with numbers corresponding to U.S. tooth numbering system. Lower half of the screen presents 3-D (upper row) and 2-D (lower two rows, top row is occlusal surface; lowest row is facial surface) graphs of the entire quadrant, with each scan approximately representing the width of a single tooth, and with 64 columns each showing normal or Agreen@ bone on the right (anterior) and damaged or Ared@ bone on the left (posterior). The shorter the column height, the more sound is attenuated. Attenuation results from LBD, dehydration or cavitation

radiographs of the same sites (Fig. 3) [16]. QUS evaluation prior to implant placement should be effective in warning the clinician of a high potential for failure, but no investigation of this has yet been published.

A Potential Therapeutic Solution

Identifying the problem is only half of the equation, of course. Ideally, regions of LBD or desiccation can be treated to bring them to a level of (healthy@ bone capable of remodeling around and holding an implant. Low-level laser therapy (LLLT) has been shown in cell culture, animal studies and human studies to affect bone in several positive ways. For example, it stimulates bone healing, increases osteoblast differentiation, increases osteoblast proliferation, decreases osteoclast numbers, enhances new bone formation, increases bone density, increases bone volume, thickens bony trabeculae, improves bone graft and implant osteointegration and even allows more rapid orthodontic movement of teeth through jawbone and reduces post-operative pain from tooth extractions [17–31].

LLLT also favorably enhances a variety of vascular and stromal activities, especially in inflamed tissues, which should be of benefit to bone health and healing. Specifically, it increases fibroblastic activity and proliferation, enhances

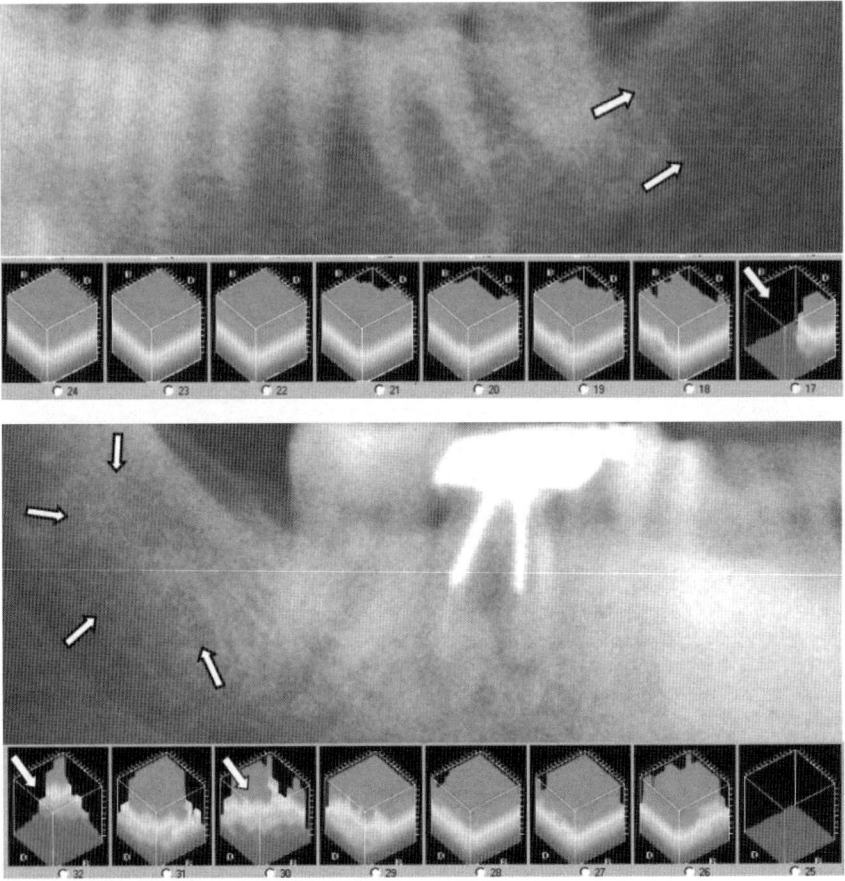

Fig. 3 Comparison of QUS scans with radiographs of the same edentulous alveolar bone sites in two patients (numbers beneath scans refer to alveolar site using U.S. tooth numbering system). The QUS scans in both very readily identify abnormal bone while the radiographic changes are very subtle and likely to be interpreted as variations of normal trabecular bone. Both patients showed bone marrow edema in QUS positive sites

endothelial cell proliferation and angiogenesis, stimulates T-cell production of angiogenetic and other factors, improves arteriole inflow, increases microcirculation flow, increases lymphatic clearance and decreases abnormally high coagulation tendencies [32–46]. It is, moreover, capable of moving inflamed tissues from the inflammatory phase into the reparative or proliferative phase, and can stimulate the photoacceptor cytochrome c oxidase to increase energy metabolism and production [47–50].

NIR-LED (near infrared – light emitting diode) photobiomodulation appears to have the same influences and the ability to use an array of LED lights which simply lies over the treated tissues (as opposed to having to constantly wave a single laser beam over a site) makes patient compliance less of an issue [45, 47–55]. As with

LLLT, the efficacy of LED therapy is dependent on the radiation wavelength, dose and distribution of light intensity in time [45, 49, 51]. It is not, however, dependent on the coherence of radiation, which is the hallmark of laser light.

Objective and Hypothesis

The above factors appealingly suggest that a combination of QUS evaluation of LED treated jawbones can provide valuable prognostic information and improve the probability of success relative to the placement of dental implants. The first step toward proving this is to determine whether or not LED therapy can increase bone density and/or hydration in presumably uninflamed, asymptomatic alveolar bone with QUS proof of LBD or desiccation. The objective of the present investigation was, therefore, to do exactly that. The study's hypothesis was that NIR-LED photobiomodulation will significantly improve LBD and/or bone hydration, as determined by QUS testing, in alveolar bone when compared to pretreatment bone.

Methods and Materials

This investigation was a retrospective record review of a cohort of 68 QUS scanned dental patients from the private practices of two co-investigators (PB, JC). All scans and treatments occurred between 2002 and 2005. The protocol was approved by the Committee for the Protection of Human Subjects of the University of Texas in Houston. Informed consent was obtained from all subjects.

Inclusion criteria included:

1. No surgical procedure at an investigative site less than 1 year prior to entry into the study
2. Pre-treatment and post-treatment QUS scans of the maxillofacial region was performed
3. LED therapy to one or more maxillofacial regions was performed

Exclusion criteria included:

1. A medical condition associated with abnormal bone growth or remodeling, such as Paget's disease of bone, fibrous dysplasia, osteopetrosis, severe systemic osteoporosis, etc.
2. A medical condition associated with poor healing, such as diabetes mellitus, immune deficiency, malnutrition, anemia, etc.
3. Unwillingness to sign informed consent form
4. Inability to perform daily LED treatments at home
5. Inability to obtain high quality QUS scans of the jaws

QUS scans were made immediately prior to LED therapy and 2 weeks after cessation of therapy. All scans were performed by the FDA cleared dental QUS device, the

Cavitat 4000 (Cavitat Medical Technologies, Inc, Aurora, Colorado). The Cavitat renders a series of three-dimensional cube images (Fig. 3) from analog signals generated when an external transmitter sends 27,000 sound pulses per microsecond through the alveolar bone at a speed of 317.6 m/s, 3.5 mHz, to an intraoral piezo screen (receiving transducer) held on the lingual aspect of the alveolus [15, 16]. The screen has 64 sensors which detect electrical changes on the screen surface as sound distorts it. The test is premised on the assumption that sound waves traveling through bone affected with LBD or dehydration (chronic ischemia) becomes attenuated, hitting the receptor screen with less intensity than sound waves which travel through normally dense, properly hydrated bone. The speed of sound is also diminished and so changes in speed are captured and accounted for by the device.

Each three-dimensional cube image is actually a 3-D graph of the 64 signals generated by the receiving transducer. Short columns represent severe attenuation of the sound signal and tall columns represent minimal loss of sound after traversing the alveolus (Figs. 2 and 3). The columns are also color coded for ease of interpretation, with green representing (Anormal@ bone and yellow, orange and red representing, respectively, more attenuated sound signals. Each column is created using a linear scale, with 252 equal sized units in a column with no loss of height.

All initial and follow-up QUS scans of alveolar bone were blindly and independently graded, after calibration, by two investigators according to an established 5-point scale (Table 1). In case of differences, a consensus grade was arrived at via discussion between the two. QUS grades of positive scans, i.e. scans with grades I–IV before LED therapy were compared with post-therapy scan grades using matched pair analysis.

Patients were treated using the investigational OsseoPulse (Version 1.0) device (Biolux Research Ltd., Vancouver, Canada). The device consists of an extra-oral

Table 1 Grading categories for individual 3-D cube images (64 columns in each) of the Cavitat QUS images[15,16]

QUS Grade*	Description**
0	AGreen bone.@ Cube shows no loss of column height and is 100% green; or mild loss of column height in less than 1/4 of columns (16 columns); and/or moderate to severe loss of column height in less than 4 non-adjacent columns.
I	Cube shows mild loss of column height in more than 1/4 of columns; and/or moderate loss of column height in 1/16 to 1/4 of the columns (5–16 columns); and/or severe loss of height in 1/16 to 1/8 of the columns (5–8 columns).
II	Cube shows moderate loss of column height in 1/4 to 2 of columns (17–32 columns); and/or severe loss of height in 1/8 to 1/4 of columns (8–16 columns).
III	Cube shows moderate loss of column height in more than 2 of columns (32 columns); and/or severe loss of column height in 1/4 to 2 of columns (17–32 columns).
IV	Cube shows severe loss of column height in more than 2 of columns (32 columns).

*high grade lesion = Grade III and IV scans; low-grade lesion = Grade I and II scans; Agreen bone@ = normal or Grade 0 scan
**definition of loss of column height: mild (crown is green, less than 1/3 loss of height); moderate (crown is yellow or brown, 1/3 to 2/3 loss of height); severe (crown is orange or red, more than 2/3 loss of height)

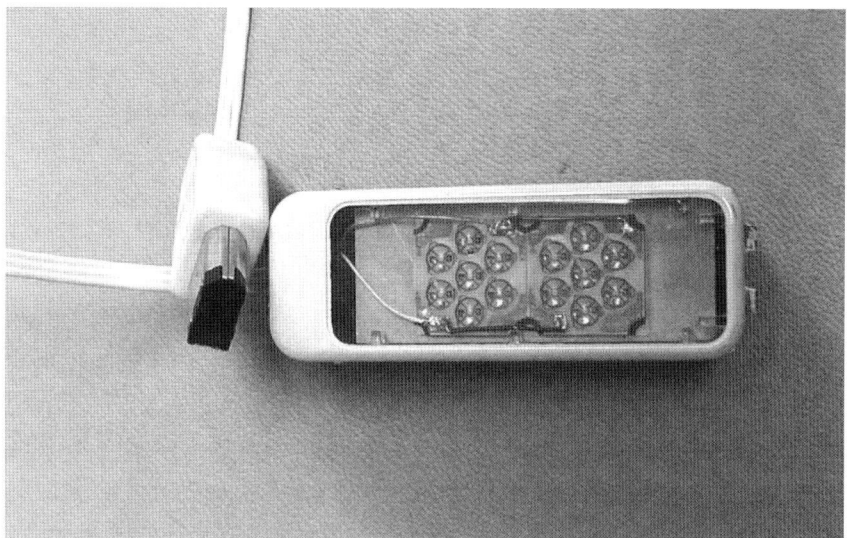

Fig. 4 LED array inside the OsseoPulse (Version 3)

array of highly-efficient light emitting diodes (LED) producing non-coherent continuous wave monochromatic light in the visible far red (660 nm @ 15 mW/cm^2) and infra-red range (840 nm @ 20 mW/cm^2), as shown in Fig. 4. In this range the light should be able to painlessly penetrate to a depth of 3–4 cm, which will effectively include the alveolar bone of the jaws. In addition, there was an integral alignment device used to ensure that the LED array was repeatably and accurately positioned directly over the treatment sites.

The OsseoPulse was placed on the facial surface for 15 minutes daily, 5 days a week for 12 weeks on each treatment side (Fig. 5). The dose per session per treatment area was approximately 200 J/in.^{2}. As can be seen in Fig. 5, the LED array irradiates almost an entire quadrant of the maxillofacial bones, not just the specific site with a positive QUS scan. This allowed a certain number of QUS-normal edentulous sites to be evaluated for random variation in scan results, but such an evaluation was not part of the present investigation.

Results

Of 1,148 pre-treatment QUS jawbone scans, each representing 11 mm^2 of alveolar bone, 294 were positive for damaged or abnormal edentulous bone. Using the 5-point scale (0 = normal; 4 = most severe), half of these sites were low grade, i.e. grades 1 or 2, while half were high grade (Table 2). The average grade for all 294 positive sites was 2.43. After LED photomodulation the average grade was 1.33, a 44.3% improvement in scan grade (Fig. 6). Additionally, almost 42% of investigated sites had returned to completely normal bone, while another 54 (18.4%) sites were grade 1 after therapy (Table 2).

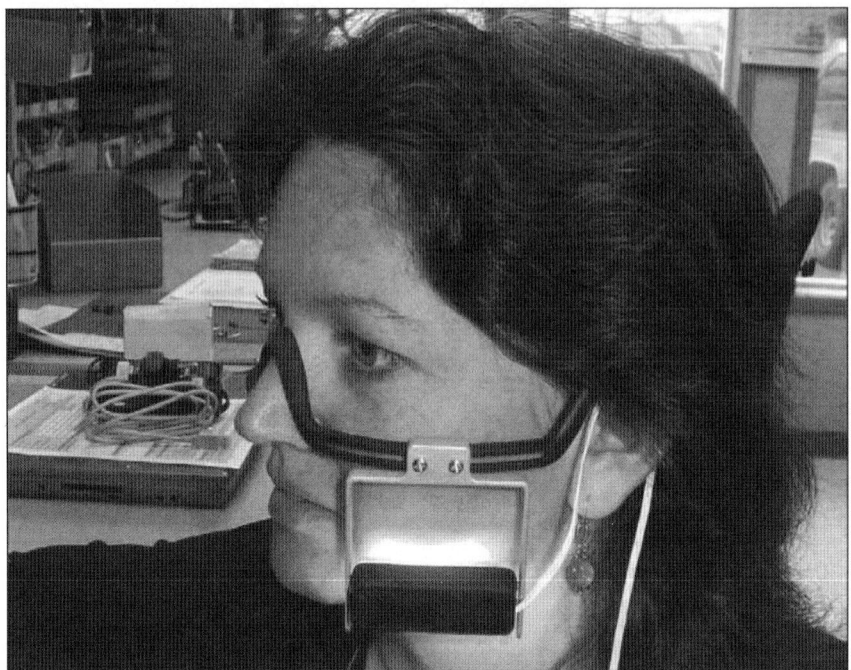

Fig. 5 OsseoPulse (version 3) providing LED photobiomodulation to the left mandible. The present investigation used version 1, which functions the same but looks somewhat different.

Table 2 Results of 294 QUS scans before and after 3 months of daily LED photobiomodulation. Numbers in right two columns represent number of sites at each grade level, but numbers in bottom row refer to average grades

Grade Level*	# at Pre-Treatment	# at Post-Treatment
1	79	120
2	69	54
3	86	53
4	61	40
Mean Grade:	2.43	1.33

*1 = mild LBD/dehydration; 4 = severe LBD/dehydration

One would expect that the lower the pre-treatment grade, the larger would be the proportion which returned to normal, since it takes a much greater grade improvement to reach normal from the higher grades. This proved to be the case: regions with pre-treatment grades of 1, 2, 3 and 4 returned to green bone 68.4%, 46.4%, 30.2% and 13.3% of the time, respectively (Table 3).

Almost 71% of the 294 treated sites demonstrated improvement of at least one QUS grade level, with most of those, 43.4%, dropping by one grade (Fig. 7). The

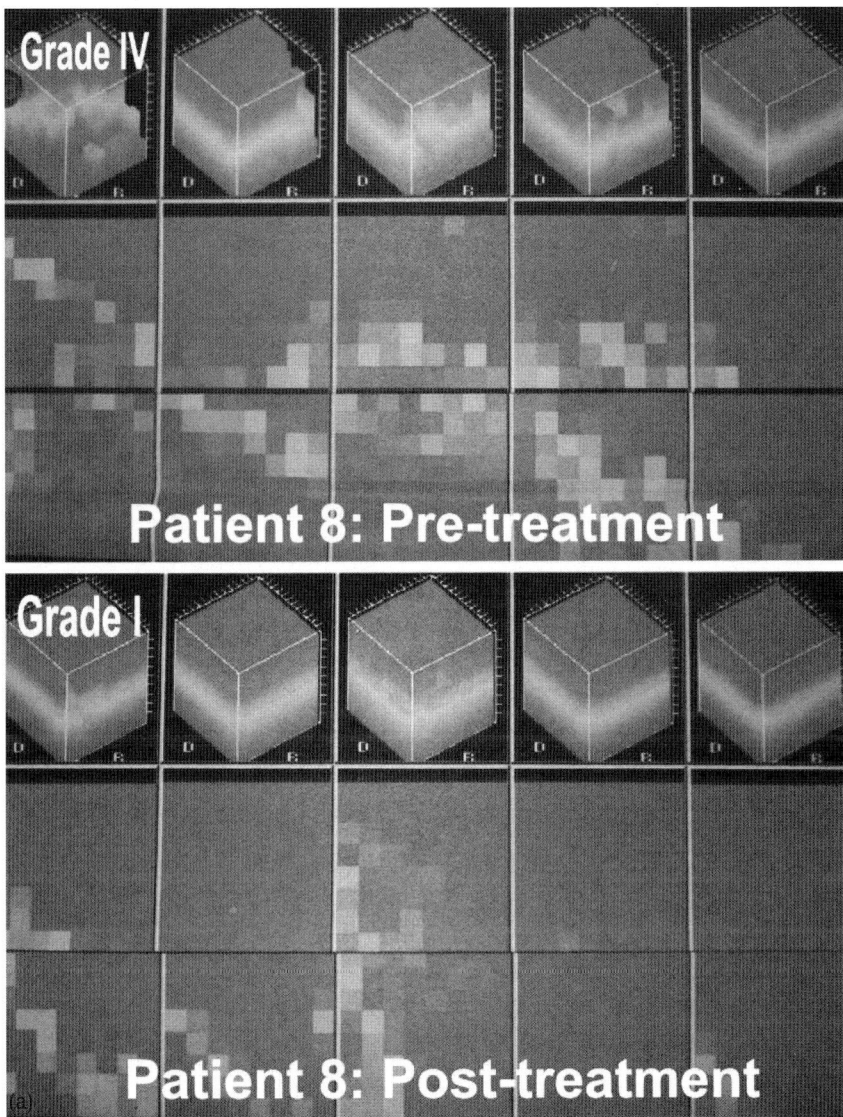

Fig. 6 Comparison of pre-treatment and post-treatment QUS scans from two typical patients: **a)** left posterior mandible of patient 8 shows a3-point improvement in the third molar area, from grade IV to grade I; **b)** left posterior mandible of patient 9 shows only a 1-point grade change in the third molar area, from Grade I to green bone (grade 0)

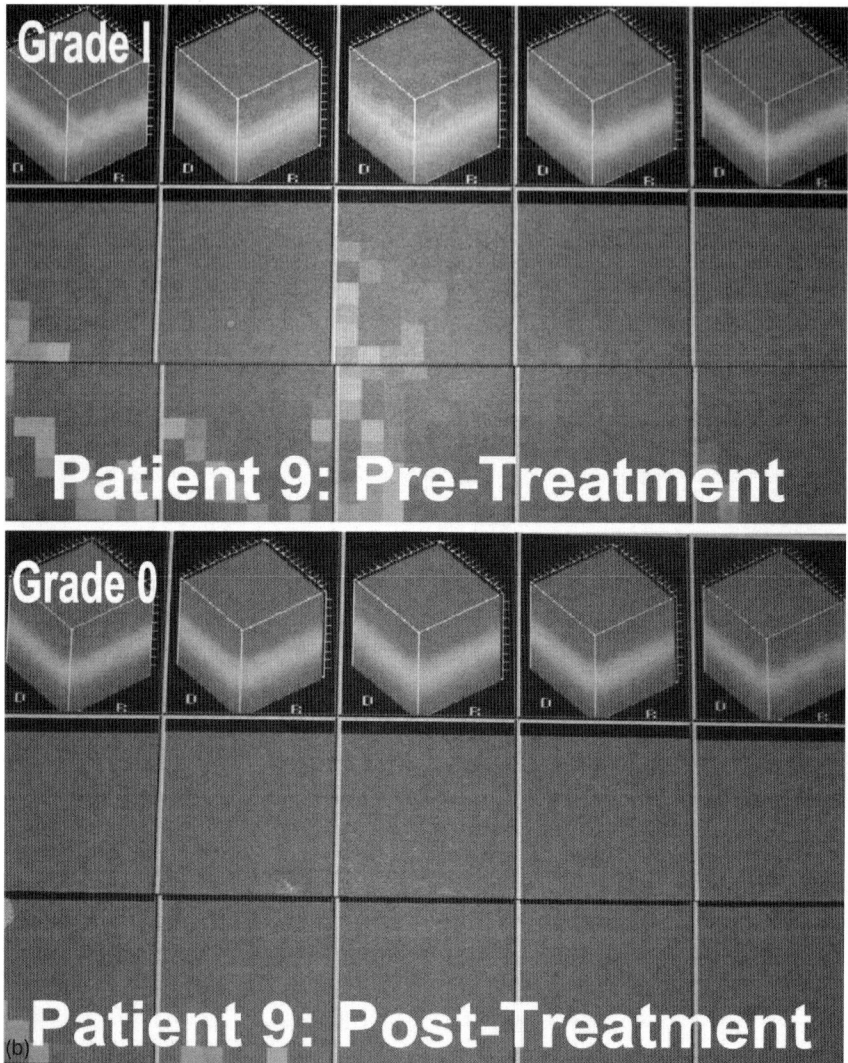

Fig. 6 (Continued)

post-treatment grade change was relatively uniform at each pretreatment QUS grade level (Table 3). Of interest is the fact that 17.7% of treated sites showed no change whatsoever and, perhaps most intriguing of all, 7.1% of sites actually showed an *increase* in the QUS grade after treatment. None of the noninvestigational sites of green bone showed a change over the 3 months of the study. Overall, for investigational sites the mean difference, i.e. improvement of bone quality, of −1.11 was very statistically significant (matched pair analysis: Std error 0.06914; t-Ratio −15.9896; DF 293; prob [t] less than 0.0001; 95% confidence interval 0.558–1.242).

Table 3 Post-treatment changes for each pre-treatment grade level, 294 QUS scans

Grade Level*	Number of pre-treatment sites @ each grade	Number of post-treatment sites @ each grade*					Average change
		0	1	2	3	4	
1	79	54	15	8	2	0	−0.54
2	69	32	19	12	4	2	−1.32
3	86	26	17	22	16	5	−1.50
4	60	8	3	10	18	21	−1.32
Mean:	2.43	120	54	52	40	28	−1.11

*1 = mild LBD/dehydration; 4 = severe LBD/dehydration (see Table 1)

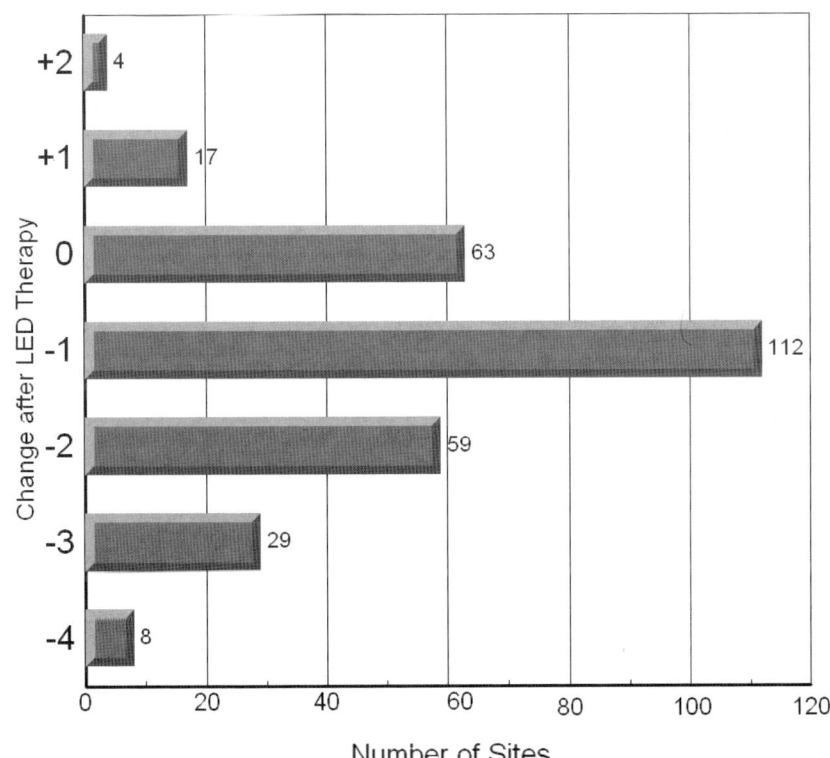

Fig. 7 Change in QUS scan grades after 3 months of LED photobiomodulation.

Discussion

Our hypothesis, based on previous animal and human studies demonstrating improved bone health and healing, presumes that QUS scanning would confirm
improvement in alveolar bone previously identified as either LBD or desiccated

via a positive scan. This hypothesis has, in fact, been confirmed reasonably well by the results of the present investigation. It should be stressed, however, that no biopsies were included in the study and so we are unaware of the *exact* condition of the bone before and after LED photobiomodulation. For the same reason, we have no way of *truly* knowing whether or not the regions of green bone, which were excluded in our initial selection of case sites and which were presumed to be normal bone, were truly normal, healthy bone [15, 16, 56].

Nevertheless, previous reports of the histopathology of QUS positive alveolar sites have demonstrated bone or marrow disease in approximately 97% of such sites, usually low grade inflammation (chronic osteomyelitis), osteoporotic bone (LBD, osteoporosis, focal osteoporotic marrow defect) or ischemic change (bone marrow edema, regional ischemic bone disease, ischemic osteonecrosis) [15, 16, 56]. These reports have also established a rather strong correlation between the microscopic presence of disease and the severity, i.e. grade, of the QUS image.

This is relatively new research and must eventually be substantiated by additional studies, of course, but it seems reasonable at this point in time to assume that our initial selection of grade 1–4 bone sites was relatively reliable as an indicator of disease or bone abnormality, and that the very high rate of specificity likewise gives us a fair degree of confidence that observed changes in the scan grade for a specific site are associated with real alterations in the bone. We assume these changes resulted from photobiomodulation but there is a slight possibility that they represent natural variation in bone hydration or density and there is at least some possibility that the ultrasound itself can increase bone density if repeated frequently [57]. Speaking against the latter possibilities is the fact that the green bone seen in noninvestigational alveolar sites in the present study did not vary over time, and that QUS was not used repeatedly but, rather, was used only once to establish a baseline assessment of bone quality.

The clinical significance of a positive QUS scan is also an issue not addressed by the present investigation. Does such a scan truly correlate with bone which is less optimally able to osteointegrate with an implant? Since bone with LBD, desiccation or intramedullary cavitations is the bone most readily able to attenuate sound transmission from cortex to cortex, it seems logical to presume that the QUS-positive sites in the present patient cohort were involved with one or more of these pathoses. And since these bone changes are known to be associated with a low rate of bone turnover, it seems logical to presume that such bone would be at risk for implant failure, or at least for less than optimal success. Additional studies are needed, however, to determine the extent of the influence of such problems on the clinical success of a dental implant. Moreover, there is an ongoing debate about the validity of determining "bone quality" as determined by a variety of imaging technologies [58]. The very definition of the term, bone quality, is, in fact, being actively debated in the orthopedic literature.

A final caution is in order. The present investigation has in no way evaluated the "staying power" of the bone which became green or which changed to a lower grade. It is possible that the sites of damaged bone might quickly revert to the original grade after cessation of LED therapy, but we suspect not. The present investigators are planning a much longer follow-up study in order to assess this possibility.

After all is said and done, and always remember the previously mentioned caveats, our preliminary results are encouraging and suggest that this combination of new technologies has the potential to significantly impact dental implant success rates. NIR-LED photobiomodulation appears to enhance bone health *in vivo* by increasing bone density and/or hydration, which accords with published *in vitro* experimentations. A myriad of other uses are also suggested, such as: improved healing for extraction sockets and jawbone fractures, slowing of alveolar bone loss in periodontitis, enhanced intraosseous circulation and bone health in chronic ischemic bone disease and bisphosphonate-associated osteonecrosis; improved healing of periapical bone infection after endodontic therapy, etc. Only time will show the extent to which appropriate light applications will be helpful in the treatment of the many inflammatory and traumatic events which so often affect the jaws and maxillofacial soft tissues.

References

1. Mish CE. Density of bone: effect on treatment planning, surgical approach, and healing. In: Misch CE (ed). Contemporary implant dentistry. St. Louis: Mosby; 1993; pp. 469–485.
2. Zarb G, Lekholm U, Albrektsson T, Tenenbaum H (eds). Aging, osteoporosis, and dental implants. Carol Stream, IL: Quintessence Publishing; 1999.
3. Lindh C, Petersson A, Rohlin M. Assessment of the trabecular pattern before endosseous implant treatment: diagnostic outcome of periapical radiography in the mandible. Oral Surg Oral Med Oral Pathol Oral Radiol Endod 1996; 82:335–343.
4. Jacobs R, Van Steenberghe D. Bone quantity. In: Jacobs R, Van Steenberghe D (eds). Radiographic planning and assessment of endosseous oral implants. Berlin: Springer; 1998; pp. 64–80.
5. McMillan PJ, Riggs ML, Bogle GC, Crigger M. Variables that influence the relationship between osteointegration and bone adjacent to an implant. Int J Oral Maxillofac Implants 2000; 15:138–145.
6. Jonasson G, Bankvall G, Kiliaridis S. Estimation of skeletal bone mineral density by means of the trabecular pattern of the alveolar bone, its interdental thickness, and the bone mass of the mandible. Oral Surg Oral Med Oral Pathol Oral Radiol Endod 2001; 92:346–352.
7. Bender JB, Seltzer S. Roentgenographic and direct observation of experimental lesions in bone. J Am Dent Assoc 1961; 62:152–160, 708–716.
8. Vanderstelt PF. Experimentally produced bone lesions. Oral Surg Oral Med Oral Pathol 1985; 59:306–312.
9. Urbaniak JR, Jones JP Jr (eds). Osteonecrosis – etiology, diagnosis, and treatment. Chicago, IL: American Academy of Orthopaedic Surgeons; 1997.
10. Bouquot JE, Rohrer M, McMahon RE, Boc T. Focal osteoporotic marrow defect (FOMD) B literature review and report of 596 new cases. Oral Surg Oral Med Oral Pathol Oral Radiol Endod 2002; 94:211.
11. Orwoll ES, Bliziotes M (eds). Osteoporosis B pathophysiology and clinical management. Totowa, NJ: Humana Press; 2003.
12. Njeh CF, Hans D, Fuerst DT, Gluer C-C, Genant HK (eds). Quantitative ultrasound: assessment of osteoporosis and bone status. London: Martin Dunitz; 1999.
13. Bouxsein ML. Skeletal assessment using quantitative ultrasound. In: Orwoll ES, Bliziotes M (eds). Osteoporosis B pathophysiology and clinical management. Totowa, NJ: Humana Press; 2003; pp. 120–147.

14. Barkmann R, Laugier P, Moser U, Dencks S, Padilla F, Haiat G, Heller M, Glüer C-C. A method for the estimation of femoral bone mineral density from variables of ultrasound transmission through the human femur. Bone 2007; 40:37–44.

15. Bouquot J, Margolis M, Shankland W, Imbeau J. Through-transmission Alveolar Ultrasonography (TAU) B A new technology for evaluation of medullary diseases. Correlation with histopathology of 285 scanned jaw sites. Oral Surg Oral Med Oral Pathol Oral Radiol Endod 2002; 94:210.

16. Bouquot JE, Margolis M, Shankland WE, II. Report to the Food & Drug Administration: Through-transmission sonography ("TTS") (A new technology for the evaluation of jawbone density and dessication. Comparison with pantographic radiographs at 167 biopsied sites and correlation with histopathology of 285 scanned alveolar sites. Washington, DC: McKenna & Cunea; November, 2001.

17. Takeda Y. Irradiation effect of low-energy laser on alveolar bone after tooth extraction. Experimental study in rats. Int J Oral Maxillofac Surg 1988; 17:388–391.

18. Clokie C, Bentley KC, Head TW. The effects of the helium-neon laser on postsurgical discomfort: a pilot study. J Can Dent Assoc 1991; 57:584–586.

19. Barushka O, Yaakobi T, Oron U. Effect of low-energy laser (He-Ne) irradiation on the process of bone repair in the rat tibia. Bone 1995; 16:47–55.

20. Yaakobi T, Maltz L, Oron U. Promotion of bone repair in the cortical bone of the tibia in rats by low energy laser (He-Ne) irradiation. Calcif Tissue Int 1996; 59:297–300.

21. Ozawa Y, Shimizu N, Kariya G, Abiko Y. Low-energy laser irradiation stimulates bone nodule formation at early stages of cell culture in rat calvarial cells. Bone 1998; 22:347–354.

22. Dortbudak O, Haas R, Mallath-Pokorny G. Biostimulation of bone marrow cells with a diode soft laser. Clin Oral Implants Res 2000; 11:540–545.

23. Guzzardella GA, Fini M, Torricelli P, Giavaresi G, et al. Laser stimulation on bone defect healing: an in vitro study. Laser Med Surg 2002; 17:216–230.

24. Silva Junior AN, Pinheiro AL, Oliveira MG, Weismann R, et al. Computerized morphometric assessment of the effect of low-level laser therapy on bone repair: an experimental animal study. J Clin Laser Med Surg 2002; 20:83–87.

25. Nicola RA, Jorgetti V, Rigau J, Pacheco MT, et al. Effect of low-power GaAlAs laser (660 nm) on bone structure and cell activity: an experimental animal study. Laser Med Sci 2003; 18:89–94.

26. Ninomiya T, Miyamoto Y, Ito T, Yamashita A, et al. High-intensity pulsed laser irradiation accelerates bone formation in metaphyseal trabecular bone in rat femur. Bone Miner Metab 2003; 21:67–73.

27. Ninomiya T, Hosoya A, Nakamura H, Sano K, Nishisaka T, Ozawa H. Increase of bone volume by a nanosecond pulsed laser irradiation is caused by a decreased osteolcast number and an activated osteoblasts. Bone 2007; 40:140–148.

28. Pinheiro AL, Limeira Junior Fde A, Gerbi ME, Ramalho LM, et al. Effect of 830-nm laser light on the repair of bone defects grafted with inorganic bovine bone and decalcified cortical osseous membrane. J Clin Laser Med Surg 2003; 21:301–306.

29. Khadra M, Ronold HJ, Lyngstadaas SP, Ellingsen JE, et al. Low-level laser therapy stimulates bone-implant interaction: an experimental study in rabbits. Clin Oral Implants Res 2004; 15:325–332.

30. Goulart CS, Nouer PR, Mouramartins L, Garbin IU, et al. Photoradiation and orthodontic movement: experimental study with canines. Photomed Laser Surg 2006; 24:192–196.

31. Khadra M. The effect of low level laser irradiation on implant-tissue interaction. In vivo and in vitro studies. Swed Dent J Suppl 2005; 172:1–63.

32. Bibikova A, Belkin V, Oron U. Enhancement of angiogenesis in regenerating gastrocnemius muscle of the toad (Bufo viridis) by low-energy laser irradiation. Anat Embryol (Berl) 1994; 190:597–602.

33. Ghamsari SM, Taguchi K, Abe N, Acorda JA, et al. Evaluation of low level laser therapy on primary healing of experimentally induced full thickness teat wounds in dairy cattle. Vet Surg 1997; 26:114–120.

34. Agaiby AD, Ghali LR, Wilson R, Dyson M. Laser modulation of angiogenic factor production by T-lymphocytes. Laser Surg Med 2000; 26:357–363.

35. Maegawa Y, Itoh T, Hosokawa T, Yaegashi K, et al. Effects of near-infrared low-level laser irradiation on microcirculation. Laser Surg Med 2000; 27:427–437.

36. Stadler I, Evans R, Kolb B, Naim JO, et al. In vitro effects of low-level laser irradiation at 660 nm on peripheral blood lymphocytes. Laser Surg Med 2000; 27:255–261.

37. Ad N, Oron U. Impact of low level laser irradiation on infarct size in the rat following myocardial infarction. Int J Cardiol 2001; 80:109–116.

38. Almeida-Lopes L, Rigau J, Zangaro RA, Guidugli-Neto J, et al. Comparison of the low level laser therapy effects on cultured human fibroblast proliferation using different irradiance and same fluence. Laser Surg Med 2001; 29:179–184.

39. Shimotoyodome A, Okajima M, Kobayashi H, Tokismitsu I, et al. Improvement of macromolecular clearance via lymph flow in hamster gingiva by low-power carbon dioxide laser-irradiation. Laser Surg Med 2001; 29:442–447.

40. Whelan HT, Smits RL Jr, Buchman EV, Whelan NT, et al. Effect of NASA light-emitting diode irradiation on wound healing. J Clin Laser Med Surg 2001; 19:305–314.

41. Pereira AN, Eduardo Cde P, Matson E, Marques MM. Effect of low-power laser irradiation on cell growth and procollagen synthesis of cultured fibroblasts. Laser Surg Med 2002; 31:263–267.

42. Schindl A, Merwald H, Schindl L, Kaun C, Wojta J. Direct stimulatory effect of low-intensity 670 nm laser irradiation on human endothelial cell proliferation. Br J Dermatol 2003; 148:224–336.

43. Mendez TM, Pinheiro AL, Pacheco MT, Nascimento PM, et al. Dose and wavelength of laser light have influence on the repair of cutaneous wounds. J Clin Laser Med Surg 2004; 22:19–25.

44. Woodruff LD, Bounkeo JM, Brannon WM, Dawes KS, et al. The efficacy of laser therapy in wound repair; a meta-analysis of the literature. Photomed Laser Surg 2004; 22:241–247.

45. Vladimirov YA, Osopov AN, Kledanov GI. Photobiological principles of therapeutic applications of laser radiation. Biochemistry 2004; 69:81–90.

46. Posten W, Wronge DA, Dover JS, Arndt KA, et al. Low-level laser therapy for wound healing: mechanism and efficacy. Dermatol Surg 2005; 31:334–340.

47. Babilas P, Karrer S, Sidoroff A, Mandthaler M, Szeimies R-M. Photodynamic therapy in dermatology B an update. Photodermal Photoimmunol Photomed 2005; 21:142–149.

48. Yeager RL, Franzosa JA, Millsap DS, Angelll-Yeage JL, et al. Effects of 670-nm phototherapy on development. Photomed Laser Surg 2005; 23:268–272.

49. Elke MV, Cagnie BJ, Cornelissen MJ, Declercq HA, et al. Increased fibroblast proliferation induced by light emitting diode and low power laser irradiation. Laser Med Sci 2003; 18:95–99.

50. Kreisler M, Christoffers AB, Al-Haj H, Willershausen B, et al. Low level 809-nm diode laser-induced in vitro stimulation of the proliferation of human gingival fibroblasts. Laser Surg Med 2002; 30:365–369.

51. Klebanov GI, Shuraeva N, Chichuk TV, Osipov AN, et al. A comparative study of the effects of laser and light-emitting diode irradiation on the wound healing and functional activity of wound exudate leukocytes. Biofizika 2005; 50:1137–1144.

52. Weiss RA, McDaniel DH, Geronemus RG, Weiss MA, et al. Clinical experience with light-emitting diode (LED) photomodulation. Dermatol Surg 2005; 31:1199–1205.

53. Desmet KD, Paz DA, Corry JJ, Eelis JT, et al. Clinical and experimental applications of NIR-LED photobiomodulation. Photomed Laser Surg 2006; 24:121–128.

54. Choi HR, Jang YY, Lim WB, Park JS, Kim OJ. Potential roles of light-emitting diode irradiation in cyclooxygenase, nitric oxide, prostaglandin E2 and reactive oxygen species

production in arachidonic acid-treated human gingival fibroblasts. J Oral Pathol Med 2006; 35:429–430.

55. Tremblay JF, Sire DJ, Lowe NJ, Moy RL. Light-emitting diode 415 nm in the treatment of inflammatory acne: an open-label, multicentric, pilot investigation. Cosmet Laser Ther 2006; 8:31–33.

56. Bouquot J, Shankland W, Margolis M. Through-transmission alveolar ultrasonography (TAU) B new technology for evaluation of bone density and desiccation. Comparison with radiology of 170 biopsied alveolar sites of osteoporotic and ischemic damage. Oral Surg Oral Med Oral Pathol Oral Radiol Endod 2002; 93:413–414.

57. Want SJ, Lewallen DG, Bolander ME, Chao EYS, Ilstrup DM, Greenleaf JF. Low intensity ultrasound treatment increases strength in a rat femoral fracture model. J Orthop Res 1994; 12:40–47.

58. Sievanen H, Kannus P, Jarvinen TLN. Bone quality: an empty term. PLOS Med 2007; 4 (e27):1–4.

Photobiomodulation by Low Power Laser Irradiation Involves Activation of Latent TGF-β1

Praveen R. Arany

Abstract Laser mediated stimulation of biological process was amongst its very first effects documented by Mester et al. but the ambiguous and tissue-cell context specific biological effects of laser radiation is now termed 'Photobiomodulation'. We found many parallels between the reported biological effects of lasers and a multifaceted growth factor, Transforming Growth Factor-β (TGF-β). This review outlines the interesting parallels between the two fields and our rationale for pursuing their potential causal correlation. We explored this correlation using an *in vitro* assay systems and a human clinical trial on healing wound extraction sockets that we reported in a recent publication. In conclusion we report that low power laser irradiation can activate latent TGF-β1 and β3 complexes and suggest that this might be one of the major modes of the photobiomodulatory effects of low power lasers.

Keywords: Low power laser therapy, photobiomodulation, latent-TGF-beta1 activation, laser medicine, wound healing, oral wounds, redox sensing.

The ability to reconstitute biological structures after injury or partial loss via the process of wound healing is amongst the central processes mediating the physiological homeostasis within an organism. This complex process involves carefully orchestrated events mediated by many distinct cell populations including resident and mobile cells from the systemic circulation. The chemical and mechanical aspects of this milieu is further intricately balance by a multitude of molecular players. The overlapping phases of healing involving hemostasis, inflammation, angiogenesis and resolution has been shown to be enhanced by exogenous agents

P.R. Arany
Molecular Reproduction Development and Genetics, Indian Institute of Science, Bangalore, Karnataka, India; Department of Oral and Maxillofacial Pathology K.L.E's Institute of Dental, Sciences Belgaum, Karnataka, India; Programs in Oral and Maxillofacial Pathology, Biological Sciences in Dental Medicine and Leder Medical Sciences, Harvard School of Dental Medicine, Boston, Massachusetts, e-mail: arany@fas.harvard.edu

R. Waynant and D.B. Tata (eds.), *Proceedings of Light-Activated Tissue Regeneration and Therapy Conference.*

varying from biomolecules like cytokines or natural products as well as physical modalities like electrical fields, hyperbaric oxygen, mechanical devices, etc. [9, 26, 38]. The fascination with laser mediated healing began with its very first biological effect observed of stimulating hair growth [23]. But more detailed studies have demonstrated their pleiotropic effects in varying cellular and tissue models as well as, perhaps more significantly, great variation in clinical therapeutic effects varying from no differences to marked improvement in healing outcomes. This led to the ambiguous and tissue-cell context specific biological effects of laser radiation being termed 'Photobiomodulation' [14].

Our extensive exploration of the laser biology literature unraveled interesting parallels with the reported biological effects of a multifaceted growth factor, Transforming Growth Factor-β (TGF-β), that has been shown to mediate various processes ranging from physiological processes like embryonic development and wound healing to pathological processes like tissue fibrosis and tumor metastasis [30]. TGF-βs are a superfamily of growth factors presently having more than forty two members including the nodals, activins, bone morphogenic proteins (BMPs), myostatin, anti-Muellerian hormone among others [22, 29] We were specifically very interested in the demonstrated effects of low power lasers on tissues that appeared to simulate similar treatment with exogenous TGF-βs. The most striking similarities were seen in the both their ability to mediate the Extracellular matrix (ECM) specifically with respect to fibronectin [7, 17] and collagen [11, 31]. More directly, the phasic pattern of healing was modulated by both agents as demonstrated by their ability to modulate the inflammatory responses [19, 34] and induce reperfusion in the healing tissues by mediating the neoangiogenesis that is central to the healing cascades [10, 32].

Further, the ability to mechanically aid wound closure inducing α-Smooth muscle actin (αSMA) in resident and infiltrating fibroblasts, termed Myofibroblast transformation, resulting in wound contraction by low power lasers [28] has been well established with TGF-βs [8]. While the effect of neural signals in mediating the neuroendocrine responses in healing tissues has been well explored, the interesting demonstration of the neurotrophic effects of low power laser [24] parallels similar observations with TGF-β to interact the Nerve Growth Factor (NGF) axis [6]. These observations clearly demonstrate the biological basis of the observed clinical efficacy of laser in promoting wound healing by well designed clinical trials [27].

The molecular basis for this photobiomodulation phenomenon still remains largely elusive. Some studies point to similarities in the heat shock response involving HSP70 with low power lasers [33] that has also been shown with TGF-β treatment [35]. There are also studies demonstrating effects of low power lasers on sub-cellular organelles [37]. The generation of free radical by high energy gamma radiation by redox is well established but laser irradiation has also been shown to induce both protease activity as well as redox in tissues [18, 20]. Interestingly, redox mediated activation of TGF-β has been well established [3] as well as conformational changes mediated by integrins [25]. A couple of recent studies have directly implicated the possible involvement of TGF-βs in the biological effects of low-power and near infrared laser in promoting wound healing [36] and recovery from spinal injury [5] in mice has also been demonstrated at the transcriptional level showing increased TGF-β mRNA.

The activation of the latent TGF-β complex is a central regulatory step in mediating its biological effects. Activation of the various latent TGF-β forms have been extensively studied and was demonstrated to be via physiochemical means like heat and altered pH [4], as mentioned previously radiation and conformation mediated as well as a large group of proteases and enzymes [15, 16, 21]. Thus, from the preceding analysis of the literature we tested our hypothesis if activatibility of the latent TGF-β complex was mediating the photobiomodulatory effects of low power laser radiation.

We used a 904 nm infrared wavelength laser at 3 J/cm^2 in our *in vivo* study while the irradiation time was varied to achieve various dosimetery in our *in vitro* activation studies [2]. We carried out a carefully designed clinical trial, using each patient as his own control, in oral tooth extraction wounds in patients undergoing complete oral rehabilitation (Complete Dentures). We first established that we could get a better healing response in our laser wounds using histological parameters including inflammation, neoangiogenesis, collagen turnover and architecture. We then analyzed TGF-β1 and β3 expressions in these wounds and found a clear correlation of increased TGF-β1 expression and in the rapid healing laser wounds. The increases were seen immediately after laser irradiation (within 15 min) and at 14 days. The latter (14 days) increase correlated with increased inflammatory infiltrate seen in the laser healing tissues but the former (within 15 min) suggested a non-transcriptional or translational event thus, suggesting a latent complex activation mechanism.

We then used an *in vitro* cell-free system analyzing various latent TGF-β complexes from various sources like cell conditioned medium (cell secreted), Recombinant (CHO) and serum. The serum was of particular interest as the wound milieu immediately after wounding is usually serum-rich post hemostasis. The experiments *in vitro* demonstrated the specificity of the TGF-β isoform in serum that have further characterized the distinct roles of TGF-β1 and TGF-β3 kinetics in the healing scenario. Our data suggests that activation of latent TGF-β1 by low power laser irradiation could via a combinatorial effect of direct activation and 'priming' of latent complexes of TGF-β making them amenable to routine physiochemical activation. We have observed these two effects might be largely due to conformational modulation of the latent complex as has been shown to be effective in the integrin mediated activation [25]. In summary, we demonstrated a novel molecular mechanism for the photobiomodulatory effects of low power laser irradiation.

Our data suggests the mechanism of activation of the latent TGF-β complex is a non-thermal, conformational change mediated event probably involving a redox-mediated process. The direct biological effects of low power laser irradiation might be a redox generating process as has been demonstrated previously due to simple hydrolysis of water or affecting other free radical species (Superoxide, NO, etc.) [18, 20], the downstream biological effects of these transient and locally acting molecular species might be limiting. Also, the effects of the redox generation in an intracellular environment would be most efficacious while it would be rapidly neutralized in the extracellular compartment. We believe the same redox generation mechanism might be involved with the conformational modulation of the Latent TGF-β complex and specific residues on the Latency Associated Peptide (LAP) have been shown to be a 'Redox sensor' [13]. Thus, now a secondary potent cytokine is activated in the ECM to induce various intracellular biological responses for a sustained period on multiple targets. This possibly explains why laser photobiomodulation is capable of eliciting cellular responses to intracellular and extracellular cues for both short as well as prolonged durations, with a variety of possible outcomes like proliferation, fibrosis, cytostasis or cell death.

One of the greatest concerns in the clinical use of Lasers to stimulate healing is the possible risk of promoting the proliferation and progression of pre-malignant and transformed cells in the vicinity of the wounds. While this risk needs to be definitely carefully addressed in patient selection and pretreatment evaluations, we and others have as demonstrated that stimulating healing might be beneficial in reducing tumor burden while a chronic 'persistent' poor healing response would be conducive to tumor growth [1, 12]. This leads me to speculate that the beneficial effects of laser mediated healing may not be detrimental in terms of promoting tumor growth *in vivo* as it inherently reduces the overall 'active' healing phase, thus resolving the probability that the wound milieu may stimulate malignant cellular characteristics. The photobiomodulatory effects of low power irradiation on tumor cells would, in themselves, be fascinating future area of research to see if these errant cells can be retrained by specific cues to modify their cell behavior.

As our present study only analyzed a single wavelength and limited dosimetry, we are currently analyzing the latent TGF-β activation mechanism using different wavelengths and fluence rates and plan to validate these effects *in vivo* using transgenic mice models. The elusive photoreceptor molecule capable of absorbing at this wavelength is another goal of our future investigations.

References

1. Arany, P.R., Flanders, K.C., Kobayashi, T., Kuo, C.K., Stuelten, C., Desai, K.V., Tuan, R., Rennard, S.I., and Roberts, A.B. (2006). Smad3 deficiency alters key structural elements of the extracellular matrix and mechanotransduction of wound closure. Proceedings of the National Academy of Sciences of the United States of America *103*, 9250–9255.
2. Arany, P.R., Nayak, R.S., Hallikerimath, S., Limaye, A.M., Kale, A.D., and Kondaiah, P. (2007). Activation of latent TGF-beta1 by low-power laser in vitro correlates with increased TGF-beta1 levels in laser-enhanced oral wound healing. Wound Repair and Regeneration *15*, 866–874.
3. Barcellos-Hoff, M.H., and Dix, T.A. (1996). Redox-mediated activation of latent transforming growth factor-beta 1. Molecular endocrinology (Baltimore, MD) *10*, 1077–1083.
4. Brown, P.D., Wakefield, L.M., Levinson, A.D., and Sporn, M.B. (1990). Physicochemical activation of recombinant latent transforming growth factor-beta's 1, 2, and 3. Growth factors (Chur, Switzerland) *3*, 35–43.

5. Byrnes, K.R., Waynant, R.W., Ilev, I.K., Wu, X., Barna, L., Smith, K., Heckert, R., Gerst, H., and Anders, J.J. (2005). Light promotes regeneration and functional recovery and alters the immune response after spinal cord injury. Lasers in Surgery and Medicine *36*, 171–185.
6. Chalazonitis, A., Kalberg, J., Twardzik, D.R., Morrison, R.S., and Kessler, J.A. (1992). Transforming growth factor beta has neurotrophic actions on sensory neurons in vitro and is synergistic with nerve growth factor. Developmental Biology *152*, 121–132.
7. Clark, R.A., McCoy, G.A., Folkvord, J.M., and McPherson, J.M. (1997). TGF-beta 1 stimulates cultured human fibroblasts to proliferate and produce tissue-like fibroplasia: a fibronectin matrix-dependent event. Journal of Cellular Physiology *170*, 69–80.
8. Desmouliere, A., Geinoz, A., Gabbiani, F., and Gabbiani, G. (1993). Transforming growth factor-beta 1 induces alpha-smooth muscle actin expression in granulation tissue myofibroblasts and in quiescent and growing cultured fibroblasts. Journal of Cell Biology *122*, 103–111.
9. Fu, X., Li, X., Cheng, B., Chen, W., and Sheng, Z. (2005). Engineered growth factors and cutaneous wound healing: success and possible questions in the past 10 years. Wound Repair and Regeneration *13*, 122–130.
10. Garavello, I., Baranauskas, V., and da Cruz-Hofling, M.A. (2004). The effects of low laser irradiation on angiogenesis in injured rat tibiae. Histology and Histopathology *19*, 43–48.
11. Ignotz, R.A., Endo, T., and Massague, J. (1987). Regulation of fibronectin and type I collagen mRNA levels by transforming growth factor-beta. Journal of Biological Chemistry *262*, 6443–6446.
12. Iyer, V.R., Eisen, M.B., Ross, D.T., Schuler, G., Moore, T., Lee, J.C., Trent, J.M., Staudt, L. M., Hudson, J., Jr., Boguski, M.S., et al. (1999). The transcriptional program in the response of human fibroblasts to serum. Science *283*, 83–87.
13. Jobling, M.F., Mott, J.D., Finnegan, M.T., Jurukovski, V., Erickson, A.C., Walian, P.J., Taylor, S.E., Ledbetter, S., Lawrence, C.M., Rifkin, D.B., et al. (2006). Isoform-specific activation of latent transforming growth factor beta (LTGF-beta) by reactive oxygen species. Radiation Research *166*, 839–848.
14. Karu, T.I., and Kolyakov, S.F. (2005). Exact action spectra for cellular responses relevant to phototherapy. Photomedicine and Laser Surgery *23*, 355–361.
15. Kojima, S., Nara, K., and Rifkin, D.B. (1993). Requirement for transglutaminase in the activation of latent transforming growth factor-beta in bovine endothelial cells. Journal of Cell Biology *121*, 439–448.
16. Kojima, S., and Rifkin, D.B. (1993). Mechanism of retinoid-induced activation of latent transforming growth factor-beta in bovine endothelial cells. Journal of Cellular Physiology *155*, 323–332.
17. Latvala, T., Tervo, K., Mustonen, R., and Tervo, T. (1995). Expression of cellular fibronectin and tenascin in the rabbit cornea after excimer laser photorefractive keratectomy: a 12 month study. British Journal of Ophthalmology *79*, 65–69.
18. Lavi, R., Sinyakov, M., Samuni, A., Shatz, S., Friedmann, H., Shainberg, A., Breitbart, H., and Lubart, R. (2004). ESR detection of 1O2 reveals enhanced redox activity in illuminated cell cultures. Free Radical Research *38*, 893–902.
19. Letterio, J.J. (2000). Murine models define the role of TGF-beta as a master regulator of immune cell function. Cytokine & Growth Factor Reviews *11*, 81–87.
20. Lubart, R., Eichler, M., Lavi, R., Friedman, H., and Shainberg, A. (2005). Low-energy laser irradiation promotes cellular redox activity. Photomedicine and Laser Surgery *23*, 3–9.
21. Lyons, R.M., Keski-Oja, J., and Moses, H.L. (1988). Proteolytic activation of latent transforming growth factor-beta from fibroblast-conditioned medium. Journal of Cell Biology *106*, 1659–1665.
22. Massague, J., and Gomis, R.R. (2006). The logic of TGFbeta signaling. FEBS Letters *580*, 2811–2820.
23. Mester, E., Nagylucskay, S., Tisza, S., Mester, A., Toth, J., and Laczy, F.I. (1977). [Current studies on the effect of laser beams on wound healing–immunologic effects]. Zeitschrift fur experimentelle Chirurgie *10*, 301–306.

24. Miloro, M., and Repasky, M. (2000). Low-level laser effect on neurosensory recovery after sagittal ramus osteotomy. Oral surgery, oral medicine, oral pathology, oral radiology, and endodontics *89*, 12–18.

25. Munger, J.S., Huang, X., Kawakatsu, H., Griffiths, M.J., Dalton, S.L., Wu, J., Pittet, J.F., Kaminski, N., Garat, C., Matthay, M.A., et al. (1999). The integrin alpha v beta 6 binds and activates latent TGF beta 1: a mechanism for regulating pulmonary inflammation and fibrosis. Cell *96*, 319–328.

26. Nuccitelli, R. (2003). A role for endogenous electric fields in wound healing. Current Topics in Developmental Biology *58*, 1–26.

27. Posten, W., Wrone, D.A., Dover, J.S., Arndt, K.A., Silapunt, S., and Alam, M. (2005). Low-level laser therapy for wound healing: mechanism and efficacy. Dermatologic Surgery *31*, 334–340.

28. Pourreau-Schneider, N., Ahmed, A., Soudry, M., Jacquemier, J., Kopp, F., Franquin, J.C., and Martin, P.M. (1990). Helium-neon laser treatment transforms fibroblasts into myofibroblasts. American Journal of Pathology *137*, 171–178.

29. Roberts, A.B., and Sporn, M.B. (1990). The Transforming Growth Factor-Betas. Vol. 1 (Heidelberg, Springer).

30. Roberts, A.B., and Wakefield, L.M. (2003). The two faces of transforming growth factor beta in carcinogenesis. Proceedings of the National Academy of Sciences of the United States of America *100*, 8621–8623.

31. Skinner, S.M., Gage, J.P., Wilce, P.A., and Shaw, R.M. (1996). A preliminary study of the effects of laser radiation on collagen metabolism in cell culture. Australian Dental Journal *41*, 188–192.

32. Smith, E.A., and LeRoy, E.C. (1990). A possible role for transforming growth factor-beta in systemic sclerosis. Journal of Investigative Dermatology *95*, 125S–127S.

33. Souil, E., Capon, A., Mordon, S., Dinh-Xuan, A.T., Polla, B.S., and Bachelet, M. (2001). Treatment with 815-nm diode laser induces long-lasting expression of 72-kDa heat shock protein in normal rat skin. British Journal of Dermatology *144*, 260–266.

34. Tadakuma, T. (1993). Possible application of the laser in immunobiology. Keio Journal of Medicine *42*, 180–182.

35. Takenaka, I.M., and Hightower, L.E. (1992). Transforming growth factor-beta 1 rapidly induces Hsp70 and Hsp90 molecular chaperones in cultured chicken embryo cells. Journal of Cellular Physiology *152*, 568–577.

36. Toyokawa, H., Matsui, Y., Uhara, J., Tsuchiya, H., Teshima, S., Nakanishi, H., Kwon, A.H., Azuma, Y., Nagaoka, T., Ogawa, T., et al. (2003). Promotive effects of far-infrared ray on full-thickness skin wound healing in rats. Experimental Biology and Medicine (Maywood, NJ) *228*, 724–729.

37. Wilden, L., and Karthein, R. (1998). Import of radiation phenomena of electrons and therapeutic low-level laser in regard to the mitochondrial energy transfer. Journal of Clinical Laser Medicine & Surgery *16*, 159–165.

38. Zamboni, W.A., Browder, L.K., and Martinez, J. (2003). Hyperbaric oxygen and wound healing. Clinics in Plastic Surgery *30*, 67–75.

Part VI
Diabetes

The Role of Laser in Diabetes Management

Leonardo Longo

Abstract

Background and Objectives: For over 20 years a few authors have reported hypoglycaemic effects of non surgical laser in type 1 and 2 diabetes. Different types of laser were used in association with other physical sources producing magnetic fields and electricity, and different treatment procedures were utilized. A few authors have published on this topic, following the famous "Helsinki" rules of clinical research: Declaration and other similar protocols. The aim of our study is to verify if some laser exposure parameters have effect on glycaemia and on the management of diabetes, and the persistence duration of this bio-effect.

Study Design/Materials and Methods: We started with a phase 2 clinical investigation. Fifteen patients were enrolled in this study, five with type 1 and ten with type 2 diabetes. The study comprised of both sexes with their ages ranging from 18–65 years. We used the same diode laser system coupled with magnetic fields but different procedure of treatment. Glycaemia was measured immediately before and 10 minutes after each session of treatment. Glycate haemoglobin were measured before and 3 months after the treatment. Insulin and anti-diabetic drugs were progressively reduced during the course of the treatment, when the glycaemia levels decreased.

Results: The results obtained were consistently positive. The follow up was variable depending of the type of the diabetes and the behaviour of the subject after the treatment.

Conclusions: Our data demonstrate that the current typical diabetes management protocols with anti-diabetic medicine could be totally modified in the near future, especially if these positive results of laser treatment remain consistently positive in the next phases of the studies. The laser induced mechanism of actions needs to be studied. Laser and magnetic field could be useful alone or in association with other therapies. Further experimentation could establish what kind of laser and energy will be the best for this treatment.

L. Longo
Institute for Laser Medicine, Siena University, Firenze, Italy

R. Waynant and D.B. Tata (eds.), *Proceedings of Light-Activated Tissue Regeneration and Therapy Conference.*
© Springer Science + Business Media, LLC 2008

Keywords: Diabetes laser therapy, bioresonance, magnetic fields, glycate haemo-globin, hypoglycaemia, laser dosage.

Background and Objectives

Since 20 years and more some Authors report news and data about the hypogly-caemic effects of non surgical laser in diabetes type 1 and 2 [1–4]. Different types of laser were used in association with other radiations, as magnetic field and electricity. Different procedures were followed. Few Authors published something on this topic, following the famous rules of the clinical research, as Helsinki. Declaration and other similar protocols. So these data were reported only as evidence based medicine. On the other hand, we can observe that diabetic mariners don't need hypoglycemic therapy when they are in high sea. Thus the large amount of light absorbed by these persons could have a therapeutic effect on the glucose metabolism. Perhaps this fact is caused by direct absorption of the sun reflexed by the water, with less portion of coherent light absorbed by the body.

At this moment, we know five different procedure of treatment of diabetes with light, associated with electromagnetic fields, and/or electrostimulation and/or laser rays [1–4]. We still don't know what procedure is most effective, but we know that all these procedure influence the glycaemia positively. This effect is strictly dose-dependent, as all the other effects of non surgical laser and light on the biological tissue [5].

Aim of our study is to verify if some laser beam have effect on the glycaemia and on the management of the diabetes. We never interrupt the diet and the physical activity of the patients, because these always continue to have a primary role on diabetes management.

We would like to verify the possibility to stop the esogenous insulin and the drugs in diabetes treatment.

Study Design/Materials and Methods

We are testing a range of dosages that surely do not have negative effects on irradiated tissue, as reported in the literature [1–9]. Previous experiments have demonstrated that laser irradiation (at a dosage level reported to be effective in the treatment of diabetic patients) can accelerate gastric emptying [1–3]. This gastric emptying would result in an increase in the glucose blood level, if a patient eats within 4 hours before the laser irradiation. It is absolutely necessary that the patient fasts for 4 hours prior to the laser treatment. Fortunately, laser dosages planned for these treatments are well below the dosage necessary to cause local burns. Further systemic effects have not been convincingly demonstrated.

Potential adverse effects caused by hypoglycemia induced by the laser therapy can betreated with sugar absorbed under the tongue.

Any potential hyperglycemic complications can be treated with the prompt administration of insulin. Each patient must be instructed to measure their blood glucose levels at the same hour each day. Usually diabetic patients know how to measures their blood glucose levels.

Each patient must contact their physician immediately, if he/she notes any change in their daily health or functioning.

We started with an experimentation as phase 2 of the schema of the clinical experimentation.15 patients were enrolled in this experience, 5 with diabetes type 1 and 10 type 2, both sex and age 18–65 years old (y.o.) In details, we enrolled patients selected with the following inclusion and exclusion criteria:

(a) **Inclusion criteria:**

   (i)   Type 1–2 human diabetes in compensative phase, both sex and middle-old age
   (ii) Weight normal or high
   (iii) Diabetes started more than 1 year ago

(b) **Exclusion criteria:**

   (i)   Diabetes I–II with unstable glycaemic control
   (ii) History of hypo-hyperglycaemic coma
   (iii) History of cardiovascular not treatable complication
   (iv) History of malignancy
   (v)   Previous participation in a clinical trial for hypoglycacmic drug effects control

(c) **Suspension criteria:**

   (i)   Patients with diabetes complication starting after laser treatment
   (ii) Patients who don't follow the protocol prescribed
   (iii) Patients with adverse effects to laser irradiation (local burns, allergic reaction)
   (iv) Patients who present some exclusion criteria that appear after the inclusion in the treatment protocol

Patients are informed about the investigative nature of the study and give their written informed consent prior to the treatment.

We used the same diode laser system (Table 1) coupled with magnetic fields (15 Hz fixed) but different procedure of treatment. The Glycaemia was measured immediately before and 10 minutes after each session of treatment. Glycate haemoglobin were measured before and 3 months after the treatment. Insulin and anti-diabetic drugs were progressively reduced during the treatment, when the glycaemia decreased. In details, the patients with diabetes type 1 are irradiated in average once a day for 3 weeks, then once a week until the total normalization of the Glycaemia and of the other parameters. The patients with diabetes type 2 are irradiated in average once a week until the total normalization of the Glycaemia and of the other parameters. We control the glycosilate haemoglobin immediately before and 3 months after the start if the laser treatment, then each 3 months.

The glycaemia level is monitored first in the initial visit, then 10 minutes after treatment, the day after, at same hour after the irradiation, and each day, at same hour after the irradiation.

Patients in both groups stop taking insulin or hypoglycaemic drugs 4 hours before the laser treatment. They will resume taking the insulin only if the glycemic

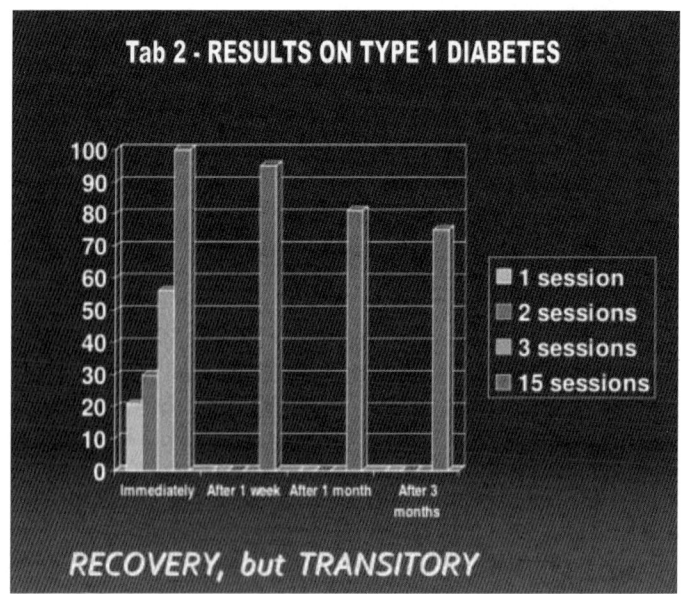

Tab 1 – Laser Dosage Parameters		
•Wavelength	nm	810
•Power density in average	mW / cm²	50
•Time/spot in average	min	5
•Emission in average	PW	50 Htz
•Mean power/spot (FLUENCE)	Joule/cm²	maximum 1
•Spot size	cm	1
•Irradiated points (zones)		Abdomen
•Nr. & rhythme of sessions		variable

Tab 2 - RESULTS ON TYPE 1 DIABETES

- 1 session
- 2 sessions
- 3 sessions
- 15 sessions

Immediately   After 1 week   After 1 month   After 3 months

RECOVERY, but TRANSITORY

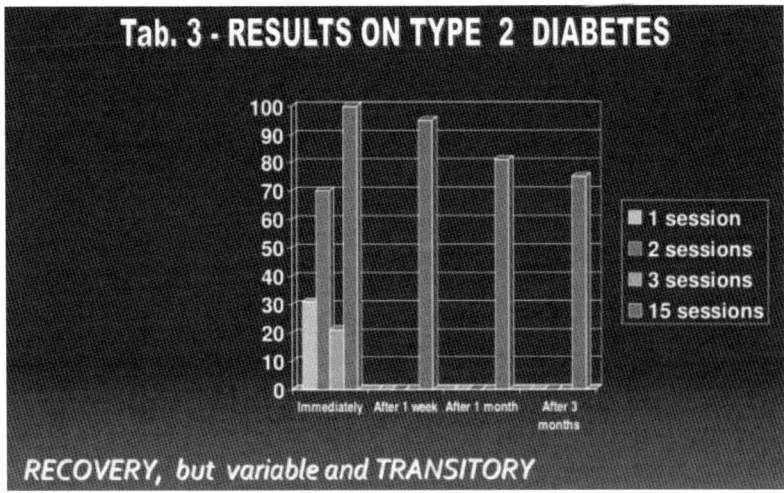

levels are not reduced by a minimum of 20% immediately after each laser treatment and persist for 1 day.

Then the insulin and hypoglycaemic drugs are decreased progressively.

Results

Our first data are positive in both types of diabetic patients (Tables 2 and 3). They stop the drugs therapy and the insulin therapy progressively, but they must continue to monitor the glycaemia once for day and more. The diet and physical activity continue to be very important. Follow up is excellent for both types of diabetes 1 month after the end of treatment, variable after 6 months, depending of the type of the diabetes and the behaviour of the patient after the treatment.

Unfortunately both types of diabetes patients require another cycle of Laser treatment 1 year after the first cycle.

Many hypothesis can be done about the action mechanisms of the light on the glucose metabolism: antinflammatory and regenerative effects on beta-cells and cromaphine cells, hyperaemic effect on the microcirculation, redistribution of the energy on the chakras, or all these effects together (table 4) In conclusion, the fact is the normalizing effect of the laser treatment. on the glucose metabolism, after a cycle of 15 sessions of irradiation in average. The causes of that must be investigated. We need further studies for maintaining the positive results of laser treatment for longer time. On the other hand, we need a comparison of the different procedures of light-based treatments to decide the best procedure and treatments.

This experimental treatment must be approved by the ethical committees of each country.

Tab 4 - Action mechanisms Hypothesis

- Anti-inflammatory effects?
- Stimulation of beta cells ?
- Stimulation of Cromaphine cells ?
- Stimulation of Microcirculation ?
- Cleaning of cell walls ?
- Hypolipemic effect ?
- Increasing of the " bioplasma" metabolism ?
- More effects associated?

Curiously this treatment was not approved by Local Ethic Committee of Firenze, Italy, but an Authority of that Committee suffering of diabetes type two required to be treated with laser and magnetic field! The mystery of the politic!

Our experimentation is conducted following the rules of the Good clinical practice established by the European Community, published on Italian Gazzetta Ufficiale, suppl n 191, 18 August 1997, n. 39 18/6/2001, and the Helsinki Declaration.

References

1. P. Ramdawon *Bioresonance Information Laser Therapy of Diabetes Mellitus A First Clinical Experience* In *Laser Florence 2001: A Window on the Laser Medicine World, Proceedings*, L. Longo, A. Hofstetter, M.L. Pascu, W. Waidelich, Editors, *Progress in Biomedical Optics and Imaging*, SPIE, Washington, Vol 4903, 2001, 146–153
2. T.V. Kovalyova, L.T. Pimenov, S.M. Denisov *Dynamics of hyperlipidemia and peripheral blood flow in patients with diabetes mellitus after the course of combined laser therapy in conditions of out-patient department*. The 2nd International Congress "Laser and Health-99": Materials. Moscow, 1999, 313–316
3. A.M. Makela *Clinical Observations on Effects of Laser Light on Blood Glucose Levels in Diabetics*, Laser Florence 2005 Abs, Suppl. *Laser in Medical Science*, Springer, London
4. L. Longo *First Experience of Laser Treatment of Diabetes in Italy*, Laser Florence 2004 Abs, Suppl. *Laser in Medical Science*, Springer, London
5. L. Longo *Terapia Laser*, USES, Firenze, 1986
6. L. Longo *Laser Therapy of Diabetes Type Two: Phase Two Of Clinical Experimentation*, Laser Florence 2005 Abs, Suppl. *Laser in Medical Science*, Springer, London
7. L. Longo *Diabetes and Lasers*, 16th International Congress of the International Society for Laser surgery and Medicine, ISLSM and 2nd Congress of World Federation of Laser Medicine and Surgery Societies, WFLSMS, Tokyo, 3–7 September, 2005
8. L. Longo *Laser Treatment of The Diabetes: Clinical Experimentation*, American Society for Laser Medicine and Surgery, 26th Annual Congress, Boston, April 5–9, 2006, Abs on *Laser in Surgery and Medicine*, Suppl. 2006, Wiley, London
9. L. Longo *Laser Treatment of the Diabetes* In *Private Hospital and Healthcare*, Campden, London, 2006

He-Ne Laser Irradiation Stimulates Proliferation and Migration of Diabetic Wounded Fibroblast Cells

Nicolette Houreld and Heidi Abrahamse

Abstract Diabetes mellitus is characterized by hyperglycemia which results in damage to the body, in particular to blood vessels and nerves. Atrophy of fat and muscle tissue shared with sensory loss can result in pressure sores, combined with delayed wound healing which can lead to lower-limb amputation.

A diabetic wound model was achieved by growing WS1 fibroblast cells in 22.6 mmol/l glucose and a wound was simulated by scraping the cell sheet. Diabetic wounded (DW) cells were irradiated on day 1, day 1 and 2 or day 1 and 4 with a He-Ne laser (632.8 nm) with 5 or 16 J/cm^2. Unirradiated DW cells were used as controls. The effect due to irradiation was determined by looking at cellular migration, viability (Trypan blue) and proliferation (IL-6 and bFGF expression and neutral red retention).

DW cells irradiated with 5 J/cm^2 on day 1 as well as day 1 and 4 showed an increase in cellular migration with complete wound closure by day 4. There was no negative effect on cellular viability and cells were able to normalize their cellular function, with an increase in IL-6, bFGF expression and neutral red retention. DW cells irradiated with 16 J/cm^2 showed an inhibition in cellular migration and a decrease in cellular viability and proliferation. The cumulative effect of irradiation on 2 consecutive days with both 5 and 16 J/cm^2 could be seen by the decrease in viability. Morphologically these cells showed an inhibition in migration and showed signs of stress. There was a decrease in proliferation in irradiated and unirradiated DW cells over time.

It is evident from this study that there is a cumulative effect, and that enough time between irradiations should be given when irradiating *in vitro*. A fluence of 5 J/cm^2 encouraged wound healing, while 16 J/cm^2 was inhibitive. The expression of IL-6 and neutral red retention was dependent on the stage of wound healing, as there was a decrease over time, thus it is important when choosing what parameters

N. Houreld and H. Abrahamse
Laser Research Group, Faculty of Health Sciences, University of Johannesburg, Doornfontein, South Africa, 2028, e-mail: habrahamse@uj.ac.za or nhoureld@uj.ac.za

R. Waynant and D.B. Tata (eds.), *Proceedings of Light-Activated Tissue Regeneration and Therapy Conference.*
© Springer Science + Business Media, LLC 2008

221

to evaluate and when to measure a response as there may be little or no observable effect. Phototherapy using the correct laser parameters appear to be beneficial to diabetic wound healing. The next phase of this study will be to link the *in vitro* findings with a 3D skin model which closely reproduces an *in vivo* situation.

Keywords: Diabetic wound healing, Laser induced stimulation, Laser light dose dependence on biological response.

Introduction

Wound healing is a complex procedure that involves the interaction of different cell types and growth factors. The events of wound healing reflect a finely balanced environment, leading to uncomplicated and rapid wound healing. Chronic wounds, for many reasons, have lost this fine balance [24]. It is well known and understood that diabetes mellitus (DM) has a negative effect on wound healing and it is accepted that diabetics are a higher surgical risk than non-diabetic individuals, their healing time is prolonged and they are at higher risk of infection. Some of these effects on wound healing include retardation of closure, delayed contraction, effects on granulocytes, defects in chemotaxis, interference with collagen synthesis and effects on red blood cells [29]. Diabetic patients suffer for months with various chronic wounds, particularly foot ulcers, which can become debilitating and lead to amputation. In 2000, the prevalence of diabetes worldwide was at 171 million; of this 814,000 were in South Africa [35]. This figure is estimated to go up to 1,286 million by the year 2030.

In an attempt to discover effective alternative treatments, Low Level Laser Therapy (LLLT), commonly known as photo-biostimulation or laser phototherapy, has emerged. Phototherapy is a highly effective therapeutic armamentarium for tissue repair and pain relief [10]. A large amount of research in laser phototherapy has been undertaken since the 1970s, and despite the advocacy of this treatment, there has been a reluctance to accept it due to a lack of an understanding of the underlying mechanisms of action [13]. The wide diversity in experimental protocols and parameters such as cell line, dose, waveform, treatment time, penetration distance, treatment area and treatment frequency make comparison of these studies difficult [2], and no therapeutic window for dosimetry and mechanism of action has been determined at the level of individual cells types [4].

Literature indicates that laser irradiation accelerates wound epithelization, granulation tissue formation, reduction of edema and inflammation, re-establishment of arterial, venous and lymph microcirculation, increased matrix synthesis, enhanced neovascularization, modulates the level of prostaglandin, enhances the action of macrophages, promotes fibroblast proliferation, facilitates collagen synthesis, fosters immunity, activates cellular metabolism and has an analgesic effect [7, 9, 27, 33]. Proteases and growth factors play an important role in regulating the fine balance of wound healing (destructive and reparative processes), and if disrupted in favor of degradation then delayed healing ensues, which is a trait of chronic wounds [6]. It has been shown that low level laser irradiation at certain fluences and wavelengths can enhance the release of growth factors and stimulate cell proliferation [3, 12, 15, 17, 19, 26, 28, 37, 39, 73].

A number of published studies, both *in vitro* and *in vivo* have reported that laser phototherapy has been found to effectively promote wound healing in diabetic models and patients (for a review see [18]). Laser treatment increases blood flow and raises local temperature and since no evidence has been found that laser therapy could aggravate diabetic symptoms it is recommended as an additional treatment modality for diabetic foot problems [23]. Diabetic patients have a 22-fold higher risk for non-traumatic foot amputation compared with the non-diabetic population [31]. In light of this data, any measure that could lower the rate of limb loss should be considered worthwhile.

Materials and Methods

Model: There is evidence that multiple mammalian cells respond to phototherapy. For wound healing studies, endothelial cells [26], fibroblasts [16, 17], keratinocytes [38], macrophages [36] and neutrophils [11] have been used. During wound healing, fibroblasts are critically involved in producing cytokines and are key players in the formation of granulation tissue through proliferation, migration, deposition and remodeling of the extracellular matrix and are involved in wound closure and contraction [34].

All materials and cells used were supplied by Adcock Ingram South Africa, unless specified otherwise. Normal human skin fibroblast cells, WS1 (CRL1502) purchased from the American Type Cell Culture (ATCC), were grown to 90% confluence in Eagles minimal essential medium (EMEM, 12-136F) containing 10% fetal bovine serum (FBS, 14-501A1), 2 mM L-glutamine (17-605F), 0.1 mM non-essential amino acids (NEAA, 13-14E), 1 mM sodium pyruvate (13-115E), and 1% Pen-Strep fungizone (17-745E). Cells were incubated at 37°C in 5% carbon dioxide and 85% humidity. A diabetic model was achieved *in vitro* by continuously growing WS1 cells in complete media containing additional D-glucose (17 mmol/l), [5, 14, 25, 32], thus simulating a diabetic condition. The media had a basal glucose concentration of 5.6 mmol/l, thus diabetic induced cells were grown in a total glucose concentration of 22.6 mmol/l.

To determine the effects of laser irradiation, cells were detached by trypsinization (1 ml/25 cm^2 0.25% trypsin-0.03% EDTA) and approximately 6×10^5 cells in 3 ml culture media were seeded into 3.3 cm diameter culture plates. Plates were incubated overnight to allow the cells to attach and a wound was simulated according to [14] and [30] by pressing a sterile pipette down onto the culture dish to scrape the cell sheet (Fig. 1a), thereby exposing a cell free zone in the center of the culture dish. The media was replaced and the cells were left at 37°C for 30 min before irradiation [30].

Fig. 1 A diabetic wound model was simulated by growing WS1 human fibroblast cells continuously in a total glucose concentration of 22.6 mMol D-glucose, and a wound was induced by the central scratch method (**a**), whereby the cell sheet is scratched with a sterile 1 ml pipette. 30 min post-wounding, cells were irradiated using a He-Ne laser (632.8 nm), (**b**), with 5 or 16 J/cm^2. The laser beam, which had a power density of 3 mW/cm^2, was expanded, reflected down towards the cells and clipped producing a Truncated Gaussian Beam with a spot size of 9.08 cm^2 (**c**)

Laser Irradiation: All lasers were supplied and set up by the National Laser Center (NLC) South Africa. Cells were irradiated in the dark from above with a Helium-Neon (He-Ne, 632.8 nm) laser (Spectraphysics, Model 127). The laser beam, which had a power density of approximately 3 mW/cm^2, was expanded, reflected and clipped producing a Truncated Gaussian Beam with a spot size of 9.08 cm^2 (Fig. 1b). Control cells were treated the same as irradiated cells, but the cells were not irradiated (0 J/cm^2). Output power was measured before each exposure and used to determine the duration of irradiation. Due to the expansion, reflection and clipping of the beam, approximately 21.3% of the power output was lost. Duration of irradiation was calculated at approximately 45 min or 2 h and 42 min for 5 or 16 J/cm^2 respectively. Cells were irradiated once either on 1 day, on 2 consecutive days, or on 2 non-consecutive days with 2 days between irradiations (irradiated on days 1 and 4). All experiments were repeated four times on different cell populations (n = 4).

Cellular Morphology and Viability: Post irradiation, various cellular and biochemical tests were done both on the culture media and on the cells themselves. Cellular morphology was examined by inverted microscopy (Olympus S.A., CKX 41). Cellular structure and activity was looked at by evaluating colony formation, cellular migration across the central scratch and haptotaxis (change in direction of growth). Following detachment of cells from the culture dish by trypsinization, cellular viability was determined using Trypan blue staining. Trypan blue is a vital stain which does not penetrate viable cells with an intact membrane. Non-viable cells have a damaged membrane and the stain enters the cells, staining them blue. The numbers of viable and non-viable cells are then counted and percentage viability calculated.

Cellular Proliferation: Proliferation of cells was determined by examining expression of interleukin-6 (IL-6) and basic fibroblast growth factor (bFGF), and retention of neutral red. bFGF plays an important role in cellular proliferation and differentiation during tissue regeneration and wound healing. bFGF was determined using the indirect Enzyme Linked ImmunoSorbent Assay (ELISA), using 1:10,000 monoclonal anti-fibroblast growth factor primary antibody (Sigma Aldrich S.A., F6162) and 1:4,000 goat anti-mouse IgG (Fab specific), horse-radish peroxidase-conjugated secondary antibody (Whitehead Scientific S.A., D0705). IL-6, a pleiotropic cytokine, expressed by a wide range of tissues and cells, including fibroblasts, plays a crucial role in the pathophysiology of wound healing and is involved in cellular proliferation. IL-6 was determined using the Human IL-6 ELISA Kit II (Scientific Group S.A., BD, 550799), which utilizes monoclonal anti-IL-6 primary antibody and streptavidin-horseradish peroxidase biotinylated anti-human IL-6 secondary antibody. The retention of neutral red was determined by a colourmetric assay which was read spectrophotometrically ($A_{550\ nm}$). This assay is based on the ability of living proliferating cells to take up and retain the dye in their lysozomes, and was done according to [1].

Hydroxyurea Treated Cells: To determine the true effect of laser irradiation on cellular migration and proliferation, hydroxyurea (Sigma, H8627), an anti-proliferative agent, was added at a concentration of 5 mM to normal wounded WS1 cells. Cells were irradiated on days 1 and 4 with a He-Ne laser with 5 J/cm^2. Post laser irradiation, cellular migration was assessed by light microscopy and proliferation by the XTT assay (Roche, 1465015). Mitochondrial succinate dehydrogenase in metabolically active proliferating cells cleaves the yellow tetrazolium salt, XTT, forming an orange formazan dye, which is measured spectrophotometrically ($A_{450\ nm}$).

Results

Morphology: DW cells irradiated with 5 J/cm^2 on day 1 showed signs of colony formation, migration and haptotaxis, while cells irradiated with 16 J/cm^2 showed no sign of migration and cells appeared stressed (Fig. 2a–c). Unirradiated control cells (0 J/cm^2) showed normal morphology and no signs of migration. When irradiated on 2 consecutive days (Fig. 2d–f), unirradiated control cells showed colony formation, haptotaxis and there was migration towards the central scratch by day 2. DW cells irradiated with 5 J/cm^2 showed signs of cellular migration and haptotaxis, although less than control cells. Cells irradiated with 16 J/cm^2 showed very little migration after irradiation on day 2. When irradiated on days 1 and 4, cells could be seen in the central scratch by day 4 and in most areas have made contact and the 'wound' is now closed (Fig. 2e–g). In unirradiated wounded cultures many cells had migrated into the central scratch, however there were still areas of incomplete wound closure, while cultures irradiated with 5 J/cm^2 showed complete wound closure. Cells irradiated with 16 J/cm^2 were less dense than unirradiated cultures, and few cells migrated into the central scratch, however cells still retained their basic morphology.

Viability: Percentage viability as determined by Trypan blue staining (Fig. 3a) showed a significant decrease in DW cells irradiated with 16 J/cm^2 on day 1 ($P < 0.05$) as well as on day 1 and 4 ($P < 0.001$). Irradiation with 5 J/cm^2 on day 1 or day 1 and 4 had no negative effect on cellular viability, with cells remaining above 90% viability, and there was an increase compared to cells irradiated with

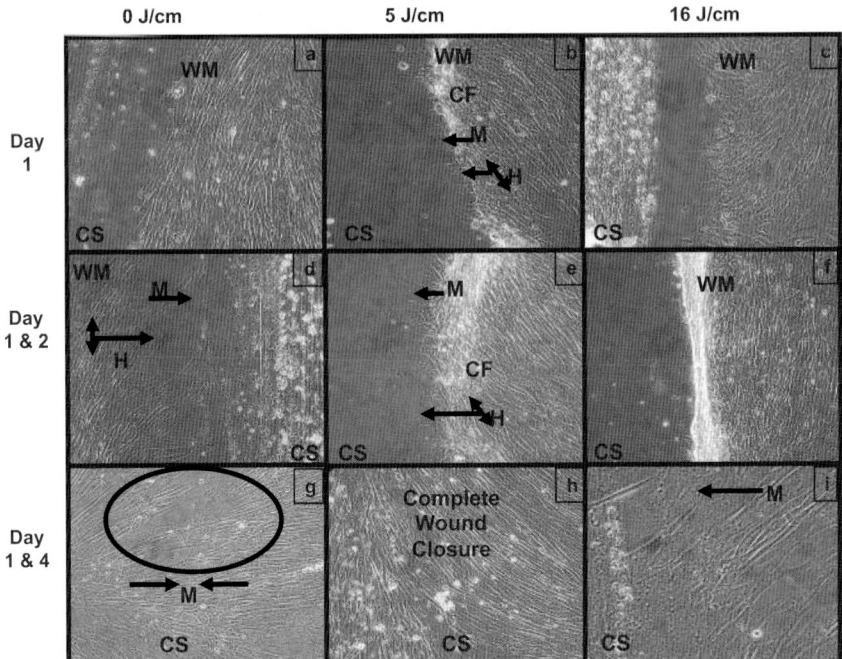

Fig. 2 Morphology of diabetic wounded cells irradiated at 632.8 nm with 5 or 16 J/cm² on various days. A wound was induced in cultures by the central scratch method; the central scratch (CS), wound margin (WM), colony formation (CF), migration of cells (M) and haptotaxis (H) could clearly be seen. Irradiation on day 1 with 5 J/cm² (**b**) stimulated colony formation, haptotaxis and migration of WS1 cells, while there was no migration in control cells (**a**) and cells irradiated with 16 J/cm² (**c**). When irradiated on day 1 and 2, control cells (**d**) showed more migration and haptotaxis than cells irradiated with 5 (**e**) or 16 J/cm² (**f**). Irradiation on day 1 and 4 with 5 J/cm² (**h**) stimulated cell migration and proliferation to the point that there was complete wound closure, while control cells (**g**) and DW irradiated with 16 J/cm² (**i**) showed incomplete wound closure

16 J/cm² ($P = 0.07$ and $P < 0.01$ respectively). All DW cells irradiated on day 1 and 2 showed a decrease in viability, pointing to a cumulative effect.

Proliferation: Cellular proliferation was determined through assessment of IL-6 (Fig. 3b), bFGF (Fig. 3c) and neutral red retention (Fig. 3d). Overall, DW cells irradiated with 5 J/cm² showed an increase in cellular proliferation, while cells irradiated with 16 J/cm² showed a decrease. There was a significant increase in IL-6 and neutral red retention in cells irradiated with 5 J/cm² on day 1 ($P < 0.05$ and $P < 0.01$ respectively). There was a delayed increase in bFGF which showed a significant increase by day 2 ($P < 0.05$), while IL-6 expression and neutral red retention showed no significant change. When irradiated on 2 non-consecutive days, there was a significant increase in IL-6 and bFGF ($P < 0.05$ and $P < 0.01$ respectively). Irradiation with 16 J/cm² produced a significant decrease in bFGF and NR retention when irradiated on day 1 ($P < 0.05$ and $P < 0.001$ respectively) as well as on day 1 and 2 ($P < 0.001$ and $P < 0.05$ respectively). Irradiation on day 1 and 4 with 16 J/cm² produced a significant decrease in IL-6 expression ($P = 0.05$).

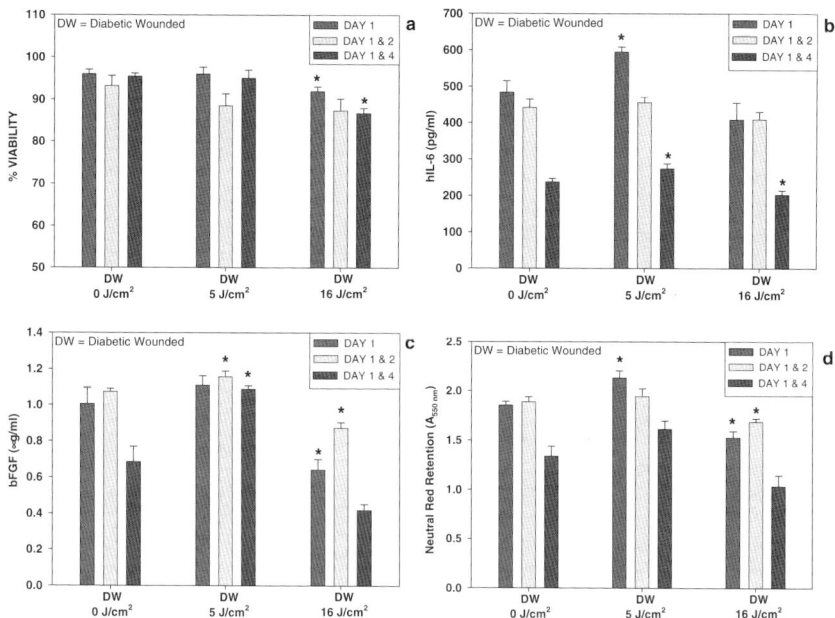

Fig. 3 Cellular viability (**a**) and proliferation (**b-d**) was determined in diabetic wounded (DW) cells irradiated at 632.8 nm with 5 or 16 J/cm^2 on various days. Viability was assessed by Trypan blue staining, and the percentage viability calculated, while the expression of IL 6 (**b**), bFGF (**c**) and the retention of neutral red (**d**) assessed cellular proliferation. Irradiation of DW cells to 5 J/cm^2 had no negative effect on cellular viability and stimulated proliferation, while cells irradiated with 16 J/cm^2 showed the opposite, there was a decrease in viability and proliferation

When comparing cells irradiated with 5 J/cm^2 to cells irradiated with 16 J/cm^2, all DW cells irradiated with 16 J/cm^2 showed a significant decrease in NR retention (P < 0.001 for day 1, P < 0.05 for day 2, and P < 0.01 for day 1 and 4) and bFGF expression (P < 0.001), while there was only a significant decrease in the expression of IL-6 when irradiated on day 1 and on day 1 and 4 (P < 0.01). As time increased, there is a decrease in the expression of IL-6 and bFGF and neutral red retention, and this decrease was significant by day 4 (with the exception of bFGF expression in DW cells irradiated with 5 J/cm^2).

Hydroxyurea Treated Cells: Hydroxyurea (HU), at a concentration was of 5 mM was added to normal, non-diabetic wounded cells and cellular migration and proliferation was determined post laser irradiation [40]. HU treated cells (Fig. 4c–d) had fewer cells in the central scratch compared to non treated cells (Fig. 4a–b). When HU treated cells were irradiated with 5 J/cm^2, there was more cells in the central scratch than non irradiated treated cells. There was complete wound closure in irradiated non-HU treated cells, again indicating a stimulatory effect of such a fluence. The XTT proliferation assay (Fig. 5) showed a significant decrease in HU treated cells compared to non-HU treated cells (P = 0.01). XTT ration such a fluencetetic wounded cells and cellular migration and proliferation was determined post laser irradiation. Irradiated HU treated cells showed an increase compared to

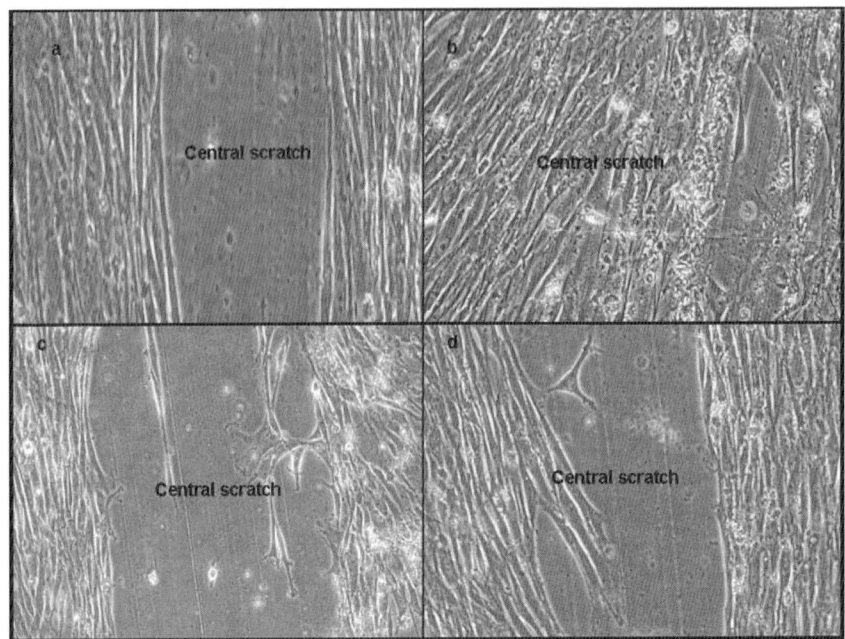

Fig. 4 Morphology of wounded hydroxyurea (HU) treated WS1 cells. Cells were then irradiated with 5 J/cm^2 on days 1 and 4 with a He-Ne laser. There were few cells in the central scratch of the HU treated non irradiated cells (**c**) compared to irradiated cells (**d**). There was complete wound closure in irradiated HU non treated cells (**b**) as opposed to the same cells which received no laser irradiation (**a**)

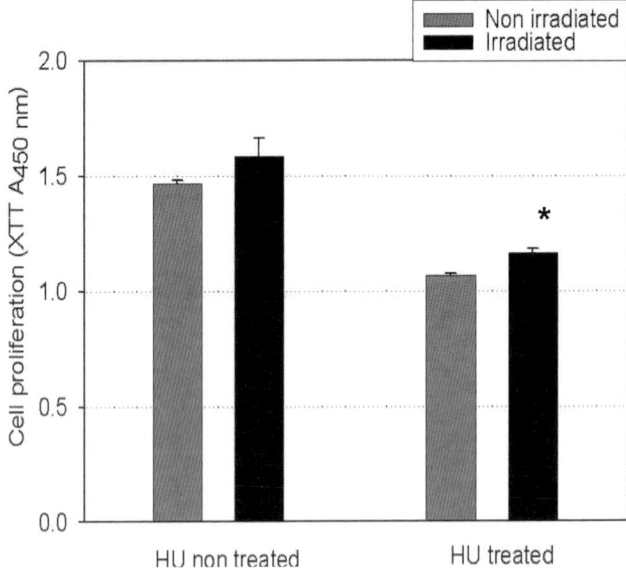

Fig. 5 The effect of laser phototherapy (632.8 nm at 5 J/cm^2) on 5 mM hydroxyurea (HU) treated cells was assessed using XTT. HU treated cells showed a decrease in cell proliferation compared to non treated cells. HU treated irradiated cells showed a significant increase compared to non irradiated cells (P = 0.01)

non irradiated treated cells (P = 0.01), while the increase in irradiated non treated cells showed an insignificant increase [40].

Conclusion

There is no hard and fast rule concerning treatment intervals, however this study demonstrated that appropriate time between treatments was more effective than irradiations that were close together [20]. This appears to be because of the cumulative effect, and what remains from the previous irradiation is added to the current irradiation, such that the accumulated dose may be above the biostimulating range entering the inhibitory range. When enough time is given between irradiations, cells have time to respond and adapt, such that they respond better to subsequent irradiations. Diabetic wounded human skin fibroblast cells exposed to 5 J/cm^2 *in vitro* to a He-Ne laser responded better than unexposed control cells, there was a higher rate of migration across the central scratch, with complete wound closure by day 4 and there was an increase in proliferation. Cells exposed to 16 J/cm^2 showed little migration from the wound margin with incomplete wound closure and a decrease in proliferation.

Irradiating diabetic induced WS1 cells to a He-Ne laser to lower fluences was stimulatory on cellular migration and proliferation, important factors in wound healing, while higher fluences such as 16 J/cm^2 was inhibitory. A dose of 16 J/cm^2 may not sound very high, however, when exposing cells *in vitro* the energy absorbed by the cells is not dispersed and passed on to neighboring cells as occurs *in vivo* leading to systemic effects. This energy is limited to the culture environment and can become above the threshold limit [21]. There was a decrease in proliferation as time increased, thus not only is it important in choosing the correct laser and irradiation parameters, but also what parameters you going to measure and when they will be measured [22]. Overall, this study showed that phototherapy at the correct parameters is beneficial to diabetic wound healing in vitro and the future role of phototherapy is as an adjuvant treatment aiming at reducing costs and raising patients' quality of life.

HU-mediated anti-proliferative effect on human skin fibroblast cells can be used as a controlling agent in assessing cellular proliferation and migration to determine the effect of laser phototherapy. Although there was a decrease in the number of cells in the central scratch, corresponding with a decrease in the XTT assay, cells still appeared morphologically normal and were able to migrate towards the central scratch. Thus HU inhibited cell proliferation. The few cells across the central scratch of the non irradiated cells can be explained as due to lack of stimuli required to stimulate an increase in the rate of cell migration compared to cells irradiated with 5 J/cm^2. HU treated cells were limited in their response to the injury as cell proliferation was inhibited. This suggests that the presence of the cells in the central scratch would only be attributed to migration as the effect of cell number or proliferation could be excluded. Despite the presence of an anti-proliferative agent, cells irradiated with 5 J/cm^2 were able to migrate and there was an increase in cell proliferation. This study showed that phototherapy using the correct parameters promotes cell migration and proliferation of wounded cells *in vitro* [40].

In vitro studies cannot provide full information about how healthy volunteers and patients will respond due to the variety of uncontrollable variables [8]. Optimal parameters can be determined *in vitro*, which can then be applied and further optimized *in vivo*. Thus the next stage of this study is to link these *in vitro* findings with an artificial skin construct which closely reproduces an *in vivo* situation.

References

1. Abe T., Hara Y., Abe Y., Aida Y. and Maeda K. (1998) Serum or growth factor deprivation induces the expression of alkaline phosphatase in human gingival fibroblasts. *J Dent Res* **77:** 1700–1707
2. Abrahamse H., Hawkins D. and Houreld N. (2006) Effect of wavelength and fluence on morphology, cellular and genetic integrity of diabetic wounded human skin fibroblasts. *Proc. SPIE* **6140:** 41—53
3. Almeida-Lopes L., Rigau J., Zangaro R.A., Guidugli-Neto J. and Jaeger M.M. (2001) Comparison of the low level laser therapy effects on cultured human gingival fibroblasts proliferation using different irradiance and same fluence. *Laser Surg Med* **29(2):** 179–184
4. Coombe A.R., Ho C.T., Darendeliler M.A., Hunter N., Philips J.R., Chapple C.C. and Yum L.W. (2001) The effects of low-level laser irradiation on osetoblastic cells. *Clin Orthod Res* **4(1):** 3–14
5. Coughlan M.T., Oliva K., Georgiou H.M., Permezel J.M. and Rice G.E. (2001) Glucose-induced release of tumour necrosis factor-alpha from human placental and adipose tissues in gestational diabetes mellitus. *Diabetic Med* **18(11):** 921–927
6. Cullen B., Watt P.W., Lundqvist C., Silcock D., Schmidt R.J., Bogan D. and Light N.D. (2002) The role of oxidized regenerated cellulose/collagen in chronic wound repair and its potential mechanism of action. *Int J Biochem Cell Biol* **34(12):** 1544–1556
7. Dyson M. (1991) Cellular and Sub-cellular aspects of low level laser therapy (LLLT). *Progress in Laser Therapy: Selected papers from the October 1990 ILTA Congress.* Published by Wiley. New York and Brisbane pp 221–222
8. Dyson M. (2006) Primary, secondary and tertiary effects of phototherapy: a review. *Proc. SPIE* **6140:** 614005-1–614002-12
9. Enwemeka C.S. (1988) Laser biostimulation of healing wounds: specific effects and mechanisms of action. *J Orthopaed Sports Physiother* **9(10):** 333–338
10. Enwemeka C.S., Parker J.C., Dowdy D.S., Harkness E.E., Sanford L.E. and Woodruff L.D. (2004) The efficacy of low-power lasers in tissue repair and pain control: a meta-analysis study. *Photomed Laser Surg* **22(4):** 323–329
11. Fujimaki Y., Shimoyama T., Liu Q., Umeda T., Nakaji S. and Sugawara K. (2003) Low level laser irradiation attenuates production of reactive oxygen species by human neutrophils. *J Clin Laser Med Surg* **21:** 165–170
12. Gavish L., Asher Y., Becker Y. and Kleinman Y. (2004) Low level laser irradiation stimulates mitochondrial membrane potential and disperses subnuclear promyelocytic leukemia protein. *Laser Surg Med* **35(5):** 369–376
13. Hamblin M.R. and Demidova T.N. (2006) Mechanisms of low level laser therapy. *Proc. SPIE* **6140:** 614001-1–614001-12
14. Hamuro M., Polan J., Natarajan M. and Mohan S. (2002) High glucose induced nuclear factor kappa B mediated inhibition of endothelial cell migration. *Atherosclerosis* **162(2):** 277–287

15. Hawkins D. and Abrahamse H. (2004) The release of interleukin-6 after low level laser therapy and the effect on migration and proliferation of human skin fibroblasts. *Med Technol S Afr* **18(1):** 11–15
16. Hawkins D. and Abrahamse H. (2005) Biological effects of helium-neon laser irradiation on normal and wounded human skin fibroblasts. *Photomed Laser Surg* **23(3):** 251–259
17. Hawkins D. and Abrahamse H. (2006) The role of laser fluence in cell viability, proliferation and membrane integrity of wounded human skin fibroblasts following helium-neon laser irradiation. *Laser Surg Med* **38:** 74–83
18. Houreld N. and Abrahamse H. (2005) Low-level laser therapy for diabetic wound healing. *Diabet Foot* **8(4):** 182–193
19. Houreld N. and Abrahamse H. (2007a) Irradiation with a 632.8 nm helium-neon laser with 5 J/cm^2 stimulates proliferation and expression of interleukin-6 in diabetic wounded fibroblast cells. *Diabetes Technology & Therapeutics* **9(5)** (accepted for publication)
20. Houreld N. and Abrahamse H. (2007b) Frequency of helium-neon laser irradiation on viability and cytotoxicity of diabetic wounded fibroblast cells. *Photomed Laser Surg* **25(6):** 474–481
21. Houreld N. and Abrahamse H. (2007c) *In Vitro* exposure of wounded diabetic fibroblast cells to a helium-neon laser at 5 and 16 J/cm^2. *Photomed Laser Surg* **25(2):** 78–84
22. Houreld N. and Abrahamse H. (2007d) Laser light influences cellular viability and proliferation in diabetic-wounded fibroblast cells in a dose and wavelength-dependent manner. *Lasers Med Sci.* DOI 10.1007/s10103-007-0445-y
23. Kleinman K., Simmer S. and Braksma Y. (1996) Low power laser therapy in patients with diabetic foot ulcers: early and long term outcome. *Laser Ther* **8(2):** 205–208
24. Komarcevic A. (2000) The modern approach to wound treatment. *Med Pregl* **53(7–8):** 363–368
25. McDermott A.M., Kern T.S. and Murphy C.J. (1998) The effect of elevated extracellular glucose on migration, adhesion and proliferation of SV40 transformed human corneal epithelial cells. *Curr Eye Res* **17(9):** 924–932
26. Moore P., Ridgway T.D., Higbee R.G., Howard E.W. and Lucroy M.D. (2005) Effect of wavelength on low-intensity laser irradiation-stimulated cell proliferation in vitro. *Laser Surg Med* **36:** 8–12
27. Moshkovska T. and Mayberry J. (2004) It is time to test low level laser therapy in Great Britan. *Post Grad Med J* **81(957):** 436–441
28. Novoselova E.G., Cherenkov D.A., Glushkova O.V., Novoselova T.V., Chudnovskii V.M., Iusupov V.I. and Fesenko E.E. (2006) Effect of low-intensity laser irradiation (632.8 nm) on immune cells isolated from mice [abstract]. *Biofizika* **51(3):** 209–518
29. Pearl S.H. and Kanat I.O. (1988) Diabetes and healing: a review of the literature. *J Foot Surg* **27(3):** 268–270
30. Rigau J., Sun C., Trelles M.A. and Berns M. (1995) Effects of the 633 nm laser on the behavior and morphology of primary fibroblasts in culture, Karu T. and Young A. (Eds) *Effects of low power light on biological systems.* Barcelona Spain, Progress in Biomedical Optics, pp 38–42
31. Trautner C., Haastert B., Giani G. and Berger M. (1996) Incidence of lower limb amputations and diabetes. *Diabetes Care* **19:** 1006–1009
32. Vinck E.M., Cagnie B.J., Cornelissen M.J., Declercq H.A. and Cambier D.C. (2005) Green light emitting diode irradiation enhances fibroblast growth impaired by high glucose level. *Photomed Laser Surg* **23(2):** 167–171
33. Walsh L.J. (1997) The current status of low level laser therapy in dentistry. Part 1. Soft tissue applications. *Aust Dent J* **42(4):** 247–254
34. Wong L.S. and Martins-Green M. (2004) Firsthand cigarette smoke alters fibroblast migration and survival: implications for impaired healing. *Wound Repair and Regeneration* **12(4):** 471–484
35. World Health Organization (2006) www.who.int/diabetes/facts/en/
36. Young S., Bolton P., Dyson M., Harvey W. and Diamantopoulos C. (1989) Macrophage responsiveness to light therapy. *Laser Surg Med* **9:** 497–505

37. Yu W., Naim J.O. and Lanzafame R.J. (2001) Effect of photostimulation on wound healing in diabetic mice. *Laser Surg Med* **20(1):** 56–63

38. Yu H.S., Wu C.L. Yu C.L., Kao Y.H. and Chiou M.H. (2003) Helium-neon laser irradiation stimulates migration and proliferation in melanocytes and induces repigmentation in segmental-type vitiligo. *J Invest Dermatol* **120:** 56–64

39. Zhevago N.A. and Samoilova K.A. (2006) Pro- and anti-inflammatory cytokine content in human peripheral blood after its transcutaneous (in vivo) and direct (in vitro) irradiation with polychromatic visible and infrared light. *Photomed Laser Surg* **24(2):** 129–139

40. Zungu I.L., Mbene A.B., Hawkins Evans D.H., Houreld N.N. and Abrahamse H. (2007) Phototherapy promotes cell migration in the presence of hydroxyurea. *Laser Med Sci* (submitted)

The Role of Colostrum Proline-Rich Polypeptides in Human Immunological and Neurological Health

Andrew Keech, John I. Buhmeyer, and Richard Kolt

Abstract Mr. Kolt describes his work with colostrum in the United States, Mexico and in Africa where he treats AIDs, autism and many other diseases. Mr. Kolt also uses photobiomodulation with LEDs.

Keywords: Colostrum, AIDs.

Colostrum, the first milk produced by a mother after her child is born, is a rich source of beneficial components, including immunological factors, anti-inflammatory factors, gastrointestinal health factors, growth factors, antioxidant and anti aging factors. It has been used for thousands of years for its health-promoting benefits. As colostrum is mother's milk, the first meal a baby has in this world, it is completely safe and without the dangerous side effects common to pharmaceuticals. For older children and adults, bovine colostrum has proven to be the most efficacious as it contains much higher levels of Immunoglobulin-G (IgG) than human colostrum (which predominantly contains IgA) and because it is produced in large volumes by cows above and beyond what their calves require.

One of the components of colostrum that is of particular interest is PRP, or Proline-Rich Polypeptides. PRP is also known as colostrinin, colostrinine, transfer factor, and other names, but they are all essentially the same fraction of colostrum. Actually a group of related polypeptides, PRPs from colostrum have demonstrated remarkable immunologic and neurologic properties [1]. PRPs immunological function relate to their ability to modulate [2] and stabilize many biological processes in the body including cytokine [3] and immune activity. A polypeptide is a simple string of greater than ten amino acids with no secondary structure like a true protein would have, such as folding, helices, sheets, and so forth. Proline Rich Peptides also exist in colostrum (PRP3,4,5), these are peptides less than ten amino acids. Their small sizes and uncomplicated structure make it easy for it to pass through membranes in the body

R. Kolt
Rejuvalight LED Technology, 220 E. Wetmore #110, Tucson, AZ 85705 USA,
e-mail: mkolt07@aol.com

R. Waynant and D.B. Tata (eds.), *Proceedings of Light-Activated Tissue Regeneration and Therapy Conference.*
© Springer Science + Business Media, LLC 2008

and skin cells without being altered or digested. Thus it can be readily absorbed directly into the bloodstream, unlike most of the other components of colostrum which must pass through the digestive tract where they are exposed to powerful gastric acids and digestive enzymes. Proline-Rich Polypeptides are so named because they contain a higher percentage of the amino acid proline than most other peptides and proteins.

Brief Review of Immune Function

The immune system is made up of two basic parts:

Innate immune system – nonspecific immunity, including barrier defenses (skin, intestinal lining, brain-blood barrier, macrophages, neutrophils, dentritic cells, etc.), airway cilia (remove antigens and small particles that are breathed in), mucus, chemicals such as lactoferrin, lactoperoxidase and lysozyme which are potent killers of pathogens, strong acids in the stomach, and so forth [4].

Adaptive immune system – system which consists of immune and scavenger cells (B- and T-lymphocytes, NK cells, mast cells, and other blood cells) and which reacts to specific antigens.

The adaptive immune system furthermore has two response modes.

Humoral Immunity – production of specific antibodies in response to an antigen, mediated by B-cells.

Cell-Mediated Immunity – production of cytotoxic lymphocytes, activated macrophages and NK cells, and cytokines in response to an antigen, mediated by T-cells.

Lymphocytes come in several different varieties. B-lymphocytes (B-cells) are produced in the bone marrow, and T-lymphocytes (T-cells) are produced in the thymus gland. T-cells are further broken down into:

Cytotoxic or killer T-cells (CD8+) – the cells which actually kill invading pathogens.
Helper T cells (CD4) – cells which direct the immune response through the secretion of cytokines [5, 6].
Suppressor T-cells – inhibit the production of cytotoxic T-cells when no longer needed to prevent excess tissue damage.
Memory T-cells – retain memory of an encountered antigen so that if encountered again the response time will be much shorter.

Helper T-cells secrete various cytokines to stimulate the production and differentiation of cytotoxic T-cells and B-cells, which produce antibodies. They also attract neutrophils (white blood cells) and stimulate macrophages to engulf and destroy pathogens.
Helper T-cells also have subsets [8]:

TH1 cells – secrete the cytokines tumor necrosis factor-alpha (TNF-a) and interleukin 12 (IL-12) which control cell-mediated immunity; TH1 activation can inhibit TH2 cell activation.

TH2 cells – secrete the cytokines IL-4, IL-5 which control humoral immunity; TH2 cell activation can inhibit TH1 cell activation.

Regulatory TH3 (Tr) cells [9] – at least three different type of regulatory TH cells exist:

Type 1 (TO) – secrete large amounts of IL-10 and low-to-moderate amounts of transforming growth factor-beta (TGF-a), may help terminate TH1-related inflammatory responses.
Type 3 (*Tr3*) – primarily secrete TGF-p, regulates multiple facets of immune response.
CD4+CD25+ – inhibit immune responses through direct cell-to-cell contact.

Immunological Properties of PRP

PRPs, are like all cytokines, and are molecular signaling device. Cytokines can act on the same cell that produced them (autocrine), on other cells near them (paracrine), and/or on distant cells (endocrine). The same cytokine may be produced by different types of cells, and often the activity of different cytokines is redundant. They often are produced in a cascade, such as is the case with TNF-a, which is the primary cytokine in the inflammatory response cascade. Cytokines interact with receptors on target cell membranes "telling" the target cell to start doing something or to stop doing something, such as synthesizing proteins.

The immunological properties of the PRP fraction of colostrum were first identified in Poland by Janusz and Lisowski et al., in 1979 [10–13]. They noted that PRP from ovine colostrum acted as a hormone in the thymus gland by stimulating thymocytes (lymphocytes which originate in the thy,15,1mus6gland) to differentiate and become activated 4 as either helper or suppressor T-cells, and furthermore that this effect was reversible [17–19]. PRP also induces the growth and differentiation of B-cells [20]. PRP increases the permeability of blood vessels in the skin, a typical pro-inflammatory action. PRP has also induces leukocyte proliferation and mitogenic stimulation of peripheral blood cells to produce a variety of cytokines [21, 22]. PRPs have also been shown to be potent stimulators of Natural Killer (NK) cell activity and induce the differentiation and maturation of monocytes and macrophages [23].

Further studies have elucidated the actual molecular mechanisms by which PRP is able to modulate immunity. A study by Inglot et al., in 1996 showed that PRP could stimulate the production of two major cytokines, tumor necrosis factor-alpha (TNF-a) and interferon-gamma (INF-y) in white blood cells [24], peritoneal cells [25], and the placenta and amniotic membrane [26]. Later it was discovered that PRP stimulates other cytokines as well, such as IL-6 and IL-10. PRP is non-species specific, meaning that PRP from bovine colostrum works on humans just as well as human PRP [27].

PRP acts on specific surface membrane receptors on target cells [29]. These receptors then release intracellular signaling peptides which act to either stimulate or inhibit actions of the cell, such as stimulating a thymocyte to differentiate into a mature lymphocyte or a lymphocyte to produce TNF-a.

PRPs obtained from colostrum can be classified into five sub-classes; PRP1 (inactive), PRP2a & 2b (very active interferon modulator), PRP3a & 3b (very active pro-inflammatory cytokine modulator), PRP4 (less active, lower concentration, unstudied), and PRP5 (again less active, lower concentration, unstudied) [28]. Chemically PRPs show amino acid sequence homology to three protein precursors: annexin, beta casein, and a hypothetical beta-casein homolog [14]. Further study may make it possible to selectively utilize PRP fractions to produce certain effects, such as selectively stimulating T-cells or selectively turning on or off the production of certain cytokines or cytokine cascades.

Some efforts to that end have already been attempted using active fragments of PRPs [29–32].

The use of PRP against viral infections has demonstrated some success [33]. In experimental systems, PRP has shown to induce immunity to vesicular stomatitis virus (VSV) [34, 35] (a close relative of the rabies virus used in experimental systems to study the properties of Rhabdoviruses), herpes viruses [36–40], Epstein-Barr virus [41], **HIV** [42–55], influenza viruses [57, 58], rotavirus gastroenteritis [59–68] child polio [69], other viral **diseases** [70–80] other immune disorders [81, 82] and measles [83]. PRP may have some promise in helping to modulate the bodies cytokine response, thus regain its balance in the dangerous and often fatal "cytokine storm" produced by H5N1 influenza virus (bird flu) in which the inflammatory response of the immune system is stimulated out of control [84].

PRPs have also shown some promise in the treatment of autoimmune disorders, conditions where the body begins to produce antibodies against its own tissues [85]. In one experiment, PRPs were used to reverse an experimentally induced autoimmune response to red blood cells [86]. Research has shown varied success in several common **allergies** [87–93], autoimmune disorders, including rheumatoid arthritis [94–98] asthma [99, 100], multiple sclerosis [101, 102] and chronic fatigue syndrome [103]. The mechanism of action was thought to be either an inhibition of the immune system to return to normal levels [41] or the targeting of specific viruses associated with the autoimmune disorder, such as Epstein-Barr Virus and Human Herpes Virus-6 in the case of chronic fatigue syndrome [43]. However, modulation of the human cytokine system could be a mechanism of action. PRPS have also been investigated as an adjunctive supplement for cancer [104, 105]

PRP An Neurological Health

One of the most exciting areas of PRP research has been in the area of neurological health, particularly in its use with Alzheimer's Disease. Preliminary research has shown marked improvement in cognitive skills in patients with mild to moderate Alzheimer's [106–108]. PRP has also been shown to boost cognitive skills in both young and aged rats [109] and was shown to both boost spatial learning and memory in the aged rats [100]. Long-term memory retention in one-day old chicks was similarly enhanced [111].

How PRP accomplishes this is still not clear. Several studies have shown that PRP inhibits nitric oxide (NO) production in cells? [102, 103]. NO and oxidative stress may enhance neurodegenerative processes. Another study showed that PRP reduced intracellular of reactive oxygen species (ROS) through down-regulation of 4HNE-mediated lipid peroxidation [9]. ROS also contributes to cellular degenerative processes. However, another study found that PRP reduced the aggregation of amyloid-beta fibers *in vitro* [14]. Amyloid-beta is the main constituent of the amyloid plaques associated with Alzheimer's Disease.

Colostrinin, a PRP product from ReGen Therapeutics, is currently undergoing Phase II clinical trials in England for use in Alzheimer's.

PRPs also activate Natural Killer (NK) Cells, which are nonspecific cytotoxic cells which kill any infected cell not recognized as "self". In fact, in experimental systems PRPs increased the activity of NK cells from about 20 lytic units (standard American male reading today) to 248 lytic units [15, 6] (strongly balanced immune system) over the normal immune response, which was five times higher than any other product tested. This is important in disease states particularly as lymphocytes and NK cells are typically suppressed in these conditions, including cancer and infectious diseases, so can help restore the normal balance of these cells.

Commercial Availability of PRPS

PRPs extracted from raw bovine colostrum can be taken either as an oral liquid or convenient mouth spray. These PRP solutions have a pleasant slight vanilla flavor. Traditional forms of PRPs exist as Transfer Factor tablet or powder by 1ife Research LLC., Sandy. UT. However uptake is much faster with liquid PRP products. The liquid PRP products can are available under the following names: Cytolog, Cytolox, FluNox, Immulox, Isamune, Receptol, Viralox, and Virasol. All are manufactured by Advanced Protein Systems LLC., Phoenix, AZ, USA. Careful processing of raw colostrum is very important to maintain the bioactivity of these PRPs [17, 118–124]

References

1. Steven J. Bock, PRPS and its clinical applications. *International Journal of Integrative Medicine.*
2. Solomons, NW, Modulation of the immune system and the response against pathogens with bovine colostrum concentrates. *European Journal of Clinical Nutrition* 56(Suppl. 3):S24–S28 (2002).
3. Zimecki, M, Artym, J. Therapeutic properties of proteins and peptides from colostrum and milk. *PostOpy Higieny i Medcyny Dos°wiadczalne j* 59:309–323 (2005).

4. Chaplin, DD. Overview of the human immune response. *Journal of Allergy and Clinical Immunology* 117(Suppl. 2):S430–S435 (2006).

5. Mosmann, TR, Cherwinski, H, Bond, MW, et al. Two types of murine helper T cell clone. I. Definition according to profiles of lymphokine activities and secreted proteins. *Journal of Immunology* 136:2348–2357 (1986).

6. Mosmann, TR, Coffman, RL. TH1 and TH2 cells: different patterns of lymphokine secretion lead to different functional properties. *Annual Reviews of Immunology* 7:145–173 (1989).

7. Mosmann, TR, Sad, S. The expanding universe of T-cell subsets: Th1, Th2 and more. *Immunology Today* 17:138–146 (1996).

8. Abbas, AK, Murphy, KM, Sher, A. Functional diversity of helper T lymphocytes. *Nature* 383:787–793 (1996).

9. McGuirk, P, Mills, KH. Pathogen-specific regulatory T cells provoke a shift in the Th1/Th2 paradigm in immunity to infectious diseases. *Trends in Immunology* 23:450–455 (2002).

10. Wieczorek, Z, Zimecki, M, Janusz, M, Staroscik, K, Lisowski, J. Proline-rich polypeptide from ovine colostrum: its effect on skin permeability and on the immune response. *Immunology* 36(4):875–881 (1979)

11. Staroscik, K, Janusz, M, Zimecki, M, Wieczorek, Z, Lisowski, J, Immunologically active nonapeptide fragment of a proline-rich polypeptide from ovine colostrum: amino acid sequence and immunoregulatory properties. *Molecular Immunology* 20(12):1277–1282 (1983).

12. Janusz, M, Staroscik, K, Zimecki, M, Wieczorek, Z, Lisowski, J. A proline-rich polypeptide (PRP) with immunoregulatory properties isolated from ovine colostrum. Murine thymocytes have on their surface a receptor specific for PRP. *Archivum immunologiae et therapiae experimentalis (Warszava)* 34(4):427–436 (1986).

13. Janusz, M, Lisowski, J. Proline-rich polypeptide (PRP)–an immunomodulatory peptide from ovine colostrum. *Archivum immunologiae et therapiae experimentalis (Warszava)* 41(5–6):275–279 (1993).

14. Zimecki, M, Staroscik, K, Janusz, M, Lisowski, J, Wieczorek, Z. The inhibitory activity of a proline-rich polypeptide (PRP) on the immune response to polyvinylpyrrolidone (PVP). *Archivum immunologiae at therapiae experimentalis (Warszava)* 31(6):895–903 (1983).

15. Wieczorek, Z, Zimecki, M, Spiegel, K, Lisowski, J, Janusz, M. Differentiation of T cells into helper cells from immature precursors: identification of a target cell for a proline-rich polypeptide (PRP). *Archivum immunologiae at therapiae experimentalis (Warszava)* 37(3–4):313–322 (1989).

16. Zimecki, M, Pierce, CW, Janusz, M, Wieczorek, Z, Lisowski, J. Proliferative response of T lymphocytes to a proline-rich polypeptide (PRP): PRP mimics mitogenic activity of II-1. *Archivum immunologiae et therapiae experimentalis (Warszava)* 35(3):339–349 (1987).

17. Zimecki, M, Janusz, M, Staroscik, K, Wieczorek, Z, Lisowski, J. Immunological activity of a proline-rich polypeptide from ovine colostrum. *Archivum immunologiae et therapiae experimentalis (Warszava)* 26(1–6):23–29 (1978).

18. Zimecki, M, Lisowski, J, Hraba, T, Wieczorek, Z, Janusz, M, Staroscik, K. The effect of a proline-rich polypeptide (PRP) on the humoral immune response. II. PRP induces differentiation of helper cells from glass-nonadherent thymocytes (NAT) and suppressor cells from glass-adherent thymocytes (GAT). *Archivum immunologiae at therapiae experimentalis (Warszava)* 32(2):197–201 (1984).

19. Lisowski, J, Wieczorek, Z, Janusz, M, Zimecki, M. Proline-rich polypeptide (PRP) from ovine colostrum. Bi-directional modulation of binding of peanut agglutinin, resistance to hydrocortisone, and helper activity in murine thymocytes. *Archivum immunologiae et therapiae experimentalis (Warszava)* 36(4):381–393 (1988).

20. Julius, MH, Janusz, M, Lisowski, J. A colostral protein that induces the growth and differentiation of resting B lymphocytes. *Journal of Immunology* 140(5):1366–1371 (1988).

21. Kruzel, ML, Janusz, M, Lisowski, J, Fischleigh, RV, Georgiades, JA. Towards an understanding of biological role of colostrinin peptides. *Journal of Molecular Neuroscience* 17 (3):379–389 (2001).

22. Boldogh, I, Liebenthal, D, Hughes, TK, Juelich, TL, Georgiades, JA, Kruzel, ML, Stanton, GJ. Modulation of 4HNE-mediated signaling by proline-rich peptides from ovine colostrum. *Journal of Molecular Neuroscience* 20(2):125–134 (2003).

23. Kubis, A, Marcinkowska, E, Janusz, M, Lisowski, J. Studies on the mechanism of action of a prolinerich polypeptide complex (PRP): effect on the stage of cell differentiation. *Peptides* 26(11):2188–2192 (2005).

24. Inglot, AD, Janusz, M, Lisowski, J. Colostrinine: a proline-rich polypeptide from ovine colostrum is a modest cytokine inducer in human leukocytes. *Archivum immunologiae et therapiae experimentalis* (*Warszava*) 44(4):215–224 (1996).

25. Blach-Olszewska, Z, Janusz, M. Stimulatory effect of ovine colostrinine (a proline-rich polypeptide) on interferons and tumor necrosis factor production by murine resident peritoneal cells. *Archivum immunologiae et therapiae experimentalis* (*Warszava*) 45(1):43–47 (1997).

26. Domaraczenko, B, Janusz, M, Orzechowska, B, Jarosz, W, Blach-Olszewska, Z. Effect of proline rich polypeptide from ovine colostrum on virus replication in human placenta and amniotic membrane at term; possible role of endogenous tumor necrosis factor alpha. *Placenta* 20(8):695–701 (1999).

27. Khan, A. Non-specificity of transfer factor. *Annals of Allergy* 38(5):320–322 (1977).

28. Keech, A. Unpublished data (2006).

29. Staroscik, K, Janusz, M, Zimecki, M, Wieczorek, Z, Lisowski, J. Immunologically active nonapeptide fragment of a proline-rich polypeptide from ovine colostrum: amino acid sequence and immunoregulatory properties. *Molecular Immunology* 20(12):1277–1282 (1983).

30. Janusz, M, Wieczorek, Z, Spiegel, K, Kubik, A, Szewczuk, Z, Siemion, I, Lisowski, J. Immunoregulatory properties of synthetic peptides, fragments of a proline-rich polypeptide (PRP) from ovine colostrum. *Molecular Immunology* 24(10):1029–1031 (1987).

31. Siemion, IZ, Folkers, G, Szewczuk, Z, Jankowski, A, Kubik, A, Voelter, W. Peptides related to the active fragment of "proline rich polypeptide", an immunoregulatory protein of the ovine colostrum. Spectroscopic and computer modeling studies. *International Journal of Protein and Peptide Research* 36(6):506–514 (1990).

32. Sokal, I, Janusz, M, Miecznikowska, H, Kupryszewski, G, Lisowski, J. Effect of colostrinin, an immunomodulatory proline-rich polypeptide from ovine colostrum, on sialidase and beta-galactosidase activities in murine thymocytes. *Archivum immunologiae et therapiae experimentalis (Warszava)* 46(3):193–198 (1998).

33. Khan, A. Transfer factor in viral diseases. *The Lancet* 1(8059):328–329 (1978).

34. Orzechowska, B, Blach-Olszewska, Z. Acquisition of susceptibility to vesicular stomatitis virus infection by murine resident peritoneal cells during culturing in vitro. *Archivum immunologiae et therapiae experimentalis (Warszava)* 44(5–6):325–328 (1996).

35. Orzechowska, B, Janusz, M, Domaraczenko, B, Blach-Olszewska, Z. Antiviral effect of proline-rich p olypeptide in murine resident peritoneal cells. *Acta Virologica* 42(2):75–78 (1998).

36. Pizza, G, Meduri, R, De Vinci, C, Scorolli, L, Viza, D. Transfer factor prevents relapses in herpes keratitis patients: a pilot study. *Biotherapy* 8(1):63–68 (1994).

37. Pizza, G, Viza, D, De Vinci, C, Palareti, A, Cuzzocrea, D, Fornarola, V, Baricordi, R. Orally administered HSV-specific transfer factor (TF) prevents genital or labial herpes relapses. *Biotherapy* 9(1–V):67–72 (1996).

38. Meduri, R, Campos, E, Scorolli, L, De Vinci, C, Pizza, G, Viza, D. Efficacy of transfer factor in treating patients with recurrent ocular herpes infections. *Biotherapy* 9(1–3):61–66 (1996).

39. "Colostrum contains Retinoic Acid, which helps fight herpes virus. It also contains glyco-protein (kappa casein) that protects against the bacteria that cause stomach ulcers." – Dr. Raloff in *Science News*.

40. Prasad, U, bin Jalaludin, MA, Rajadurai, P, Pizza, G, De Vinci, C, Viza, D, Levine, PH. Transfer factor with anti-EBV activity as an adjuvant therapy for nasopharyngeal carcinoma: a pilot study. *Biotherapy* 9(1–3):109–115 (1996).

41. Claes-Henrik, F., Chinenye, S., Elfstrand L., Hagman, C., Ihse, I., ColoPlus, A new product based on bovine colostrum, alleviates HIV-associated diarrhea. *Scandinavian Journal of Gastroenterology*, 41:682–686 (2006).

42. Raise, E, Guerra, L, Viza, D, Pizza, G, De Vinci, C, Schiattone, ML, Rocaccio, L, Cicognani, M, Gritti, F. Preliminary results in HIV-1-infected patients treated with transfer factor (TF) and zidovudine (ZDV). *Biotherapy* 9(1–3):49–54 (1996).

43. Heaton, P. Bovine colostrum immunoglobulin concentrate for cryptosporidiosis in AIDS. *Archives of Disease in Childhood* 70(4):356–357 (1994).

44. Shield, J, Melville, C, Novelli, V, Anderson, G, Scheimberg, I, Gibb, D, Milla, P. Bovine colostrum immunoglobulin concentrate for cryptosporidiosis in AIDS. *Archives of Disease in Childhood* 69(4):451–453 (1993).

45. Rump, JA, Aarndt, R, Arnold, A, Bendick, C, Dichtelmuller, H, Franke, M, Helm, EB, Jager, H, Kampmann, B, Kolb, P, et al. Treatment of diarrhoea in human immunodeficiency virus infected patients with immunoglobulins from bovine colostrum. *Clinical Investigator* 70 (7):588–594 (1992).

46. Plettenberg, A et al. A preparation from bovine colostrum in the treatment of HIV positive patients with chronic diarrhea. *Clinical Investigator* (Jan. 1993).

47. Richie, J. Update on the management of intestinal cryptosporidiosis in AIDS. *Annals of Pharmacotherapy* 28:767–778 (1994).

48. Unger, BLP et al. Cessation of cryptosporidium-associated diarrhea in AIDS patients after treatment with hyperimmune bovine colostrum. *Gastroenterology* 98:486–489 (1990).

49. Ritchie, DJ, Update on the management of intestinal cryptosporidiosis in AIDS. *Annals of Pharmacotherapy* 28:767–778 (1994).

50. Greenberg PD, Cello JP. *Journal of Acquired Immune Deficiency Syndromes and Human Retrovirology* 13(4):348–354 (Dec. 1, 1996).

51. Plettenberg, A et al. A preparation from bovine colostrum in the treatment of HIV-positive patients with chronic diarrhea. *Clinical Investigator* 71(1):42–45 (1993).

52. Greenberg, PD, Cello, JP. Treatment of severe diarrhea caused by Cryptosporidium parvum with oral bovine immunoglobulin concentrate in patients with AIDS. *Journal of Acquired Immunodeficiency Syndromes and Human Retrovirology* 13(4):348–354 (1996).

53. "Reducing viral levels in the body and stimulating natural immune capabilities holds the most promise in helping our immune systems contain the HIV virus," according to Dr. Nowa and Dr. McMichael in *Scientific American*.

54. Shortridge, KF, Lawton, JW, Choi, EK. Protective potential of colostrum and early milk against prospective influenza viruses. *Journal of Tropical Pediatrics* 36(2):94–95 (1990).

55. Qiu, J, Hendrixson, DR, Baker, EN, Murphy, TF, St. Game, JW, III, Plaut, AG. Human milk lactoferrin inactivates two putative colonization factors expressed by *Haemophilus influenzae*. *Proceedings of the National Academy of Science* (USA) 95(21):12641–12646 (1998).

56. Hilpert, H, Brussow, H, Mietens, C, Sidoti, J, Lerner, L, Werchau, H. Use of bovine milk concentrate containing antibody to rotavirus to treat rotavirus gastroenteritis in infants. *Journal of Infectious Disease* 156(1):58–166 (1987).

57. Brussow, H, Hilpert, H, Walther, I, Sidoti, J, Mietens, C, Bachmann, P. Bovine milk immunoglobulins for passive immunity to infantile rotavirus gastroenteritis. *Journal of Clinical Microbiology* 25(6):982–986 (1987).

58. Ebina, T, Sato, A, Umezu, K, Ishida, N, Ohyama, S, Ohizumi, A, Aikawa, K, Katagiri, S, Katsushima, N, Imai, A. Prevention of rotavirus infection by cow colostrum antibody against human rotaviruses. *Lancet* 2(8357):1029–1030 (1983).

59. Ebina, T, Sato, A, Umezu, K, Ishida, N, Ohyama, S, Oizumi, A, Aikawa, K, Katagiri, S, Katsushima, N, Imai, A. Prevention of rotavirus infection by oral administration of cow

colostrum containing anti-human rotavirus antibody. *Medical Microbiology and Immunity (Berlin)* 174(4):177–185 (1985).

60. Ebina, T, Ohta, M, Kanamaru, Y, Yamamoto-Osumi, Y, Baba, K. Passive immunizations of suckling mice and infants with bovine colostrum containing antibodies to human rotavirus. *Journal of Medical Virology* 38(2):117–123 (1992).

61. Ebina, T. Prophylaxis of rotavirus gastroenteritis using immunoglobulin. *Archives of Virology Supplement* 12:217–223 (1996).

62. Sarker, SA, Casswall, TH, Mahalanabis, D, Alam, NH, Albert, MJ, Brussow, H, Fuchs, GJ, Hammerstrom, L. Successful treatment of rotavirus diarrhea in children with immunoglobulin from, immunized bovine colostrum. *Pediatric Infectious Disease Journa* 17(12):1149–1152 (1998).

63. Davidson, GPE, Nunan, H, Moore, AG, Whyte, PBD, Franklin, K, McCloud, PL, Moore, DJ. Passive Immunization of children with bovine colostrum containing antibodies to human rotavirus. *Lancet* 2:709–712 (1989).

64. Mitra, AK, Mahalanabis, D, Ashraf, H, Unicomb, L, Eeckels, R, Tzipori, S. *Acta Paediatrica* 84(9):996–1001 (Sep. 1995).

65. Yolken, RH et al. Antibody to human rotavirus in cow's milk. *New England Journal of Medicine* 312(10):605–610 (1985).

66. Sabin, AB. Anti-poliomyelitic substance in milk from human beings and certain cows. *Journal of Diseases of Children* 80:866–870 (1950).

67. Palmer, EL et al. Antiviral activity of colostrum and serum immunoglobulins A and G. *Journal of Medical Virology* 5:123–129 (1980).

68. Petschow, BW, Talbott, RD. Reduction in virus-neutralizing activity of a bovine colostrum immunoglobulin concentrate by gastric acid and digestive enzymes. *Journal of Pediatric Gastroenterology and Nutrition* 19(2):228–235 (1994).

69. Theodore, C et al. "Immunologic Aspects of Colostrum and Milk: Development of Antibody Response to Respiratory Syncytial Virus and Bovine Serum Albumin in the Human and Rabbit Mammary Gland" *Recent Advances in Mucosal Immunity* (1982) (Raven Press), New York.

70. Immunoglobulin components and anti-viral activities in bovine colostrum. *Kansenshogaku Zassh* 64(3):274–279 (Mar. 1990).

71. van Hooijdonk AC, Kussendrager KD, Steijns JM., In vivo antimicrobial and antiviral activity of components in bovine milk and colostrum involved in non-specific defence. *British Journal of Nutrition* 84(Suppl. 1):S127–134 (Nov. 2000).

72. Ushijima, H, Dairaku, M, Honnma, H, Mukoyama, A, Kitamura, T. *Kansenshogaku Zasshi* 64(3):274–279 (Mar. 1990).

73. Ushijima, H, Dairaku, M, Honnma, H, Mukoyama, A, Kitamura, T. (1990) Immunoglobulin components and anti-viral activities in bovine colostrum. 64:274–279.

74. Orzechowska, B et al. Antiviral effect of proline-rich polypeptide in murine resident peritoneal cells. *Acta Virologica* 42(2):75–78 (1998).

75. Ferrer-Argote, VE, Romero-Cabello, R, Hernandez-Mendoza, L, Arista-Viveros, A, Rojo-Medina, J, Balseca-Olivera, F, Fierro, M, Gonzalez-Constandse, R. Successful treatment of severe complicated measles with non-specific transfer factor. *In Vivo* 8(4):555–557 (1994).

76. Tzipori, S, Roberton, D, Chapman, C. Remission of diarrhoea due to cryptosporidiosis in an immunodeficient child treated with hyperimmune bovine colostrum. *British Medical Journal (Clinical Research Edition)* 293(6557):1276–1277 (1986).

77. Orzechowska, B, Janusz, M, Domaraczenko, B, Beach-Olszewska, B, Antiviral effect of proline-rich polypeptide in murine resident peritoneal cells. *Acta Virologica* 42:75–78 (1998).

78. Ferrer-Argote, VE, Romero-Cabello, R, Hernandez-Mendoza, L, Arista-Viveros, A, Rojo-Medina, J, Balseca-Olivera, F, Fierro, M, Gonzalez-Constandse, R. Successful treatment of severe complicated measles with non-specific transfer factor. *In Vivo* 8(4):555–557 (1994).

79. Chan, MC, Cheung, CY, Chui, WH, Tsao, SW, Nicholls, JM, Chan, YO, Chan, RW, Long, HT, Poon, LL, Guan, Y, Peiris, JS. Proinflammatory cytokine responses induced by influenza

A (H5N1) viruses in grimary human alveolar and bronchial epithelial cells. *Respiratory Research* 6:135 (2005).

80. Lawrence, HS. Transfer factor and autoimmune disease. *Annals of the New York Academy of Science* 124(1):56–60 (1965).

81. Hraba, T, Wieczorek, Z, Janusz, M, Lisowski, J, Zimecki, M. Effect of proline-rich polypeptide on experimental autoimmune response to erythrocytes. *Archivum immunologiae et therapiae experimentalis (Warszava)* 34(4):437–443 (1986).

82. Savilahti, E, Tainio, VM, Salmenpera, L, Arjomaa, P, Kallio, M, Perheentupa, J, Siimes, MA. Low colostral IgA associated with cow's milk allergy. *Acta Paediatrica Scandinavica* 80(12):1207–1213 (1991).

83. Delespesse, G. Polypeptide factors from colostrum. US Patent #5,371,073 (1994).

84. Collins, AM et al. Bovine milk, including pasteurized milk, contains antibodies directed against allergens of clinical importance to man. *International Archives of Allergy and Applied Immunology* 96:362–367 (1991).

85. Institute of Chemistry and Biochemistry, Helibrunner Str. 34, 5020 Salzburg, Austria. Univ.-Prof. Dr. Albert Duschl, http://tapir.sbg.ac.at/allergy.htm

86. Grohn, P, Anttila, R, Krohn, K. The *effect of* non-specifically *acting transfer factor* component *on* cellular immunity in juvenile rheumatoid arthritis. *Scandinavian Journal of Rheumatology* 5(3):151–157 (1976).

87. "The Clinical Use of Bovine Colostrum", Alejandro Nitsch, MD, Fabiola P. Nitsch, MD.

88. De Keyser, F, et al. Gut inflammation and spondyloarthropathies. *Current Rheumatology Reports* 4(6):525–532 (2002).

89. Op. cit., Fallanc, L, p. 192. Feldmann, M, Brennan, F, Maini, R. Role *of* cytokines *in* rheumatoid arthritis. *Annals of Review in Immunology* 14:397–440 (1996).

90. Feldmann, M, Brennan, F, Maini, R. Role of cytokines *in* rheumatoid arthritis. *Annual Review of Immunology* 14:397–440 (1996).

91. Khan, *A*, Sellars, W, Grater, W, Graham, MF, Pflanzer, J, Antonetti, A, Bailey, J, Hill, NO. *The* usefulness of transfer factor in asthma associated with frequent infections. *Annals of Allergy* 40(4):229–232 (1978).

92. Runa Ali, F, Barry Kay, A, Larcha, M. *The* potential *of* peptide immunotherapy *in* allergy and asthma. *Current Allergy and Asthma Reports* (2001).

93. Basten, A, McLeod, JG, Pollard, JD, Walsh, JC, Stewart, GJ, Garrick,. R, Frith, JA, Van Der Brink, CM. Transfer *factor in treatment of* multiple sclerosis. *The Lancet* 2(8201):931–934 (t980).

94. Ebina, T, et al. "Treatment of multiple sclerosis with anti-measles cow colostrum." *Med Microbiol Immunol (Berl)* 173(2):87–93 (1984).

95. De Vinci, C, Levine, PH, Pizza, G, Fudenberg, HH, Orens, P, Pearson, G, Viza, D. *Lessons from a* pilot *study of transfer factor in* chronic *fatigue syndrome*. *Biotherapy* 9(1–3):87–90 (1996).

96. Pizza, G, et al. A preliminary report *on the use of transfer factor for treating stage* D3 hormone-unresponsive *metastatic prostate cancer*. *Biotherapy* 9(1–3):123–132 (1996). Transfer *factor* (PRPs) increased the survival rate of stage D3 prostate cancer patients.

97. See, D, et al. Increased tumor *necrosis factor* alpha (TNF-alpha) *and* natural killer cell (NK) *function using an integrative* approach *in* late *stage cancers*. *Immunological Investigations* 31(2):137–153 (2002).

98. Leszek, J, Inglot, AD, Janusz, M, Lisowski, J, Krukowska, K, Georgiades, JA. Colostrinin: *a* prolinerich polypeptide (PRP) complex isolated from ovine colostrum for treatment of Alzheimer's disease. A double-blind, placebo-controlled study. *Archivum immunologiae et therapiae experimentalis (Warszava)* 47(6):377–385 (1999).

99. Leszek, J, Inglot, AD, Janusz, M, Byczkiewicz, F, Kiejna, A, Georgiades, J, Lisowski, J. Colostrinin proline-rich polypeptide complex from ovine colostrum–a long-term study of its efficacy in Alzheimer's disease. *Medical Science Monitor* 8(10):193–196 (2002).

100. Bilikiewicz, A, Gaus, W. Colostrinin (*a* naturally *occurring*, proline-rich, polypeptide mixture) *in the* treatment of Alzheimer's disease. *Journal of Alzheimer's Disease* 6(1):17–26 (2004).
101. Popik, P, Bobula, B, Janusz, M, Lisowski, J, Vetulani, J. Colostrinin, *a* polypeptide isolated *from* early milk, facilitates learning and memory in rats. *Pharmacology and Biochemistry of Behavior* 64(1):183–189 (1999).
102. Popik, P, Galoch, Z, Janusz, M, Lisowski, J, Vetulani, J. *Cognitive effects of* Colostral-Val nonapeptide in aged rats. *Behavioral Brain Research* 118(2):201–208 (2001).
103. Stewart, MG, Banks, D. Enhancement *of* long-term memory retention by Colostrinin *in* one-day-old chicks trained on a weak passive avoidance learning paradigm. *Neurobiology of Learning and Memory* prepublication (2006).
104. Mikulska, JE, Lisowski, J. A proline-rich polypeptide complex (PRP) from ovine colostrum. Studies on the effect of PRP on nitric oxide (NO) production induced by LPS in THP-1 cells. *Immunopharmacology and Immunotoxicology* 25(4):645–654 (2003).
105. Zablocka, A, Janusz, M, Macala, J, Lisowski, J. A proline-rich polypeptide complex and its nonapeptide fragment inhibit nitric oxide production induced in mice. *Regulatory Peptides* 125(1–3):35–39 (2005).
106. Schuster, D, Rajendran, A, Hui, SW, Nicotera, T, Srikrishnan, T, Kruzel, ML. Protective effect of colostrinin on neuroblastoma cell survival is due to reduced aggregation of beta-amyloid. *Neuropeptides* 39(4):419–426 (2005).
107. Stoff, JA, The use of dialyzable bovine colostrum extract in conjunction with a holistic treatment model for natural killer cell stimulation in chronic illness. Obtained by *Mato/ Botanical International (1–800–363–1890)*.
108. Stoff, J, et al. An Examination of Immune Response Modulation in Humans by AUE^{1Q}0 Utilizing A Double Blind Study. *Townsend* Letter April, 2002.
109. Godden, SM, et al. Effect of on-farm commercial batch pasteurization of colostrum on colostrum and serum immunoglobulin concentrations in dairy calves. *Journal of Dairy Science* 86(4):1503–1523 (2003).
110. Meylan, M, et al. Survival of *Mycobacterium paratuberculosis* and preservation of immunoglobulin G in bovine colostrum under experimental conditions simulating pasteurization. *American Journal of Veterinary Research* 57(11):1580–1585 (1996).
111. Li-Chan, E, et al. Stability of bovine immunoglobulins to thermal treatment and processing. *Food Research International* 28(1):9–16 (1995).
112. Dominguez, E, et al. Effect of heat treatment on the antigen-binding activity of anti-peroxidase immunoglobulins in bovine colostrum. *Journal of Dairy Science* 80(12):3182–3187 (1997).
113. Collier, RJ, et al. Factors affecting insulin-like growth factor-1 concentration in bovine milk. *Journal of Dairy Science* 74(9):2905–2911 (1991).
114. Paulsson, MA, et al. Thermal behavior of bovine lactoferrin in water and its relation to bacterial interaction and antibacterial activity. *Journal of Dairy Science* 76(12):3711–3720 (1993).
115. da Costa, RS, et al. Characterization of iron, copper and zinc levels in the colostrum of mothers of term and pre-term infants before and after pasteurization. *International Journal of Food Science and Nutrition* 54(2):111–117 (2003).
116. Holloway, NM, et al. Serum immunoglobulin G concentrations in calves fed fresh and frozen colostrum. *Journal of the American Veterinary Medicine Association* 219(3):357–359 (2001).

Part VII
Neuroscience

Phototherapy and Nerve Tissue Repair

Shimon Rochkind

Abstract

Background: Posttraumatic nerve repair continues to be a major challenge of restorative medicine. Numerous attempts have been made to enhance and/or accelerate the recovery of injured peripheral nerves. One of the methods studied is the use of laser phototherapy. Laser phototherapy was applied as a supportive factor for accelerating and enhancing axonal growth and regeneration after injury or reconstructive peripheral nerve procedure.

Methods:

I – These studies summarize our experience with 632 and 780 nm low power laser irradiation for treatment of peripheral nerve injury using a rat sciatic nerve model after crush injury, neurorraphy or neurotube reconstruction.

II – A clinical double-blind, placebo-controlled randomized study was performed to measure the effectiveness of laser phototherapy on patients who had been suffering from incomplete peripheral nerve injuries for 6 months up to several years.

III – 780 nm laser irradiation was investigated on the growth of embryonic rat brain cultures embedded in neurogel (cross-linked hyaluronic acid with adhesive molecule laminin). Neuronal cells attached to microcarriers (MCs) were laser treated, and their growth in stationary cultures was detected.

Results:

I – *Animal studies* show that laser phototherapy has a protective immediate effect, maintains functional activity of the injured nerve over time, prevents or decreases scar tissue formation at the injured site, prevents or decreases degeneration in corresponding motor neurons of the spinal cord and significantly increases axonal growth and myelinization. Moreover, direct laser irradiation of the spinal cord improves recovery of the corresponding injured peripheral nerve.

S. Rochkind
Division of Peripheral Nerve Reconstruction, Tel Aviv Sourasky Medical Center, Tel Aviv
University, 6 Weizmann st., Tel Aviv, Isrel 64239, e-mail: rochkind@zahav.net.il

R. Waynant and D.B. Tata (eds.), *Proceedings of Light-Activated Tissue Regeneration*   247
and Therapy Conference.
© Springer Science + Business Media, LLC 2008

II – *A clinical double-blind, placebo-controlled randomized study* shows that in long-term peripheral nerve injured patients low power laser irradiation can progressively improve peripheral nerve function, which leads to significant functional recovery.

III – *Cell therapy*: 780 nm laser irradiation accelerated migration and fiber sprouting of neuronal cells aggregates. The irradiated cultures contained a higher number of large neurons than the controls. Neurons in the irradiated cultures developed a dense branched interconnected network of neuronal fibers.

Conclusions:

I – The animal and clinical studies on the promoting action of phototherapy on peripheral nerve regeneration make it possible to suggest that the time for broader clinical trials has come.

II – 780 nm laser treatment of embryonic rat brain cultures embedded in neurogel and attached to positively charged cylindrical MCs, stimulated migration and fiber sprouting of neuronal cells aggregates and, therefore, can be considered as potential therapy for neuronal injury.

Keywords: Laser phototherapy, nerve injury, nerve regeneration, cell therapy.

Basic Sciences and Clinical Trial

Studies, which evaluated the effects of 632.8 and 780 nm laser irradiation on Schwann [1] and nerve cell [2] cultures and injured peripheral nerves of animals [3–7] showed positive results. Laser phototherapy induces Schwann cell proliferation [1] and affects nerve cell metabolism and induces nerve processes sprouting [2].

Peripheral Nerve Injury and Clinical Perspective

Injury of a peripheral nerve frequently results in considerable disability. In an extremity such lesions may be associated with loss of sensory and motor functions, which leads to severe occupational and social consequences.

Surgical repair is the preferred modality of treatment for the complete or severe peripheral nerve injury [8–11]. In most cases, the results can be successful if the surgery is performed in the first 6 months after injury, in comparison to long-term cases where surgical management is less successful.

Although, in related literature, there are several publications of surgical treatment of long-term injuries of the brachial plexus and peripheral nerve [12–15]. For most patients who suffered from long-term peripheral nerve injuries, the continuation of rehabilitation therapy was recommended, especially in those regions or countries, which don't have specially dedicated peripheral nerve surgeons. Unfortunately, spontaneous recovery of long-term severe incomplete peripheral nerve injury is often unsatisfactory. The usual results after such an injury are degeneration of the axons and retrograde degeneration of the corresponding neurons of the spinal cord, followed by a very slow regeneration. Recovery may eventually occur, but it is slow and frequently incomplete. Understandably, therefore, numerous attempts have been made to enhance and/or accelerate the recovery of injured peripheral nerves. One of the methods studied is the use of different wavelengths of low power laser irradiation to enhance the recovery of peripheral nerve injuries.

I – Laser Phototherapy for Treatment of Experimental Peripheral Nerve Injury

Laser phototherapy significantly improves recovery of the injured peripheral nerve [3, 4, 6, 7] and, in addition, decreases posttraumatic retrograde degeneration of the neurons in the corresponding segments of the spinal cord [5].

Our previous studies investigating the effects of low power laser irradiation 632.8 and 780 nm on injured peripheral nerves of rats have found:

1. Protective immediate effects which increase the functional activity of the injured peripheral nerve [16]
2. Maintenance of functional activity of the injured nerve over time [4]
3. Influence of the LPLI on scar tissue formation at the injured site (Fig. 1) [6]
4. Prevention or decreased degeneration in corresponding motor neurons of the spinal cord (Fig. 2) [5]
5. Influence on axonal growth and myelinization (Fig. 3) [4, 7]

Moreover, direct laser irradiation of the spinal cord improves recovery of the corresponding injured peripheral nerve [7, 17].

Our results suggest that laser phototherapy accelerates and improves the regeneration of the injured peripheral nerve.

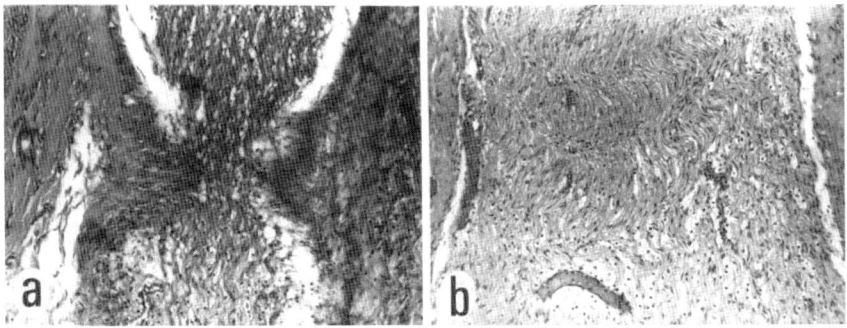

Fig. 1 Decrease or prevention of scar tissue formation at the site of injury (*Laser Surg Med* 7: 441–443, 1987). **a** – Scar in the place of the injury in the non-laser treated nerve. **b** – Prevention of scar formation after laser treatment

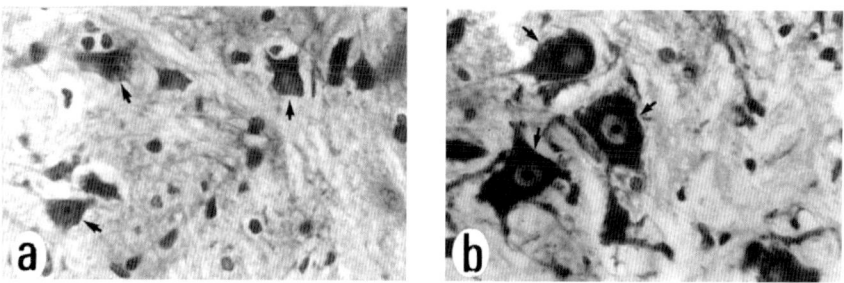

Fig. 2 (*Spine* 15: 6–10, 1990). Progressive degeneration changes in the corresponding neurons of the spinal cord after peripheral nerve injury in the control non-irradiated group (**a**). Decrease of degeneration process after laser treatment (**b**)

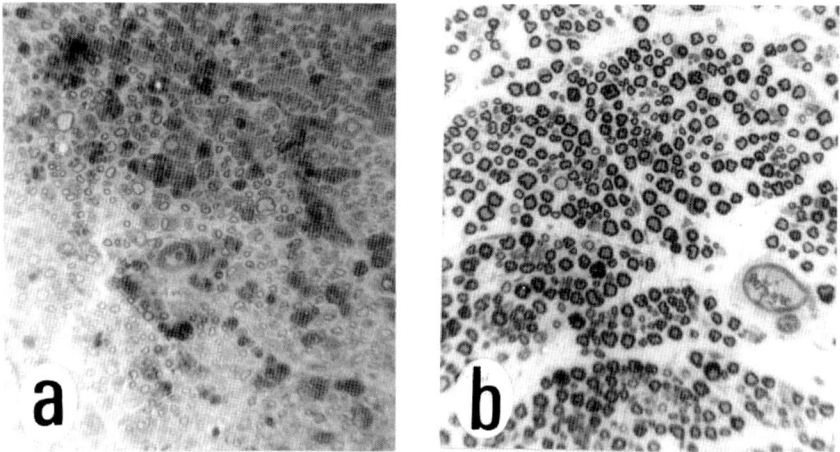

Fig. 3 (*Neurosurg* 20: 843–847, 1987). Increase in rate of axonal growth and myelinization: **a** – without treatment; **b** – laser treated nerve

780 nm Laser Phototherapy in Clinical Study

II – Clinical Double-Blind, Placebo-Controlled Randomized Trial

Since our animal studies were positive, an evaluation of the response to 780 nm laser phototherapy was in order. Therefore, a clinical double-blind, placebo-controlled randomized study was performed to measure the effectiveness of 780 nm low power laser irradiation on patients who had been suffering from incomplete peripheral nerve and brachial plexus injuries for 6 months up to several years [18]. Most of these patients were discharged from initial orthopedics, neurosurgeons and plastic surgeons without further treatment.

In this study 18 patients with a history of traumatic peripheral nerve/brachial plexus injury (at least 6 months after the injury), with a stable neurological deficit

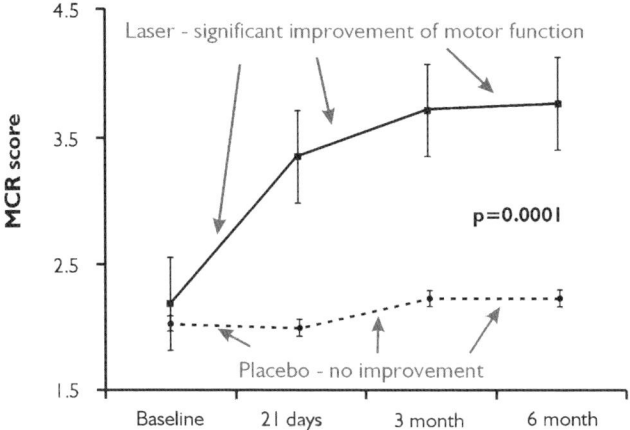

Fig. 4 (*Photomed Laser Surg* 25: 436–442, 2007)

and a significant weakness, were randomly divided to receive either 780 nm laser or placebo (non-active light) irradiation. The analysis of the results of this trial in the laser-irradiated group showed statistically significant improvement in motor function in the previously partially paralyzed limbs, compared to the placebo group, where no statistical significance in neurological status was found (Fig. 4).

Electrophysiological observation during the trial supplied us with important diagnostic information and helped to determine the degree of functional recovery in nerve-injured patients. The electrophysiological analysis also showed statistically significant improvement in recruitment of voluntary muscle activity in the laser-irradiated group, compared to the placebo group (Fig. 5)

This study is not the ultimate word and recommendations regarding 780 nm laser phototherapy in peripheral nerve injured patients. Nevertheless, this study shows that in long-term peripheral nerve injured patients 780 nm low power laser irradiation can progressively improve peripheral nerve function, which leads to significant functional recovery.

III – Further Development in Peripheral Nerve Reconstruction and Role of 780 nm Laser Phototherapy

This study was done to show the use of low power laser treatment enhances the regeneration and repair of a reconstructed injured peripheral nerve [19]. The 5 mm segment of the right sciatic nerve was removed and proximal and distal parts were inserted into a bioabsorbable neurotube (Fig. 6).

The rats were divided into two groups laser treated and non-laser treated. Postoperative low power laser irradiation was applied for 30 min. transcutaneously

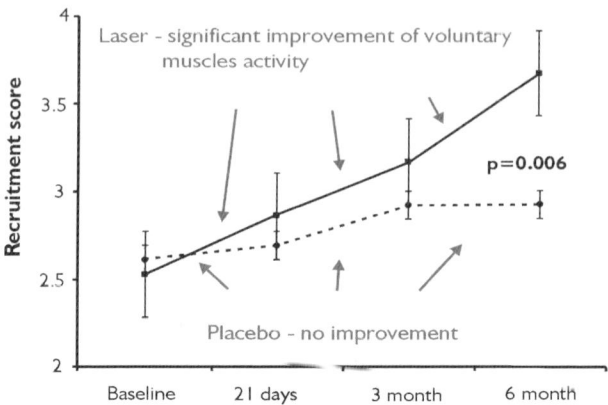

Fig. 5 (*Photomed Laser Surg* 25: 436–442, 2007)

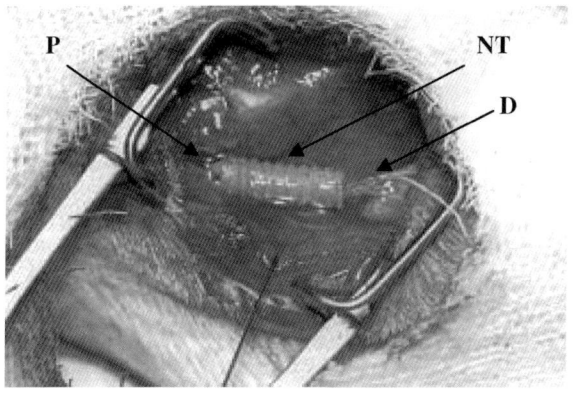

Fig. 6 (*Photomed Laser Surg* 25: 137–143, 2007) A neurotube (NT) placed between the proximal (P) and the distal (D) parts of the nerve for the reconnection of 0.5 cm nerve defect (arrows)

on the transplanted peripheral nerve area and corresponding segments of the spinal cord, during 14 consecutive days. Conductivity of the sciatic nerve was studied by stimulating the sciatic nerve and recording the somato-sensory evoked potentials (SSEP) from the scalp. Three months after surgery SSEP were found in 70% of the rats in the laser-treated group in comparison with 40% of the rats in the non-irradiated Group [19].

Morphologically, the previously transected nerve had good reconnection 4 months after surgery in both groups and the neurotube had dissolved (Fig. 7).

The immuno-histochemical staining (Fig. 8) using a monoclonal antibody-neurofilament showed more intensive axonal growth in neurotube-reconstructed and laser-treated rats (C) compared with the results of the non-laser treated group (B).

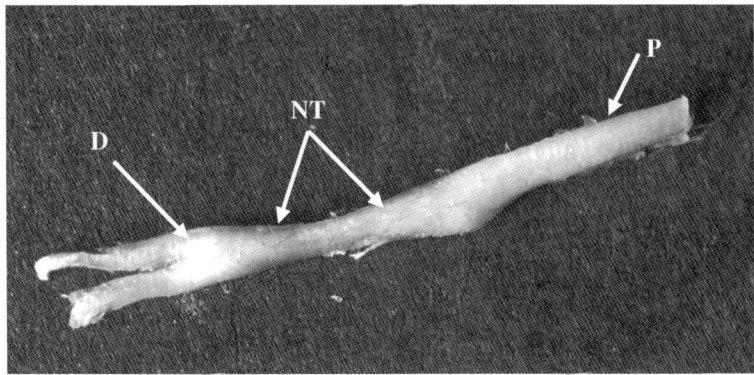

Fig. 7 (*Photomed Laser Surg* 25: 137–143, 2007). Sciatic nerve of adult rat which was reconstructed by the neurotube (see arrows: NT – neurotube area, D – distal part, P – proximal part)

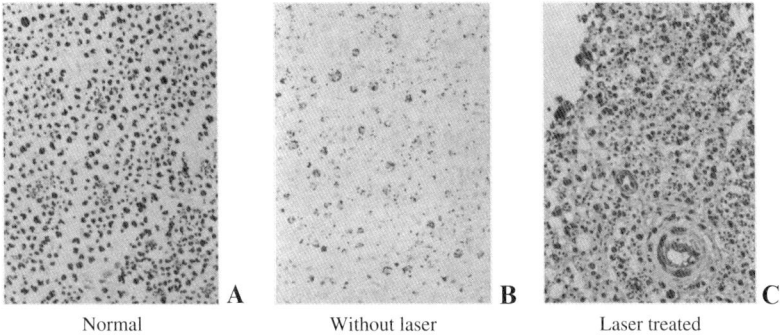

Normal   Without laser   Laser treated

Fig. 8 Influence of 780 nm low power laser irradiation on nerve cell growth in vitro

IV – Influence of 780 nm Low Power Laser Irradiation on Nerve Cell Growth In Vitro

In this work the effect of 780 nm laser phototherapy on sprouting and cell size of embryonic rat brain cells on microcarriers (MC) embedded in NVR-N-Gel in culture [20] was investigated.

Cell cultures: Whole brains were dissected from 16-day old rat embryos (Sprague Dawley). After mechanical dissociation, cells were seeded directly in NVR-N-Gel, or suspended in positively charged cylindrical MC. Single cell-MC aggregates were either irradiated with LPLI within 1 hour after seeding, or cultured without irradiation.

NVR-N-Gel (hyaluronic acid and laminin) was enriched with the following growth factors:

BDNF and IGF-1.

780 nm low power laser irradiation: Laser powers were 10, 30, 50, 110, 160, 200 and 250 mW. Dissociated cells or cell-MC aggregates embedded in NVR-N-Gel, were irradiated for 1, 3, 4 or 7 min.

Fluorescent staining: Cultures were fixed with 4% paraformaldehyde and incubated with antibodies against neural cell marker: Mouse anti Rat microtubule associated protein. Cells were then washed and incubated with Texas Red conjugated goat anti mouse IgG.

A rapid sprouting of nerve processes from the irradiated cell-MC aggregates was detected already within 24 after seeding (Fig. 9).

The extension of nerve fibers was followed by active neuronal migration. Differences between controls, and irradiated stationary dissociated brain cultures, became evident at about the end of the first week of cultivation – several neurons in the irradiated cultures exhibited large perikarya and thick elongated processes (Figs. 10 and 11).

V – Further Development in Spinal Cord Reconstruction and Role of 780 nm Laser Phototherapy

The following treatment method was developed recently in our laboratories to enhance regeneration and to repair traumatic paraplegia in rats, resulting from spinal cord transaction [21, 22]. Embryonal spinal cord cells dissociated from rat fetuses were cultured on biodegradable microcarriers (MCs) (Fig. 12A) and embedded in hyaluronic acid (HA) (Fig. 12B). Biodegradable microcarriers Embryonal spinal cord cells.

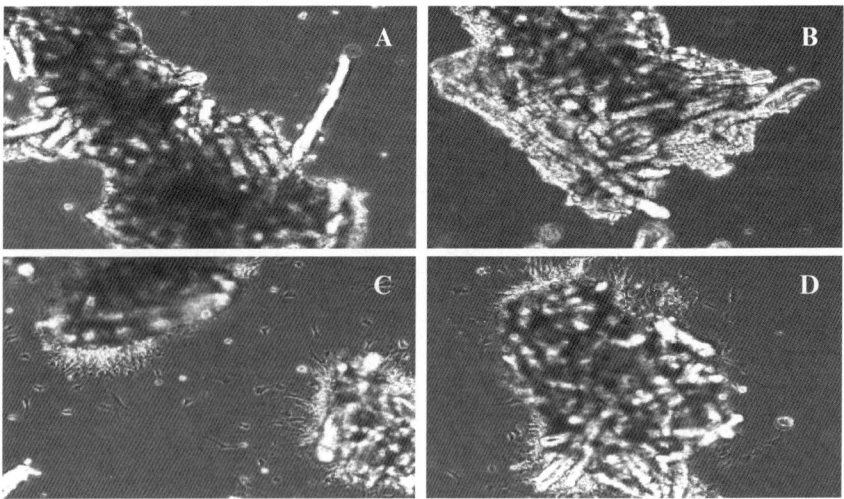

Fig. 9 Effect of 780 nm low power laser irradiation on initial sprouting and migration from DE-53 MCs in NVR-N-Gel. Initial sprouting and cellular migration is observed in irradiated cultures but not in non-irradiated control, already one day after the transfer to stationary cultures in NVR-N-Gel. **A & B**: Non-irradiated controls. **C**: Single irradiation of 250 mW, for 1 min. **D**: Single irradiation of 250 mW for 3 min. Original magnification: 200X.

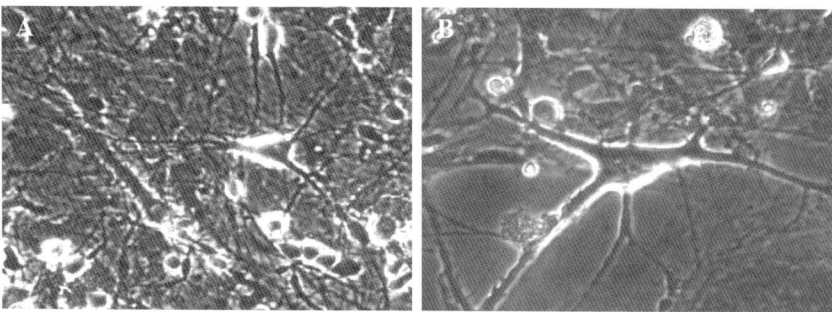

Fig. 10 Effect of LPLI treatment on perikarya and fibers of nerve cells derived from rat embryonic brain. Dissociated brain cells were embedded in NVR-N-Gel and were either exposed to single irradiation of 160 mW for 3 min (**B**), or served as non-irradiated controls (**A**). Large neural cells exhibiting thick fibers were observed in 8 days in vitro (DIV) irradiated cultures. Original magnification: 200×

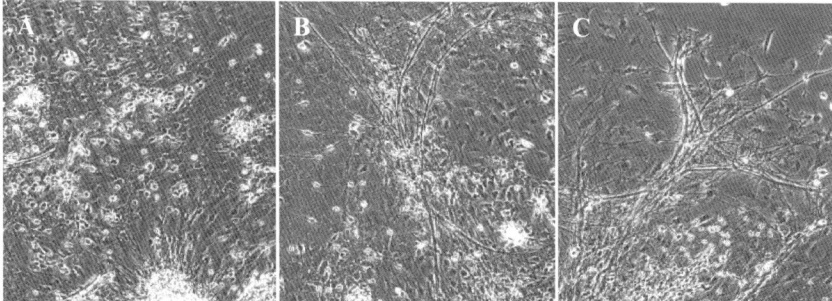

Fig. 11 Effect of LPLI treatment on sprouting of rat embryonic brain neurons. Floating cell-MC aggregates were transferred to stationary cultures and exposed to a single irradiation of 50 mW. **A**: Non-irradiated control. **B**: 4 min irradiation. **C**: 7 min irradiation. Note branching of huge processes in the laser treated cultures (B & C) after 7 days in NVR-N-Gel. Original magnification: 100×

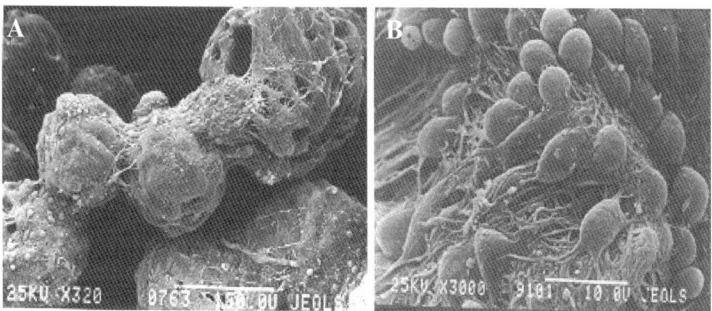

Fig. 12 A: (*Neurol Res* 24: 355–360, 2002). In vitro reconstructed composite implants containing neuronal cells attached to gelatinous microcarriers (MCs). **B**: Embryonal spinal cord cells (B)

The cell-MCs aggregates were implanted into sites of the completely transected spinal cord of adult rats. These implants served as regenerative and repair sources for reconstructing neuronal tissue. During the following 14 post-operative days, the implanted area of the spinal cord was irradiated transcutaneously, 30 min daily to enhance the neuro-regenerative repair process.

The post-operative follow-up (from 3 to 6 months) on operated rats which underwent embryonic nerve cell implantation in the transected area of the spinal cord, showed that the most effective re-establishment of active leg movements were found in the laser-treated group. The rats which underwent embryonic nerve cell transplantation and laser treatment showed that most effective re-establishment of limb function and gait performance occurred after nerve cell implantation and laser irradiation (Fig. 13A), compared to rats that were treated by nerve cell implantation alone.

The rats which underwent spinal cord transection only remained completely paralyzed in the lower extremities (Fig. 14A).

Intensive axonal sprouting was observed in the group which was implanted with embryonal nerve cells-MCs culture in HA and treated with low power laser irradi-

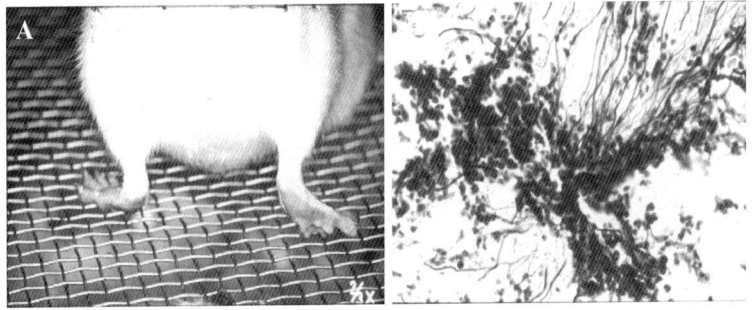

Fig. 13 (*Neurol Res* 24: 355–360, 2002). Active movement in both legs after embryonal nerve cell implantation in the transected spinal cord followed by low power laser treatment (**A**). Diffuse sprouting of axons at the site of nerve cell implantation followed by laser irradiation (Modified Bodian's stain X 400) (**B**)

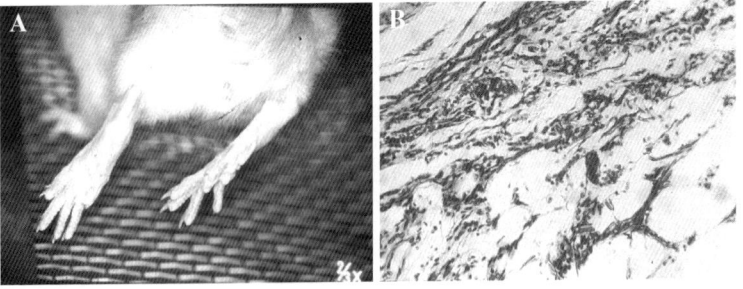

Fig. 14 (*Neurol Res* 24: 355–360, 2002). Complete paralysis of legs after spinal cord transection (**A**). Proliferating fibroblast and capillaries in a tansected spinal cord of a rat without treatment (Modified Bodian's stain X 400) (**B**)

ation (Fig. 13B), in comparison with the untreated completely transected spinal cord area that contained only proliferating fibroblasts and blood capillaries (Fig. 14B).

This study suggests that nerve cell implants which contain embryonal spinal cord cells attached to microcarriers and embedded in hyaluronic acid are a regenerative and reparative source for the reconstruction of the transected spinal cord. In addition, low power laser irradiation accelerates axonal growth and spinal cord regeneration.

In conclusion: The extensive review article, which was published in *Muscle and Nerve* in 2005 [23] revealed that most of experimental studies showed phototherapy to promote the recovery of the severely injured peripheral nerve. This review makes possible to suggest that time for broader clinical trials has come.

The significance of our experimental and clinical studies is the provision of new nerve tissue engineering technology and 780 nm laser phototherapy for treatment of severe nerve injury.

References

1. Van Breugel HH, Bar PR. HeNe laser irradiation affects proliferation of cultured rat schwann cells in a dose-dependent manner. *J Neurocytol* 22: 185–190, 1993.
2. Wollman Y, Rochkind S, Simantov R. Low power laser irradiation enhances migration and neurite sprouting of cultured rat embryonal brain cells. *Neurol Res* 18: 467–470, 1996
3. Andres JJ, Borke RC, Woolery SK, et al. Low power laser irradiation alters the rate of regeneration of the rat facial nerve. *Laser Surg Med* 13: 72–82, 1993
4. Rochkind S, Barr-Nea L, Razon N, et al. Stimulatory effect of HeNe laser low-dose laser on injured sciatic nerves of rats. *Neurosurgery* 20: 843–847, 1987
5. Rochkind S, Barr-Nea L, Volger I. Spinal cord response to laser treatment of injured peripheral nerve. *Spine* 15: 6–10, 1990
6. Rochkind S, Nissan M, Barr-Nea L, et al. Response of peripheral nerve to HeNe laser: experimental studies. *Laser Surg Med* 7: 441–443, 1987
7. Shamir MH, Rochkind S, Sandbank J, et al. Double-blind randomized study evaluating reneration of the rat transected sciatic nerve after suturing and postoperative low power laser treatment. *J Reconstruct Microsurg* 17: 133–138, 2001
8. MacKinnon SE, Dellon AL: **Surgery of the Peripheral Nerve**. New York: Thieme Medical, 1988
9. Noble J, Munro CA, Prasad VS, Midha R. Analysis of upper and lower extremity peripheral nerve injuries in a population of patients with multiple injuries. *J Trauma* 45: 6–122, 1998
10. Sunderland S. **Nerves and Nerve Injuries**. 2nd ed. Edinburgh, UK: Churchill Livingstone, 1978.
11. Terzis JK, Smith KL. **The Peripheral Nerve: Structure, Function and Reconstruction**. Norfolk, VA/New York: Hampton Press, Raven Press, 1990
12. Kline DG, Hackett ER. Reappraisal of timing for exploration of civilian peripheral nerve injuries. *Surgery* 78: 54–65, 1975
13. Narakas A. Surgical treatment of traction injuries of the brachial plexus. *Clin Orth Rel Res* 133: 71–90, 1978
14. Rochkind S, Alon M. Microsurgical management of old injuries of the peripheral nerve and brachial plexus. *J Reconstructive Microsurg* 16: 541–546, 2000

15. Rochkind S, Filmar G, Alon M, et al. Delayed microsurgery of penetrating peripheral nerve injuries: Pre, intra- and postoperative analysis and results. *Acta Neurochirurgica, Acta Neurochirurgica* (Suppl), 100: 21–24, 2007.
16. Rochkind S, Nissan M, Lubart, et al. The in vivo nerve response to direct low-energy laser irradiation. *Acta Neurochirurgica (Wien)* 94: 74–77, 1988
17. Rochkind, S, Nissan M, Alon M, et al. Effects of laser irradiation on the spinal cord for the regeneration of crushed peripheral nerves in rats. *Laser Surg Med* 28: 216–219, 2001
18. Rochkind S, Drory V, Alon M, Nissan M, Ouaknine GE. The treatment of incomplete peripheral nerve injuries using a new modality – Laser phototherapy (780 nm). *Photomed Laser Surg* 25: 436–442, 2007.
19. Rochkind S, Leider-Trejo L, Nissan M, Shamir M, Kharenko O, Alon M. Efficacy of 780-nm laser phototherapy on peripheral nerve regeneration after neurotube reconstruction procedure (double-blind randomized study). *Photomed Laser Surg* 25: 137–143, 2007.
20. Rochkind S, El-Ani D, Hayun T, Nevo Z, Shahar A. Increase of neuronal sprouting and migration using 780 nm laser phototherapy as procedure for cell therapy. *Laser Surg Med.* Accepted for publication, 2007
21. Rochkind S, Shahar A, Alon M, Nevo Z. Transplantation of embryonal spinal cord nerve cells cultured on biodegradable microcarriers followed by low power laser irradiation for the treatment of traumatic paraplegia in rats. *Neurol Res* 24: 355–360, 2002
22. Rochkind S, Shahar A, Fliss D, El-Ani D, Astachov L, Hayon T, Alon M, Zamostiano R, Ayalon O, Biton IE, Cohen Y, Halperin R, Schneider D, Oron A, Nevo Z. Development of a tissue-engineered composite implant for treating traumatic paraplegia in rats. *Eur Spine J* 15: 234–245, 2006
23. Gigo-Benato D, Geuna S, Rochkind S. Phototherapy for enhancing peripheral nerve repair: a review of the literature. *Muscle Nerve* 31: 694–701, 2005

Laser Regeneration of Spine Discs Cartilage: Mechanism, In-Vivo Study and Clinical Applications

Emil Sobol, Andrei Baskov, Anatoly Shekhter, Igor Borshchenko, and Olga Zakharkina

Abstract Laser Reconstruction of the spine discs (LRD) is a novel minimally invasive approach for the treatment of spine diseases. This approach belongs to the laser therapy, but it differs from low intensity laser therapy because LRD uses local and moderate heterogeneous laser heating that mostly does not effect directly on the cells; LRD procedure modifies the extra cellular matrix to provide better surroundings for the cells. Our main finding is that laser irradiation can activate the growth of hyaline cartilage. The predictability of the result, the locality and safety of laser effect allowed to use the technology for spine problems. LRD can be performed in an outpatient setting requiring only 30 min to complete without the need for general anesthesia. The new type of Erbium doped glass fiber laser (1.56 μm in wavelength) has been tested first on animals and then in a clinical trial. The mechanism of laser-induced tissue regeneration include: (1) formation of nanopores enchasing water permeability through end plates and annulus fibrosus of the disc that provide feeding for biological cells, and (2) activation of cell due to mechanical oscillations resulting from the periodically thermo expansion of nucleus pulposus under modulated laser irradiation. The clinical trials have shown positive results for 90% from 240 laser treated patients.

Keywords: Laser induced biostimulation, laser reconstruction, spine disc healing, spine treatment, near infrared laser therapy.

E. Sobol
Institute on Laser and Information Technologies, Russian Academy of Sciences, Troitsk, Russia,
e-mail: sobol@labr.ru

R. Waynant and D.B. Tata (eds.), *Proceedings of Light-Activated Tissue Regeneration and Therapy Conference.*
© Springer Science + Business Media, LLC 2008

Introduction: The Problem

Discogenic degenerative spine diseases are still a serious problem as they are a major cause of back pain that deteriorates the quality of life of patients and leads to disability [1]. To avoid the negative effects of surgery minimally invasive methods have been increasingly adopted during the last years [2] due to benefits, such as minimum injury, no need for general anesthesia and significant shortening of rehabilitation time. This trend includes puncture techniques conducting various physical stimuli [3–5], to the disc tissues. However, despite the clinical use of intradiscal therapy, its medical background and long-term efficacy sometimes seem doubtful [6–8].

In 2000 we introduced a novel approach to the treatment of spine disc degeneration based on the thermo mechanical effect of modulated laser irradiation without strong heating of the tissues [9–14]. The LRD procedure involves puncture of the disc, laser irradiation of the nucleus pulposus (NP) to facilitate the reparative processes in the tissue. This paper is aimed to present the results of animal study, clinical data and discussion of possible mechanisms of laser-induced activation of the generation of hyaline cartilage in spine discs.

In-Vivo Animal Studies

The effect of laser radiation on the generation of hyaline cartilage in spine disc has been demonstrated first for rabbits [9, 10]. Annulus fibrosus (AF) and NP of the intervertebral discs of rabbits have been irradiated *in vivo* using an 1.56 μm fiber laser with various pulse duration and repetition rate (from 0.3 to 2 Hz). Conventional histological technique and atomic force microscopy (AFM) have been used for examination of the new growing tissue. It has been demonstrated that laser radiation of spine discs induces metaplasia of fibrous cartilage into hyaline type cartilage (Fig. 1).

In 2 months after laser irradiation, the most pronounced signs of regeneration are seen in the inner layers of AF and in the NP. Neogenic tissue has features both of fibrous and hyaline cartilage types: the shape and ultra structure of chondrocytes are close to the cells of hyaline cartilage, but the structure of intercellular matrix possesses the random (disorderly) distribution of thin collagen fibrils like hyaline cartilage, as well as more thick and more aligned collagen fibrils like in the fibrous cartilage. This cartilaginous tissue is called as fibrous-hyaline cartilage. Apparently, its origin is poorly differentiated chondroblasts activated as a result of laser radiation. Alongside with that, there are the regions of typical hyaline type cartilage with homogeneous matrix structure and lacunas surrounded chondrocytes.

For AFM examination, samples of new tissues arisen in the intervertebral discs as a result of laser radiation were kept in 70% proof alcohol, and then cut into sections of about 300 μm in thickness perpendicular to the irradiated surface. The surface of the cartilage samples was imaged with a commercial multimode microscope NanoScope IIIa with phase extender (Digital Instruments, Santa Barbara, CA) operated in tapping

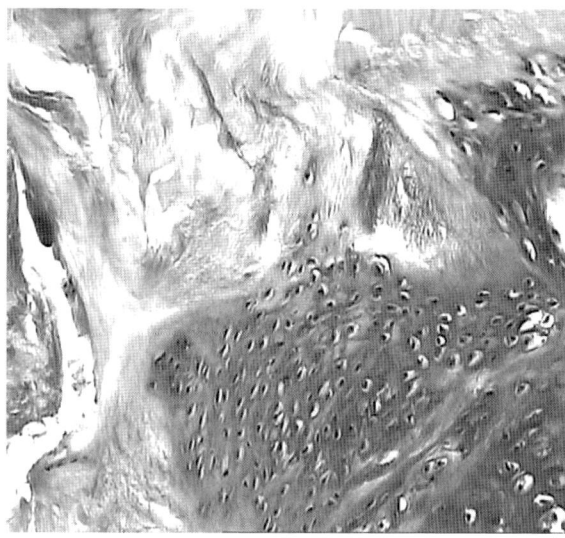

Fig. 1 Histological microphotograph of the new tissue growing in a rabbit spine disc in a month after laser treatment ($\times 200$)

mode at about 300 kHz. AFM has shown that nondestructive laser irradiation provokes formation of new nano-pores in cartilaginous matrix (Fig. 2).

In vivo studies performed on rabbits showed, that the therapeutic effect is due to spatial-temporal irregularities of temperature and mechanical stress in the tissue caused by local heating from modulated laser irradiation. Non-ablative laser treatment induces the formation of hyaline and fibrous-hyaline cartilage within the NP and the AF. Laser procedure doesn't cause any destruction of the discs, does not lead to the tissue necrosis, injury of adjacent spinal nerves and vessels.

Mechanism of Laser-Induced Regeneration of Cartilage

Mechanisms of laser-induced regeneration of hyaline type cartilage in the interveretebral discs are still under investigation [13, 14]. Possible mechanisms include:

(1) Non-destructive laser radiation leads to the formation of nano-pores in cartilage matrix (Fig. 2). Nano-pores promote water permeability and increase the feeding of biological cells.

(2) Space and temporary modulated laser beam induces non-homogeneous and pulse repetitive thermal expansion and stress in the irradiated zone of cartilage. Mechanical effect due to controllable thermal expansion of the tissue and micro and nano-bubbles formation in the course of the moderate (up to 45–50°C) heating of the NP activate biological cells (chondrocytes) and promote cartilage regeneration. The temperature dynamics during LRD is shown on the Fig. 3.

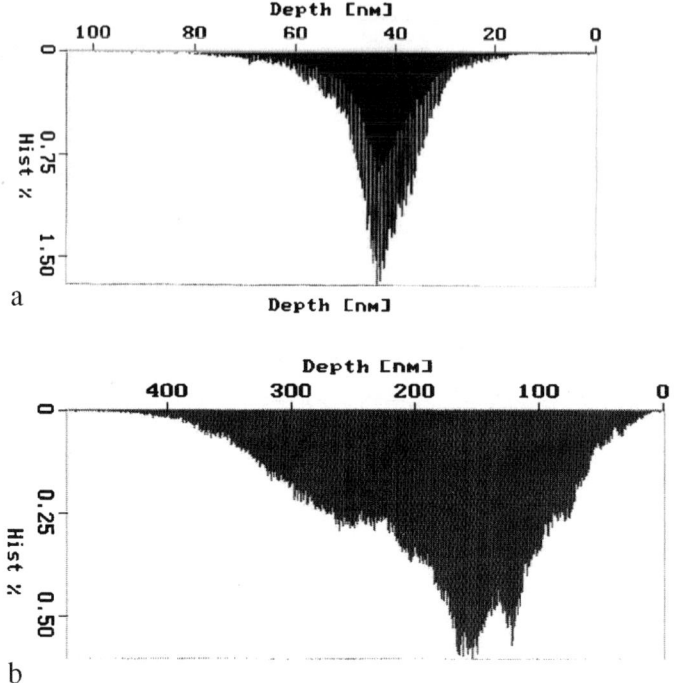

Fig. 2 Size distribution of nanopores in cartilage (**a**) before and (**b**) after laser treatment

It is known that dynamic mechanical oscillations may activate tissue regeneration. There are specific amplitudes and frequencies of mechanical oscillations which have positive effect on chondrocytes proliferation [13, 15]. Temperature and amplitude of the thermo mechanical oscillations relating to temperature oscillations shown on the Fig. 3 diminish rapidly with the distance from the laser spot centrum. This character of oscillations in the laser treated disc provides a possibility to control the locality of laser-induced regeneration.

Laser-induced heterogeneous heating and mechanical oscillations in the NP result in the formation of temporal micro bubbles shown on the Fig. 4. The formation and movement of these small gas micro bubbles provoke a creation of new pores (of sub-microns in size), which (together with mechanical oscillations) promote regeneration processes in the spine disc [13].

It is important that mechanical properties of the disc tissue do not change for the worse as a result of laser treatment, because the size of micro-bubbles arising under LRD is much less than characteristic dimensions of cavitation bubbles forming under ultrasound treatment of the tissues. Also the submicron-pores arising under laser treatment in the AF and in the end plates of the discs are much smaller than large pores and cracks which may diminish mechanical properties of the tissue and lead to disc degeneration.

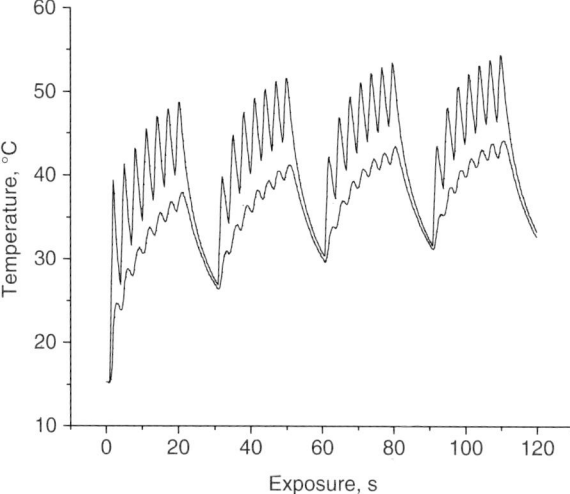

Fig. 3 Temperature in a rabbit NP in the course of laser irradiation at the centre of laser spot (*upper curve*) and 1 mm apart from the centre (*lower curve*). Laser power is 1 W; exposure time is 20 s for each set of irradiation; each set consists of seven laser pulses; duration of a laser pulse is 2 s; interval between pulses is 1 s; interval between neighboring sets is 10 s

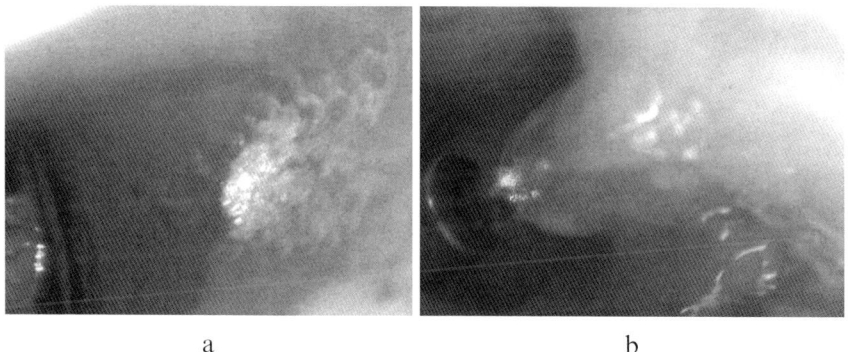

a   b

Fig. 4 Endoscopic images of a rabbit NP during laser irradiation: (**a**) Laser power is of 1.5 W, exposure time of 15 s. Micro cavitation (micro bubbles of few micrometers in size are seen). (**b**) Laser power is of 2.5 W, exposure time of 30 s Macro caviation. Periodic thermal expansion produced mechanical oscillations. The diameter of the optical fiber (on the left) is of 1 mm

Other possible mechanisms (more active motion of the disc liquid media and of ions contained in them, increasing of Ca concentration near the cells; possible activation of stem and committed cells that become a source of cartilage tissue regeneration) discussed in [13] require future studies.

Clinical Trials of LRD

Indications for percutaneous LRD were chronic pain in neck and back or chronic vertigo as a sign of vertebro-basilar insufficiency that could not be controlled by complex conservative therapy. Necessary conditions for LRD therapy were clear signs of degeneration of intervertebral discs on MRI scans (dark discs), including the appearance of a hyperintensive zone [16], and induction of highly to moderately concordant pain syndrome during CT-discographic challenge and presence of fissures in the AF [17]. Contraindications for LRD therapy were absence of signs of degeneration of intervertebral discs at MRI and CT-discography, significant protrusion or extrusion of intervertebral discs associated with apparent compression of nervous structures, significantly reduced height of intervertebral discs, local and general infections, blood coagulation disorders and uncontrolled coagulopathy, obvious psychological component in the pain syndrome, including extreme intolerance to discography.

Since 2001 the laser reconstruction of intervertebral discs have been performed at the Spine and Orthopedic Medical Center in Moscow for 240 patients with chronic symptoms of low back or neck pain who failed to improve with non-operative care. LRD was given in the out-patient setting under local anesthesia. Cervical discs were punctured with the use of anterior-lateral approach. The needle diameter was 18G while that of the laser light guide 0.6 mm. LRD therapy was conducted through the needle introduced into the disc under X-ray guidance to irradiate several zones in the disc. In cervical spine, the central zone of the NP and two zones of the transition layer of the AF were irradiated. In lumbar spine, two central zones of the NP and two zones of the transition layer of the AF were irradiated. An Er-glass fiber laser (wavelength 1.56 μm) was used for irradiation. Every zone was irradiated by three series of pulses (each series lasted for 30 s with intervals of 20 s between them; pulse duration – 2 s, interval between pulses – 1 s).

All patients obtained LRD demonstrated improvement. Substantial relief of back pain was obtained in 90% of patients treated who returned to their daily activities with many participating in recreational sports [12–14]. As our clinical observations show, discogenic pain syndrome was controlled or alleviated immediately after the procedure in the majority of patients. This can be due to dereception of pathological nerves, changes in the cartilage mechanical properties and relaxation of strain leading to disc decompression. In many patients pain syndrome alleviation was a gradual process lasting for 3–4 months during which new cartilage tissue was growing. In such cases maximal clinical improvement was observed in 6–8 months and was followed by stabilization of patients' condition, which lasted through the whole follow-up period.

Postoperative MRI examination of the patients has shown: (a) The locality of laser effect: New well hydrated tissue is growing in the laser treatment zones only. (b) The increasing of water contents in the disc (the increasing of T2 signal) was observed for 80% patients after LRD. (c) Decreasing in the protrusion size for some patients after LRD.

The method of laser treatment of spinal diseases was registered with the Federal Service of Supervision in the health and social development sector of the Russian Federation (registration certificate FS-2006/025).

LRD therapy allows to halt disc destruction and prevent the development of disc fibrosis and spinal segment instability. Thus the spine will be in a better shape and there will be no causes for pain syndrome to develop. LRD therapy is different from other puncture methods of treatment, as it is non-destructive. In experimental conditions temperature was measured in all disc zones, in the posterior longitudinal ligament and spinal canal. The temperature change of the most bulk of the disc tissue during LRD is less than $2°C$. Only small area (of about 1 mm^3 in the volume) near the laser fiber tip is heated to the 45–50$°C$. At this heating temperature disc tissue is not destroyed and the procedure is thermally safe but still laser heating of different zones is not uniform, which creates a significant temperature gradient resulting in a significant mechanical effect that modifies the fields of mechanical strains in disc tissue. The safety of the procedure is also ensured by the control system allowing to adjust delivered laser power and stop the irradiation if there appear gaps or big gas bubbles (when microgravity is transformed into cavitation that can lead to disc injury).

The mode of action of non-destructive laser irradiation on intervertebral disc tissue is multifaceted. It was shown experimentally that it can induce the formation of fibrous-hyaline and hyaline cartilage in the disc [13, 14]. Additionally to a high number of proliferating chondrocytes, including multiple multicellular cell clones indicative of regeneration activity induction by LRD, this type of tissue was found at histological examination of two biopsy specimens taken during later surgery for disc hernia recurrence in several years after previous microdiscectomy in combination with LRD. In these two cases, the disc cavity was filled with well structured fibrous-hyaline cartilage and there were no defects in the AF.

Arcuo Medical Inc. has manufactured the equipment for Laser reconstruction of spine discs. The specification of the equipment: Erbium-glass fiber laser, Wavelength 1.56 μm, Average power 0.3–3 W, Beam modulation: pulse duration from 10 ms to 2 s, Pulse repetition rate 0.3–10 Hz, Exposure time 10–30 s, A number of pulse series 3–5, Pause between pulse series 10–30 s. The equipment includes a feedback control system which provides efficacy and safety of laser procedure.

The advantages of LRD compared to other techniques of low-invasive physical treatment of spine disc diseases are:

1. LRD is directed to the reason of the disease due to activation of tissue regeneration.
2. We do not heat AF. The locality, the efficacy and safety of LRD can be controlled by the stress distribution and dynamics using space and temporal modulation of laser radiation.
3. The positive dynamics of the results. The stability of the improvement of life and pain relief.

Open Questions and Future Work Are Necessary to Study

1. Mechanisms of tissue regeneration need future investigation
2. The origin of chondrocytes producing hyaline type cartilage (stem cells, or rapid diffusion of chondrocytes from hyaline plates?)

3. The reasons of different dynamics of pain relief (including denervation of pathologic nerves in the AF, stress relaxation, growth of the new tissue decrease spine instability)
4. Long term stability (more than 5 years) of the positive results

Conclusions

Laser reconstruction of spine discs is a novel minimally invasive approach for the treatment of intervertebral disc diseases. The new type of Erbium doped glass fiber laser (1.56 μm) has been tested first on animals and then in a clinical trial. The histological studies showed that laser treatment stimulated growth of hyaline cartilage in the spine discs. Since 2001 LRD procedure has been performed for 240 patients in Moscow. Most of the patients demonstrated significant improvement, returned to the work and normal life.

Acknowledgments The authors thank the Russian Foundation for Basic Research (grant 05-02-16902), CRDF (grant RUP2–2660-MO-05), ISTC (grant 3360), and Arcuo Medical Inc. for the financial support.

References

1. Luo, X., Pietrobon, R., et al. (2004). "Estimates and patterns of direct health care expenditures among individuals with back pain in the United States." Spine 29 (1): 79–86.
2. Maroon, J.C. (2002). Neurosurgery. Nov; 51 (5 Suppl): 137–145.
3. Singh, K., Ledet, E., Carl, A. (2005). Spine; 30 (17 Suppl): 20–26.
4. Singh, V., Derby, R. (2006). Pain Physician; 9: 139–146.
5. Davis, T.T., Sra, P., Fuller, N., Bae, H. (2003). Orthopedic Clinics of North America 2003; 34 (2): 255–262.
6. Houpt, J.C., Conner, E.S., McFarland, E.W. (1996). Spine; 21: 1808–1812.
7. Bass, E.C., Nau, W.H., et al. (2006). Spine; 31 (2): 139–145.
8. Freeman, B.J.C., Fraser, R.D.F., Cain, C.M.J., et al. (2005) Spine; 30 (21): 2369–2377.
9. Sobol, E. et al. (2000). *Proceedings of SPIE*; 3907: 504–507.
10. Sobol, E. et al. (2001). Proc 12th World Congress Neurosurgery, Sydney, Australia.
11. Baskov, A. (2002). Laser thermodiscoplasty in the treatment of degenerative disc disease: experimental and clinical study. IITS Meeting, Seoul, Korea.
12. Baskov, A., Borshchenko, I., et al. (2005). Minimally invasive treatment of degenerative disc disease by Laser Reconstruction of Discs. SF-36 questionnaire and VAS analysis. 13th World Congress of Neurological Surgery, Marrakesh, Morocco.
13. Laser Engineering of Cartilage, edited by V.N. Bagratashvili, E.N. Sobol, A.B. Shekhter Fizmatlit, Moscow, pp. 423–472, 2006.
14. Sobol, E.N. et al. (2007). "Laser reshaping and regeneration of cartilage." Laser Phys Lett; 4: 488–502.
15. Heath, C.A. (2000) Biotechnol Genet Eng Rev, 17: 533–551.
16. Aprill, C., Bogduk, N. (1992). "High-intensity zone: a diagnostic sign of painful lumbar disc on magnetic resonance imaging." Br J Radiol 65 (773): 361–369.
17. Bernard, T.N., Jr. (1990). "Lumbar discography followed by computed tomography. Refining the diagnosis of low-back pain." Spine 15 (7): 690–707.

Part VIII
FDA Regulations

Requirements for FDA Approval

Sankar Basu

Abstract In the United States the Food and Drug Administration regulates the sale of medical devices. Questions concerning the approval process of light therapy devices where discussed by a distinguished FDA staff member.

Keywords: 510(k), pre-IDE, IDE, PMA.

Many of the presented talks questioned the regulatory policies of the Medical Devices, particularly on the Low Level Light Therapy Devices. What is needed for a marketing application of these devices includes when and how to submit the 510(k), Pre-IDE, IDE, or PMA applications. Since the safety and effectiveness of most of the low level light therapy (LLLT) devices are not known, a decision may depend on the indications for use of these devices. Pre-IDE and IDE applications may become prerequisites for collecting data before marketing applications are submitted. A IDE application is mandatory for initiating a clinical trial for significant risk devices. A 510(k) (lowest cost) application for a proposed device may be submitted only when the device manufacturer can present for comparison a similar device in terms of its specifications and intended uses. A PMA (highest cost) application is needed when the device presents new intended uses and safety and effectiveness are not known for the device for its intended uses. Whether a 510(k) or PMA application is appropriate for an LLLT device submission, it is likely that FDA may ask for clinical data to establish its safety and effectiveness. For small manufacturers the FDA offers help in bringing devices to FDA. The Division of Small Manufacturers, International, and Consumer Assistance Office of Communications, Education and Radiation Programs can be contacted toll free at 800 638-2041×122 for help with any questions.

S. Basu

Food and Drug Administration, Office of Device Evaluation, 9200 Corporate Boulevard, Rockville, MD 20850, USA

R. Waynant and D.B. Tata (eds.), *Proceedings of Light-Activated Tissue Regeneration and Therapy Conference.*

269

Part IX
Pain

Pain Relief with Phototherapy: Session Overview

Mary Dyson

Abstract In this overview of the session on the relief of pain with phototherapy the clinical significance of acute and chronic pain are considered. It is recommended that phototherapy be the treatment of choice for acute pain since this is associated with acute inflammation, the resolution of which by phototherapy accelerates tissue repair where this is proceeding suboptimally. Another advantage of phototherapy is the absence of harmful side effects.

Attention is drawn to the inhibition by phototherapy of nerve conduction in small and medium diameter of peripheral nerve fibres, and the possibility that this blockade may reduce nociceptive input to the dorsal horn, down-regulating the spinal and cortical tracts involved in pain, resulting in long-term analgesia. Clinical examples of pain relief by phototherapy are considered, together with the need to monitor effectiveness and to tailor treatments to the response of each patient.

Keywords: Phototherapy, low level laser therapy, low intensity laser therapy, photobiomodulation, pain relief, dentistry, craniofacial pain, acupuncture, inflammation complex regional pain syndrome.

Phototherapy, also known as low level laser therapy (LLLT) and low intensity laser therapy (LILT), has been used with considerable success to relieve acute and chronic pain for many years. It has proved particularly successful in dentistry, where pain can be an obstacle to the patient's acceptance of treatment, to prevent pain as well as to reduce it. There are, however, circumstances where acute pain has a protective value, being indicative of damage and usually resulting in the avoidance of further damage; in these conditions it is appropriate to use LLLT to reduce the level of pain but not to eliminate it. With chronic pain, the clinician's aim and the patient's wish is its complete eradication; LLLT helps with this, but the cause of the chronicity must also be removed to ensure that it does not recur. LLLT has the

M. Dyson
Emeritus Reader in the Biology of Tissue Repair, Kings College London (KCL), University of London, London, e-mail: md41139@aol.com

R. Waynant and D.B. Tata (eds.), *Proceedings of Light-Activated Tissue Regeneration and Therapy Conference.*
© Springer Science + Business Media, LLC 2008

additional advantage, not shared with many medications administered for pain relief, of accelerating the resolution of acute inflammation and thus accelerating tissue repair where this is proceeding suboptimally [1, 2]. Pain is one of the signs of inflammation, generally ending when inflammation ends. LLLT should therefore be the treatment of choice for pain associated with recent injuries. Another advantage of LLLT is the absence of harmful side-effects when used correctly.

Understanding of the mechanisms involved in the relief of pain by means of LLLT has increased in recent years. Once considered to be primarily a means of stimulating cell activity, there is a growing awareness of our ability to use LLLT to inhibit, as well as stimulate, cell activity. This duality has led to LLLT now being referred to as a modulator of cell activity. There is evidence, presented in this session by Dr. Chow, that the inhibitory effects of LLLT on certain neurons underlies clinical pain relief. The inhibition of nerve conduction has been observed in small and medium diameter peripheral nerve fibers; it is hypothesized that this blockade may reduce nociceptive input to the dorsal horn, down-regulating the spinal and cortical tracts involved in pain, resulting in long-term analgesia.

The clinical value of the pain relieving attributes of LLLT is illustrated by the other presenters in this session.

- Dr. Bjordal translates the observation that LLLT can modulate acute inflammation in the laboratory to analogous clinical situations such as the inflammation and consequent pain associated with extraction of the third molar in dentistry. His conclusion, following a systematic review of randomized controlled trials (RCT) with meta-analysis of pain up to 24 hours post surgery, is that LLLT with red or infrared wavelengths and energy doses of 6–7.5 J is effective in reducing the pain associated with acute inflammation. He recommends that future research should focus on the determination of the most effective clinical doses for LLLT. It should be appreciated that these will vary with the disorder and clinical attributes of each patient including their other disorders and the treatments being received for them.
- Dr. Gardiner presents a retrospective case series demonstrating the efficacy of bandwidth limited visible light, combined with real-time physiologic monitoring, in the treatment of Complex Regional Pain Syndrome (CRPS). This is characterised by severe pain associated with swelling and vasomotor changes in an injured limb which may feel either cold or hot. There is a decrease in the numbers of superficial small diameter nerve fibers in the limbs of patients with CRPS. Following preliminary indications that LLLT may have a beneficial effect on autonomic tone and be effective in reducing pain in these patients, a retrospective analysis was done to find out is localized application of visible light to patients with CRPS could induces changes in regional skin temperatures and reduction in pain. Because neurons respond to change, not to static conditions, the light used as a stimulus was varied in wavelength and in rate of change of wavelength during each treatment, a logical though novel approach to the application of LLLT. When modulated low intensity light was applied to skin sites remote from the area of pain for 2 minutes per site 75% of patients reported a reduction in pain in the affected limb and there was a greater change in the

temperature of the skin of the limb than in controls. Because the light was applied, not to the injured limb, but to skin at a distance from it, it is considered that pain relief and associated, tertiary, physiological responses to the light are under central rather than local control.

When a patient is treated with LLLT primary, secondary and tertiary effects occur, summating to produce clinically meaningful changes in the patient. Primary (molecular) and secondary (cellular) effects are predictable and reasonably well understood. However tertiary events are less predictable due to variation in the metabolic status of the patient and are less well understood. To optimise treatment, the therapist must be aware of and respond to this variation, tailoring treatment to fit the patient. One size doesn't fit all. Analysis of the mechanisms of LLLT at the primary and secondary levels of activity has been very successful; what is now required is synthesis, informed by the results of this analysis, to improve clinical outcome.

References

1. Young SR, Dyson M. 'The effect of light on tissue repair'. Acupuncture Med 11:17–20, 1993.
2. Baxter D. 'Low-intensity laser therapy'. In 'Electrotherapy: Evidence-Based Practice' Edited by S. Kitchen, Churchill Livingstone, Edinburgh, pp. 171–189, 2002.

Is Relief of Pain with Low-Level Laser Therapy (LLLT) a Clinical Manifestation of Laser-Induced Neural Inhibition?

Roberta Chow

Abstract Dr. Çhow looks at the mechanism of pain and asks the question regarding the effects on nerves.

Keywords: LLLT, inhibition, nerves, inflammation.

Introduction

Low-Level Laser Therapy (LLLT) has been used in the treatment of painful conditions for more than thirty years. Increasingly rigourous evidence for the clinical effectiveness of laser therapy has steadily accumulated, particularly in the last five years. Though clinical benefits have been established for a number of conditions, the mechanisms underlying these effects are not yet well understood. The lack of a plausible mechanism or mechanisms of action has been a factor in the slow acceptance of LLLT by mainstream medicine.

Proposed Mechanisms for Pain Relief

Much of the focus of early discussion and research of laser-mediated pain relief centred on stimulatory effects of laser irradiation on cell cultures and wound healing models. Cells involved in wound repair such as fibroblasts, neutrophils and macrophages, show a generally reproducible pattern of stimulation at low doses and inhibition at higher doses of laser irradiation. While tissue repair, mediated by activity of such cells, is critical to recovery from acute injury and ultimately pain relief, the time course of activation occurs over days to weeks. This does not reflect

R. Chow
Central Clinical School, Department of Medicine, The University of Sydney, Australia,
e-mail: rtchow@bigpond.net.au

R. Waynant and D.B. Tata (eds.), *Proceedings of Light-Activated Tissue Regeneration and Therapy Conference.*

277

the immediate change in pain, often seen within minutes of laser application, in patients and in animals.

More recent research has clearly delineated specific, anti-inflammatory effects of laser irradiation in acute inflammatory models [2]. The complex cascade of events occurring in acute injury, which initiates tissue repair, is associated with the release of pro-inflammatory mediators, such as prostaglandins E_2, which sensitize nociceptors and cause pain [7]. Laser irradiation causes decreased production of PGE_2 along with other inflammatory markers and is a plausible mechanism for pain relief in acute injury. The mechanism whereby PGE_2 production is suppressed by laser irradiation remains unknown.

Neural Inhibition as a Mechanism for Pain Relief

Laser-mediated inhibition of neural activity has provided a parallel hypothesis for laser-induced pain relief over the last twenty-five years, supported by researchers in this domain of laser therapy [6; 1; 4].

The time frame for laser-induced neural effects occurs on a more rapid scale (minutes to hours) compared to that of cellular effects (days to weeks). Moreover, the rapidity of response is also compatible with the rapid reduction of PGE_2 seen in experimental studies. In contrast to models of laser action where stimulatory effects are seen as most relevant, inhibition of neural activity provides a model by which both acute and chronic pain relief can occur.

The identification of mechanisms for laser-induced pain relief is critical for its acceptance into mainstream medicine. The literature is explored for evidence to support the hypothesis that inhibitory effects of laser irradiation on nerves provide a plausible mechanism for clinical pain relief.

Method

Randomised, controlled studies of any wavelength of laser irradiation to peripheral, mammalian nerve cells in culture and to nerves in animals and humans *in vivo* were evaluated following a search of sensitizatio databases and relevant literature, using multiple key words. Electrical, functional and structural changes in nerves or related structures in response to laser irradiation, when compared to baseline or a control group, were evaluated.

Results

Conduction Velocity

In 44 experiments from 20 studies, conduction velocity (CV) was reduced in 39% of studies when results of all wavelengths were grouped together. Visible wavelength ($\lambda = 632.8$ and 670 nm) irradiation caused a reduction in CV in 33% of experiments and infrared wavelengths ($\lambda = 780, 820, 830, 904, 1{,}064$ nm) reduced CV in 42% of experiments.

Amplitude of Compound Action Potentials (CAP)

Visible wavelengths caused a reduction in the amplitude of the CAP in 59% of 34 experiments from 15 studies. From the same group of studies, infrared wavelengths caused a reduction in 63% of studies.

Selective Inhibition of Aδ and C Fibres

830 and 1,064 nm laser irradiation selectively inhibited thinly myelinated Aδ and unmyelinated C fibre nociceptors. There were no studies of visible wavelengths on nociceptors.

Suppression of Noxious Stimulation

Visible and infrared wavelengths caused inhibition of noxious stimuli in both inflammatory and non-inflammatory models of noxious stimulation.

Other Neural Functions

Laser irradiation caused inhibition of several other structural and functional activities of nerves such as suppression of Na^+-K^+-ATPase, at high doses [5] and fast axonal flow [3].

Discussion

Studies of laser-induced changes in conduction velocity failed to show an overall inhibitory effect when all wavelengths were grouped together for analysis. When wavelength-specific effects were evaluated slowing of conduction velocity was more likely to occur with infrared laser irradiation of several wavelengths than visible laser. Reduction in the amplitude of compound action potentials was more sensitive an indicator of inhibitory effects and occurred with both infrared and visible wavelengths. As the nature of the experimental stimulus became more noxious, infrared wavelengths were seen to have a specific inhibitory effect on nociceptors, the thinly myelinated Aδ and unmyelinated diameter neurons C fibers, implying a direct analgesic effect. No studies of visible wavelengths were retrieved so direct nociceptive effects of these wavelengths cannot be commented upon. How such inhibitory effects influence pain relate to the nature of the nociceptive process.

In acute injury with tissue damage, laser irradiation may cause inhibition of peripheral terminals of Aδ and C fibres in the most superficial layers of the epidermis, which become damaged and sensitized by locally released pro-inflammatory mediators. Suppression of neurogenic inflammation arising from damage to these terminals and associated with peripheral sensitization is consistent with the decrease in PGE_2 observed following laser irradiation in animal and human models. Direct inhibition of peripheral afferent terminals suppresses peripheral sensitization and limits further release of neurokinins, such as Substance P and CGRP and may be a mechanism associated with acute pain relief.

In chronic pain, sensitization is more important than inflammation, in the absence of a specific disease process. Nociceptors become sensitized by changes occurring in the spinal cord and higher centres, as a result of neural plasticity, causing the phenomenon of long-term potentiation, a form of pain memory. Sensitized peripheral nociceptors may gradually be "de sensitized" by repeated application of laser irradiation to peripheral sites of tenderness. Blockade of fast axonal flow by reversible, laser-induced disruption of the microtubule structure in peripheral afferent terminals disrupts both anterograde and retrograde transport in neurons. Laser blockade of anterograde transport may block the release of neurotransmitters centrally by inhibition of transmission of signals from the sensitized nociceptors in the region of pain. Similarly, inhibition of retrograde transport of neurotransmitters, which are sensitizati in the body of the neuron at the dorsal root ganglia, will also suppress nociceptor activity in the periphery.

The elongated structure of the neuron, up to 1 m in the length, with no ATP production occurring along the axon, makes it uniquely vulnerable to disruption of its energy transport system by interference at any point along this continuous transport system. The superficial location of peripheral terminals in the skin is associated with minimal attenuation of laser irradiation, especially when applied with a contact pressure technique, thus sensitizat the potential for inhibition of neural function at this level.

Conclusion

Evidence from the literature demonstrates that laser inhibits several critical aspects of nerve function, though change in conduction velocity is a poor measure of potential, anti-nociceptive effects of laser irradiation. Based on the overall body of literature, it is proposed that in painful conditions, application of laser causes inhibition of nerve activity in small and medium diameter peripheral afferent fibres. In acute tissue injury, suppression of the neurogenic component of peripheral sensitization may limit the extent of inflammation, thereby relieving pain. In persistent pain, laser-induced neural blockade may result in reduced nociceptive input to the dorsal horn, thereby facilitating down-regulation of spinal and cortical tracts involved in pain perception, resulting in long-term analgesia. Inhibitory effects of laser irradiation are therefore important to the application of laser in both acute and chronic pain relief.

References

1. Baxter GC, Walsh DM, Allen JM, Lowe AS, Bell AJ. Effects of low intensity infrared laser irradiation upon conduction in the human median nerve in vivo. Experimental Physiology 1994; 79:227–234.
2. Bjordal JM, Johnson MI, Iverson V, Aimbire F, Lopes-Martins RAB. Photoradiation in acute pain: A systematic review of possible mechanisms of action and clinical effects in randomized placebo-controlled trials. Photomedicine and Laser Surgery 2006; 24(2):158–168.
3. Chow R, David M, Armati P. 830-nm laser irradiation induces varicosity formation, reduces mitochondrial membrane potential and blocks fast axonal flow in small and medium diameter rat dorsal root ganglion neurons: implications for the analgesic effects of 830-nm laser. Journal of the Peripheral Nervous System 2007; 12:28–39.
4. Jimbo K, Noda K, Suzuki H, Yoda K. Suppressive effects of low-power laser irradiation on bradykinin evoked action potentials in cultured murine dorsal root ganglia cells. Neuroscience Letters 1998; 240(2):93–96.
5. Kudoh C, Inomata K, Okajima K, Ohshiro T. Low-level laser therapy pain attenuation mechanisms. Laser Therapy 1990; 2:3–6.
6. Maeda T. Morphological demonstration of low reactive laser therapeutic pain attenuation effect of the gallium aluminium arsenide diode laser. Laser Therapy 1989; 1(1):23–30.
7. Siddall PJ, Cousins MJ. Neural Blockade in Clinical Anesthesia In: Introduction to Pain Mechanisms – Implications for Neural Blockade. M Cousins, P Bridenbaugh, editors. Philadelphia, PA: Lippincott-Raven, 1998. pp. 675–713.

Complex Regional Pain Syndrome: A New Approach to Therapy

Allan Gardiner, Robert E. Florin, and Constance Haber

Abstract The principal symptoms in Complex Regional Pain Syndrome (CRPS) are severe pain and vasomotor changes in an injured extremity.

This ongoing observational case study demonstrates that localized application of variable wavelength light to patients with CRPS can induce changes in regional skin temperatures and reduce the pain.

Initially, case histories for 12 adult patients with CRPS of 2 to 25 years duration and 11 normal controls were included in the study. Modulated low intensity light in the visible spectrum was applied to specific skin sites remote from the affected part for 2 minutes at each site. Skin temperatures of the affected extremities were recorded during the light sessions from infrared thermograms. Changes in pain intensity were measured from pre- and post treatment numeric pain scores.

Following light applications, the average maximal change in skin temperature in the affected limbs of the CRPS group was $4.3°C$ (range $+11.2 - 6.9°C$) relative to the pre-treatment temperature. In normal controls, the average change was $1.3°C$ (range $0.3 - 2.3°C$). In the same period, 75% reported a reduction in pain intensity.

The rapid response to light at sites remote from the affected extremities suggests a photobiological interaction with the subcutaneous network of peripheral sensory receptors. The bilateral temperature changes observed support the idea that the vasomotor responses are mediated via central autonomic pathways. The association of pain reduction with autonomic responses suggests a common mechanism in some of the patients.

Keywords: Complex regional pain syndrome, reflex sympathetic dystrophy, light therapy, infrared thermography, vasomotor changes.

A. Gardiner

PhotoMed Technologies, Kensington, California; 65 Franciscan Way, Kensington, 94707, USA, e-mail: allan@photomedtech.com

R. Waynant and D.B. Tata (eds.), *Proceedings of Light-Activated Tissue Regeneration and Therapy Conference.*
© Springer Science + Business Media, LLc 2008

Introduction

Complex Regional Pain Syndrome (CRPS) is a disorder of the peripheral and central nervous system that appears following trauma to a limb with symptoms disproportionate in intensity, duration and distribution to the degree of injury. Clinically, it manifests as severe spontaneous pain, hyperalgesia, impaired motor function, swelling, changes in skin temperature, and abnormal sweating. The symptoms are maximum distally in a non-dermatomal distribution and tend to spread to other body regions over time.

CRPS is classified into two types; type I is also known as reflex sympathetic dystrophy (RSD) while type II is associated with injury to a major nerve and is also called causalgia.

The severity of symptoms and their resistance to most medical and surgical interventions results in many cases lasting for years to decades. In a Canadian survey of RSD patients, only 16% were employed while 60% were completely disabled and 20% had attempted suicide [1]. Because there is no treatment that is consistently able to cure CRPS, most therapy is designed to relieve the pain. This usually includes a variety of pharmacological, manipulative and surgical interventions that are frequently ineffective and carry significant risk and costs associated with their use. Many require inpatient care during periods of extreme pain.

This report describes a new method that applies variable wavelength visible light locally to the skin of CRPS patients that has two novel effects. The first is a change in the skin temperature of the affected limbs. The second effect is a beneficial reduction in the intensity of pain experienced following the same exposures to light.

Use of variable wavelength visible light locally applied to a patients' skin as an autonomic stimulus is a new method of eliciting changes in skin temperature of affected limbs in patients with CRPS. The amount of change is comparable to that seen with external thermal challenges, and is significantly different from the response of normal controls to the same stimulation.

Measurement of Skin Temperature Changes

Infrared thermal imaging (TI) is a useful non-contact tool for measuring skin temperature. TI allows large areas to be included in one view for easy comparison and documentation.

Normal people have symmetrical skin temperatures in matched areas on each side of the body [10] (Uematsu 1988). In patients with neuropathic pain, it is common to observe abnormal thermal patterns in the affected extremity compared to the opposite limb. These changes in temperature are linked to abnormal sympathetic vasomotor activity that controls perfusion of the cutaneous vascular network [2, 3]. Since the side differences in temperature are not static but can change depending on the amount of sympathetic activity, measurement of resting differences in temperature in paired sites were used but proved to be an unreliable diagnostic measure in CRPS patients with involvement of both limbs. However, when techniques designed to induce changes in sympathetic vaso constrictor activity were applied to CRPS patients, significant changes in skin temperature during stimulation could be mapped that were not evoked in normal controls or other patients with limb pain who did not have CRPS [12]. This involved applying external thermal challenges to the patients to activate vasomotor reflexes.

Types of Abnormal Sympathetic Vasomotor Activity in CRPS

In patients with early CRPS (Acute Stage), generally within the first 6 months, the affected limb appears swollen and with reddish or mottled skin, and is warm relative to the opposite limb, These findings represent an absence of inhibition of vasoconstriction in that limb, with increased perfusion associated with what has been termed neurogenic inflammation. Beyond 6 months, there is a gradual transition to cold CRPS characterized by cold pale skin, continued pain and hypersensitivity and with continued changes in motor functions as well early trophic changes in the skin and nails. In some cases, spread of symptoms and autonomic changes appear in the opposite limb, increasing with the duration of the CRPS. These changes reflect overactive vasoconstriction with severely restricted cutaneous perfusion due to a combination of central and peripheral mechanisms [12].

Methods of Stimulating Sympathetic Vasomotor Reflexes *Cold Water Immersion*

Immersion of an unaffected hand or foot in cold water was shown to result in a paradoxical warming of the affected limb in CRPS compared to cooling as observed in normal controls [4]. This method is easily applied in an office setting and has had utility in supporting the clinical diagnosis of CRPS (Fig. 1) [9].

Body Suit Dynamic Stimulation of Thermoregulatory Cycle

Wasner et al. [11, 12] refined this method using a thermal body suit to sequentially cool and then warm the patient while monitoring skin temperatures bilaterally. They demonstrated significant dynamic changes in skin temperature asymmetry with a maximal side difference of $>2.2°C$, and a sensitivity of 75% and specificity of 100% that support it's use as a diagnostic tool for CRPS.

The results of this method in warm CRPS and cold CRPS are illustrated in Fig. 2, which shows the type and degree of the temperature asymmetries during the thermoregulatory cycle. The downside to this method is the need to apply a body suit and thermal control equipment and then monitor distal limb temperatures for a rather extended period of time while cycling from the maximal cold response to the peak warm response.

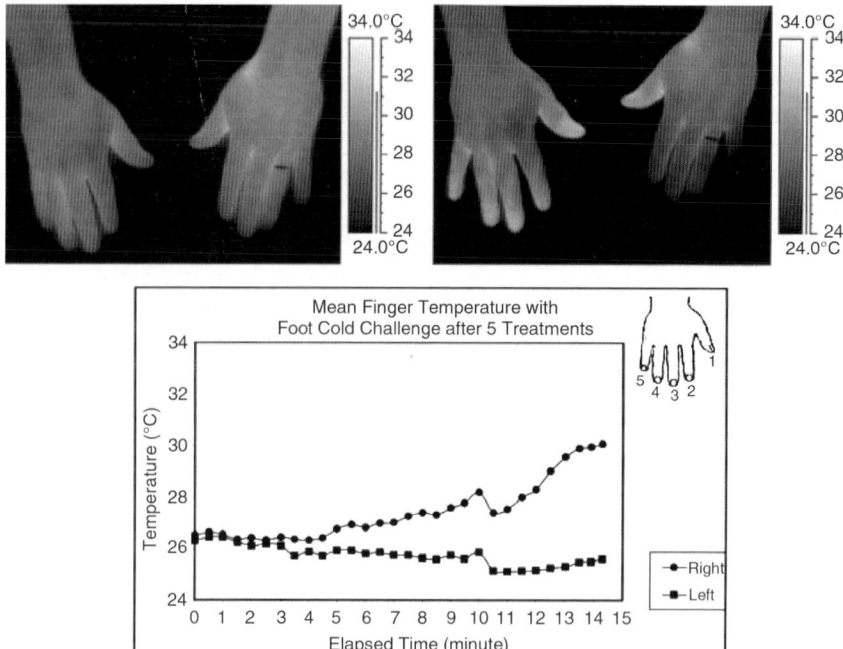

Fig. 1 Shows hands responding to a cold challenge. The left hand is responding nearly normally while the right hand (CRPS side) responds by warming abnormally

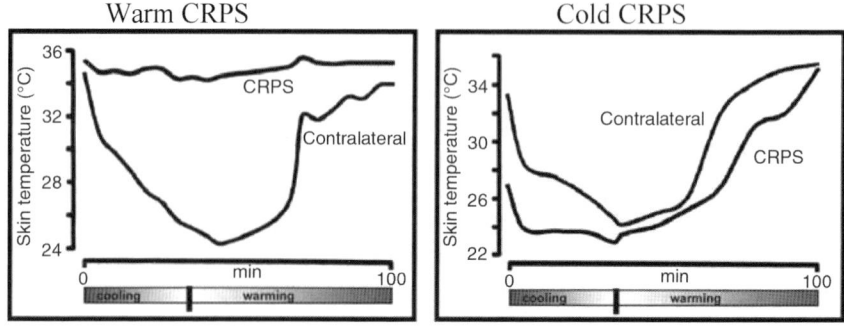

Fig. 2 Body suit response for cold CRPS and warm CRPS [12] (With permission)

Photon Cutaneous Stimulation of Sympathetic Vasomotor Reflexes

In 2002, use of light in the visible spectrum with a variable wavelength was applied to the skin at selected sites with comparable effects (C. Haber). These patients were monitored with IR thermography during and following the light applications which permitted objective measurement of the thermal changes observed.

The changes in skin temperature with variable wavelength light were greater in CRPS patients than in patients with pain of other causes. The changes were also greater in the distal portion of the affected extremities.

Methods

CRPS Case Series Treated with Light

In 2005, a group of 12 patients with a history of 2–25 years of CRPS I were treated by a single clinician (author CH) with localized applications of variable wavelength light stimulation (VWLS) during late 2005 and early 2006 at an outpatient clinic in Pennsylvania. The patients came from several RSD support groups in Pennsylvania, Delaware and New Jersey. All had been diagnosed with CRPS by previous treating physicians. The diagnosis was confirmed using International Association for the Study of Pain (IASP) criteria [5, 6] which includes continuing regional pain with abnormal sensory, vasomotor, motor, and/or trophic changes.

The mean duration of pain in the group was 7.3 years with a range from 2 to 25 years. Six patients had CRPS for more than 5 years while only two had it for less than 3 years. All 12 patients were suffering from CRPS Type I.

Normal Subjects

Eleven subjects without clinical evidence of CRPS, peripheral vascular disease, arthropathy or peripheral neuropathy served as a control group. Each had three applications of light to a site in the upper lateral neck where the CRPS patients often exhibited a significant response.

Measurement of Pain

All patients who presented for treatment with VWLS were provided an informed consent which they signed. Records of their past medical/surgical history and demographic parameters were reviewed. Visual Analog Pain scales were done at entry and Numeric Rating Scales at later visits. Pain scores were recorded at the beginning of each visit during the first 5 days.

Thermographic Monitoring

For patients with upper extremity/body CRPS, the finger pads (or nail beds) were used as the targets for measurement of surface temperature, while the toe pads (or nail beds) were used for lower body/extremity CRPS. A digital infrared thermographic camera (FLIR A40M) was focused on an area that included both distal extremities up to approximately mid-forearm (or mid-calf.) A pre-treatment image was recorded for baseline temperature measuresi. Monitoring at 10 images per minute was continued for up to 15–20 minutes after initiation of light applications and the images were stored on digital media.

Sites of Application

The sites of application conformed to points on acupuncture meridians, and usually were remote from the distal affected extremity. Several very effective sites were identified over the upper lateral neck, the supraclavicular fossa and the posterior midline over spinal processes.

Protocol

A halogen lamp with a motorized filter provided variable wavelength visible light ranging in wavelength from 400 to 700 nm from two hand-held applicators. Infrared and ultraviolet were removed from the light by filters located in the light path. A motorized shutter interrupted the light delivered through fiber optic cables to the applicators at frequencies from zero to 400 kHz. The maximum power delivered by each applicator is approximately 0.050–0.150 W (0.07–0.11 J/cm^2), depending on wavelength.

The treatments typically included three or four 2-minute applications of light having the center wavelength of the output beam vary between two fixed wavelengths (typically, 60 nm apart) at about 1 nm per second while excluding IR and ultraviolet wavelengths with appropriate filters. This was repeated for each of at least three visits per patient. Patients who manifested significant changes in regional temperature (over 2°C) or beneficial changes in symptoms (reduced pain and allodynia, reduced use of analgesics, improved sleep) were invited to return for additional sessions after the initial three visits. The elapsed time for the series of initial visits was usually 4–6 days with subsequent visits and VWLS dependent on the patient's requests for additional therapy. The intervals between visits varied widely during the follow up period, ranging from several days to several months.

Measures of Regional Temperature Changes

Spot temperatures of targeted sites in the images were measured by software analysis of the recorded thermographic data. The digit pads or nail beds of both extremities were the sites measured and the mean temperature of all five digits on each side was calculated to represent the temperature of the affected side and the contralateral side.

The maximal change in skin temperature on either side was used to represent the effect of VWLS for that session. Since some of the changes observed were cooling rather than warming, we converted the maximal values on each side into absolute values before calculating the maximal change in temperature for either extremity.

For the CRPS group, we selected only those VWLS sessions that had thermographic recordings of the affected limbs that permitted measurement of temperatures in all of the ten digits. The normal control group had thermographic measures of all ten digits for each subject.

Measures of Pain Changes

Pain intensity was measured initially with the Visual Analog Pain Scale (VAS) and the baseline pain intensity was calculated from the initial VAS as the average of the current pain plus the average pain in the past week plus the least pain and their worst pain.

Since pain is a predominant symptom in CRPS, the quality of any intervention designed to reduce pain needs to be judged in the context of how much improvement the patient perceives as a result of that invention. Farrar et al. [8] showed that a 30% reduction from baseline in chronic pain was judged to be a clinically important difference by patients. We used a 30% reduction in pain score from baseline as the threshold for significant pain relief in this report.

Results and Discussion

Changes in Skin Temperature

Significant changes in skin temperature ($>2°C$) following VWLS during the first 5 days of visits were observed in 11 of the 12 patients (92%).

An example of skin temperature change during and following light applications in a bilateral CRPS patient is depicted in Fig. 3 which illustrates the thermographic patterns in grayscale over the dorsum of both hands during a VWLS session. The chart plots the mean temperature of each hand during the session.

The thermal changes observed were both rapid and substantial, reflecting the rapid response to the applied light stimulus.

In the group of 12 CRPS patients, the mean maximal change in either limb was 4.3°C. Nine of the twelve had a significant temperature change within three visits. When a group of normal controls were stimulated using the same parameters and sites as the CRPS patients, the mean maximal skin temperature change in either limb was 1.3°C. The mean maximal changes with VWLS were significantly greater in CRPS patients than in normal controls ($p < .001$).

Comparison of Body Suit Versus VWLS Stimulation

Range of temperature changes in CRPS I patients:

Thermal body suit stimulation $= -9.5°C$ to $+10.5°C$
CVWLS stimulation $= -6.9°C$ to $+11.3°C$

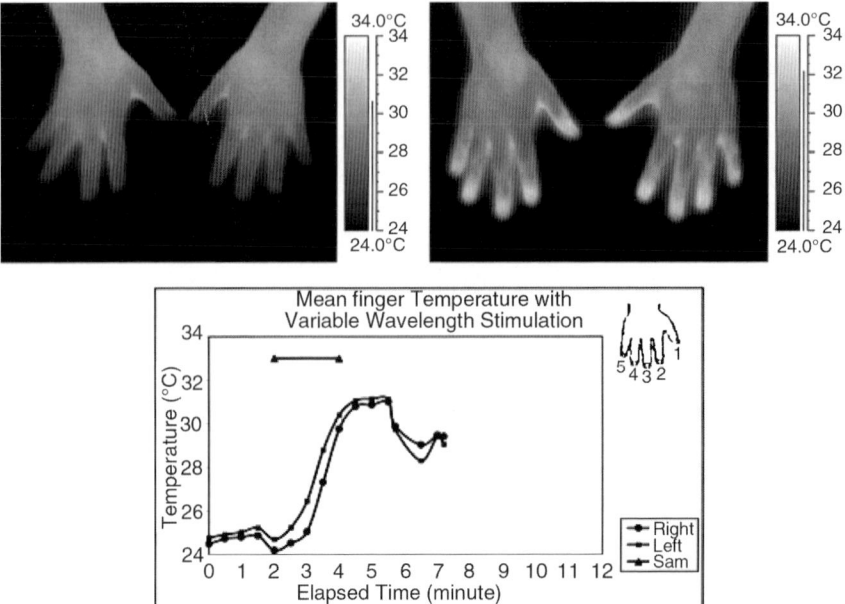

Fig. 3 Thermographic monitored skin temperature changes in a patient with bilateral cold CRPS stimulated by VWLS during the period marked by (▲)

In CRPS patients, temperature changes from the resting level, including positive and negative changes around the baseline temperature were comparable with both VWLS and thermal stimulation.

Value of Measures of Thermal Changes with VWLS

Testing dynamic sympathetic vasomotor reflexes with VWLS evoked changes in skin temperature with a potential diagnostic specificity of 91% and sensitivity of 75% for this small sample. This is equivalent to the results of body suit thermal stimulation of the same vasomotor reflexes in CRPS [12], with a method that is significantly more convenient to utilize and more comfortable for the patient. Applying VWLS as a method of influencing the sympathetic dysregulation in patients with CRPS suggests it's potential use as a tool to supplement the clinical criteria used in diagnosis of this disorder.

The advantage of using absolute values of change in either limb rather than the net change around a baseline is that it permits measurement of significant changes from resting levels when both extremities are affected by CRPS as in chronic CRPS that has spread.

Value of Real Time Monitoring During VWLS

Skin temperature changes during VWLS are objective evidence of a significant physiological response to the light stimulus. As such, temperature change indicates that the patients' autonomic dysregulation is reacting to the stimulus and suggests to both the patient and treating physician that continued efforts may be worthwhile.

Pain Changes in CRPS I Patients Treated with VWLS

The mean baseline numeric (0–10) pain score for the group of 12 patients was 6.9+/−0.5 before therapy, calculated from Visual Analog Scores. The pain scores after the first three visits with VWLS reduced by 50% for 6/12 and by 22–40% for five additional patients. Using a 30% reduction as a significant change, 9 of the 12 patients met this level of change [7, 8]. This matches the 75% of patients that had a significant skin temperature change during therapy. However, four patients in this group had either a thermal response without a significant reduction in pain, or pain reduction without a corresponding thermal change during the initial three visits.

In the thermal suit dynamic stimulation of vasomotor reflexes, no significant changes in pain were reported subsequent to the thermal changes observed [11, 12].

Conclusions

Use of VWLS compares favorably to thermal challenges as a means of modulating sympathetic vasomotor reflexes in patients with CRPS. Changes in skin temperature recorded by IR thermography provide objective measures of physiological changes in regional perfusion of affected limbs in patients with CRPS. Measurement of the degree of abnormal thermoregulation represented by the amount of skin temperature change from baseline may be considered in support of the clinical diagnosis of CRPS. Significant pain reduction occurs in some CRPS patients treated with VWLS. The association of pain reduction with sympathetic responses suggests a common mechanism in some of these patients.

The rapid response to VWLS at sites remote from the affected extremities suggests the photobiological interaction initially may involve the subcutaneous network of peripheral sensory receptors. The bilateral temperature changes observed with VWLS support the idea that the vasomotor responses are mediated via central autonomic pathways.

References

1. Agarwal S, Broatch J, Raja S. Web-based epidemiological survey of complex regional pain syndrome-1. Anesthesiology 2005; 103:A902.
2. Birklein F, Riedl B, Claus D, Neundorfer B. Pattern of autonomic dysfunction in time course of complex regional pain syndrome. Clin Auton Res 1998a; 8:79–85.

3. Birklein F, Riedl B, Claus D, Neuundorfer B, Handwerker HO. Sympathetic vasoconstrictor reflex pattern in patients with complex regional pain syndrome. Pain 1998b; 75:93–100.

4. Bej M, Schwartzman R. Abnormalities of cutaneous blood flow regulation in patients with reflex sympathetic dystrophy as measured by laser Doppler fluxmetry. Arch Neurol 1991, 48:312–915.

5. Breuhl S, Lubenow TR, Nath H, Ivankovich O. Validation of thermography in the diagnosis of reflex sympathetic dystrophy. Clin J Pain 1996; 12:316–325.

6. Breuhl S, Steger H, Harden R. Complex Regional Pain Syndrome. In Turk D, Melzack R. eds. Handbook of Pain Assessment, 2nd ed. Guilford Press, New York, 2001.

7. Cepeda M, Africano J, Polo R, Alcala R, Carr D. What Decline in Pain Intensity Is Meaningful to Patients with Acute Pain? In: Dostrovsky J, Carr D, Koltzenburg M, eds. Proceedings of the 10th World Congress on Pain, Progress in Pain Research and Management, Vol 24, IASP Press, Seattle, WA, 2003.

8. Farrar J, Portenoy R, Berlin J, Kinman J, Strom B. Defining the clinically important difference in pain outcome measures. Pain 2000; 88:287–294.

9. Gulevich SJ, Conwell TD, Lane J, Lockwood B, Schwettmann RS, Rosenberg N, Goldman LB. Stress infrared telethermography is useful in the diagnosis of complex regional pain syndrome, type I (formerly reflex sympathetic dystrophy). Clin J Pain 1997 Mar; 13(1):50–59.

10. Uematsu S, Edwin D, Jankel W, Kozikowski J, Trattner M. Quantification of thermal asymmetry. J Neurosurg 1988; 69:552–555.

11. Wasner G, Schattschneider J, Baron R. Skin temperature side differences – a diagnostic tool for CRPS? Pain 2002; 98:19–26.

12. Wasner G, Schattschneider J, Heckmann K, Maier C, Baron R. Vascular abnormalities in reflex sympathetic dystrophy (CRPS I): mechanisms and diagnostic value. Brain 2001, 124:587–599.

Part X
Electric Field Interactions

Introduction

Martin J.C. van Gernert

Abstract Stimulation of biological tissue has been observed from electromagnetic fields from several studies. Recent questions have been raised as to whether such fields may be similar in cause to light therapy. This session looks at some experiments where fields do appear to trigger similar results.

Keywords: rf studies, mid-IR results, refractive index extension.

The session on Electric Field Interactions covers three topics where electromagnetic fields of various frequencies, varying between bursts of eight square pulses of electric fields of 250 μs, 0.8 to 1.2 GHz, to 405 nm wavelengths, produce unexpected and so far unexplained biological effects.

The first is the effect of light illumination on the refractive index of ATP dissolved in water, first discovered by the Spanish physician Albert Amat I Genis and described in his Ph.D. thesis [1], showing an increasing refractive index proportional to the illumination time. The largest effect was at the shorter wavelengths, 405 nm was the shortest used, and this effect disappeared at 969 nm. In that wavelength band, the excess refractive index decreased continuously. Interestingly, ATP solutions have exceedingly small if any absorption in those wavelengths. Unfortunately, Dr. Amat was unable to attend the conference but a summary of his results will be given by Dr. Darrell B Tata, FDA, White Oak, USA. Subsequently, Dr. Bernhard J Hoenders, from the Center of Theoretical Physics, University of Groningen, The Netherlands, will summarize the theory of refractive index, both the classical and quantum dynamics description.

The second is the application of pulsed electric fields and continuous wave broad band light source on suppressing the metabolic activity of malignant human brain cancer (glioblastoma) cells. Dr. Ron Waynant will present the results that, although not exhaustive in range of values of dose, indicate that electric fields in the kilohertz range and far infrared radiation in the 20–50 μm wavelength produce stimulation and inhibition results much like those in the visible and near infrared. As yet tests

M.J.C. van Gernert
Laser Center, Academic Medical Center, University of Amsterdam, The Netherlands

R. Waynant and D.B. Tata (eds.), *Proceedings of Light-Activated Tissue Regeneration and Therapy Conference.*

for and measurements of hydrogen peroxide have not been done, but the results led the investigators to speculate that the generation of hydrogen peroxide may extend over a broad region of the electromagnetic spectrum.

The third is on the effects of electromagnetic radiation emitted by a mobile phone (mobile phone electro smog) on the activity of the human brain. The possibility of health risk particular for children will be discussed by the German-Dutch physician Dr. Careen A Schroeter, Department of Laser Therapy, Medical Center Maastricht, The Netherlands.

Our aim is to identify the underlying mechanisms that cause these intriguing biological responses to electromagnetic interactions.

The Painful Derivation of the Refractive Index from Microscopical Considerations

Bernhard J. Hoenders

Abstract The derivation of the refractive index from the microscopical structure of matter is analysed in detail. In particular the many various assumptions leading to the basic Clausius- Mosotti (Lorentz-Lorenz) equation are carefully stated. The most general formulation of the second order correlation theory for the refractive index, the so-called Yvon-Kirkwood theory, is given. These considerations will facilitate the explanation of a very peculiar effect observed by Amat.

Keywords: Refractive index, ATP solution, polarisation, Clausius-Mosotti equation, Lorentz-Lorenz law, effective field.

Introduction

When measuring the refractive index of an ATP solution in water Amat observed that the refractive index of this solution was increasing during the illumination and even kept on on increasing when the illumination was stopped! A similar effects, though less pronounced was observed in a solutions of ADP. A solution of AMP did not show the effect. References to his work are [1, 2, 3]. An explanation of this effect is still lacking. Because the refractive index is the key quantity which is measured, the key thoughts leading to the concept "refractive index" have to be clearly stated in order to get an understanding of the Amat effect. This contribution to the conference provides all the essential thoughts which, starting from the basic microscopical quantity, viz. the microscopical polarisability, leads to the macroscopical concept "refractive index". The polarisation of a medium originates from the effect of an electric field on a electrically neutral atom or molecule: The electric field displaces the negatively charged electron cloud from the positively charged nucleus, thus inducing a net electric field as the field of the positive- and negative

B.J. Hoenders

University of Groningen, Institute for Theoretical Physics, and Zernike Institute for Advanced Materials, Nijenborgh 4, NL-9747, AG Groningen, The Netherlands

R. Waynant and D.B. Tata (eds.), *Proceedings of Light-Activated Tissue Regeneration and Therapy Conference.*
297

charges no longer compensate each other outside the atom. The resulting field is the dipole field far away from the atom. Near the atom corections must be added known as the quadripole- octipole- etc. fields.

Nucleus (charge Ze)

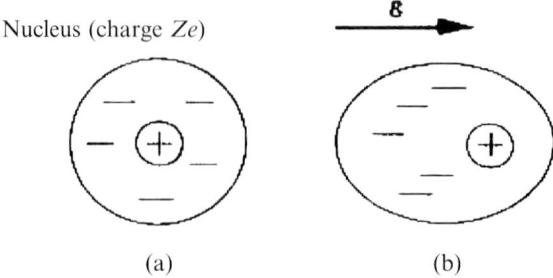

(a)   (b)

(**a**) The unpolarised atom, (**b**) The polarised atom

In the macroscopic theory of the electrodynamics of continuous media, the electromagnetic state of a medium is described by four vector fields, the electric and magnetic field vectors **E** and **B**, and the electric and magnetic displacement fields, **D** and **H**, which are all functions of the space and time coordinates, **r**. The space and time derivatives of these fields are related to each other and to the macroscopic charge and current densities, ρ_e and **j** by the *Maxwell equations* which are suppose to hold at each point in space and time where the derivatives of the field exist. They read as:

$$\nabla \wedge \mathbf{E} + \dot{\mathbf{B}} = 0, \quad \nabla \cdot \mathbf{B} = 0$$
$$\nabla \wedge \mathbf{H} - \dot{\mathbf{D}} = \mathbf{j}, \quad \nabla \cdot \mathbf{D} = \rho_e \tag{1a, b}$$

The field vectors **B,H,D** and **E** are related by the constitutive relations:

$$\mathbf{B} = \mu_0 \mu_r \mathbf{H}, \text{ and } \mathbf{D} = \varepsilon_0 \varepsilon_r \mathbf{E}. \tag{2}$$

The functions $\varepsilon_{0,r}$ and $\mu_{0,r}$ denote the dielectrical- and magnetical (relative) permeabilities resp., and are tensors in general. The physical significance and the unambiguous definition of the fields **E, B, D**, and **H** on the macroscopic level are based, on the one hand that in vacuo the fields satisfy $\mathbf{D} = \varepsilon_0 \mathbf{E}$ and $\mathbf{B} = \mu_0 \mathbf{H}$ and give the force on a unit test charge or current, and on the other hand on the fact that the polarisation and magnetisation **P** and **M**, *defined* by the equations:

$$\mathbf{D} = \varepsilon_0 \mathbf{E} + \mathbf{P}, \quad \mathbf{B} = \mu_0 \mathbf{H} + \mathbf{M} \tag{3}$$

are resp. the electric- and magnetic dipole moments per unit volume of the medium. These quantities should be derived from the underlying microscopical equations, and it was exactly this derivation which was initiated as early as 1878 by Lorentz [4] (for an English translation see [5, 6]). Further basic articles see: [7, 8]. Once it has become clear how the macroscopical Maxwell equations can be derived from the microscopical Maxwell equations the relation between the polarisation, the magnetisation, and the microscopical structure of the medium is established. This then enables the analysis of the connection between refractive index and the microscopical material properties. It will be shown in particular that the Lorentz-Lorenz relation, also known as the Clausius-Mosotti equation, follows immediately by the transition from the microscopical- to the macroscopical equations. Moreover, the influence of statistical fluctuations of the density or on the refractive index will be calculated.

The outline of this paper is as follows: We start with the microscopical Maxwell equations and will then derive the corresponding macroscopical equations, viz. (1a, b) by the appropriate ensemble averaging procedures. The underlying physical assumptions will be discussed at length, as they essentially determine the final expression for the refractive index in terms of the microscopical properties of the medium.

The Calculation of the Polarisation

In this section we will derive the macroscopical Maxwell equations (1a, b) from the microscopical Maxwell equations:

$$\nabla \wedge \mathbf{e} + \dot{\mathbf{b}} = 0, \quad \nabla \wedge \mathbf{b} - \dot{\mathbf{e}} = \mathbf{j} = \sum_{i,k} e_{k,i} \dot{\mathbf{R}}_{\mathbf{k},\mathbf{i}} \delta(\mathbf{R}_{k,i} - \mathbf{R})$$

$$\nabla \cdot \mathbf{b} = 0, \qquad \nabla \cdot \mathbf{e} = \rho_e = \sum_{i,k} e_{k,i} \delta(\mathbf{R}_{k,i} - \mathbf{R}) \tag{4a, b}$$

where ρ_e and $\mathbf{j}$ denote the microscopic source and current distributions. Next we derive the the so-called atomic field equations, i.e. The Mawell equations in which the existence of atoms (stable groups of point particles) has been taken into account. The position vector $\mathbf{R}_{k,i}$ of a stable group k of particles is written as:

$$\mathbf{R}_{k,i} = \mathbf{R}_i + \mathbf{r}_{\mathbf{k},\mathbf{i}}, \tag{5}$$

where the $\mathbf{r}_{k,i}$ denote the internal coordinates, which specify the positions of the constituent particles k, i with respect to the position $\mathbf{R}_k$ of the privileged group. Then, introducing the multipole expansions of the group of point charges:

$$\rho_e = \nabla \cdot \sum_{k,i} e_{k,i} \sum_{n=1}^{\infty} \frac{(-1)^{n-1}}{n!} (\mathbf{r}_{k,i} \cdot \nabla)^n \delta(\mathbf{R}_k - \mathbf{R}) \tag{6}$$

and

$$\mathbf{j} = \sum_{k,i} e_{k,i} (\dot{\mathbf{R}}_k + \dot{\mathbf{r}}_{k,i}) \sum_{n=0}^{\infty} \frac{(-1)^{n-1}}{n!} (\mathbf{r}_{k,i} \cdot \nabla)^n \delta(\mathbf{R}_k - \mathbf{R}) \tag{7}$$

and

$$\rho_e^{mono} = \sum_{k,i} e_{k,i} \delta(\mathbf{R}_k - \mathbf{R}) \tag{8}$$

we obtain:

$$\nabla \cdot \mathbf{e} = \rho_e^{mono} - \nabla \cdot \mathbf{p}, \quad \nabla \wedge \mathbf{h} - \dot{\mathbf{e}} = \mathbf{j} + \dot{\mathbf{p}} + \nabla \wedge \mathbf{m}$$
$$\nabla \cdot \mathbf{b} = 0, \quad \nabla \wedge \mathbf{e} = -\dot{\mathbf{b}} \tag{9}$$

with

$$\mathbf{m} = \sum_k \mathbf{p}_k \nabla \dot{\mathbf{R}}_k + \sum_i e_{k,i} \sum_{n=1}^{\infty} \frac{(-1)^n n}{(n+1)!} \mathbf{r}_{k,i} \wedge \dot{\mathbf{r}}_{k,i} (\mathbf{r}_{k,i} \cdot \nabla)^{n-1} \delta(\mathbf{R}_k - \mathbf{R}) \tag{10}$$

and

$$\mathbf{p} = \sum_k \mathbf{p}_k; \quad \mathbf{p}_k = \sum_i e_{k,i} \sum_{n=1}^{\infty} \frac{(-1)^{n-1}}{n!} \mathbf{r}_{k,i} (\mathbf{r}_{k,i} \cdot \nabla)^n \delta(\mathbf{R}_k - \mathbf{R}). \tag{11}$$

The model we take for the medium is that we suppose that the medium consists of a randomly oriented collection of scatterers each of which experiences the field radiated by the excited other scatterers. For a macroscopic description to be valid the system has to satisfy certain general conditions. One of these is that it must be possible to divide the system into "physically infinitesimal" volume elements, of diameter Δ, which on one hand are small compared with distances over which the macroscopical variables vary appreciably, and which, on the other hand, contain so many molecules that the principles of statistical mechanics are applicable, and quantities like the entropy and the temperature can be defined for each volume element. We characterise these conditions by the basic inequalities:

$$a << \Delta << \lambda, \tag{12}$$

where a is a molecular dimension and λ is a length characterising the spatial variations of the macroscopic variables in the system. *This approximation is known as the local field approximation and constitutes the key assumption of all the theories of molecular optics!* It is then assumed that these scatterers are so widely separated that the various fields they experience from the other scatters are *dipole* fields. This assumption, usually justified in practice, simplifies the calculations enormously. However, once the calculations have been set up using this approximation, higher order multipole fields like quadripole-octipole etc. can be dealt with similarly [9]. The local field approximation and the dipole field approximation are exemplified in Fig. 1. The dipole at some place inside the medium experiences the field due to the contributions of all the dipoles outside a sphere, the so-called Lorentz sphere. These dipoles lead then to a depolarising field of magnitude $1/3\varepsilon_0 \mathbf{P}(\mathbf{r})$ which makes that the dipole inside the sphere experiences a *local* field

$$\mathbf{P} = \rho a (\mathbf{E} + \frac{1}{3\varepsilon_0} \mathbf{P}), \tag{13}$$

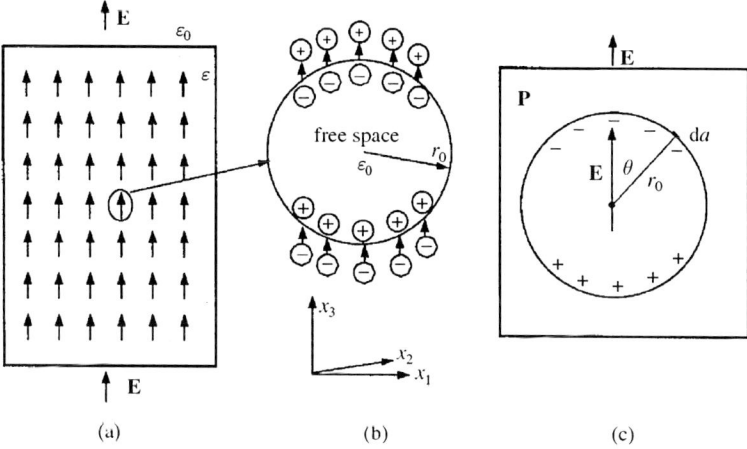

Fig. 1 (**a**) Macroscopical medium, (**b**) The Lorentz sphere, (**c**) The Lorentz sphere inside the medium.

If the dielectric were simple it would also be true that, see (2) and (3):

$$\mathbf{P} = (\varepsilon_0 \varepsilon_r - 1)\mathbf{E} \tag{14}$$

where ε_r denotes the relative dielectrical permeability. Combination of (13) and (14) then yields the Clausius-Mosotti or Lorentz-Lorenz law:

$$\varepsilon_0 \frac{\varepsilon_r - 1}{\varepsilon_r + 2} = \frac{1}{3}\rho a, \tag{15}$$

where ρ denote the density of the dipoles with "dipole strength" a. We would like to stress the point that the choice of the particular form of the excluding geometry, viz. the Lorentz *sphere*, determines the numerical factor of the depolarizing field: $\frac{1}{3}$ for this case. If we had chosen a different geometry, another numerical factor would have occurred! This observation may well be of interest if we consider non-spherically symmetrical molecules, in which case another type of cavity e.g. an ellipsoidal one has to be chosen.

Now consider again a collection of dipoles, where each dipole is characterised by its polarisability tensor $\mathbf{a}_i$ and give rise to a dipole with strength $\mathbf{p}_i$, if subject to an electric field $\mathbf{E}_0(\mathbf{r}_i)$

$$\mathbf{p}_i = \mathbf{a}_i \cdot \mathbf{E}_0(\mathbf{r}_i) \tag{16}$$

The polarisability tensor $\mathbf{a}_i$ is the basic molecular quantity to be derived either classically or quantum mechanically from the structure of the constituents of the material under consideration. For a detailed derivation see [10].

Next, a dipole with dipole moment $\mathbf{p}_i$ generates a field distribution:

$$\mathbf{p}_i \cdot \mathbf{F}_{i,j}, \tag{17}$$

with

$$\mathbf{F}_{i,j}(\mathbf{r}_i, \mathbf{r}_j) = (\nabla_i \nabla_i + k^2) \frac{exp(-ikr_{i,j})}{r_{i,j}}, \tag{18}$$

where $r_{i,j}$ denotes the distance between the coordinates $\mathbf{r}_i$ and $\mathbf{r}_j$, and the symbol denotes the unit tensor of order and rank 3. The theory, and in fact *any* theory enabling the transition of microscopical- to macroscopical Maxwell equations is then based on the following integral equation:

$$\mathbf{p}_i = \mathbf{a}_i \cdot \left[\mathbf{E}_0(\mathbf{r_i}) - \sum_{j=1, j \neq i}^{N} \mathbf{F}_{i,j}(\mathbf{r}_i, \mathbf{r}_j) \cdot \mathbf{p}_i \right]. \tag{19}$$

The average dipole moment $\bar{\mathbf{p}}_i$, the average polarisibility $\bar{\mathbf{a}}_i$, and the polarisation $\mathbf{P}(\mathbf{r})$ are given by:

$$\bar{\mathbf{p}}_i = \frac{\overline{\mathbf{p}_i \delta(\mathbf{r}_i - \mathbf{r})}}{\overline{\delta(\mathbf{r}_i - \mathbf{r})}} \tag{20}$$

and

$$\bar{\mathbf{a}}_i = \bar{\mathbf{a}} = \rho^{-1} \sum_i \overline{\mathbf{a}_i \delta(\mathbf{r}_i - \mathbf{r})} \tag{21}$$

and

$$\mathbf{P}(\mathbf{r}) = \sum_i \overline{\mathbf{p}_i \delta(\mathbf{r}_i - \mathbf{r})} = \sum_i \bar{\mathbf{p}}_i \overline{\delta(\mathbf{r}_i - \mathbf{r})} = \rho \bar{\mathbf{p}}_i \tag{22}$$

respectively, and the quantity ρ denotes the molecular density:

$$\rho = \sum_i^{N} \overline{\delta(\mathbf{r}_i - r)}. \tag{23}$$

Then, combination of equations (19)–(23) yields:

$$\mathbf{P}(\mathbf{r}) = \rho \bar{\mathbf{a}} \left[\mathbf{E}_0(r) - \int_v \mathbf{F}(\mathbf{r}, \mathbf{r}') \mathbf{P}(\mathbf{r}') g^{(2)}(\mathbf{r}, \mathbf{r}') dr' \right] + \mathbf{J}(\mathbf{r}), \tag{24}$$

if

$$\mathbf{J(r)} = \int_v \sum_{i,j,i\neq j} \overline{\left[\overline{\mathbf{a}} \cdot \mathbf{F}_{i,j} \cdot \overline{\mathbf{p}}_j - \mathbf{a}_i \cdot \mathbf{F}_{i,j} \cdot \mathbf{p}_j\right] \delta(\mathbf{r}_i - \mathbf{r})\delta(\mathbf{r}_j - \mathbf{r}')} d\mathbf{r}' \tag{25}$$

and $g^{(2)}(\mathbf{r}, \mathbf{r}')$ denotes the so-called pair correlation function

$$g^{(2)}(\mathbf{r}, \mathbf{r}') = \overline{\sum_{i,j,i\neq j} \delta(\mathbf{r}_i - r)\delta(\mathbf{r}_j - r)}. \tag{26}$$

The first of these equations, (24) describes the dominant interaction, whereas the second one, (25) describes the deviations from the main process because its integrand denotes the difference between the mean (macroscopical) and the microscopical quantities. These equations are the basic equations of molecular optics and lead to the classical laws connected with the refractive index such as the Clausius-Mosotti (Lorentz-Lorenz) law see

$$(15): \rho^{-1}\frac{n^2-1}{n^2+2} = \frac{1}{3}\mathbf{a}_0(\omega)$$

A straightforward but tedious calculation [11] then leads to:

$$\rho^{-1}\frac{n^2-1}{n^2+2} = \frac{1}{3}\mathbf{a}_0\frac{[1++G(\rho,T,,\omega)]}{1+\rho\mathbf{a}_0(\omega)D}\left[1 + \frac{\Delta\overline{\mathbf{a}}}{\mathbf{a}_0(\omega)} + G(\rho,T,,\omega)\right] \tag{27}$$

if

$$G(\rho,T,\omega) = \frac{\overline{\mathbf{a}}}{\mathbf{a}_0(\omega)}R, \quad \Delta\overline{\mathbf{a}} = \overline{\mathbf{a}} - \mathbf{a}_0(\omega). \tag{28}$$

The relation (27) has been for the first time derived by [12], see also [13, 14]. The functions $D(\rho,T,\omega)$ and $R(\rho,T,\omega)$ are defined by:

$$D(\rho,T,\omega) = \int_{v(\mathbf{r})}^{V} \mathbf{u} \cdot \mathbf{F}(\mathbf{r},\mathbf{r}') \cdot \mathbf{u} \, exp(-in\mathbf{k} \cdot (\mathbf{r}' - \mathbf{r}))[g^2(\mathbf{r},\mathbf{r}') - 1]d\mathbf{r}', \tag{29}$$

$$R(\rho,T,\omega) = \int_{v(\mathbf{r})}^{V} \mathbf{u} \cdot \mathbf{K}(\mathbf{r},\mathbf{r}') \cdot \mathbf{u} \, exp(-in\mathbf{k} \cdot (\mathbf{r}' - \mathbf{r})) \tag{30}$$

resp. The wave vector $\mathbf{k}$ is equal to $\frac{\omega}{c} \times$ the unit vector in the propagation direction. The vector $\mathbf{u}$ denotes the unit vector in the direction of $\mathbf{P(r)}$. The basic result (27) contains the complete physical information of our system!! The fluctuations of the medium are described in terms of the average polarisability $\overline{\mathbf{a}}$ and the polarizability $\mathbf{a}_0$ of the free molecule. The tensor function $\mathbf{K}(\mathbf{r},\mathbf{r}')$ is determined by the basic equations (24) and (25) and describes the fluctuations of the medium due to whatever course, such as density- temperature fluctuations.

The assumptions on which this refined model for the calculation of the refractive index equations is based together with the result (27) and the remarks made below

(15) constitute the basic tools for an in depth analysis of the results obtained by Albert Amat i Genñis [1, 2, 3], showing an increasing refractive index proportional to the illumination time for a solution of ATP in water.

Discussion

In the previous sections we developed the theory of the refractive index and carefully stated various assumptions involved. These pertinent assumptions involved are listed below, together with a short statement concerning the implications for the values of the refractive index:

(a) A Lorentz ellipsoid could replace the sphere, then the depolarisation field takes a different value: its magnitude becomes $\beta\varepsilon_0\mathbf{P}(\mathbf{r})$ instead of $1/3\varepsilon_0\mathbf{P}(\mathbf{r})$, where β can be any number between zero and one. The Lorentz-Lorenz relation (15) then changes because the local field equation (13) then becomes:

$$\mathbf{P} = \rho a(\mathbf{E} + \beta\varepsilon_0\mathbf{P}), \tag{31}$$

(b) The polarizability a is in general a tensor. This is probably true for the case considered by Albert. Amat and accounts for asymmetry effects.
(c) Should only dipole-dipole interactions be considered or are quadripole- and octupole- etc. interactions important as well? These interactions become important if the interaction between the various dipoles is such that not only nearest neighbors interactions are important but also further neighbors interactions become important.
(d) Are non-linear effects important? These effects become important if the field strength is sufficiently large.
(e) The effect of the polarity of water was not considered. The large permanent dipole moment of water influences the value of the refractive index and may well account partly for the Amat effect as a changing geometrical shape of the ATP molecule leads to another distribution of the water molecules surrounding the molecule. An effective theory for such solutions is given by Onsager [15] and Böttcher [16–18].

These issues listed above may be of great value for the analysis and further understanding of the Amat effect.

References

1. A. Amat i Genñis, *Effect of visible and near-infrared light on adenosine triphosphate (atp).*, Ph.D. thesis, University of Rovira i Virgili, Reus, Spain, April 2005.
2. A. Amat, J. Rigau, R.W.Waynant, I.K. Ilev and J.J. Anders, *The electric field induced by light can explain cellular responses to electromagnetic energy: a hypothesis of mechanism*, J Photochem Photobiol B 82 (2006), no. 2, 152–160.

3. M. Willemse, E. Janssen, F. de Lange, B. Wieringa and J. Fransen, *ATP and FRET a cautionary note*, Nat Biotechnol 25 (February 2007), no. 2, 170–172.
4. H.A. Lorentz, Verh. Kon. Akad. v. Wetensch. Amsterdam 18 (1878), no. 1, see for a reprint: *Collected Papers*, vol. 2, p. 1.
5. H.A. Lorentz, *Collected Papers*, vol. 1, Martinus Nijhoff, The Hague, 1936.
6. H.A. Lorentz, *The Theory of Electrons*, Dover, New York, 1952.
7. J. van Kranendonk and J.E. Sipe, *Foundations of the Macroscopical Electromagnetic Theory of Dielectric Media*, vol. 15, ch. 4, pp. 247–350, North Holland, Amsterdam/Oxford, 1977.
8. H.A. Lorentz, *Enzyklopädie der Mathematischen Wissenschaften, Physik, Weiterbildung der Maxwellschen Theorie, Elektronentheorie*, vol. 5, ch. 14, pp. 145–280, 1903.
9. L. Jansen, *Molecular theory of the dielectric constant*, Phys Rev 112 (1958), no. 2, 434–444.
10. Y.R. Shen, *The Principles of Nonlinear Optics*, Wiley, New York, 1984.
11. A. Münster, *Statistische Thermodynamik kondensierter Phasen*, vol. 13, ch. 1, pp. 251–267, Springer, Berlin, 1962.
12. J. Yvon, *La propagation et la diffusion de la lumière*, Actualités scientifiques et industrielles Nr. 543 (1937), no. 543.
13. J.G. Kirkwood, *On the theory of dielectric polarization*, J Chem Phys 4 (1936), 592–601.
14. B.U. Felderhof, *On the propagation and scattering of light in fluids*, Physica 76 (1974), 486–502.
15. L. Onsager, *Electric moments o molecules in liquids*, J Am Chem Soc 58 (1936), no. 8, 1486–1493.
16. C.J.F. Böttcher, *Zur Theorie der Inneren Elektrischen Feldstärke*, Physica 9 (1942), 937–944.
17. C.J.F. Böttcher, *Die Druckabhangigkeit der Molekularpolarisation von Dipolfreien Gasen und Flüssigkeiten*, Physica 10 (1942), 945–953.
18. C.J.F. Böttcher, *Theory of Dielectric Polarisation*, Elsevier, Amsterdam, 1952.

Independent Applications of Near-IR Broadband Light Source and Pulsed Electric Potential in the Suppression of Human Brain Cancer Metabolic Activity: An In-Vitro Study

Darrell B. Tata and Ronald W. Waynant

Abstract The roles of two different non-conventional techniques in suppressing the metabolic activity of a malignant human brain cancer (glioblastoma) cell line were explored through the application of (i) pulsed electric field, or (ii) independent application with a continuous wave broadband near infrared light channeled through a fiber-optic bundle exposing cancer cells within growth medium. Human glioblastomas were grown in T-75 flasks and were utilized when the cells were 50–70% confluent. The cells were either transferred into 96 well plates for exposures through a fiber bundle, or into 1.4 ml sterile eppendoff tubes for exposures to pulsed electric potential. The glioblastomas within wells were light exposed through a fiber bundle at an average intensity of 0.115 W/cm^2 from the underside of the well, with the light dose (fluence) values ranging from 0.115–50 J/cm^2. Glioblastomas exhibited a maximal decline in the metabolic activity (down 80%) relative to their respective sham exposed control counterparts between the fluence dose values of 5.0–10 J/cm^2. The cellular metabolic activities for various treatment doses were measured through the colorimetric MTS metabolic assay 3 days after the broadband near infrared light exposure. Interestingly, the metabolic activity was found to return back to the (sham exposed) control levels as the fluence of exposure was increased up to 50 J/cm^2. Glioblastomas in suspension within sterile eppendoff tubes were exposed to pulsed electrical potential fluctuations: rectangular pulse width = 250 μs with pulse amplitude = 100 V, with 8 square pulses per burst, 2 bursts per second. A time course study of treatment exposure revealed a complete obliteration of glioblastomas for in-vitro treatment duration beyond 7 min.

Keywords: rf irradiation, glioblastoma cells.

D.B. Tata
U.S. Food and Drug Administration, Center for Devices and Radiological Health, Silver Spring, Maryland, USA, e-mail: Darrell.Tata@fda.hhs.gov

Introduction

Novel therapeutic devices are streaming into regulatory agencies seeking review and approval for their usage in clinical settings for diverse medical indications ranging from cellular/tissue engineering and wound healing to disease therapy. Identification and characterization through basic science research of novel/unconventional treatment modalities and strategies for fighting aggressive disease such as cancers is of great importance.

One potential and novel treatment strategy is to allow cancer cells to interact with optical or near infrared radiation at low intensities to induce desired bioeffects. Reports on low level light bio-effects on cancer cells are sparse and have primarily entailed monitoring proliferation/mitotic rates from a limited number of cancer cell lines (such as the He-La cervical cancer cell lines, the KB cell line, the melanoma cell lines) with several different and discrete exposure parameters in wavelength, intensity, and the amount of energy deposition on the cancer cells (findings summarized in [1]).

Wavelength, fluence and intensity have been noted as important light exposure parameters through previous investigations playing an important role in biomodulations which bring about various biological effects [2]. A feature in low level light exposure, as noted through past research, is a "biphasic" biological response in intensity and most notably in the light energy dose, i.e., the fluence [3]. Consequently, for a specified optical or infrared wavelength and for a specified low intensity there exists an optimal fluence value for light exposure to produce a maximum modulation for a specified biological response/effect. A few noteworthy bio-effects due to low level light exposure which have been reported in the literature include: (i) modulation in gene expressions [4], (ii) increase in the intracellular calcium levels [5], (iii) increase in the mitochondrial metabolic activity [6] and in enhanced production levels of ATP [7], (iv) with a concomitant increase in cellular proliferation [8].

Short electrical pulses having durations in the microsecond to the millisecond range have been utilized over the past 2 decades for passively importing genomic or proteomic material into cells due to the induction of nano-scale pores in the cellular membrane – a technique known as electroporation. If the delivered electric pulse amplitude is "weak" and the pulse duration is "short", the nano-scale pores are believed to reseal and the cell continues to function normally. However, if the electrical pulse amplitude is "strong" with long duration, the induced pores are beyond repair and the exposed cells die through necrosis.

Taken in the context of targeted cancer therapy, such a technique could be utilized to kill cancer locally through the applications of strong amplitude electrical fields. Unfortunately, the strong amplitude electrical fields in turn would induce electrical currents and joule heating would be expected to ensue which would diffuse out into the normal healthy tissue and would be expected to either damage or destroy it. The goal of a targeted site specific cancer obliteration, which leaves normal healthy tissues undamaged due to the lack of thermal heating effects, necessarily requires basic science exploratory research. A fine balance needs to be realized between the electric field strength and the duration of the pulse in addition to an understanding of the biological electrical properties of the tissue. To be sure, if several electrical pulses are involved in the treatment, the intermediate duration between pulses would also become an important parameter. Recent basic science research has shown that such a strategy could potentially be achieved through the applications of short (micro-second) pulses with moderately large electric field strengths [9].

In this communication, we report on our recent findings on the suppression of the metabolic activity in human glioblastoma, due to in-vitro exposure to either a low intensity broadband near infrared light or exposure to pulsed electric fields. The cellular metabolic activities for all treatment conditions were measured through the colorimetric MTS metabolic assay. The findings revealed that broadband visible and near infrared low level light exposures could potentially be a viable tool in reducing the metabolic activity of glioblastomas; however, a "biphasic" response in the energy of deposition (i.e., fluence) was realized with the maximum suppression occurring at the fluence value of 10 J/cm^2. Interestingly, the metabolic activity was found to return back to the sham exposed control levels as the fluence of exposure was increased beyond 10 J/cm^2 and up to 50 J/cm^2. Unlike the low intensity visible light findings, an in-vitro treatment study with 250 μs pulsed electrical potentials on glioblastomas has revealed a remarkable 100% brain cancer cells obliteration for treatment durations greater than 7 min. The end point results of our findings are consistent with the recent results of a 100% liver cancer obliteration in-vitro, with approximately the same pulse duration [10], although in our case the applied peak electric field amplitude was only 15% of the strength value utilized for the liver cancer in-vitro experiments.

Materials and Methods

Cell Line Maintenance

Human malignant (brain cancer) glioblastoma was purchased from American Type Culture Collection (Rockville, MD) and grown in monolayer and maintained in T-75 flasks under incubation conditions of 5% CO_2 at 37°C. The glioblastoma cells were maintained in ATCC formulated DMEM/F12 growth medium with 10% of fetal bovine serum (FBS) and 50 units/ml of penicillin and streptomycin (Pen/Strep) antibiotics.

Cell Preparation for Broadband Light Exposure

When the adherent glioblastoma cells reach 50–70% confluence within the T-75 flasks, the glioblastomas were trypsinized and brought into suspension. The cells were spun down and the (trypsin) supernate was discarded. The cells were re-suspended in fresh growth medium with 10% FBS and 50 units/ml of Pen/Strep at an initial working concentration 75,000 cells/ml. The cell suspension was then transferred into single well's of the 96 well plates with a transfer volume of 0.2 ml or 15,000 cells seeded per selected well. The cells were seeded into every other well in order to ensure no possible overlap in the broadband light exposure. The cells within the 96 well plates were returned back into the incubator for approximately 16 h before the light treatments.

Cell Exposure to Broadband Light

A continuous wave broadband light source (Tungsram, Inc.) was positioned far removed from the 96 well plate exposure platform, and its unfocused light was channeled through a fiber-optic bundle (5 mm in diameter). The other end of the fiber bundle was position in the center at the underside of the well, and the cancer cells were exposed with an intensity of 0.115 W/cm^2 (as measured through the Spectra-Physics, Inc. model # 407A power meter). Figure 1 below exhibits the normalized emission characteristics of the broadband lamp source after passing through the fiber-optic bundle (at the site of exposure) measured through Ocean Optics, Inc. high resolution spectrophotometer HR4000 CG UV-NIR.

Each alternating column of the 96 well plate (with the exception of column#1 which served as the control/sham exposed condition) received a pre-determine fluence dose of exposure for those wells within the column which contained cells. Fluence levels were varied by keeping the intensity of exposure invariant in time and through varying the irradiation time of exposure. The selected times of broadband light exposure ranged from 1 s to 7 min and 15 s with the corresponding fluence range between 0.115 and 50 J/cm^2. Experiments were done in quadruplicates

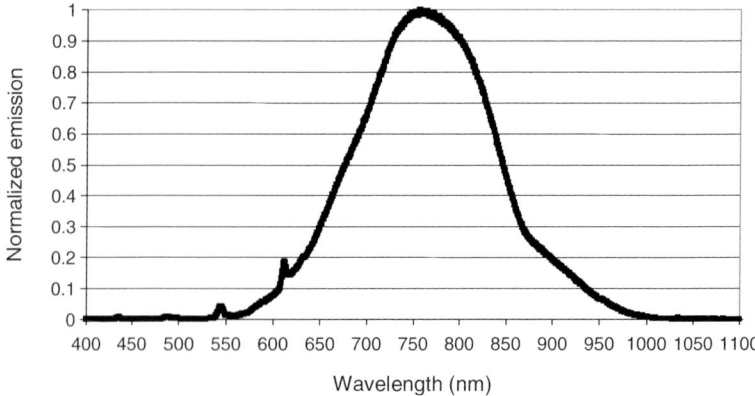

Fig. 1 Normalized emission through a fiber bundle of a broadband lamp source. Location of the central peak: 757 nm

for each fluence value. All exposures were done at room temperature ($\sim$23°C), and the average duration that the 96 well plates were left out at room temperature for laser treatment was approximately 2 h. Upon completion of the experiment, the 96 well plates were immediately returned to the incubator.

Cell Preparation for Pulsed E-Field Exposure

Identical preparation protocol as above, except the cells were re-suspended in fresh growth medium with 10% FBS and 50 units/ml of Pen/Strep at a working concentration 350,000 cells/ml and 1.4 ml of cell suspension was directly transferred into 1.4 ml sterile eppendoff tubes for pulsed electric potential experiments.

Time Varying Electric Potential Exposure

Glioblastoma cell suspension within sterile eppendoff tubes was exposed to pulsed electrical potential fluctuations through direct physical contact of immersed leads within the cellular growth medium. The electrical potential temporal wave form generation was achieved through a transcutaneous electrical nerve stimulator model: BioMed 2000 (Biomedical Life Science, Inc.) operated under a burst mode setting with the following parameters (as verified with an oscilloscope): temporal pulse width = 250 μs, pulse amplitude δV = +100 V, with 8 square pulses per burst, and a repetition frequency of 2 burst per second.

Just prior to field exposures and fluctuations, the positive and ground leads were sterilized through 70% ethanol immersion and thoroughly dried within the sterile laminar flow hood.

The stainless steel positive and ground leads were identical in dimensions: length = 11.43 mm, cylindrical diameter = 2.032 mm, and distance of separation between them within the sterile eppendoff tube was 4.445 mm.

All E-field exposures were conducted within the sterile laminar flow hood.

Measurement of Cellular Metabolic Activity

The metabolic response of glioblastoma cells to various broadband light fluences were assessed with a non-radioactive colorimetric cell metabolic tetrazolium compound (MTS) assay (Promega, Madison, WI) in four independent (control and exposed) replicates, 3 days after the light treatments. On the day of measurement, the 96-well plates were removed from the incubator and 20 μl of the MTS solution was added to each cell containing well. Thereafter, the plates were immediately returned to the incubator for a 2 h incubation period.

Functionally, the MTS readily permeates through the cell membrane and is metabolized and converted into formazan by living cells. Conversion into formazan induces a maximum change in absorption at 490 nm.

Two hours after the addition of MTS, absorption measurements were made at 490 nm with a 96 well plate reader (Perkin Elmer 1420 Multilabel counter: VICTOR^3), and the average absorbance value at 490 nm of laser treated cell's metabolic activity was computed with standard deviations and compared to the sham laser exposed average absorbance value with its standard deviation. The percentage of laser treated cell's metabolic activity was computed relative to the sham exposed metabolic activity.

Results and Discussion

The human glioblastoma cells exhibited a decline in metabolic activity relative to their control (sham exposed) counterparts between the fluence values of 0.115–10 J/cm^2. Maximal suppression in metabolic activity was noted at the fluence value of 10 J/cm^2. See Fig. 2 below. As the near infrared light dose was further increased beyond 10 J/cm^2 the metabolic activity was found to return towards the control levels. Thus, due to the biphasic response characteristics in the metabolic activity we deduce that there is a window of opportunity of the fluence level at which maximum suppression occurs. Although the bulk growth medium temperature immediately after the laser irradiation did not appreciably change ($\Delta T \sim 0.5°C$), one plausible and speculative mechanism for the metabolic activity trend to return

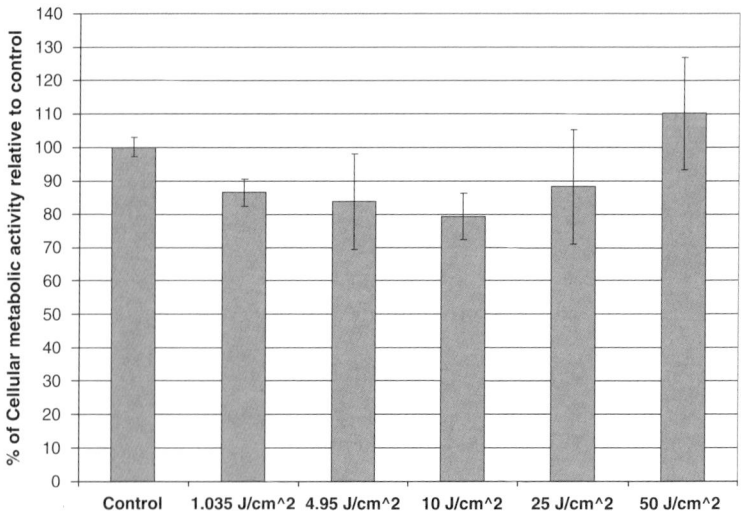

Fig. 2 Malignant human glioblastoma cellular metabolic activity measured 72 h after the broadband light exposure through a fiber bundle. Average ± SD, n = 4

towards the control condition could be attributed to the *intra-cellular* temperature which may potentially reach hyperthermic levels in which case the cellular constitutive heat shock proteins would be immediately called into action to protect the integrity of cellular and mitochondrial proteins and the DNA from thermal damage. Alternatively, elevated and sustained levels of long lived reactive oxygen species, such as hydrogen peroxide, could be expected to mediate similar pro-active responses from the constitutive heat shock proteins.

An important question which needs to be addressed in the pre-clinical/clinical setting is that of the light dosing scheme: Would repetitive light exposures with a specified fluence dose, separated by a specified time interval yield a greater photo-modulation in a bio-effect than a single exposure dose?

Strategic applications of micro-second pulsed electric fields to irreversibly electroporate at a specified target site within the tissue is expected to be of great clinical value, due to the fact that the electric field effects are predominantly at the cell membrane level and does not generate thermal damaging effects which could alter other healthy surrounding cells and tissue structures such as blood vessels or connective tissues [9].

Figure 3 below exhibits the metabolic activity results from our exploratory experiments with glioblastoma treatments using pulsed electric field on glioblastomas in suspension within sterile eppendoff tubes with the following exposure parameters: rectangular pulse width = 250 μs, pulse amplitude $\delta V = 100$ V, with 8 square pulses per burst, 2 bursts per second. A time course study of treatment exposure revealed a complete obliteration of glioblastomas for in-vitro treatment duration beyond 7 min.

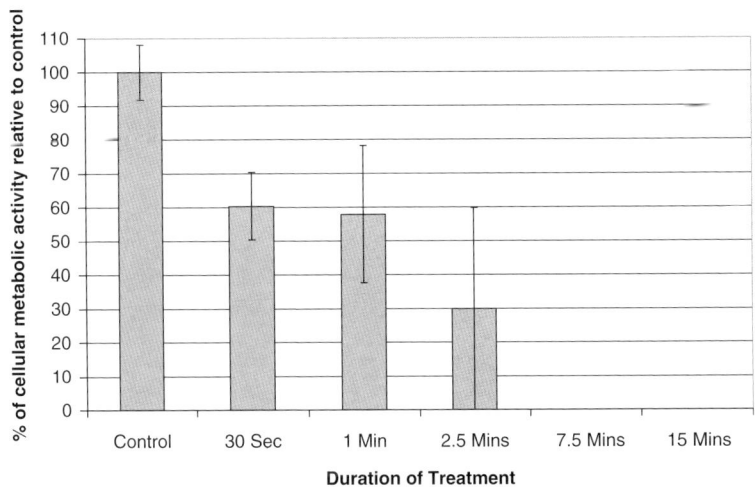

Fig. 3 Glioblastoma growth 3 days after pulsed electric potential exposures for several treatment durations. Exposure Parameters: Pulse width = 250 us, with 8 pulses/burst, 2 burst per second. [Glios] = 350,000/ml in 1.4 ml of sterile eppendoff tubes

Conclusion

These in-vitro findings have suggested further scrutiny into pre-clinical tumor bearing animal models. Of particular interest are the possible light induced growth modulations in aggressive surface cancers, such as the melanomas, due to their convenient accessibility to sunlight and to the conventional and non-conventional low level light exposures.

Selective time varying electrical fields which induce irreversible electroporation/ablation could prove to be a valuable tool in the clinical settings for minimally invasive cancer surgery treatments.

References

1. Tuner, J., Hode, L. 2002, Laser Therapy – Clinical Practice and Scientific Background. Prima Books AB, Grangesberg, Sweden. Biostimulation, Chapter 4.1.7: Cancer 130–134, and see table listing in Chapter 11 on page 349.
2. Tuner, J., Hode, L. 2002, Laser Therapy – Clinical Practice and Scientific Background. Prima Books AB, Grangesberg, Sweden. Chapter 3: Biostimulation, Chapter 4: Medical Indications.
3. Hamblin, M., Demidova, T.N., Mechanisms of low light therapy. Proc. SPIE. Vol. 6140, pp 1–12, 2006.
4. Zhang, Y., Song, S., Fong, C.C., Tsang, C.H., Yang, Z., Yang M. cDNA microarray analysis of gene expression profiles in human fibroblast cells irradiated with red light. J. Invest. Dermatol., Vol. 120, pp 849–857, 2003.
5. Lubart, R., Lavi, R., Friedmann, H., Rochkind, S. Photochemistry and photobiology of light absorption by living cells. Photomed. Laser Surg., Vol. 24, pp 179–185, 2006.
6. Yu, W., Naim, J.O., McGowan, M., Ippolito, K., Lanzafame, R.J. Photomodulation of oxidative metabolism and electron chain enzymes in rat liver mitochondria. Photochem. Photobiol. Vol. 66, pp 866–871, 1997.
7. Passarella, S. Helium – neon laser irradiation of isolated mitochondria. J Photochem Photobiol B, Vol. 3, pp 642–643, 1989.
8. Hawkins, D., Abrahamse, H. Biological effects of helium – neon laser irradiation on normal and wounded human skin fibroblast. Photomed Laser Surg, Vol. 23, pp 251–259, 2005.
9. Rubinsky, B., Onik, G., Mikus, P. Irreversible electroporation: a new ablation modality – clinical implications. Technol Cancer Res Treat, Vol. 6, pp 37–48, 2007.
10. Miller, L., Leor, J., Rubinsky, B. Cancer cells ablation with irreversible electroporation. Technol Cancer Res Treat, Vol. 4, pp 699–705, 2005.

Disclaimer

Electroencephalogram Changes Caused by Mobile Phones a Protective Device

Ngozy Mbonu, Elmar Weiler, and Careen Schroeter

Abstract The authors have seen electroencephalograms that appear to be modified by cell phone radiation. They present their evidence and have created a small device to protect the brain.

Keywords: Alpha waves, mobile phones, protective device; quantitative electroencephalogram.

Introduction

Wireless phones fall broadly into two categories: cellular phones and personal communication system (PCS) phones. Both categories use radiofrequency radiation (RFR) at a specific wave frequency to transmit voice messages via an antenna located in the phone. The radio frequency of mobile phones varies from 800 to 1,990 MHz. Analogue telephones apply frequencies between 800 and 900 MHz while digital telephones use frequencies between 1,850–1,990 MHz. The exposure to it is characterized through the specific absorption rate (SAR) expressed as watts per kilogram [1].

The use of cellular and cordless telephones is widespread and increasing in the society. The appearance and evolution of cellular phones have been one of the fastest in the history of innovation. The popularity of this technology cannot be underestimated yet the current science is not definitive about what type of health risks the use of mobile phones cause. The potential health risks of radiofrequency electromagnetic fields emitted by mobile phones are of considerable public interest. A potential association between cellular and cordless telephones and health effects is of concern and has been discussed in several articles during recent years [2–7].

Mobile phones and brain tumors are an issue which was first raised in the Scandinavian countries, as these have the most experience in using mobile phones

C. Schroeter
Department of Lasertherapy, Becanusstraat 17, Maastricht, Limburg 6216BX,
The Netherlands, e-mail: lasermed.schroeter@planet.al

R. Waynant and D.B. Tata (eds.), *Proceedings of Light-Activated Tissue Regeneration and Therapy Conference.*
© Springer Science + Business Media, LLC 2008

since the 1980s [8–11]. The brain is a sensitive organ and highly susceptible to the electromagnetic field. There is public concern that using mobile phones with a latency period of more than 10 years could increase the risk of brain tumors [12]. In another study on the effect of high frequency electromagnetic fields on the human EEG it was concluded that EMF emitted by cell phones may be harmful to the human brain since the delta waves are pathological if seen in awake subjects [13]. There is some evidence to suggest that exposure to mobile phones can affect neuronal activity particularly in response to auditory stimuli [14]. Another study showed slow response time in subjects performing specific cognitive tasks [15]. Despite all this information some studies have also shown that mobile phones have no effect on cognitive function [16] and another study carried out on human brain activity during exposure to radio frequency fields emitted by cellular phones suggested that exposure to radiofrequency fields emitted by cellular phones has no abnormal effects on human EEG activity [17].

Not much information is available on the effect of mobile phones on the electroencephalographic activity in the human body with and without masking the effect. In one study using GSM mobile phones it was shown that a combination of 64 electrodes electro-cap and 64 electro quick cap caused a peak specific absorption rate reduction of 14–18% respectively in both the whole head and in the temporal region [18].

In this preliminary study, brain mapping technology was applied in order to investigate the effects of mobile phones on the electro-encephalographic activity in humans with and without a protective device.

Methods

Quantitative electroencephalography: Technological advances increased the ability of encephalography to read brain activity data from the entire head simultaneously. Quantitative EEG (QEEG) applies multichannel measurements which can better determine spatial structures and localize areas with brain activity or abnormality. The results are often used for topographic brain mapping represented by color maps in 2D and/or 3D to enhance visualization. We acquired the brain waves (EEG signals) employing a Neurosearch 24 instrument (Lexicor Medical Technology, Inc., Boulder, CO) by placing 19 electrodes on the scalp in a standard international (10–20) pattern.

The QEEG was taken using the Electro-cap (Electro-cap International, Inc., Eaton, OH). Electro-Caps are an EEG electrode application technique. They are made of an elastic spandex-type fabric with recessed, pure tin electrodes attached to the fabric. The electrodes on the standard caps are positioned by the International 10–20 method of electrode placement (Fig. 1). Each lead was checked separately. Impedance was judged acceptable when electrode impedance registered below 5,000 Ω. The EEG signals from each electrode were independently amplified by matched differential amplifiers with less than 2-mV peak-to-peak noise, input impedance of more than 70-MΩ differential, common mode rejection of more than 90 dB at 60 Hz, high-pass filter of 2 Hz, and low-pass filter of 32 Hz.

Fig. 1 The "10–20 System" of Electrode Placement: Each point on this figure indicates the electrode positions employed. Each site has a letter (to identify the lobe) and a number or another letter to identify the hemisphere location. The letters F, T, C, P, and O stand for Frontal, Temporal, Central, Parietal and Occipital. Even numbers (2, 4, 6, 8) refer to the right hemisphere and odd numbers (1, 3, 5, 7) refer to the left hemisphere. The z refers to an electrode placed on the midline

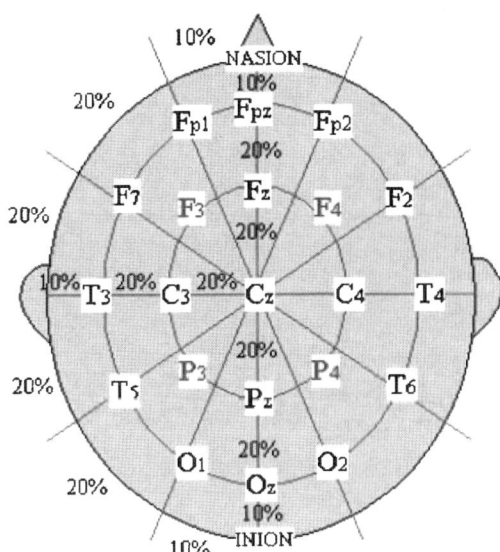

Analog-to-digital conversion of the signal was achieved with a 12-bit A/D converter, the sampling interval of which was governed by a 50-kHz crystal oscillator.

All EEG data were visually inspected for artifacts due to movements and to muscular activity before the records were subjected to quantitative analysis and interpretation. Frequency analysis was performed using a fast Fourier transform. The QEEG frequency bands chosen were delta (2–4 Hz), theta (4–7 Hz), alpha (8–13 Hz), beta (14–21 Hz) beta (19–23 Hz) and beta 21–30 Hz), with sub-alpha bands of alpha$_1$ (7–9 Hz), alpha$_2$ (9–11 Hz) and alpha$_3$ (11–13 Hz). Statistical calculations were performed using only artifact-screened data.

All EEG data were collected under controlled conditions with subjects reclining comfortably in an armchair with eyes closed and then eyes open in a sound-attenuated, electrically shielded room. Unless stated otherwise, the data used have been manually screened for eye-blink movements, and only eye-blink-free epochs were used in the preparation of quantitative results.

Experimental design

Condition	Eyes closed	Eyes open
Baseline condition	X	X
Listening to announcement mobile phone without protective system	X	X
Listening to announcement mobile phone with (E.M.I.) protective system	X	X

Prior to testing the effects of the mobile phone on the EEG pattern in humans, a baseline EEG was recorded with eyes closed and then with eyes open. Subsequently, the subject listened to time announcements for 1 minute while holding the mobile phone (with and without the protective E.M.I. system) with the left hand close to the ear, however, avoiding a direct contact with the EEG cap. One minute EEG samples were employed for an in depth EEG analysis.

Mobile phone: The tests were performed with a mobile phone, Motorola C115.

Protective chip (EuroMedInnovation = E.M.I.): made out of a polyester film, based on acryl which is able to store data. This polyester film data base contains biological products that send and receive information to and from the mobile phone. This information is read by the chip consequently responding to the electromagnetic radiation neutralizing the effect by absorption of the emitted radiation of the mobile phones.

Results

Demographics: All participants were healthy subjects without any neurological diseases. The EEG data collected from 30 normal subjects (females n = 21 and males n = 9) were subjected to a computer assisted EEG analysis. The average age of the male subjects was 57 ± 5 years and of the female control group 50 ± 4 years. No significant difference was calculated.

Results of the Computer Assisted EEG Analysis

Average total power: The average total power for a subject is calculated by averaging the total power from each of the 19 electrode leads.

The average total power for the female group (n = 21: baseline: eyes *closed*) $18.8 \pm 0.6 \, \mu V^2$, *MP without protective system:* $17.9 \pm 0.6 \, \mu V^2$, *MP with protective system:* $21.1 \pm 0.7 \, \mu V^2$.

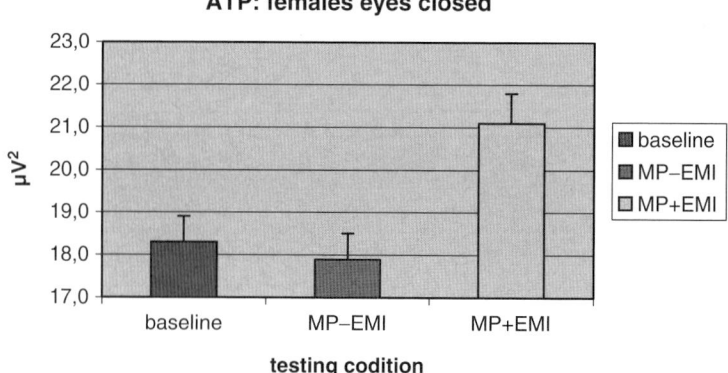

The average total power for the female group (n = 21: baseline: eyes *open*) 11.2 ± 0.2 μV², *MP without protective system*: 11.5 ± 0.2 μV², *MP with protective system*: 11.7 ± 0.3 μV².

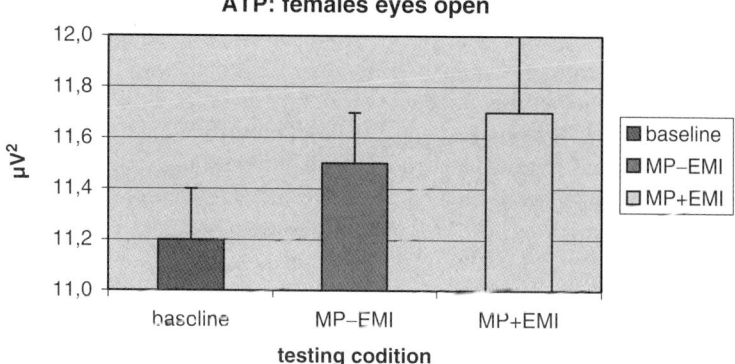

The average total power for the male group (n = 9: baseline: eyes *closed*) 19.2 ± 0.9 μV², *MP without protective system*: 14.4.2 ± 0.5 μV², *MP with protective system*: 15.4.5 ± 0.6 μV².

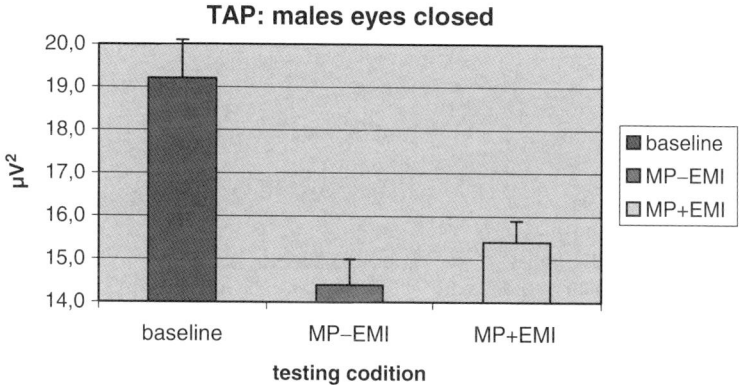

The average total power for the male group (n = 9: baseline: eyes *open*) 9.1 ± 0.2 μV², *MP without protective system*: 9.2 ± 0.2 μV², *MP with protective system*: 8.8 ± 0.2 μV².

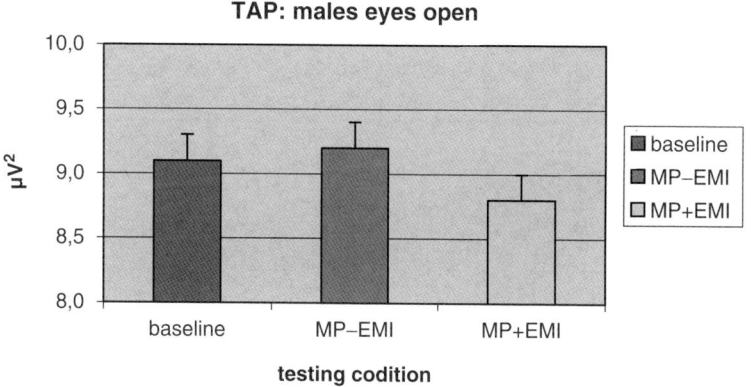

Absolute Power Values

Subsequently, absolute power values (μV^2) are calculated and subjected to the analysis of variance (ANOVA, post hoc tests Fisher's PLSD).

In Table 1 absolute power values (with eyes open) are calculated for baseline and for both testing conditions. Comparing baseline EEG data with protective E.M.I. chip EEG data none of the frequency bands revealed significant differences. However, comparing baseline EEG data without protective E.M.I. chip EEG data a significant difference was calculated for delta power values only ($p = 0.002$).

Comparing protective chip EEG data to EEG data without protective E.M.I. chip a significant difference was noted for the delta frequency band ($p = 0.05$).

In Table 2 power values (with eyes closed) are calculated for baseline and for both testing conditions. Comparing baseline EEG data with protective E.M.I. chip EEG data a significant increase in delta power values was noted ($p = 0.007$). Similar results were obtained for the delta power values when comparing baseline EEG data to EEG data without protective chip ($p = 0.02$).

Comparing the protective chip EEG data to the EEG data without protective chip significant differences were noted for the power values of the theta frequency band ($p = 0.03$) and for the power values of the beta frequency bands ($p = 0.05$).

Subsequently, the effects of the mobile phone with and without the protective systems were calculated for all frequency bands according to the following formula:

EEG-power values (%) for a defined frequency band obtained in the presence of the protective system minus EEG-power values (%) obtained in the absence of the protective system.

Table 1 Comparison of absolute power values. EEG data were recorded with eyes open while using a mobile phone with (+E.M.I.) and without (−E.M.I.) protective chip

Condition	Baseline $\mu V^2 \pm$ SEM	+EMI, chip $\mu V^2 \pm$ SEM	−EMI, without $\mu V^2 \pm$ SEM
Delta	14.4 ± 0.4	15.1 ± 0.5	16.5 ± 0.5
Delta/theta	24.5 ± 0.9	25.1 ± 1.1	26.7 ± 1.0
Theta	11.8 ± 0.6	11.8 ± 0.6	12.2 ± 0.7
Alpha	18.1 ± 1.1	19.4 ± 1.9	15.8 ± 1.3
Beta	9.9 ± 0.3	10.2 ± 0.4	10.7 ± 0.4

Table 2 Comparison of absolute power values. EEG data were recorded with eyes closed while using a mobile phone with (+E.M.I.) and without (−E.M.I.) protective chip

Condition	Baseline $\mu V^2 \pm$ SEM	+EMI, with $\mu V^2 \pm$ SEM	−EMI, without $\mu V^2 \pm$ SEM
Delta	14.9 ± 0.4	16.7 ± 0.5	16.4 ± 0.5
Delta/theta	27.8 ± 1.1	31.1 ± 1.6	27.1 ± 1.0
Theta	14.8 ± 0.8	18.4 ± 1.2	12.7 ± 0.6
Alpha	51.0 ± 3.1	51.9 ± 3.5	42.9 ± 3.2
Beta	11.2 ± 0.3	11.8 ± 0.4	10.6 ± 0.4

Negative values represent the effects of the mobile phone on the electro-encephalo-graphic activity in the *absence* of the protective system. Positive values represent the effect of the mobile phone on the electro-encephalographic activity in the *presence* of the protective system (Fig. 2a).

The subtraction analysis of the males patients eyes open revealed in the presence of the protective system the EEG strongest increase of the alpha power values at P4, T6 and FZ. In the absence of the protective system an increase of delta power was registered at the following sites: P4, O2 and T6.

The subtraction analysis of the female patients eyes open revealed in the presence of the protective system an overall increase of the alpha power with maximal increases in: T6, T5, O1,O2, P3 and FP2. In the absence of the protective system an increase of delta power was registered. Strongest effects were noted at FP1, FP2, F4, FZ, CZ and C3 (Fig. 2b).

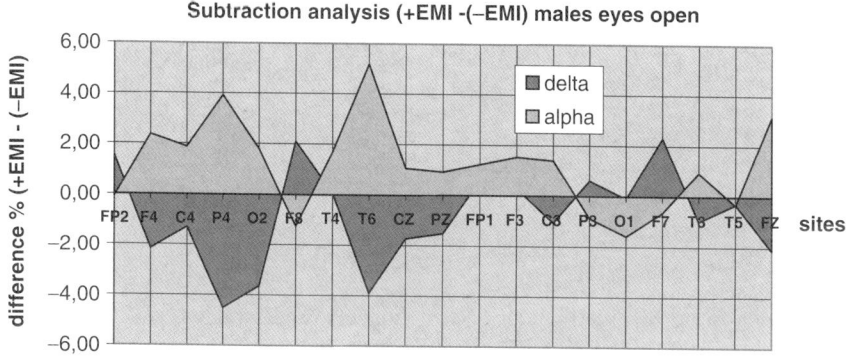

Fig. 2a Illustration of the calculated EEG data for the male subjects

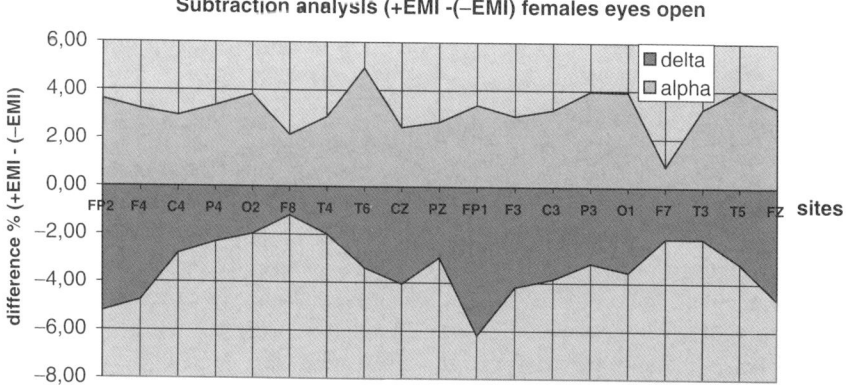

Fig. 2b Illustration of the calculated EEG data for the female subjects

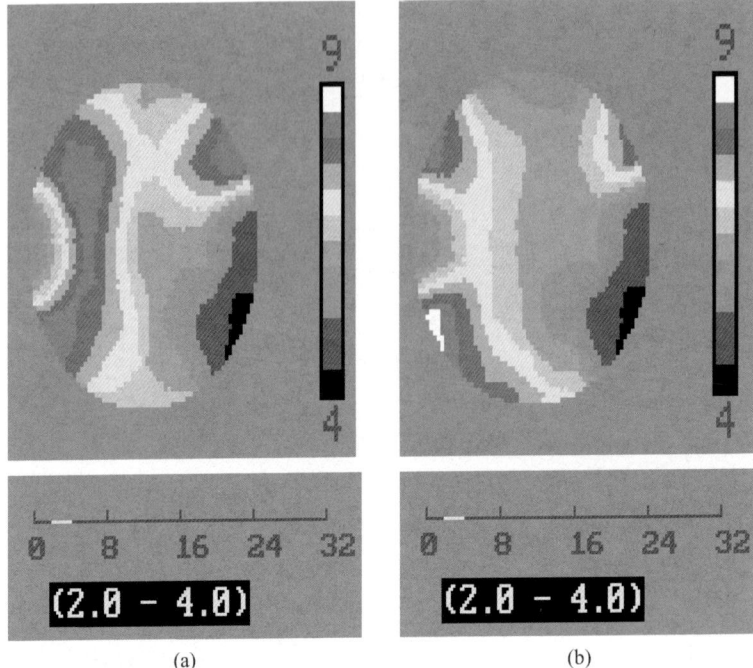

(a)   (b)

Fig. 3

EEG data were recorded while the test subjects used a mobile phone with and without a protective system. Artifact – free data were subjected to a Fast Fourier Transformation and subsequently converted into topographic maps.

As illustrated in Fig. 3a an increased delta activity was demonstrated within the left hemisphere and in F4/F8 site without the protective system. In Fig. 3b a significant reduction of the delta activity could be demonstrated with the protective E.M.I. chip.

Interpretation

This study investigated the effect of electromagnetic frequencies of mobile phones on the brain applying quantitative electroencephalography (QEEG) with and without the protective device (E.M.I. chip). Our study shows generally high electromagnetic frequency alpha values with eyes closed compared to eyes open, as there is no visual input and no suppression of alpha waves. With eyes closed, the baseline value of the absolute power of the alpha frequency is $51 \pm 3.1 \ \mu V^2$, but using mobile phones the absolute power of the alpha frequency decreased to $43 \pm 3.1 \ \mu V^2$, Applying the protective E.M.I. chip, the alpha frequency has returned back even to higher values than the baseline value ($51.9 \pm 0.4 \ \mu V^2$).

Generally, alpha waves are physiological during relaxation, which shows that alpha waves in patients with eyes closed generally show a relaxed state. The baseline value for the alpha frequency of the absolute power with eyes open was 18.1 ± 0.6 μV^2, but using mobile phones, the absolute power of the alpha waves was significantly reduced to 15.8 ± 1.3 μV^2. The application of the protective chip brought the value of the alpha waves back to more than the baseline value, which means that the application of the protective chip reverses the effect of mobile phones in the EEG and patients get more relaxed. With respect to the QEEG with eyes open, the electromagnetic frequency values of alpha waves are generally lower than with eyes closed, due to a visual input and suppression of alpha waves. Patients with eyes open are less relaxed than with eyes closed. Using the mobile phone without the protective chip, alpha waves are enhanced in the back of the head, simultaneously, alpha power was reduced at the right side of the head.

Delta waves are present during sleep, dreaming and light meditation. Using mobile phones puts the patient in a less relaxed state during sleep, but again, applying the protective chip, patients sleep more relaxed.

Concerning our results, with eyes closed, delta waves, which are found in deep sleep did not show a significant difference between using mobile phones with and without the protective chip. Delta waves with eyes open and mobile phones without protection showed an increase in delta wave absolute power (16.5 ± 0.5 μV^2) compared to the baseline value absolute power (14.4 ± 0.4 μV^2). This finding was improved with the application of the protective chip (15.1 ± 0.5 μV^2) which was a significant difference. Delta waves are pathological when they are increased in the awake condition [19].

These preliminary data indicate that mobile phones have an influence on the brain function as delta power was increased without the protective system and alpha waves were decreased. The strongest increase in delta waves, in female patients was noted in FP2, FP1, F4, F3, FZ and C3, which means the frontal brain, the limbic system and the precentral gyrus were involved. In our study, delta waves, if increased, make the patient more sleepy and tired using mobile phones.

Furthermore, if there is an increase in delta power, also mental performance is impaired. Our result agrees with another study which showed some statistical evidence that mobile phones may cause headache, extreme irritation, increase in carelessness, forgetfulness, decrease of the reflex and clicking sound in the ears (13). Alpha waves are increased in the presence of the protective chip, which was shown in the subtraction analysis of the whole frequency band. The strongest increase was noted in FP2, O2, O1 (optic radiation and visual cortex, where alpha waves presented strongest together with occipital area movements) furthermore in T6, T5 (referring to speech and hearing processing) and P3 (movements in three-dimensional space). Increase in alpha power means improved relaxation, which means all these procedures are taking place more relaxed. Without using the protective chip alpha power is reduced which makes the person more susceptible to stress and nervous with decreased relaxation. Our finding confirms previous findings which investigated the effect of the electromagnetic field (EMF) exposure on the EEG waking activity and its temporal development. That study showed that under real exposure as compared to baseline and sham conditions, EEG spectral power was influenced in some bins of the alpha band [18].

Another study noted that for all brain tumors an increased risk for analogue phones was found which was most pronounced in the group with >10 year latency period, odds ratio (OR) 1.6, 95% confidence interval 1.1–2.5 (1). While using a mobile phone with the protective E.M.I. chip in our study these negative changes have been reversed and the normal brain function can be maintained. Alpha waves are also improved.

Our study shows that while an increased number of hours and long latency period is needed to get a stable risk estimate, yet it is necessary that the protective chip is used to reverse the effects seen on the human brain, but more investigations are needed on the mechanisms of damage to the human brain. One of the hypotheses is oxidative stress, which was confirmed by Irmak et al. Studying the brain of rats, in which he found an increase of the serum SOD activity [20].

The conclusion out of these preliminary data indicates that using mobile phones produces a lot of EEG changes. Applying the protective E.M.I. device is a way of increasing alpha waves which means patients are in a more relaxed state. The protective chip can preserve normal brain function while using a mobile phone by increasing the alpha power which improves mental functioning. The present data lend further support with other published articles that pulsed high frequency electromagnetic fields can affect normal brain functioning (18). The usage of mobile phones with the EMI protective chip allowed normal brain function to be maintained as alpha waves are improved.

Further studies are needed to investigate the long term follow-up.

References

1. Hardell, L, Carlberg, M (2006) Pooled analysis of two case control studies on use of cellular and cordless telephones and the risk for malignant brain tumours diagnosed in 1997–2003. Int Arch Occup Environ Health 79:630–639
2. Colonna A (2005) Cellular phones and cancer: current status. Bull cancer 92(7):637–643 (ISSN:1769–6917)
3. Van Rongen E, Roubos EW, Van Aernsbergen, LM, et al. (2004) Mobile phone and children: is precaution warranted? Bioelectromagnetics 25:142–144
4. Braune S, Riedel A, Schulte-Monting, J, et al. 2002 Influence of a radiofrequency electromagnetic field in cardiovascular and hormonal parameters of the autonomic nervous system in healthy individual. Radiat Res 158(3):352–356 (ISSN: 0033–7587)
5. Morissey J. State of the science: RF genetic toxicology (2001) In: Carlo GL, ed. Wireless phone and health State of the Science. Norwell, MA: Kluwer
6. Moulder JE, Erdreich LS, Malyapa RS, et al. (1999) Cell phones and cancer: what is the evidence for a connection? Radiat Res 151:513–531
7. Blettner M, Berg G (2000) Are mobile phones harmful? Acta Oncol 39:927–930
8. Christensen HC, Schuz J, Kosteljanetz M, et al. (2004) Cellular telephone use and risk of acoustic neurinoma. Epidemiology 159:277–283
9. Lonn S, Ahlbom A, Hall P, et al. (2004) Mobile phone use and the risk of acoustic neurinoma. Epidemiology 15:653–659

10. Weinberger Z, Richter ED (2002) Cellular telephones and effects on the brain: the head as an antenna and brain tissue as a radio receiver. Med Hypotheses 59:703–705
11. Hansson MK, Hardell L, Kundi M, et al. (2003) Mobile telephones and cancer: is there really no evidence of an association? (review). Int J Mol Med 12:67–72
12. Lahkola A, Tokola K, Auvinen A (2006) Meta-analysis of mobile phone use and intracranial tumors. Scand J Work Environ Health 32(3):171–177
13. Repacholi, MH (1998) Low level exposure to radio-frequency electromagnetic fields: health effects and research needs. Bioelectromagnetics 19:1–19
14. Hamblin DL, Croft RJ, Wood AW, et al. (2006) The sensitivity of human event-related potentials and reaction time to mobile phone emitted electromagnetic fields. Bioelectromagnetics 27(4):265–273 (ISSN:0197–8462)
15. Eliyahu I, Luria R, Hareuveny R, et al. (2006) Effects of radiofrequency radiation emitted by cellular telephones on the cognitive functions of humans. Bioelectromagnetics 27(2):119–126
16. Besset A, Espa F, Dauvilliers Y, et al. (2005) No effect on cognitive function from daily mobile phone use. Bioelectromagnetics 26(2):102–108 (ISSN:0197–8462)
17. Hietanem M, Kovala T, Hamalainen AM (2000) Human brain activity during exposure to radiofrequency fields emitted by cellular phones. Scand J Work Environ Health 26(2):87–92 (ISSN: 0353-3140)
18. Curcio G, Ferrara M, Moroni F, et al. (2005) Is brain influenced by a phone call? An EEG study of wakefulness. Neuroscience Res 53(3):265–270
19. Kramarenko AV, Tan U (2003) Effects of high-frequency electromagnetic field on human EEG: a brain mapping study. Int J Neurosci 113(7):1007–1019 (ISSN:0020–7454)
20. Irmak MK, Fadilhoglu E, Guelec M, et al. (2002) Effects of electromagnetic radiation from a cellular phone on the oxidant and antioxidant level of rabbits. Cell Biochem Funct 20:279–283

Appendix 1

A 3D Dose Model for Low Level Laser/LED Therapy Biostimulation and Bioinhibition

James D. Carroll

Abstract There have been numerous reports describing the phenomena of biostimulation/bioinhibition using low-level laser therapy (LLLT) and other light and IR sources within the laboratory and in clinical trials. Stimulation or inhibition employed correctly has been shown clinically to reduce pain, improve tissue repair, resolve inflammation and stimulate an immune response. All these effects are sensitive to irradiance and total energy delivered. A 3D Arndt Schulz style model is proposed to illustrate possible 'dose sweet spots' for the intended clinical effects.

Keywords: Low level light therapy, Arndt-Schultz Law, biphasic response model, fluence, intensity, laser biostimulation, laser bioinhibition.

Introduction

The Arndt-Schultz Law is frequently quoted as a suitable model to describe dose dependant effects of LLLT [1–4]. The Arndt-Schultz Law states that "weak stimuli increases physiologic activity, moderate stimuli inhibit activity, and very strong stimuli abolish activity" [5–7]. Simply put; a small stimulus may have no biological effect, a moderate stimulus may have a biostimulatory effect, a large stimulus may have a toxic or bioinhibitory effect. In the context of LLLT the increasing "stimulus" may be irradiation time or it may be increased beam intensity (irradiance). A non-linear, s-curve response has been demonstrated many times in LLLT research [8–10] This non-linear effect contradicts the Bunsen-Roscoe rule of reciprocity which predicts that if the products of time of exposure and irradiance are equal, then the quantities of material undergoing change will be equal. This inverse linear relationship between intensity and time has frequently failed in LLLT research [4, 11].

J.D. Carroll
THO Photomedicine Ltd, Chesham

R. Waynant and D.B. Tata (eds.), *Proceedings of Light-Activated Tissue Regeneration and Therapy Conference.*
© Springer Science + Business Media, LLC 2008

The 3D Model

At the NAALT 2007 conference, Anders [12] hypothesized that "If we had extended the ranges I believe we would have found an island in the middle of the data table indicating the effective combinations". Karu [11] plotted the rate of DNA synthesis dependence on light intensity and time. Taking this template for both increases in time and intensity it has been possible to speculate and redraw this in a three-dimensional graph to illustrate how a dose "sweet spot" for biostimulation may appear when the appropriate combinations of time and irradiance are achieved at the target tissue.

Whilst this model could help explain and explore dose rate effects it must be remembered that there are many other factors affecting LLLT dosimetry including wavelength, depth of target tissue, attenuation, target cell type, state of target cells, pulse regimes and treatment intervals. Other confounding factors include a lack of beam measurement standardization in this field of study and some unstable laser and LED product performance. These factors mean that much published data is not robust and should be referred to cautiously. When conducting LLLT research all parameters should be recorded and published including wavelength, average power (pulse width, pulse and peak power if any) and beam area (including the beam measurement method). The recommended method for expressing beam area is the $1/e^2$ point [13].

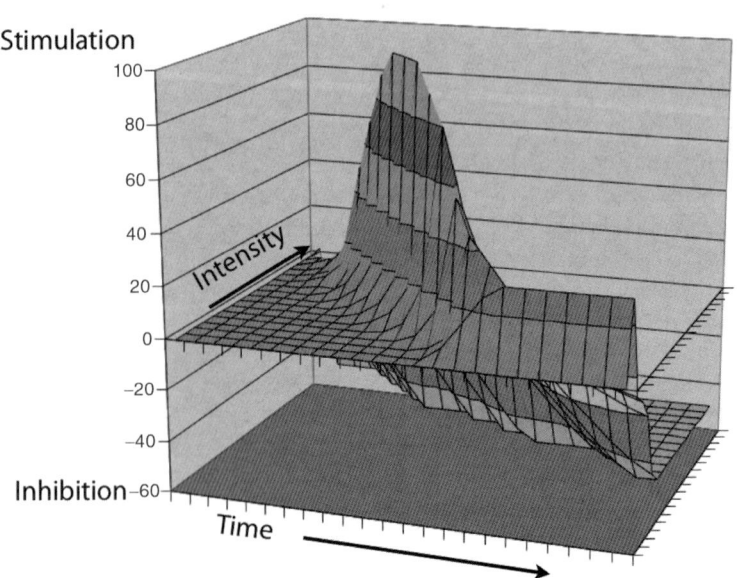

References

1. Chow RT, David MA, Armati PJ. (2007) 830 nm laser irradiation induces varicosity formation, reduces mitochondrial membrane potential and blocks fast axonal flow in small and medium diameter rat dorsal root ganglion neurons: implications for the analgesic effects of 830 nm laser. J Peripher Nerv Syst. Mar; 12(1):28–39.
2. Sommer AP, Pinheiro AL, Mester AR, Franke RP, Whelan HT. (2001) Biostimulatory windows in low-intensity laser activation: lasers, scanners, and NASA's light-emitting diode array system. J Clin Laser Med Surg. Feb; 19(1):29–33.
3. Hawkins D, Abrahamse H. (2006) Effect of multiple exposures of low-level laser therapy on the cellular responses of wounded human skin fibroblasts. Photomed Laser Surg. Dec; 24 (6):705–714.
4. Lubart R, Lavi R, Friedmann H, Rochkind S. (2006) Photochemistry and photobiology of light absorption by living cells. Photomed Laser Surg. Apr; 24(2):179–185.
5. Dorland's Illustrated Medical Dictionary.
6. Schulz, H. (1888) "Uber Hefegiste", Pflügers Archiv Gesammte Physiologie, Vol. 42 pp. 517
7. Zur Lehre von der Arzneiwirkung. (1887) [Virchows] Archiv für pathologische Anatomie und Physiologie und für klinische Medizin, Berlin; 108:423–445.
8. Oron U, Yaakobi T, Oron A, Hayam G, Gepstein L, Rubin O, Wolf T, Ben Haim S. (2001) Attenuation of infarct size in rats and dogs after myocardial infarction by low-energy laser irradiation. Lasers Surg Med.; 28(3):204–211.
9. Lanzafame RJ, Stadler I, Kurtz AF, Connelly R, Peter, AT, Sr, Brondon P, Olson D. 2007 Reciprocity of exposure time and irradiance on energy density during photoradiation on wound healing in a murine pressure ulcer model. Lasers Surg Med. July; 39(6):534–542.
10. Michtchenko, A, Hernandez, M. (2006) Photobiostimulation Effects Caused for Low Level Laser Radiation with 650 nm in the Growth Stimulus of Biological Systems. Electrical and Electronics Engineering, 2006 3rd International Conference on Volume, Issue, Sept. Page(s):1–4.
11. Karu TI, Kolyakov SF. (2005) Exact action spectra for cellular responses relevant to phototherapy. Photomed Laser Surg. Aug; 23(4):355–361.
12. Anders, J, Romanczyk, T, Moges, H, Ilev, I, Waynant, R, Longo, L (2007) Light Interaction with Human Central Nervous System Progenitor Cells, NAALT Conference Proceedings.
13. Carroll, J (2006) The Characteristics of a LLLT/Photobiomodulation Light Source Defines the Medicine. The Exposure Time Defines the Dose. A Collapsed Distinction. SPIE BiOS Proceedings 2006.

Appendix 2

Does Body Contouring Need to Be Painful?

Photomology™ – A Dynamic Combination of Dual-Band Low Intensity Light Therapy with Vacuum and Mechanical Massage for Non-destructive Treatment of Cellulite and Subcutaneous Fat Reduction

Michail M. Pankratov

Abstract Cellulite is a skin condition characterized by hypodermal "engorged" fat lobules and fibrosis of fibrous strands (septae) frequently aggravated by skin laxity. Current non-invasive technologies for treatment of cellulite include massagers (mechanical manipulation), bulk tissue heaters (lasers, IPL, RF) and combination of both.

Photomology module of the SmoothShape device delivers Low Intensity Light Therapy (LILT) and moderated temperature elevation ($\leq 41°C$) combined with vacuum and mechanical massage for restorative, non-destructive treatment of cellulite, fat reduction and skin rejuvenation. It improves blood circulation and lymphatic drainage in the treated tissue and gently stimulates natural cell processes that modify cell activity leading to: (1) alteration in cells' behavior, (2) restoration of normal cell activity, and (3) improvement of tissue functionality. Photomology simultaneously addresses the physical manifestation of cellulite and alleviates some underlying causes of cellulite. This two level approach improves existing condition and promises longer lasting results.

Keywords: Cellulite, low intensity light therapy, LILT, LLLT, photomology, SmoothShapes.

M.M. Pankratov
Elemé Medical, Merrimack, NH 03054

R. Waynant and D.B. Tata (eds.), *Proceedings of Light-Activated Tissue Regeneration and Therapy Conference.*
331
© Springer Science + Business Media, LLC 2008

Introduction

Cellulite is a skin condition characterized by a dimpling, uneven, or "cottage cheese" skin texture occurring almost exclusively in post-pubertal females. It is most commonly seen in the thighs and buttocks, but may also be apparent in the breasts, inner knees, hips, lower abdomen or upper arms. It occurs even in slim and fit or otherwise healthy women, but the appearance of cellulite is naturally exacerbated by an excess of fat. Cellulite is made of adipocytes residing in the skin in the subcutaneous fat layer, and is not affected by diet or exercise. There is no uniform theory for the origin of cellulite and purported etiologies vary from morphological differences between male and female skin anatomy (the hypodermal connective tissue strands run diagonally in male skin and vertically in female skin), to hormonal involvement (cellulite is frequently present in androgen deficient males and not seen in pre-pubertal females), to lipodystrophy (gynoid lipodystrophy is one of the synonyms for cellulite) and circulatory disorders, to hereditary factors [4, 5, 8, 15, 31, 43, 49, 61, 66, 67, 71, 74, 78]. It is very likely that all these factors play some role in the origin of cellulite.

The regional fat accumulation is responsible for vertical stretching of the superficial fat lobules protruding into the lower reticular dermis producing irregularities and dimpling of the skin at the surface. Advanced stages of cellulite are associated with deterioration of dermal vasculature, deposition of hydrophilic byproducts [49], and retention of excess fluid within the dermis and adjacent interstitial structures. Ensuing edema leads to vascular compression, cellular changes and tissue hypoxia. In time, continuing tissue congestion leads to thickening and sclerosis of fibrous septae causing a dimpled appearance. Some investigators reported the presence of chronic inflammatory cells, e.g., macrophages and lymphocytes, in fibrous septae of cellulite biopsies [43].

Although our understanding of cellulite has grown exponentially in the last few years, we still rely on highly subjective measures such as digital photography, or other somewhat primitive and outdated methods of classification. For example, the Nürnberger and Müller Severity Scale, proposed more than 30 years ago [61], is used to assign patients to one of four clinical stages (0–III) based on clinical alterations observed at rest and with pinch test or muscular contraction. This severity scale leaves too much room for subjectivity in evaluation and comparison of changes over time, making settled but appreciable changes in cellulite appearance hard to correlate with changes on the severity scale. Several attempts to develop a new classification system or to modify and expand the existing severity scale to incorporate some objective biometric measurements did not gain wide acceptance and remains a domain of individual users [6, 25, 32, 33].

There are a few objective methods for demonstrating and monitoring of cellulite conditions and its changes over time: specialized 3-D photography, high-resolution MRI and high frequency ultrasound [23, 25, 50, 55, 65, 72, 73, 77]. Unfortunately, these tests are not readily available, cost-effective, or easily justified when treating a cosmetic, non-life threatening condition.

Advanced stages of cellulite involve multi-level pathologic changes in affected tissue, i.e., cellular, metabolic, circulatory, physiologic, and biomechanical; and therefore treatment modalities require interventions capable of dealing with many if not all of these pathologic changes in order to deliver effective and long lasting improvement.

Treatments for Cellulite

Currently available treatments for cellulite include surgical techniques (liposuction, subcision, and laser- and ultrasound-assisted liposuction) [29, 30, 47], mechanical manipulation (endermologie, lymphatic massage, etc.), topical creams, energy-based devices (light, ultrasound, acoustic shock, radio frequency, electrical stimulation, etc.), mesotherapy and injection lipolysis [11, 48] as well as combinations of the above mentioned techniques.

The benefits of the many available treatments for cellulite vary from marginal to somewhat effective, but are not without drawbacks and limitations. For example, liposuction and laser- and/or ultrasound-assisted liposuction are unlikely to "cure" cellulite and in many cases require additional interventions to improve appearance of cellulite [4, 82]. Mesotherapy and injection lipolysis are capable of inducing local lipolysis and improve skin condition and cellulite appearance but are invasive and associated with numerous side effects, including pigmentary changes and fibroses.

There are a myriad of energy sources (laser light, ultrasound, acoustic shock, radiofrequency, electrical stimulation, low temperature (cold), mechanical massage and vacuum, etc.) that are being used or proposed for ablative as well as non-ablative intervention in the treatment of cellulite. Techniques employing energy sources for ablative fat reduction and cellulite treatment are subject to the same complications and side effects as any surgical intervention. Non-ablative energy-based techniques are safe and nearly free of side effects when administered by a well-trained operator, but the efficacy and longevity of the therapeutic benefits are unpredictable.

Endermology[®] (LPG Systems[®], Valence, France), TriActive[TM] (Cynosure[TM], Westford, MA), and Velasmooth[TM] (Syneron[TM] Medical Ltd, Yokneam Illit, Israel) and several alike devices are among the most widely used non-invasive devices for the treatment of cellulite.

The Endermologie device incorporates a mechanical massager with controlled suction that pulls skin into a hand-held motorized device with adjustable rollers. The skin is manipulated, compressed and rolled to increase blood and lymphatic flow and to modify the underlying connective tissue. This therapy is administered in 35–45 minutes sessions over 14–20 visits [7, 14, 21].

TriActive technology combines a weak version of the endermologie (mechanical massage and suction) with topical cooling and low intensity 808 nm diode laser light (6 W max). It was designed to increase lymphatic drainage, tighten the skin by stimulating underlying muscles and fascia, and increase superficial blood flow thereby reducing the appearance of cellulite. The TriActive treatment regimen is similar to the endermologie with greater emphasis on the microcirculation impairment theory of cellulite formation [60, 64].

The Velasmooth device incorporates a proprietary combination of bi-polar radiofrequency and broadband infrared light (700–2,000 nm) – ELŌS^{TM} (electro-optical synergy) technology – with a mechanical massager and suction for deep penetration into adipose tissue for treatment of cellulite. Infrared light provides bulk tissue heating and conducted RF energy is designed to penetrated up to 15-mm deep into the tissue causing heating of subcutaneous fat. The combination of mechanical manipulation and IR/RF energies is expected to increase metabolism of adipose tissue, homogenize sub-dermal fat and increase skin elasticity. The recommended treatment regimen is 8–10 sessions lasting for 40 minutes once or twice a week [2, 44, 60, 82, 84].

The effectiveness of these treatments and longevity of therapeutic benefits is dependant upon many variables, including frequency, duration and total number of treatments, size of the treatment area, power levels of employed energy sources, and intensity of vacuum and mechanical effort as well as operator's technique [44]. A patient's age, severity of cellulite and contributory medical conditions may also affect the outcome of the treatment regiment. Cellulite management with devices employing mechanical massage and suction require monthly maintenance treatments following the original treatment regiment in order to sustain the therapeutic benefits; otherwise, these benefits tend to disappear by 3 months.

Patients treated with the devices employing light and/or light and radiofrequency as well as other energy sources find that their therapeutic benefits begin to diminish around 3 months post-therapy with almost all benefits reversed by 6 months. Accordingly, they are also advised to return for maintenance treatments not later than 6 months.

The chase for a better non-invasive treatment of cellulite and fat reduction continues with more and more players entering the market utilizing established modes and energy sources and introducing new ones: acoustic shock, low temperatures, microwaves, etc.

Clinical Study Utilizing Low Intensity Light Therapy

Anecdotal observations of successful treatment of cellulite using Low Intensity Light Therapy (LILT) combining He-Ne (632 nm) and GaAs (904 nm) laser radiation were discussed at various symposia but never published in a refereed journal. However, some *in vitro* studies have demonstrated that exposure of adipose tissue samples collected from lipectomy to 635 nm wavelength light from a 10 mW diode laser for 6 minutes (3.6 J/cm^2) produced near total emptying of fat from these cells through, what the authors postulated, transitory pores in the membranes of adipocytes [35, 56–59]. In addition, they observed that these transitory changes didn't cause destruction of adipocytes or other interstitial structures.

These findings led to an inception of laser-assisted liposuction where LILT is administered transcutaneously for 6–15 minutes (in an hour prior to liposuction). It became an adjunct, intraoperative procedure. Practitioners of this technique mostly agree that there are significant benefits to patients in post-operative recovery: fewer side effects, less swelling, less bruising, less discomfort, decrease in post-operative use of pain medication, etc., but there is no consensus on whether it also simplifies liposuction procedure in terms of ease of performing liposuction, reduced time in surgery, or enhanced emulsification of extracted fat [12, 35]. Although some studies have failed to reproduce Dr. Neira's results [12, 54] the hypothesis of "transitory pores" was not overturned because there were significant differences between the device, wavelength, and method of application used by Dr. Neira and the other studies.

For the purpose of non-destructive treatment of cellulite and fat reduction, another wavelength that can penetrate relatively deep into the skin and has preferential affinity for lipids (i.e., absorbed better in fat than surrounding tissue) was added to the mix in order to compliment the pore-inducing effect of 635-nm wavelength and assist in fat evacuation from adipocytes. It has been established for some time [51] and recently reconfirmed by the Rox Anderson, M.D. and his team [3] that four infrared wavelength bands – 915, 1,205, 1,715 and 2,305 nm – have about 50% more absorption in lipid-rich tissue than in aqueous tissue.

Combining the visible 630 ± 20 nm wavelength for its adipocyte membrane pore-inducing properties and the infrared 915 ± 20 nm wavelength for its lipid-liquefying properties creates a dual-band light source capable of positively affecting skin adipose tissue. Augmenting this dynamic distinct light duo with suction (vacuum) and mechanical massage produces entirely new process that was termed *photomology*^{TM} and defined as:

> Exclusive process that treats cellulite and subcutaneous fat by combining dynamic light and laser energy along with vacuum massage to gently stimulate natural cell processes and modify cell activity.

The study under discussion was designed to evaluate the effectiveness of the low intensity dual-band (630 and 900 nm) laser light for the improvement of appearance of cellulite and removal of discrete subcutaneous deposits of adipose tissue from the thighs of otherwise healthy volunteer subjects. It was conducted in two phases: first a pilot study of 12 subjects, followed by a second, much larger multi-center study. Pre- and post-treatment MRI scans were used to accurately document changes in fat and photography, physical examination and questionnaires were used to assess and document changes in cellulite. In the pilot study, the bi-dimensional MRI fat images demonstrated remarkable reduction in fat thickness by as high as 35% relative to the control group. According to the investigators:

> weight loss per se appeared to be random and did not correlate to findings of changes found in a single thigh.... However, the use of MRI was found to be a reliable cross indicator of fat change in the treated legs. Furthermore, the use of MRI was found to be a reliable means of monitoring the longevity of the results, which was confirmed to have persisted as far as four years post treatment. [45]

Encouraged by a success of the pilot study, the investigators set to replicate these findings in a large group of volunteer subjects in the multi-center, IRB-approved, clinical investigation [46].

Subjects

One hundred two healthy female volunteers between the ages of 18 and 50 with mild-to-moderate cellulite ($\leq 15\%$ overweight from a standard actuarial table) on the thighs were enrolled in the IRB-approved study with 74 subjects completing the treatments and follow-up. Prior to treatment, cross-sectional MRI scans of both thighs were taken. Subjects were randomly assigned to have one thigh treated with the combination of laser and massage, while the contra lateral thigh served as "control" and was subjected to massage treatment only. Subjects were "blinded" as to which leg received laser plus massage treatment *vs.* massage alone.

Device

904-nm 1 W laser diode with the spot size of 3 mm and 0.5 W 632-nm He-Ne laser were scanned over large treatment area (30.5×15 cm) at a frequency of 30 Hz, thus covering the entire area in 2 seconds.

Pixel densitometry of each layer of each MRI image was employed to objectively and consistently calculate changes in the fat pad. The pixel counts of the respective anatomic elements identified in each image were calculated. Selective outlining of the entire thigh, fat, muscle, and bone was performed. The total cross-section of the thigh was calculated in square centimeters using standard software algorithms. The pixel count of the thigh minus the fat layer was calculated and subtracted from the complete thigh surface area to produce an objective representation of fat thickness.

Treatment

Each subject underwent 12–14 treatment sessions—two to three times a week over a 4–6 week period. MRI scans of both thighs were performed pre- and post-treatment. Subcutaneous peri-muscular fat pads were measured on MR images by a "blinded" independent radiologist who had no knowledge of the laser vs. non-laser treated side. Volumetric determination technique was used to quantify the difference in fat deposits based on pixel densitometry of the respective anatomic constituents identified in the MRI. Changes in skin condition were evaluated using digital photographs by blinded observers and subjects' questionnaires.

Results

Only subjects – 65 – with complete pre- and post-treatment data were included in the analyses. As measured by the MRI, there was a statistically significant difference over time between the laser plus massage treated thigh vs. the massage-only treated thigh (p < 0.001), with the average fat thickness in the experimental leg decreasing by an average of 1.19 cm^2. vs. an increase of an average 3.82 cm^2 in the control leg. There was a high degree of patient satisfaction with the treatment and therapeutic results:

Notice a difference between treated and control legs – 83%
Pleased with the results – 89%
Would participate in the treatment again – 99%
Would recommend this treatment to a friend – 93%

Five subjects returned for an additional follow-up examination including MRI evaluation 13 months or longer after completion of the study. Four of these subjects maintained some improvements on the laser plus massage treated thigh, while the control thigh continued to demonstrate an increase in the circumference based on the MRI measurements.

Discussion

The therapeutic effect of Low Intensity Light Therapy is no longer in question [19, 26–28, 34, 52, 53, 70]. However, the question remains as to how the light works at the cellular, tissue, and organ levels, and what are the optimal parameters for the most effective therapeutic effects for various medical conditions. It is believed that, on the cellular level, light first interacts with mitochondria leading to an increase in ATP production, modulation of reactive oxygen species, and induction of transcription factors [1, 16, 18, 19]. These effects lead to increased cell proliferation and migration, especially, fibroblasts, modulation in levels of cytokines, growth factors and inflammatory mediators, and increased tissue oxygenation. The ability of monochromatic light to modify cellular function of living tissue, enhance healing and restore its normal function is a basis for the treatment of numerous dermatological, musculoskeletal, and neurological conditions [1, 13, 19, 26, 28, 34, 52, 53, 69, 70, 75, 76, 80, 81, 83, 85, 86, 88].

Light traversing living tissue is subjected to scattering and absorption which are both wavelength-dependent phenomena. Absorption of monochromatic visible and near infrared light at the cellular level by components of the cellular respiratory chain is at the heart of the beneficial therapeutic effect [20, 24, 62]. Mitochondria are conceivably the principal unit within the cell controlling the LILT response and by being at the core of a chain reaction within the cell that originates with improving respiratory and ion transporting activity in the cell and culminating in beneficial therapeutic reaction on cellular, tissue, and systemic levels [9, 10, 27].

Phototherapy produces primary, secondary and tertiary effects which collectively enhance tissue repair and produce pain relief [17, 38, 63, 75, 76, 79, 87]. **Primary effects** in the cells, often referred to as photoreception, occur through direct interaction of photons with cytochromes. **Secondary effects**, e.g., cell proliferation, protein synthesis, growth factors secretion, myofibroblast contraction, neurotransmitter modification, etc., occur in the same cells as primary and can be initiated by light as well as other stimuli. **Tertiary effects**, e.g., tissue repair and pain relief, are the indirect responses of distant cells to photon induced changes in other cells. These effects are the least predictable because they are influenced by cellular as well as environmental factors.

It is believed that the most beneficial effects are achieved in the 600–950 nm wavelength range with the exception of 700–770 nm wavelengths range which is considered ineffective. Hence the choice of a monochromatic wavelength is an important consideration when considering treatment of superficial or deeply located tissues: shorter wavelengths for more superficial applications and longer wavelengths for deeper penetration [9, 27, 36, 37, 39].

SmoothShapes 100 – Mechanism of Action

LILT therapy produces no significant temperature elevation and no immediate, visible structural changes in the cells, tissues, or organs. It modulates the level of cellular activity that eventually improves local blood circulation and metabolism, reduces inflammatory response to injury and improves cell survival, modifies perception of pain (nociception) and improves tissue condition that causes pain. Although photobiostimulating effects of LILT can occur in the absence of temperature elevation there is evidence that moderate temperature elevation to the level of 'high-fever' in the treated tissue ($\leq 40°C$) would facilitate many of the same functions and activities that LILT alone is attempting to improve: local blood circulation, metabolic activity, chemical reaction, etc. [89].

Photomology process for non-destructive, modulating action of LILT in combination with moderate temperature elevation in the irradiated tissue is at the root of the therapeutic effects of dual-band light application for improvement in the appearance of cellulite and fat reduction by the **SmoothShapes®100** technology. This improvement appears to be the result of the tertiary effects of phototherapy that together with primary and secondary effects enhance tissue repair [36–41].

SmoothShapes®100 device (SmoothShapes, Merrimack, NH) utilizes the **photomology** process that combines dual-band light therapy 630 ± 20 nm red light from laser diodes and LEDs with near infrared 900 ± 20 nm from LED with mechanical massager and suction to reduce discrete subcutaneous adipose tissue and improve the appearance of cellulite (Fig. 1). The laser & LED light sources operate in continuous mode and deliver 0.56 W of combined optical power (Fig. 2). The 650-nm wavelength was selected for its action in modifying permeability of the adipocytes' membranes and enhancement of fat emulsification. This wavelength

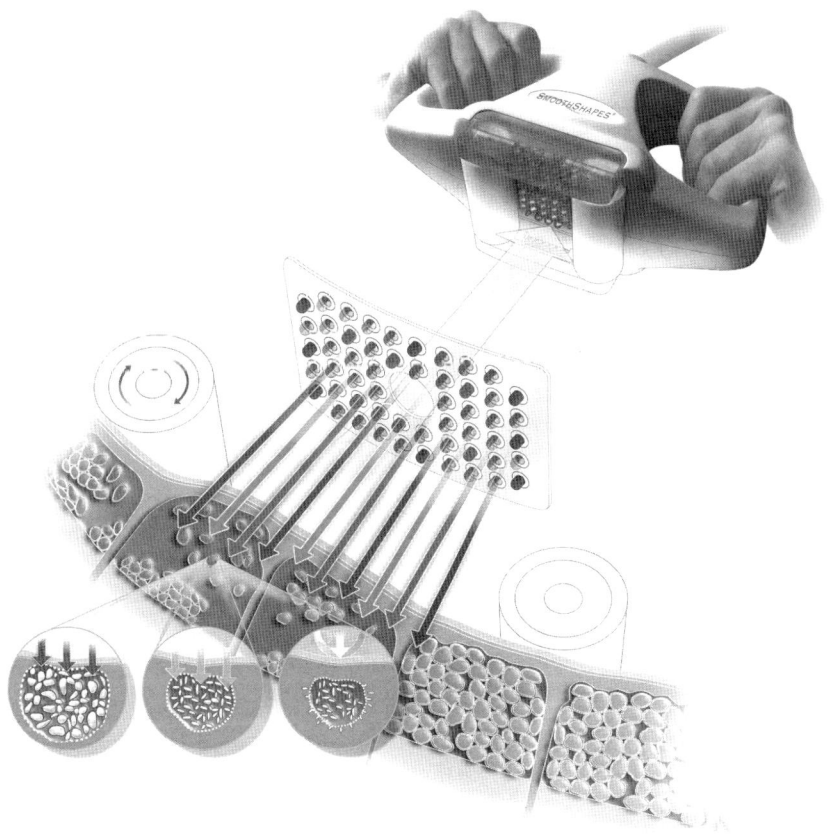

Fig. 1 Dynamical action of SmoothShapes 100 device

falls into a so called optical window where there is relatively minimal absorption in hemoglobin, water and other skin components, except for melanin, which allows for adequate penetration to the subcutaneous fat level.

The choice of 900 nm wavelength was based on its preferential absorption in lipids. This wavelength penetrates well into the tissue with even less scattering than 650 nm but gets absorbed by the lipids in fat. The temperature inside the adipocytes gets slightly elevated causing fat liquefaction. Evacuation of liquefied fat from the inside of adipocytes into intracellular space is facilitated by continuous vacuum massage.

In addition, both wavelengths are implicated in producing other beneficial therapeutic effects discussed earlier:

− Improved local blood and lymphatic circulation
− Improved metabolic activity of treated tissue at the tissue and cellular level

Fig. 2 Cellulite treatment with the SmoothShapes100 device

– Induce mild inflammatory reaction thus attracting fibroblasts that leads to new
  collagen production and deposition manifesting in thicker dermis

Mechanical massager with suction has proven track of:

– Improved blood and lymphatic circulation
– Modification of underlying connective and muscular tissue leading to tissue
  tightening
– Modification of subcutaneous fat layer

Conclusion

There was a high degree of the participants' satisfaction with the treatment which
they found virtually painless, pleasant and relaxing. Photomology – a combined and
synergetic effects of four energy modalities (visible and infrared light, mechanical
massage and suction) – produce greater therapeutic results than the simple addition
of individual components in short term. There is some indication that cellular and
tissue changes incited by the dual-band LILT therapy may have produced long
lasting effect. The latter observation requires further investigation and confirmation
by a larger study.

Acknowledgments Author is thankful to Elliot Lach, MD and Tracey Corby, RN for sharing with me their knowledge, data and experience and to Mary Stoll, RN for her valuable comments and corrections.

References

1. Abrahamse H, Hawkins D, Houreld N (2006) Effect of wavelength and fluence on morphology, cellular and genetic integrity of diabetic wounded human skin fibroblasts. In: Hamblin MR, Waynant RW, Anders J (eds) Mechanisms for low-light therapy, Proceedings of SPIE vol 6140, 6140006:1–13
2. Alster TS, Tehrani M (2006) Treatment of cellulite with optical devices: an overview with practical considerations. Laser Surg Med 38:727–730
3. Anderson RR, Farinelli W, Laubach H, Manstein D, Yaroslavsky AN, Gubeli J, Jordan K, Neil GR, Shinn M, Chandler W, Williams GP, Benson SV, Douglas DR, Dylla HF (2006) Selective photothermolysis of lipid-rich tissues: a free electron laser study. Laser Surg Med 38:913–919
4. Avram MA ((2004) Cellulite: a review of its physiology and treatment. J Cosmet Laser Ther 6:181–185
5. Bacci PA (2006) Anatomy of cellulite and interstitial matrix. In: Goldman MP, Bacci PA, Leischoff G, Hexsel D, and Angelini F (eds) Cellulite: pathophysiology and treatment. Taylor & Francis, New York, pp 29–40
6. Bacci PA (2006) Clinical-therapeutic classification: BIMED-TCD. In: Goldman MP, Bacci PA, Leischoff G, Hexsel D, and Angelini F (eds) Cellulite: pathophysiology and treatment. Taylor & Francis, New York, pp 115–141
7. Bacci PA (2006) The role of endermologie in treatment of cellulite. In: Goldman MP, Bacci PA, Leischoff G, Hexsel D, and Angelini F (eds) Cellulite: pathophysiology and treatment. Taylor & Francis, New York, pp 171–187
8. Bacci PA, Leisbaschoff G (2006) Pathophysiology of cellulite. In: Goldman MP, Bacci PA, Leischoff G, Hexsel D, and Angelini F (eds) Cellulite: pathophysiology and treatment. Taylor & Francis, New York, pp 41–74
9. Bisland SK, Wilson BC (2006) To begin at the beginning: the science of bio-stimulation in cells and tissues. In: Hamblin MR, Waynant RW, Anders J (eds) Mechanisms for low-light therapy, Proceedings of SPIE vol 6140, 6140002:1–10
10. Bortoletto R, Silva NS, Zangaro RA, Pacheco MT, Da Matta RA, Pacheco-Soares C (2004) Mitochondrial membrane potential after low-power laser irradiation. Laser Med Sci 18:204–206
11. Braun M (2006) Lipodissolve for body sculpting. In: Goldman MP, Bacci PA, Leischoff G, Hexsel D, and Angelini F (eds) Cellulite: pathophysiology and treatment. Taylor & Francis, New York, pp 301–322
12. Brown SA, Rohrich RJ, Kenkel J, Young VL, Hoopman J, Coimbra M (2004) Effect of low-level laser therapy on abdominal adipocytes before lipoplasty procedure. Plast Reconstr Surg 113:1796–1804
13. Capon A, Souil E, Gauthier B, Sumian C, Bachelet M, Buys B, Polla BS, Mordon S (2001) Laser-assisted skin closure (LASC) by using 815-nm diode laser system accelerates and improves wound healing, Laser Surg Med 28:168–175
14. Chang P, Wiseman J, Jacoby T, Salisbury AV, Ersek RA (1998) Noninvasive mechanical body contouring: (endermologie) a one year clinical outcome study update. Aesthetic Plast Surg 22 145–153
15. Draelos ZD, Marenus KD (1997) Cellulite. Etiology and purported treatment. Dermatol Surg 23:1117–1181

16. Dyson M (1991) Cellular sub-cellular aspects of low level laser therapy (LLLT). In: Oshiro T, Calderhead RG (eds) Progress in laser therapy: selected papers from the October 1990 ILTA Congress. Wiley, New York, pp 221–222

17. Dyson M (2006) Primary, secondary and tertiary effects of phototherapy: a review. In: Hamblin MR, Waynant RW, Anders J (eds) Mechanisms for low-light therapy, Proceedings of SPIE vol 6140, 6140005:1–12

18. El Said SO, Dyson M (1990) Comparison of the effect of multi-wavelength light produced by a cluster of semiconductor diodes and of each individual diode on mast cell number and degranulation in intact and injured skin. Laser Surg Med 10:559–568

19. Enwemeka CS (1988) Laser biostimulation of healing wounds: specific effects and mechanism of action. J Orthop Sports Physiotherapy 9:333–338

20. Enwemeka CS (2001) Attenuation and penetration of visible 632 and invisible 904 infra-red light in soft tissue. Laser Ther 13:95–101

21. Fodor PB (1997) Endermologie (LPG) does it work? Aesthetic Surg J 21:68

22. Gan L, Tse, C, Pilliar RM, Kandel RA (2007) Low-power laser stimulation of tissue engineering cartilage tissue formed on a porous calcium polyphosphate scaffold. Laser Surg Med 39:286–293

23. Gniadecka M (1997) Potential for high-frequency ultrasonography, nuclear magnetic resonance, and Raman spectroscopy for skin studies. Skin Res Tech 3:139–146

24. Grimblatov V, Rubinshtein A, Rubinshtein M (2006) Spectral dosimetry in low light therapy. In: Hamblin MR, Waynant RW, Anders J (eds) Mechanisms for low-light therapy, Proceedings of SPIE vol 6140, 614000S:1–6

25. Guillard G, Lagarde JM (2005) Skin lesions segmentation and quantification from 3D body's models. Skin Res Tech 11:123–131

26. Gupta AK, Filonenko N, Salansky N, Sauder DN (1998) The use of low energy photon therapy (LEFT) in venous leg ulcers: a double-blind, placebo-controlled study. Dermatol Surg 24:1383–1386

27. Hamblin MR, Demidova TN (2006) Mechanisms of low level light therapy. In: Hamblin MR, Waynant RW, Anders J (eds) Mechanisms for low-light therapy, Proceedings of SPIE vol 6140, 6140001:1–12

28. Hawkins D, Houreld N, Abrahamse H (2005) Low level laser therapy (LLLT) as an effective therapeutic modality for delayed wound healing. Ann NY Acad Sci 1056:486–493

29. Hexsel DM, Mazzuco R (2000) Subcison: a treatment for cellulite. Int J Dermatol 39:539–544

30. Hexsel D, Mazzuco R (2006) Subcision®. In: Goldman MP, Bacci PA, Leischoff G, Hexsel D, and Angelini F (eds) Cellulite: pathophysiology and treatment. Taylor & Francis, New York, pp 251–262

31. Hexsel D, Dal'Forno T, Cignachi S (2006) Definition, clinical aspects, associated conditions, and differential diagnosis. In: Goldman MP, Bacci PA, Leischoff G, Hexsel D, and Angelini F (eds) Cellulite: pathophysiology and treatment. Taylor & Francis, New York, pp 7–28

32. Hexsel D, Dal'Forno T, Hexsel C, Rodrigues T (2007) Cellulite severity scale. J Am Acad Dermatol 56 (Suppl. 2): AB59

33. Hexsel D, Dal'Forno T, Hexsel C (2007) Severity scale of cellulite. J Europ Acad Dermatol Venerol 91:FC3.9

34. Horwitz LR, Burke TJ, Carnegie D (1999) Augmentation of wound healing using monochromatic infrared energy. Adv Wound Care 12:35–40

35. Jackson RF, Roche G, Butterwick KJ, Dedo DD, Slattery KT (2004) Low-level laser-assisted liposuction: A 2004 clinical study of its effectiveness for enhancing ease of liposuction procedures and facilitating the recovery process for patients undergoing thigh, hip, and stomach contouring. Am J Cosmet Surg 21:191–198

36. Karu, T (1989) Photobiology of low-power laser effects. Health Phys 56:691–704

37. Karu T (1989) Laser biostimulation: a photobiological phenomenon. J Photochem Photobiol B3:638–640

38. Karu T (1999) Primary and secondary mechanisms of action of visible to near-IR radiation on cells. J Photochem Photobiol B 49:1–17

39. Karu, T. and Afanas'eva, NI (1995) Cytochrome c oxidase as the primary photoacceptor upon laser exposure of cultured cells to visible and near IR-range light. Dokl Akad Nauk 342:693–695

40. Karu TI, Pyatibrat LV, Kalendo GS (1993) The effect of He-Ne laser irradiation on the adhesive properties of the cell membranes. Int J Radiat Biol Med 115:622–623

41. Karu TI, Pyatibrat LV, Kalendo GS, Esenaliev RO (1996) Effects of monochromatic low-intensity light and laser irradiation on adhesion of HeLa cells in vitro. Laser Surg 18:171–177

42. Kasai S, Kono T, Yamamoto Y, Kotani H, Sakamoto T, Mito M (1996) Effect of low-power laser irradiation on impulse conduction in anesthetized rabbits. J Clin Laser Med Surg 14:107–109

43. Kligman AM (1997) Cellulite: facts and fiction. J Geriatr Dermatol 5:136–139

44. Kulick M (2006) Evaluation of the combination of radio frequency, infrared energy and mechanical rollers with suction to improve skin surface irregularities (cellulite) in limited treatment area. J Cosmet Laser Ther 8:180–195

45. Lach E, et al. Unpublished data.

46. Lach E, Pap S (2004) Laser treatment for cellulite: a non-invasive alternative to liposuction. Laser Surg Med (Suppl. 16):32

47. Leibaschoff G (2006) Surgical treatment. In: Goldman MP, Bacci PA, Leischoff G, Hexsel D, and Angelini F (eds) Cellulite: pathophysiology and treatment. Taylor & Francis, New York, pp 211–250

48. Leibaschoff G (2006) Mesotherapy in the treatment of cellulite. In: Goldman MP, Bacci PA, Leischoff G, Hexsel D, and Angelini F (eds) Cellulite, pathophysiology and treatment. Taylor & Francis, New York, pp 263–286

49. Lotti T, Ghersetich I, Grappone C, Dini G (1990) Proteoglycans in so-called cellulite. Int J Dematol 29:272–274

50. Lucassen GW, van der Sluys, van Herk JJ, Nuijs AM, Weirenga PE, barel AO, Lambrecht R (1997) The effectiveness of massage treatment on cellulite as monitored by ultrasound imaging. Skin Res Tech 3:154–160

51. Manstein D, Erofeev AV, Altshuler GB, Anderson RR (2001) Selective photothermolysis of lipid-rich tissue. Laser Surg Med (Suppl. 13):6

52. Marovino T (2004) Cold lasers in pain management. Practical pain management, Sept/Oct

53. McDaniel DH, Geronemous RG, Weiss RA, Weiss M, Newman J (2004) LED photomodulation reverses acute UV induced skin damage. Laser Surg Med (Suppl. 16):30

54. Medrado AP, Trindale E, Reis SRA, Andrade ZA (2006) Action of low-level laser therapy on living fatty tissue of rats. Laser Med Sci 21:19–23

55. Mirrashed F, Sharp JC, Krause V, Morgan J, Tomanek B (2004) Pilot study of dermal and subcutaneous fat structures by MRI in individuals who differ in gender, BMI, and cellulite grading. Skin Res Tech 10:161–168

56. Neira R, Ortiz-Niera C (2002) Low level laser assisted liposculpture: clinical report of 700 cases. Aesthetic Surg J 22:451–455

57. Niera R, Jackson R, Dedo D, Ortiz CL, Arroyave JA (2001) Low-level laser-assisted lipoplasty appearance of fat demonstrated by MRI on abdominal tissue. Am J Cosmet Surg 18:133–140

58. Neira R, Arroyave J, Ramirez H, Ortiz CL, Solarte E, Sequeda F, Gitierrez MI (2002) Fat liquefaction: effect of low-level laser energy on adipose tissue. Plast Reconstruct Surg 110:912–922

59. Neira R, Toledo L, Arroyave J, Solarte E, Isaza C, Gutierrez O, Criollo W, Ramirez H, Gutierrez MI, Ortiz-Neira CL (2006) Low-level laser-assisted liposuction: the Neira 4 L technique. Clin Plast Surg 33:117–127

60. Nootheti PK, Magpantay A, Yosowitz G, Calderon S, Goldman MP (2006) A single center, randomized, comparative, prospective clinical study to determine the efficacy of the Velas-

mooth system versus the TriActive system for the treatment of cellulite. Laser Surg Med 38:908–912

61. Nürnberger F, Müller G (1978) So-called cellulite: an invented disease. J Dermatol Surg Oncol 4:221–229
62. Nussbaum EL, van Zuylen J (2006) Transmission of phototherapy through human skin: dosimetry adjustment for effects of skin color, body composition, wavelength and light coupling. In: Hamblin MR, Waynant RW, Anders J (eds) Mechanisms for low-light therapy, Proceedings of SPIE vol 6140, 614000H:1–8
63. Ohno T (1997) Pain suppressive effect of low power laser irradiation: a quantitative analysis of substance P in the rat spinal dorsal root ganglion. Nippon Ika Daigaku Zasshi 64:395–400
64. Pabby A, Goldman MP (2006) The use of TriActive^{TM} in the treatment of cellulite. In: Goldman MP, Bacci PA, Leischoff G, Hexsel D, and Angelini F (eds) Cellulite: pathophysiology and treatment. Taylor & Francis, New York, pp 189–195
65. Perin F, Perrier C, Pittet JC, Beau P, Schnebert S, Perrier P (2000) Assessment of skin improvement treatment efficacy using the photograding of mechanically-accentuated macrorelief of thigh skin. Int J Cosmet Sci 22:147–156
66. Piérard GE (2005) Commentary on cellulite: skin mechanobiology and the waist-to-hip ratio. J Cosmetic Dermatol 4:151–152
67. Piérard GE, Nizet JL, Pierard-Franchimont C (2000) Cellulite: from standing fat herniation to hypodermal stretch marks. Am J Dermatopathol 22:34–37
68. Porgel MA, Chen JW, Zhang K (1998) Effects of low-energy gallium-aluminum-arsenide laser irradiation on cultured fibroblasts and keratinocytes. Laser Surg Med 20:426–432
69. Posten W, Wrone DA, Dover JS, Arndt KA, Alam M (2004) Low level laser therapy for wound healing: mechanism and efficacy – a review. Laser Surg Med (Suppl. 16):30
70. Posten W, Wrone DA, Dover JS, Arndt KA, Silapunt S, Alam M (2005) Low-Level laser therapy for wound healing: mechanism and efficacy. Dermatol Surg 31:334–340
71. Quatresooz P, Xhauflaire-Uhoda E, Piérard-Franchimont C, Piérard GE (2006) Cellulite histopathology and related mechanobiology. Int J Cosmet Sci 28:3, 207–210
72. Querleux B (2006) Cellulite characterization by high-frequency ultrasound and high-resolution magnetic resonance imaging. In: Goldman MP, Bacci PA, Leischoff G, Hexsel D, and Angelini F (eds) Cellulite: pathophysiology and treatment. Taylor & Francis, New York, pp 105–114
73. Querleux B, Cornillon C, Jolivet O, Bittoun J (2002) Anatomy and physiology of subcutaneous adipose tissue by in vivo magnetic resonance imaging and spectroscopy: relationships with sex and presence of cellulite. Skin Res Technol 8:118–124
74. Rosenbaum M, Prieto V, Hellmer J, Boschmann M, Krueger J, Leibel RL, Ship AG (1998) An exploratory investigation of the morphology and biochemistry of cellulite. Plast Reconstr Surg 101:1934–1939
75. Schindl A, Schindl M, Pernerstorfer-Schon H, Schindl L (2000) Low intensity laser therapy: a review. J Invest Med 48:312–326
76. Schindl A, Schindl M, Schon H, Knobler R, Havelec L, Schindl L (1998) Low-intensity laser irradiation improves skin circulation in patients with diabetic microangiopathy. Diabetes Care 21:580–584
77. Smalls LK, Lee CY, Whitestone J, Kitzmiller WJ, Wickett RR, Visscher MO (2005) Quantitative model of cellulite: three dimensional skin surface topography, biophysical characterization and relationship to human perception. J Cosmet Sci 56:105–120
78. Smalls LK, Hicks M, Passeretti D, Gerson K, Kitzmiller WJ, Bakhsh A, Wickett RR, Whitestone J, Visscher MO (2006) Effect of weight loss on cellulite: gynoid lypodystrophy. Plastic Reconstr Surg 118:510–516
79. Steinlechner CWR, Dyson M (1993) The effect of low-level laser therapy on macrophage-modified keratinocyte proliferation. Laser Ther 5:65–73
80. Tam, G (1999) Low power laser therapy and analgesic action. J Clin Laser Med Surg 17:29–33

81. Thomasson TL (1996) Effects of skin-contact monochromatic infrared irradiation on tendonitis, capsilitis, and myofascial pain. J Neurol Orthop Med Surg 16:242–245
82. Van Vliet M, Ortiz A, Avram MM, Yamauchi PS (2005) An assessment of traditional and novel therapies for cellulite. J Cosmet Laser Ther 7:7–10
83. Wakabayashi H, Hamba M, Matsumoto K, Tachibana H (1993) Effect of irradiation by semiconductor laser on response evoked in trigeminal caudal neurons by tooth pulp stimulation. Laser Surg Med 13:605–610
84. Wanitphakdeedecha R, Manuskiatti W (2006) Treatment of cellulite with a bipolar radiofrequency, infrared heat, and pulsatile suction device: a pilot study. J Cosmet Dermatol 5:284–288
85. Webb C, Dyson M, Lewis WHP (1998) Stimulatory effect of 660 nm low laser energy on hypertrophic scar-derived fibroblasts: possible mechanisms for increase in cell counts. Laser Surg Med 22:294–301
86. Weiss RA, McDaniel DH, Geronemus RG, Weiss M, Newman J (2004) Non-ablative, nonthermal light emitting diode (LED) phototherapy of photoaged skin. Laser Surg Med (Suppl. 16):31
87. Yu W, Naim JO, McGowan H, Ippolito K, Lanzafame RJ (1997) Photomodulation of oxidative metabolism and electron chain enzymes in rat liver mitochandria. Photochem Photobiol 66:866–871
88. Zitzlsperger K, Counters J, Tasi S, Zelickson B (2004) Photomodulation using Gentlewaves LED with and without rejuvenating masque products. Laser Surg Med (Suppl. 16):30
89. United States Patent Application (2004) Publication No. US 2004/0162596

Appendix 3

Effect of Far Infrared Therapy on Inflammatory Process Control After Sciatic Crushing in Rats

Carolina L.R.B. Nuevo, Renata Amadei Nicolau, Renato Amaro Zângaro, Aldo Brugnera, Jr., and Marcos Tadeu Tavares Pacheco

Abstract

Background – Peripheral nerves are often traumatized, and as a consequence there is loss or decrease in sensitivity. Radiation in the far infrared region (Far-IR) between 4–16 μm of the electromagnetic spectrum, was found to be efficient in treating inflammatory processes. However, it had not been previously studied in the neuronal reparation process.

Objective – This study investigated the initial process of sciatic nerve recuperation in rats treated with Far-IR treatment, after lesion by crushing.

Material and Methods – The experiment comprised twelve Wistar rats, divided into control group and treated group. Animals in both groups had their sciatic nerves crushed using Kelly tweezers and underwent pressure of about 6 N for 30 seconds. The Far-IR therapy was performed in the treated group 30 minutes after surgery, and repeated after 24, 48, and 72 hours. All animals were sacrificed on the 21st day after surgery (DAS), so that histomorphometrical analyses could be carried out. Functional recuperation was assessed in the before and after surgery periods, respectively on the 7th, 14th, 21st days after surgery (DAS) using the Sciatic Functional Index (SFI) calculations. Although results showed a progressive improvement in both groups, the most expressive values were found in the treated group. There was a significant difference in the SFI of the treated group between the 7th and 21st DAS ($p < 0.05$), showing a significant recuperation of this group between the two lesion

Renato Amadei Nicolau

Institute of Research and Development – Universidade do Vale do Paraíba (UNIVAP) Biomodulation Tissue Laboratory and Lasertherapy and Phototherapy Center, São José dos Campos, São Paulo, 12244-000, Brasil, e-mail: rain@univap.br

R. Waynant and D.B. Tata (eds.), *Proceedings of Light-Activated Tissue Regeneration and Therapy Conference.*
© Springer Science + Business Media, LLC 2008

phases. The index evolution in the control group was not statistically significant between the different evaluation periods. The myelin sheath mean area did not show significant difference between treated and control groups. Results showed that even though no statistical differences were noted between the control and treated groups regarding the myelin area and the SFI, acceleration in the reparation process between the 7th and 21st DAS was observed in the latter.

Keywords: Far infrared, sciatic nerve, crushing.

Introduction

Peripheral nerve lesions constitute an important functional problem. Although there is some recovery in most cases of such lesions, it occurs slowly and incompletely. Difficulties faced by lesioned patients while executing their daily life activities became prime elements to determine the goals for patients' recovery [9].

In axonotmesis lesions, the distal part endures a process called Wallerian degeneration, in which the axoplasm breaks up and disappears, and the myelin gradually dissolves. The nerve's connective tissues are kept almost intact and there are good chances of spontaneous recovery [22]. Recovery in humans is usually slow (from few months to more than a year) [24].

Experimental and clinical studies have been done in search for therapy that could help in neuronal restitution and the Far infrared radiation (Far-IR) has been among the newest therapeutic modalities.

According to Toyokawa et al. [23], irradiation with Far-IR on tissue undergoing an inflammatory process can cause a quick temperature rise with a notable increase in cell activity. Theories indicate that energy produced by a rise in temperature is absorbed by the cell membrane, improving local circulation. Venous capillary vasodilatation, cell metabolism elevation and increase in interstitial fluid lymphatic drainage favour the tissue cicatrization process [10].

When using wavelength between 4–16 μm, propagation and absorption of the Far-IR by biological tissues promote biostimulation with the following therapeutic outcomes: analgesia, edema reductions, microcirculation and cell metabolism elevation, collagen and elastin synthesis, water molecule agglomerate reduction, free radicals and impurities elimination, muscular spasm and hematoma reductions [5, 6]. So far no study using a Far-IR (4–16 μm) exposure in nervous tissular restitution has been found in literature reviews. The objective of this study was to investigate the initial process of the sciatic nerve restitution in rats treated with Far-IR after a lesion by crushing, and assess the Sciatic Functional Index (SFI) in addition to histomorphometrical evaluation of the myelin sheath.

Material and Methods

Twelve Wistar male rats (*Rattus norvegicus* albinus variety), weighing approximately 300 g (90 days) were used. Animals were obtained from Anilab farm and kept at Camilo Castelo Branco University-UNICASTELO/Fernandópolis – SP vivarium. Animals were kept in cages (two animals per cage), at room temperature and the day/night cycle was of 12 hours with Labina® standard diet and water *ad libitum*. All of them were conditioned to walk on a gait belt, according to De Medinaceli et al. [3]. After conditioning, pre-operative register of footprints was obtained, which was the initial evaluation parameter to compare with the after-surgery register.

Surgical Procedure

The pre-anesthetic Butorfanol (Turbogesic®, 2 mg/kg) was used in association with Acepromazin (Acepran®, 1 mg/kg), both administered via a unique intramuscular dose. Zolazepan and Tiletamin (Zoletil 50®, 40 mg/kg) were given to the animals 15 minutes later. After that, they were trichotomyzed at the right dorsal region. Antisepticism of this region with iodized alcohol was followed by a longitudinal incision on the lateral face of the right thigh from the greater trochanter to the knee, exposing it from its emergence at the inferior edge of the bigger gluteal until its bifurcation into three distal branches. Suture (Prolene 3/0, Ethicon®) was passed through the epineurium, about 3 mm distally to its emergence, to facilitate the visualization of the crushed region, whose size was of about 50 mm. The crush was done with a Kelly forceps (Kinelato) exerting pressure of about 6 N for 30 seconds. Muscular tissue and skin were positioned and sutured with simple stitches of monofilament nylon (Mononylon 3/0, Ethicom®) [11]. The region was washed with iodized alcohol antiseptic solution and only one dose of antibiotic (Penicillin associated with estreptomicin) to prevent infections was administered. Fentanil® drug via intraperitoneal (0.03 mg/kg, 12/12 hours) was used for after-surgery analgesia for 48 hours DAS.

Therapy with Far-IR

Animals were divided into two groups. Group 1 (n = 6) was the control group. The other group (n = 6) was exposed to infrared radiation (Far-IR – Dome Face – Invel), 30 minutes after surgery, and repeated 24, 48, and 72 hrs post surgery. Temperature inside the equipment was of 36°C, and animals of both groups were positioned in right lateral decubitus, in order to receive 30 minutes of irradiation (Fig. 1).

Fig. 1 Therapy with Far-IR.
(**A**) Far-IR emission
equipment, (**B**) animal in
lateral decubitus, (**C**)
thermometer

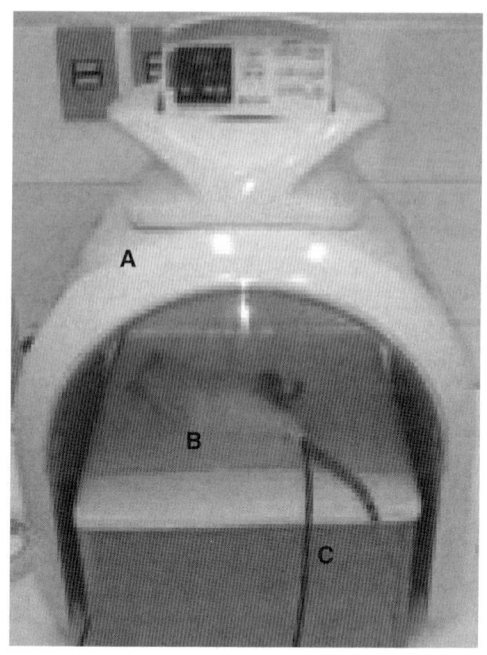

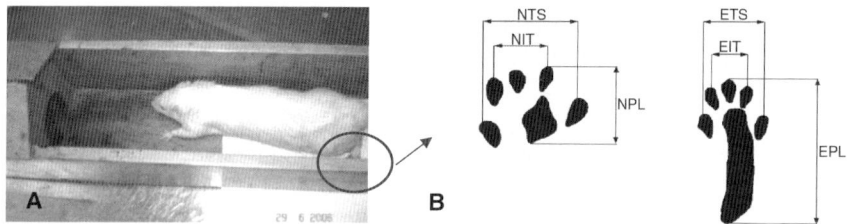

Fig. 2 (A) Platform for footprint registration pre and post-surgery. (**B**) Footprint registration example

Footprint Registers

Animal footprints were obtained based on the De Medinaceli et al. [3] method, modified by Lowdon et al. [8] (Fig. 2). Their back paws were sunk into black ink, and placed to walk on the gait belt covered with a sheet of paper, to obtain the footprint register. Papers were left to dry and stored once again, until measurements using the SFI, modified by Bain et al. [2] were done.

$$\text{SFI} = -38.3 \times \frac{\text{EPL} - \text{NPL}}{\text{NPL}} + 109.5 \times \frac{\text{ETS} - \text{NTS}}{\text{NTS}} + 13.3 \times \frac{\text{EIT} - \text{NIT}}{\text{NIT}} - 8.8$$

Footprints were obtained prior to surgery and also the on the 7th, 14th and 21st DAS. On the 21st DAS, animals were sacrificed in order to have the sciatic nerves removed for later analysis through histomorphometric studies.

Histomorphometric Analysis

Histomorphometric analysis was done with the aid of *Image Pro Plus* program (version 4.5.0.69). A total of four cuts were obtained from each animal, and five of their quadrants in the crushed area (four peripheral and one central) were analyzed. Total area of the myelin sheath was calculated from a rectangular area of 350 × 350 μm, totaling an area of 122,500 μm^2 in each quadrant.

Gait Functional Analysis

Rat footprints were obtained at every stage of the experiment. Measurements of footprint length (PL), total toe spread (TS) and spread of intermediate toes (IT) enabled the SFI calculation proposed by Bain et al. [2]. The obtained results express the percentage of functional loss: 0 (zero) represents the normal function and the −100 value represents the total loss of nerve function.

Results

In the control group, the SFI mean was of −73.12 on the 7th DAS, improving to −36.76 on the 14th DAS and −12.10 on the 21st DAS (Table 1). In the treated group, these means were: −103.66 on the 7th DAS, −63.32 on the 14th DAS and finally −29.34 on the 21st DAS (Table 2). Results demonstrated that no significant statistical differences related to myelinization between the control and treated groups ($p > 0.05$) were observed.

Table 1 SFI values obtained in the control group

Control	Pre-operative	7th DAS	14th DAS	21st DAS
RAT 1	−0.85	−125.26	−13.74	−20.44
RAT 2	−15.80	−139.26	−121.44	2.79
RAT 3	−6.73	−17.28	−44.62	−34.85
RAT 4	−11.19	−11.38	27.36	3.51
RAT 5	−10.08	−137.98	−55.53	−19.46
RAT 6	−17.19	−7.60	−24.64	−4.16
Mean	−10.31	−73.13	−38.77	−12.10
Standard deviation	6.01	67.12	49.70	15.30

DAS – days after surgery

Table 2 SFI values obtained in the treated group

Treatment	Pre-operative	7th DAS	14th DAS	21st DAS
RAT 1	−26.13	−120.71	−7.93	−4.96
RAT 2	−9.66	−119.99	−125.03	−26.15
RAT 3	−1.87	−111.25	−64.48	−7.35
RAT 4	−18.06	−118.30	−135.08	−124.39
RAT 5	−18.16	−13.74	−38.37	1.11
RAT 6	6.93	−138.02	−9.07	−14.35
Mean	−11.16	−103.67	−63.33	−29.35
Standard deviation	12.14	44.93	55.84	47.48

DAS – days after surgery

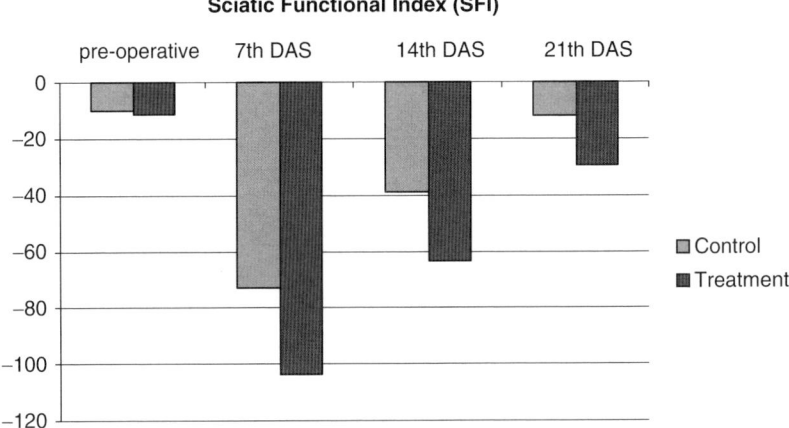

Fig. 3 SFI of the control and treated groups. DAS – days after surgery

The treated group showed expressive difference ($p < 0.05$) between the 7th and 21st DAS, which did not occur among the control group animals at any evaluated time, demonstrating that recovery of the sciatic nerve function was better in the treated group (Fig. 3).

Discussion

The objective of this study was to test the electromagnetic radiation efficacy with wavelength between 4–16 μm (Far-IR) in the initial process of crushed sciatic nerve reparation in rats.

The methodology applied to this study was based on literature reviews in which Wistar rats are usually employed due to their fast natural nervous regeneration.

Their peripheral nerve morphologic structures, as well as the physiology and regeneration process, are very similar to humans [11, 13–18].

The crush lesion (axonotmesis) was chosen because it preserves the nerve's support structure even when subjected to application of certain pressure, which in this study was approximately 6N. Preservation of the support structure helps regeneration, respecting the fact that neural tubes are in continuance ([11, 13–18]).

Evaluations was done using the SFI measurement, which has proved to be a reliable instrument to evaluate the peripheral nerve regeneration process, as it supplies a function value that allows statistical analysis of results. Measurement of the myelin sheath area, which lines each neuron was another used parameter.

With regard to the SFI, results obtained in this study showed there was no significant difference between the control group and the one treated with Far-IR. However, there was a significant difference in the intra-group evaluation of the treated group in the period between the 7th and 21st DAS showing the Far-IR had a beneficial effect. In the inter-group analysis, the treated group showed an improvement 66% higher in relation to the 7th and 21st DAS of the control group. Both groups, control and treated, showed differences among the before and after surgery values. There was a significant difference in the treated group intra-group evaluation in the post-operative comparison between the 7th and 21st DAS. These results showed that the Far-IR had a beneficial effect upon the peripheral nervous structures during the initial process of restitution, corroborating results obtained with infrared radiation [4, 7, 23].

The tendency of neuronal regeneration was observed using the SFI calculation and was more evident in the treated group between the 7th and 21st DAS. According to existing literature and patients' reports, when treating patients with peripheral nervous compromising in the clinical practice, usually low back pain, there is little improvement after the first week of treatment. Thus, future studies with animals for longer evaluation periods should be done.

On the 21st DAS, which had an expressive difference with regard to the 7th DAS, results demonstrated probable continuance of the Far-IR action upon the tissue. Assia et al. [1] demonstrated that irradiation effects at different wavelengths last for up to three weeks, being more effective after two weeks. The authors also showed that electromagnetic irradiation effects in the visible and close infrared regions are observed after a week of irradiation.

It is known that the radiation emitted in the Far-IR region increases cell membrane permeability, favouring metabolic exchanges with the extracellular medium, thus feeding the membrane. It also acts on the mitochondria promoting ATP synthesis, which is used in muscular contraction [23]. The skeletal muscle system is benefited by Far-IR effects, thus its application promotes loss of muscular spasms. Skin stimulation reduces the gamma fibers, resulting in excitability reduction, constituting the physiological base for the slackness of muscular spasms after the use of heat [4]. These events may have contributed to the relative intra-group improvement observed in the treated group. Temperature rise can change tissue properties reducing resistance of muscles and tendons. Temperature influences the degree of resistance, so that low temperature infers bigger resistance to the movement, and higher temperature reduces resistance, favouring articular mobility [7]. An increase in temperature also culminates in blood perfusion and increase of neuronal activity, justifying the increase in animals' functional capacity in the treated group.

The response to low power laser stimulus results from the amount of total energy applied, and consequently from the stimulation time upon the structure was observed by Nissan et al. [11]. Given that studies of Far-IR therapy dosimetry are still incipient, there is need for further studies in this area, using different irradiation times and intervals between applications.

Histomorphometric analysis showed there were no statistically significant values to compare the control and the treated groups, taking into consideration the mean area of the myelin sheath, which lines each neuron. These results, associated with functional results, call for future studies to be carried out using a longer analysis period.

Conclusion

Results showed that although no significant statistical differences between the control and treated groups regarding the myelinization area and SFI were observed, acceleration in the restitution process between the 7th and 21st DAS was observed in the irradiated group. Therefore, further studies with increased number of animals and longer evaluation periods are necessary to clarify the possible mechanisms of Far-IR therapy with regard to the initial process of the crushed sciatic nerve recuperation in rats.

References

1. Assia, E; Rosner, M; Belkin, M; Solomon, A; Schwatz, M. Temporal parameters of low energy laser irradiation for optimal delay of post-traumatic degeneration of rat optic nerve. **Brain Res.**, v. 476, pp. 205–212, 1989.
2. Bain, J.R; Mackinnon, S.E; Hunter, R.T. Functional evaluation of complete sciatic, peroneal posterior tibial nerve lesions in the rat. **Plast. Reconstr. Surgery**, v. 83, pp. 129–138, 1989.
3. De Medinaceli, L.; Freed, W.J.; Wyatt, R.J. An index of the functional condition of rat sciatic nerve based on measurements made from walking tracks. **Exp. Neuro.**, v. 77, pp. 6634–643, 1982.
4. Guyton, A. C.; Hall, J. E. H. **Fisiologia humana e mecanismos das doenças**. 6 ed. Rio de Janeiro: Guanabara Koogan, pp. 181–190, 514–519, 1998.
5. Inoué, S.; Kabaya, M. Biological activities causes by far-infrared radiation. **Biometerol.**, v. 33, pp. 145–150, 1989.
6. Kitchen, S.S.; Patridge, C.J. Infra-red therapy. **Physiotherapy**, London, v. 7, n. 4, pp. 249–254, 1991.
7. Kitchen, S.; Bazin, S. **Eletroterapia de Clayton**. 10 ed. São Paulo: Manole, 1998.

8. Lowdon, I M.R.; Scabei, V.; Urbaniak, J.R. An improved method of recording rat tracks for measurement of the sciatic functional index of Dc Medinaceli. **J. Neuroscience Meth.**, v. 24, pp. 279–281, 1988.

9. Machado, A.B.M. **Neuroanatomía Funcíonal**. 2 ed. São Paulo: Atheneu, 1993.

10. Maeda, K. **Far infrared-trcatment and medical revolution**. Tokyo: Toppan Insatsu Kabushiki, 1997.

11. Nissan, M.; Rochkind, S.; Razon, N.; Bartal, A. HeNe Laser irradiation delivered transcutaneously: its effect on the sciatic nerve of rats. **Laser Surg. Med.**, v. 6, pp. 435–438, 1986.

12. Rigau, J. **Accion de la luz laser a baja intensidad em la modulacion de la funcion celular. Réus**. Tese (doutorado em histologia)- Facultat de Medicina i Ciência de la salut Univ. rovira i Virgili, 1996.

13. Rochkind, S.; Nissan, M.; Razon, N.; Schwartz, M.; Bartal, A. Electrophysiolgical eftect of HeNe laser on normal and injured sciatic nerve in the rat. **Acta Neurochir.**, v. 83, pp. 125–130, 1986.

14. Rochkind, S.; Nissan, M.; Barr-Nea, L.; Razon, N.; Schwartz, M.; Bartal, A. Response of peripheral nerve to He-Ne laser: experimentalstudies. **Laser Surg. Med.**, v. 7, pp. 441–443, 198/a.

15. Rochkind, S.; Barr-Nea, L.; Razon, N.; Bartal, A.; Schwartz, M. Stimulatory effect of HeNe dose laser on injured sciatic nerves of rats. **Neurosurgery**, v. 20, n. 6, pp. 843–847, 1987b.

16. Rochkind, S.; Nissan, M.; Lubart, R.; Avram, J.; Bartal, A. The in-vivo-nerve response to direct low-energy-laser irradiation. **Acta Neurochir.**, v. 94, pp. 74–77, 1988.

17. Rochkind, S.; Rousso, M.; Nissan, M.; Vilarreal, M.; Barr-Nea, L.; Rees, D.G. Systemic effects of low-power laser irradiation on the peripheral and central nervous system. cutaneous wounds and burns. **Laser Surg. Med.**, v. 9, pp. 174–182, 1989.

18. Rochkind, S.; Vogler, I.; Barr-Nea, L. Spine cord response to laser treatment of injured peripheral nerve. **Spine**, v. 15, n. 1, pp. 6–10, 1990.

19. Rochkind, S.; Nissan, M.; Alon, M.; Shamir, M.; Salame, K. Effects of laser irradiation on the spinal cord for the regeneration of crushed peripheral nerve in rats. **Laser Surg. Med.**, v. 28, pp. 216–219, 2001.

20. Simoneau, L.A. **Etude de L'efficate clinique cosmétique & ds effects biometrologique**: d''une methode associant produit cosmétique, soins esthétique, & utilisation du DOME PROFESSIONAL VISAGE INVEL. s.l.: s.n., 2002 [20] p.

21. Sonnewend, D.; Oliveira, J.L.R.; Nicolau, R.A.; Magalhães, R.G.; Conrado, L.A.; Zângaro, R. A.; Pacheco, M.T.T. O efeito da radiação infravermelho longo e microcorrentes sobre o processo de reparação de feridas em ratos. In: XIX Congresso Brasileiro de Engenharia Biomédica de João Pessoa, n° 2696, 2004. **Anais**. João Pessoa: Universidade do Federal da Paraíba, 2004.

22. Stokes, M. **Neurologia para Fisioterapeutas**. São Paulo: Premier, 2000.

23. Toyokawa, H. et al. Promotive effects of far-infrared ray on full-thickness skin wound healing in rats. **Exp. Biol. Med.** v. 228, n. 6, pp. 724–729, 2003.

24. Umphread, D.A. **Fisioterapia Neurológica**. São Paulo: Manole, 1994.

Appendix 4

Effects of Diode Laser Therapy on the Acellular Dermal Matrix

Livia Soares, Marília de Oliveira, Sílvia Reis, and Antônio Pinheiro

Abstract Acellular dermal matrix (ADM) has been used for facial reconstruction, and for soft tissue augmentation. During the past 10 years, this material has been extensively evaluated in primary and secondary burn reconstruction, numerous facial soft tissue augumentation procedures, and intraoral mucosal and gingival replacements.

Keywords: Acellular dermal network, rats, 685 nm, 4 J/cm^2, accelerated repair.

Aim

The purpose of this study was to define the effects of Low- level laser therapy on acellular dermal matrix (ADM), from a histological point of view.

Study Design/Materials and Methods

ADM (Alloderm – LifeCell Corporation, Woodlands, TX) in 5 × 5 mm squares was subcutaneously implanted into calvarian skin of male Wistar rats (n = 40). Low-level laser (InGaAlP, $\lambda = 685$ nm, 4 J/cm^2) was locally applied in experimental group (n = 20) above the skin flap. Grafts were harvested at 1, 3, 7 and 14 days after implantation and underwent histological analyses.

Serial sections were then cut at 5-μm and colored with hematoxylin and eosin (H&E) and Sirius Red. An analysis of two different regions of each sections were

L. Soares
Pontifica Universidade Católica do Rio Grande do Sul, Porto Alegre, RS Brasil

R. Waynant and D.B. Tata (eds.), *Proceedings of Light-Activated Tissue Regeneration and Therapy Conference.*
© Springer Science + Business Media, LLC 2008

made. The field near the basal lamina of the allograft was region A and the opposite side was region B.

Changes of collagen and inflammatory features were semi-quantitatively evaluated in coded slides and registered as absent (0), mild (1), moderate (2) and marked (3). The differences between the control group and experimental groups were analyzed by using the Student paired t test.

Results

In the control group of dead animals within 1 day and 3 days, an intense edema and polymorphonuclear inflammatory infiltrate was observed, although in the irradiated group the same features were seen the average was significantly smaller (Tables 1 and 2). The amount of collagen in graft treated with low level laser were significantly higher than those of controls and were statistically more prominent on the 14th day after surgery (Table 3).

Table 1 Distribution of edema in the control and laser treated groups at 1, 3, 7 and 14 days

Intervals	Group	Region A		Region B	
		Mean ± SD	p	Mean ± SD	p
1 day	Control	1.40 ± 0.54	0.0203*	2.20 ± 0.44	0.0125*
	Irradiated	0.40 ± 0.54		1.00 ± 0.70	
3 days	Control	0.40 ± 0.55	0.1411	1.00 ± 0.00	0.04*
	Irradiated	0.00 ± 0.00		0.40 ± 0.55	
7 days	Control	0.00 ± 0.00	1.00	0.20 ± 0.45	0.3466
	Irradiated	0.00 ± 0.00		0.00 ± 0.00	
14 days	Control	0.00 ± 0.00	1.00	0.80 ± 0.45	0.0039*
	Irradiated	0.00 ± 0.00		0.00 ± 0.00	

*The mean difference is significant at the 0.05 level.

Table 2 Distribution of polymorphonuclear inflammatory cells in the control and laser treated groups at 1, 3, 7 and 14 days

Intervals	Groups	Region A		Region B	
		Mean ± SD	p	Mean ± SD	p
1 day	Control	1.20 ± 0.44	0.0943	2.00 ± 0.70	0.172
	Irradiated	0.60 ± 0.54		1.40 ± 0.54	
3 days	Control	0.00 ± 0.00	1.00	0.20 ± 0.45	0.3466
	Irradiated	0.00 ± 0.00		0.00 ± 0.00	
7 days	Control	0.00 ± 0.00	1.00	0.00 ± 0.00	1.00
	Irradiated	0.00 ± 0.00		0.00 ± 0.00	
14 days	Control	0.00 ± 0.00	1.00	0.00 ± 0.00	1.00
	Irradiated	0.00 ± 0.00		0.00 ± 0.00	

Table 3 Distribution of collagen fibers in the control and laser treated groups at 1, 3, 7 and 14 days

Intervals	Group	Region A		Region B	
		Mean ± SD	p	Mean ± SD	p
1 day	Control	0.00 ± 0.00	0.0003*	0.80 ± 0.45	0.3466
	Irradiated	1.20 ± 0.45		1.00 ± 0.00	
3 days	Control	1.00 ± 0.00	0.04*	0.80 ± 0.44	0.0353*
	Irradiated	1.60 ± 0.55		1.60 ± 0.54	
7 days	Control	1.00 ± 0.70	0.04*	0.40 ± 0.54	0.0140*
	Irradiated	2.20 ± 0.83		1.80 ± 0.83	
14 days	Control	1.40 ± 0.54	0.0085*	0.80 ± 0.44	0.0109*
	Irradiated	2.60 ± 0.54		2.20 ± 0.83	

*The mean difference is significant at the 0.05 level.

Conclusion

The results of the effect of Low-level laser therapy on acellular dermal implants show an accelerated process of repair of the dermis and implant integration. These findings extend the theory of the positive influence of Low-level laser therapy on the healing of skin wounds to its use on acellular dermal implants.

Appendix 5

Laser Therapy in Inflammation: Possible Mechanisms of Action

Rodrigo Alvaro B. Lopes Martins

Presented by:
Jan M. Bjordal
Bergen University College & University of Bergen

R.A.B.L. Martins
Department of Pharmacology, Institute of Biomedical Sciences, University of Sao Paulo, Brazil

R. Waynant and D.B. Tata (eds.), *Proceedings of Light-Activated Tissue Regeneration and Therapy Conference.*
© Springer Science + Business Media, LLC 2008

Pathophysiology in arthritis
and tendinopathies

Inflammation

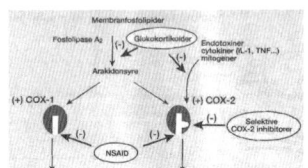

Inflammation in **synovia**

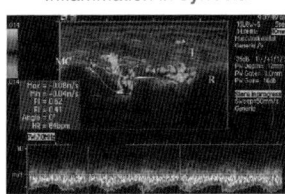

Terslev et al. Ann Rheum Dis 2005

COX expression in all layers of **cartilage**

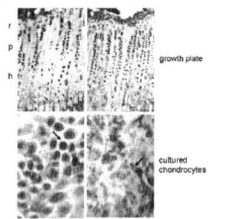

Brochausen et al. Arthr Ther Res 2006

Analysis of Cochrane reviews of randomized
placebo-controlled trials with pain therapy

Physical agents patients n=5482
Drug patients n= 6284

Reviews - patient samples characterisics

- Physical
- Pharmacological

	Total trials	Positive trials	Total reviews	Mean review sample	Median review sample
Physical	101	58	19	260	259
Pharmacological	71	41	26	313	113

Patient numbers

Analysis of Cochrane reviews of randomized placebo-controlled trials with pain therapy

Methodological quality for pain RCTs with positive results by two scales used in the Cochrane Library

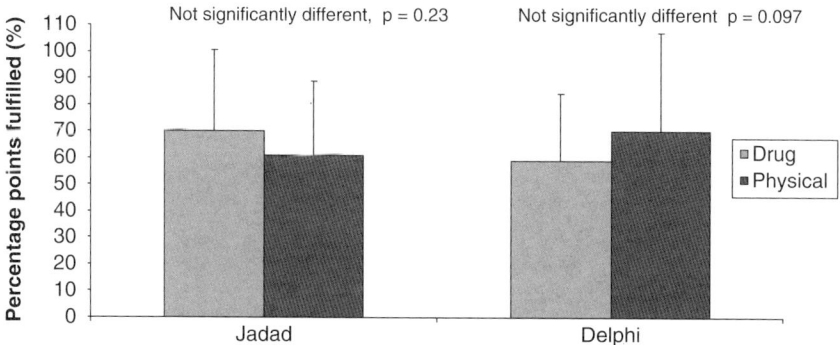

Analysis of Cochrane reviews of randomized placebo-controlled trials with pain therapy

Grading of review conclusions (percentage in each category)

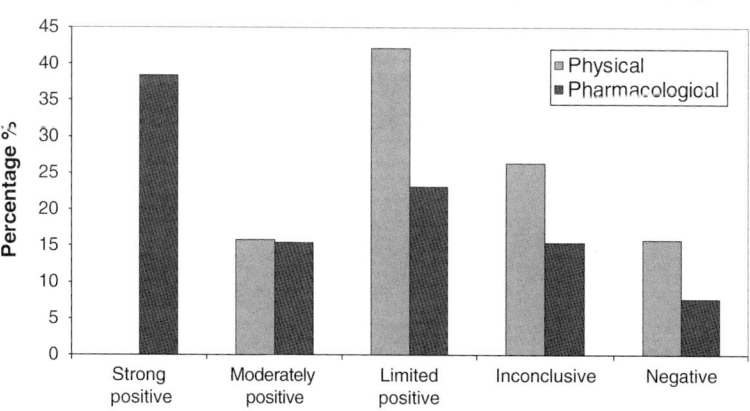

Inflammation
Definitions

"A Defensive Reaction against an Aggression"

Characterized by the generation of inflammatory mediators, cell migration and fluids movements.

Phases of Inflammation

• Initiation
• Amplification
• Termination

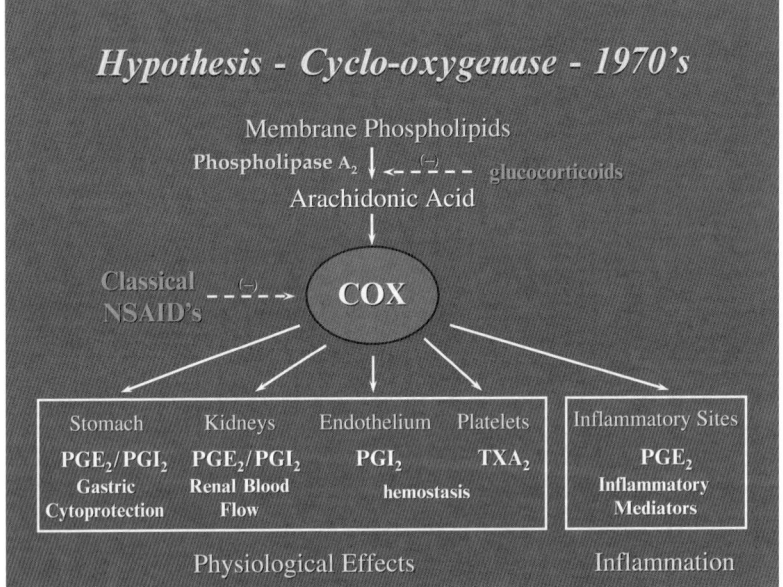

Hypothesis - Cyclo-oxygenase - 1970's

Membrane Phospholipids

Phospholipase A$_2$ $\downarrow$ $\leftarrow$ (-) - - - glucocorticoids

Arachidonic Acid

Classical NSAID's - - (-) - -> **COX**

Stomach	Kidneys	Endothelium	Platelets	Inflammatory Sites
PGE$_2$/PGI$_2$	PGE$_2$/PGI$_2$	PGI$_2$	TXA$_2$	PGE$_2$
Gastric Cytoprotection	Renal Blood Flow	hemostasis		Inflammatory Mediators

Physiological Effects   Inflammation

COX-2 Discovery

"Cells can express two pools of COX, one constitutive and a second inducible isoform expressed during pathological conditions"

(Fu et al., 1990)

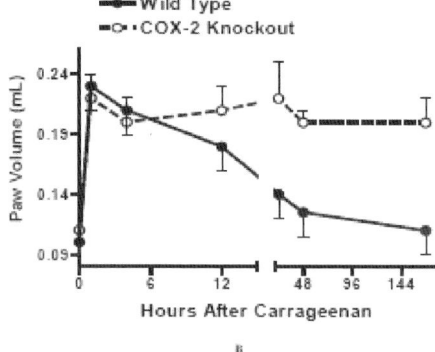

Deletion of COX-2 Gene delays the resolution of Inflammation

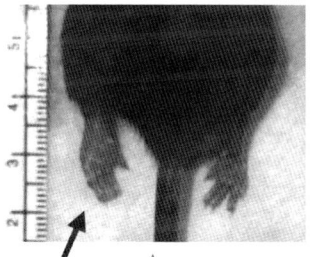

TheScientificWorldJOURNAL (2006) 6, 577–588
ISSN 1537-744X; DOI 10.1100/tsw.2006.122

TABLE 1
Comparison of Load to Failure (Newtons) Among Control, Indomethacin, and Celecoxib Groups at 2-Week, 4-Week, and 8-Week Time Periods

	Control	Indomethacin	Celecoxib
2 weeks	8.4 ± 3.0	5.1 ± 2.6[a]	5.3 ± 3.8[a]
4 weeks	20.0 ± 4.7	10.5 ± 4.8[b]	9.3 ± 6.1[b]
8 weeks	34.8 ± 7.3	26.4 ± 9.2[c]	24.8 ± 5.9[c]

[a]Compared with control group, $P < .006$.
[b]Compared with control group, $P < .0001$.
[c]Compared with control group, $P < .001$.

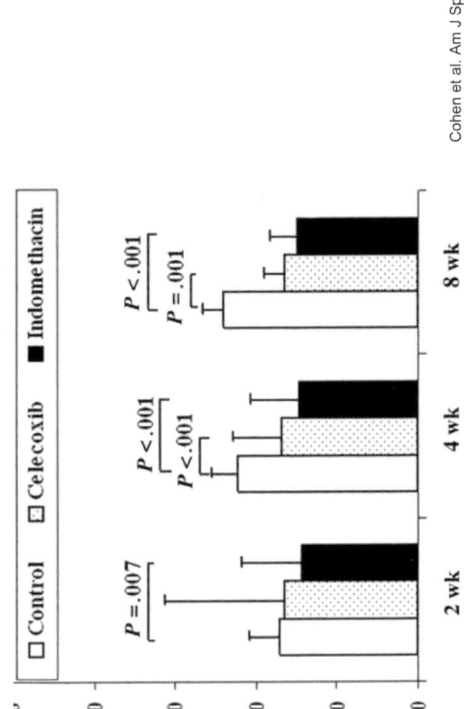

Figure 1. Custom-designed uniaxial testing system for bio mechanical testing.

Cohen et al. Am J Sports Med. 2006

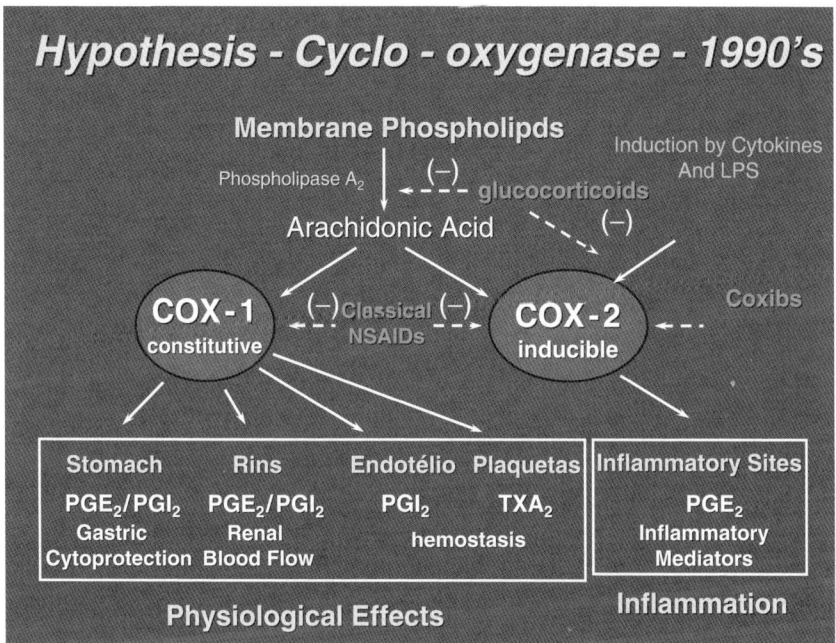

A New Hypothesis from the discovery of the 1990's

• COX-1 inhibition by NSAIDs produces important side-effects

• COX-2 inhibition produces antiinflammatory effects.

• The ideal antiinflammatory drug for systemic use should probably inhibit COX-2 selectively.

Cox 1 and Cox 2

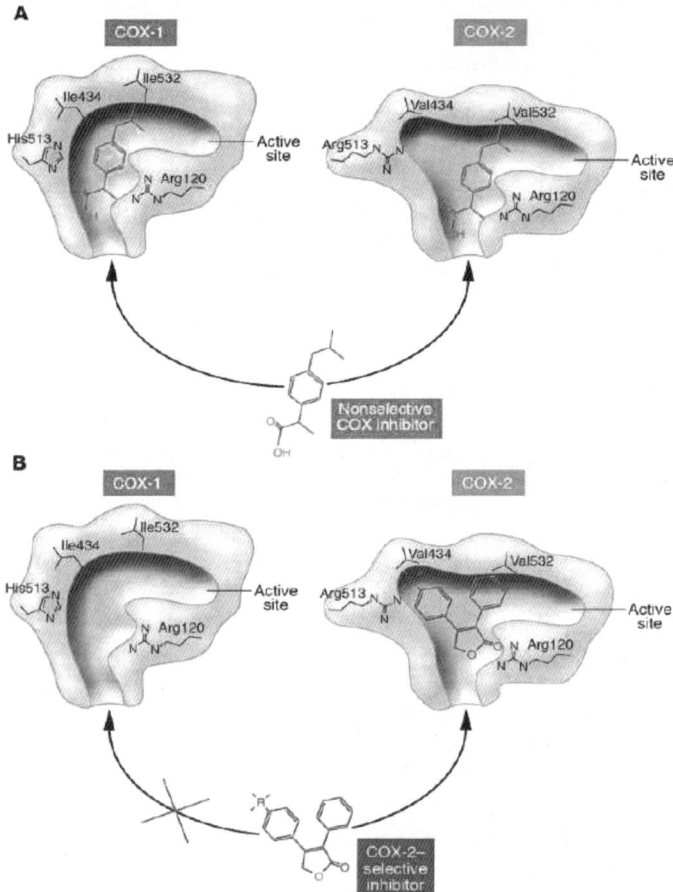

Figure 1

Schematic depiction of the structural differences between the substrate-binding channels of COX-1 and COX-2 that allowed the design of selective inhibitors. The amino acid residues Val434, Arg513, and Val523 form a side pocket in COX-2 that is absent in COX-1. (**A**) Nonselective inhibitors have access to the binding channels of both isoforms. (**B**) The more voluminous residues in COX-1, Ile434, His513, and Ile532, obstruct access of the bulky side chains of COX-2 inhibitors.

Consequences of COX inhibition for prostacyclin and TXA2 production in normal and atherosclerotic arteries

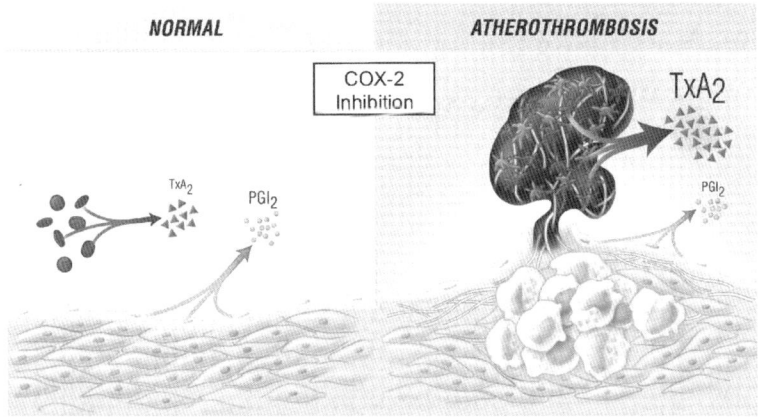

Antman. E. M. et al. Circulation 2007:115:1634-1642

Comparison of effects of different selective COX-2 inhibitors vs placebo on myocardial infarction

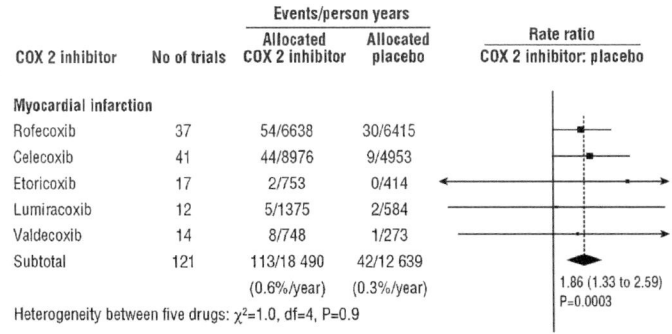

| | | Events/person years | | Rate ratio |
COX 2 inhibitor	No of trials	Allocated COX 2 inhibitor	Allocated placebo	COX 2 inhibitor: placebo
Myocardial infarction				
Rofecoxib	37	54/6638	30/6415	
Celecoxib	41	44/8976	9/4953	
Etoricoxib	17	2/753	0/414	
Lumiracoxib	12	5/1375	2/584	
Valdecoxib	14	8/748	1/273	
Subtotal	121	113/18 490	42/12 639	1.86 (1.33 to 2.59)
		(0.6%/year)	(0.3%/year)	P=0.0003

Heterogeneity between five drugs: χ^2=1.0, df=4, P=0.9

Antman, E. M. et al. Circulation 2007;115:1634-1642

Adverse effects of Coxibs vs non-specific NSAIDs in placebo-controlled trials

COX 2 inhibitor versus:	No of trials	Events/person years		Rate ratio COX 2 inhibitor: NSAID
		Allocated COX 2 inhibitor	Allocated NSAID	
Vascular events				
(a) Naproxen	42	185/16 360 (1.1%/year)	81/10 978 (0.7%/year)	1.57 (1.21 to 2.03) P=0.0006
Ibuprofen	24	46/5848	47/5160	
Diclofenac	26	101/10 886	79/6913	
Other non-naproxen	7	8/166	4/274	
(b) Any non-naproxen	51	155/16 900 (0.9%/year)	130/12 347 (1.1%/year)	0.88 (0.69 to 1.12) P=0.3
Any NSAID	91	340/33 260 (1.0%/year)	211/23 325 (0.9%/year)	1.16 (0.97 to 1.38) P=0.1

Heterogeneity between (a) and (b): χ^2=10.2, df=1, P=0.001
Between non-naproxen NSAIDs: χ^2=2.6, df=2, P=0.3

Implication of the relative degrees of selectivity

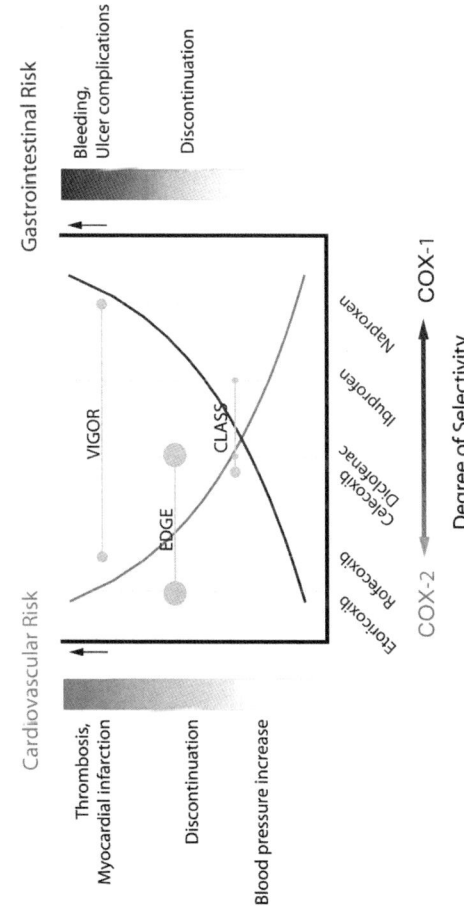

Antman, E. M. et al. Circulation 2007;115:1634-1642

Editorials represent the opinions
of the authors and JAMA and not those of
the American Medical Association.

COX-2 Inhibitors, Other NSAIDs, and Cardiovascular Risk
The Seduction of Common Sense

David J. Graham, MD, MPH

" . . . (A) long habit of not thinking a thing wrong, gives it a superficial appearance of being right, and raises at first a formidable outcry in defence of custom. But the tumult soon subsides. Time makes more converts than reason."

Thomas Paine, *Common Sense*, 1776

Experimental Models

- Rat and Mouse Paw Edema
- Rat Achilles Tendinitis
- Mice Pleurisy
- Mice Peritonitis
- Rat airway smooth muscle
- Mice Temporomandibular Osteoarthritis
- Rat knee Osteoarthritis

Mouse Paw Edema induced by Carrageenan

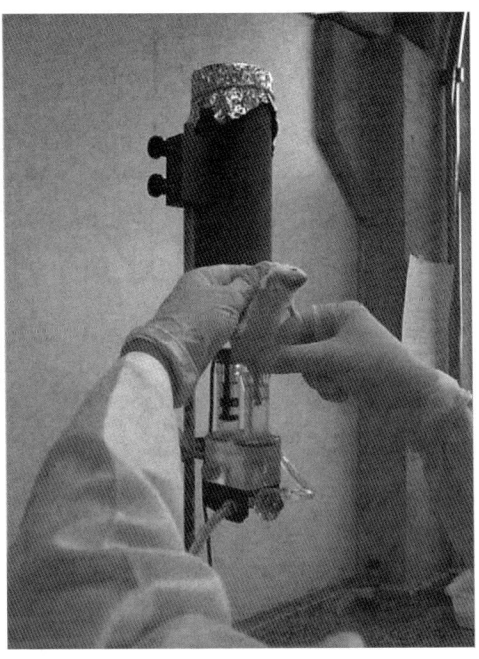

Effects of different protocol doses of low power
gallium aluminum arsenate (Ga Al As) laser radiation (650 nm)
on carrageenan induced rat paw ooedema

R. Albertini [a], F.S.C. Aimbire [a], F.I. Correa [b], W. Ribeiro [b], J.C. Cogo [b], E. Antunes [c], S.A. Teixeira [c], G. De Nucci [c], H.C. Castro-Faria-Neto [d], R.A. Zángaro [a], R.A.B. Lopes-Martins [b,*]

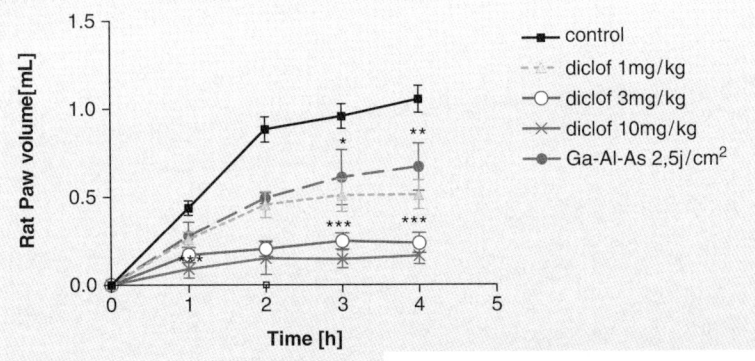

Journal of Photochemistry and Photobiology B: Biology 74 (2004) 101–107

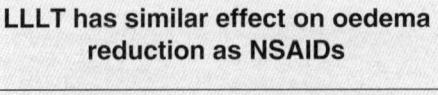

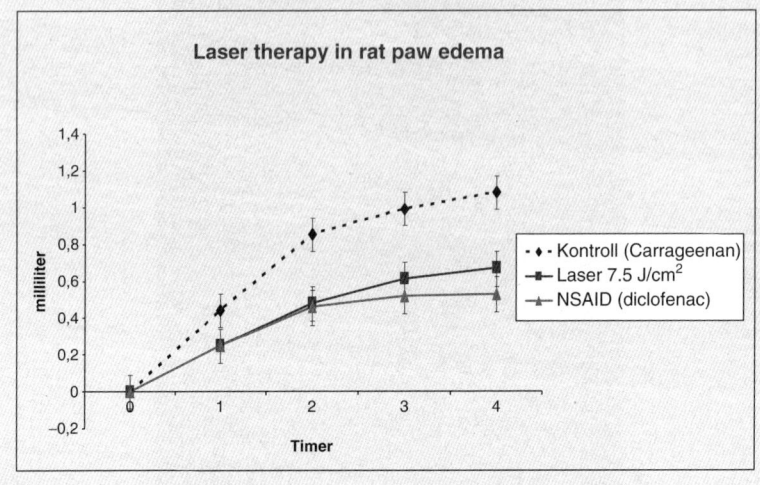

LLLT has similar effect on oedema reduction as NSAIDs

Albertini et al. 2004 Photochem PhotoBiol

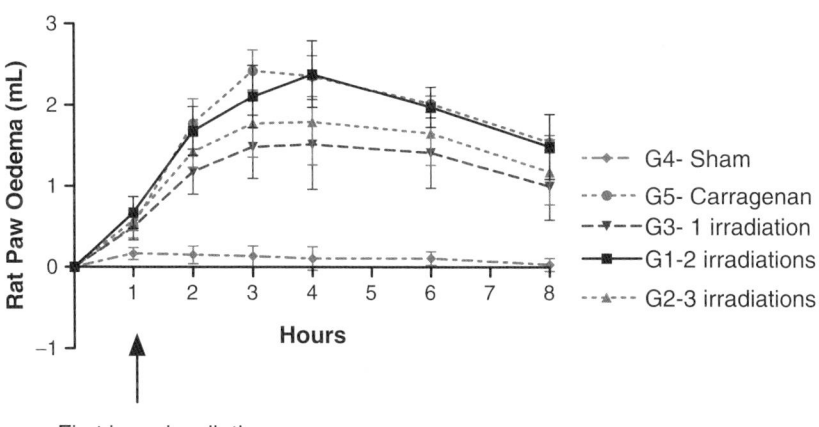

820nm 200 mW experiment in tendon inflammation

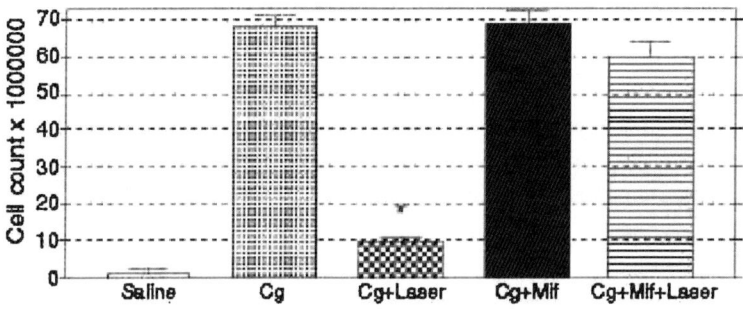

Steroids may block the modulation of inflammatory processes byLLLT

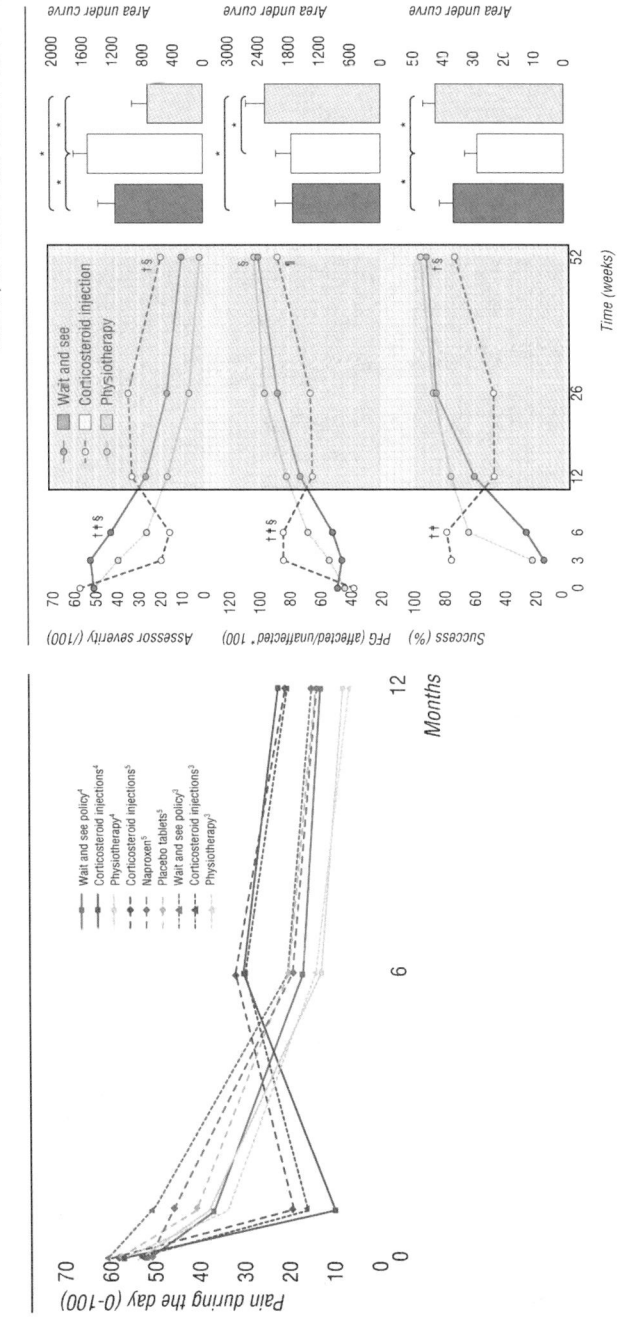

Long-term efficacy of steroid injections in tendinopathies

Severity of pain during one year of follow-up in three randomised controlled trials

Smidt, N. et al. BMJ 2006;333:927-928

Achilles tendinopathy

Results from our new RCT with eccentric exercises and LLLT

Pain intensity

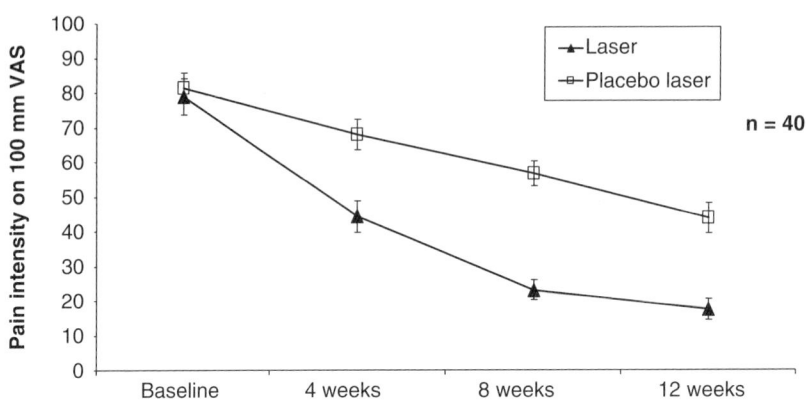

Spontaneous Effects of Low-Level Laser Therapy
(650 nm) in Acute Inflammatory Mouse
Pleurisy Induced by Carrageenan

RODRIGO ALVARO BRANDÃO LOPES-MARTINS, Ph.D.,[1] REGIANE ALBERTINI, M.Sc.,[1]
PATRÍCIA SARDINHA LEONARDO LOPES MARTINS, M.Sc.,[1] JAN MAGNUS BJORDAL, Ph.D.,[2]
and HUGO CAIRE CASTRO FARIA NETO, M.D., Ph.D.[3]

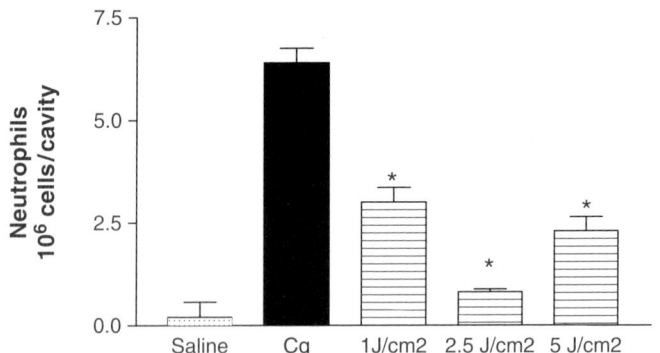

**Laser therapy gives a dose-dependent reduction of
neutrophil influx after < 4 hours**

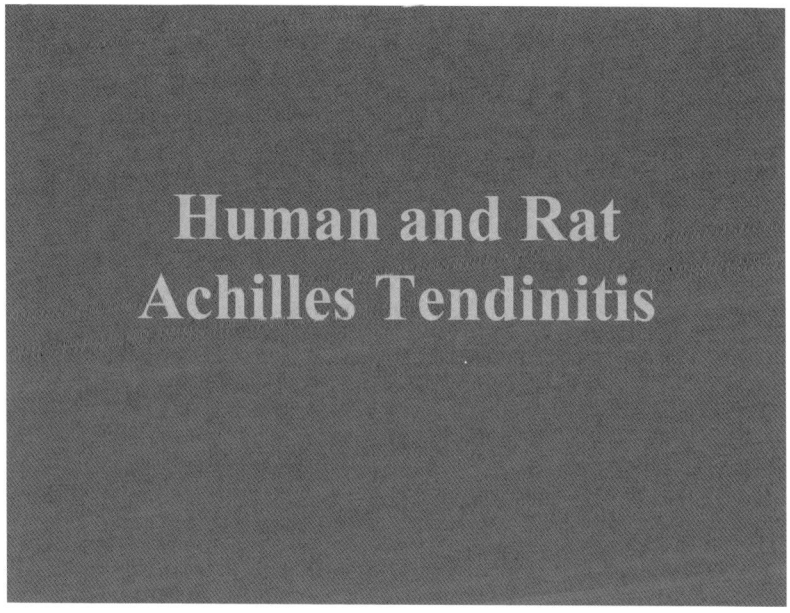

Human and Rat Achilles Tendinitis

Injecting the membrane

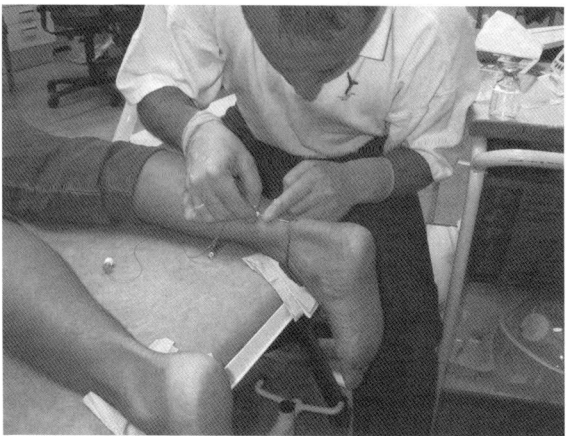

Membrane in position

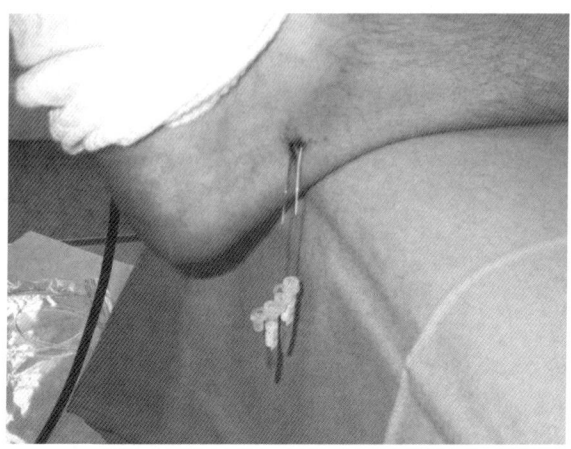

The set-up

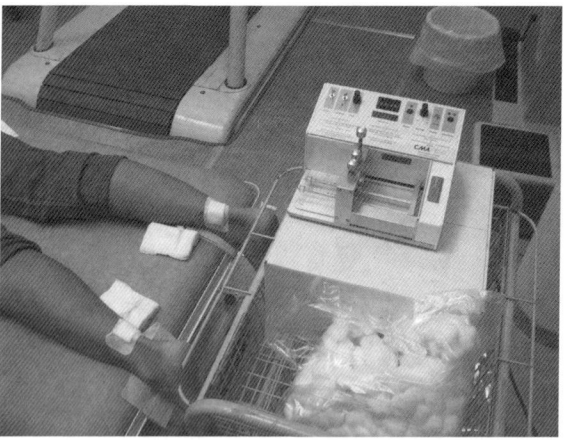

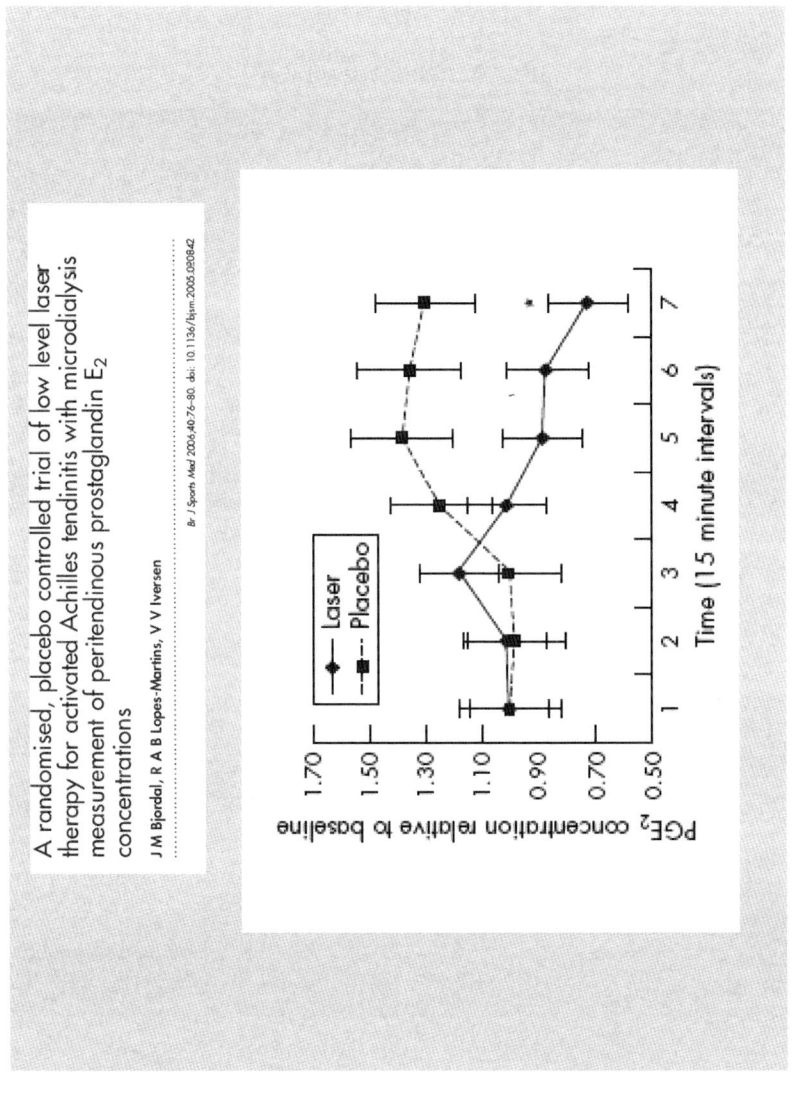

A randomised, placebo controlled trial of low level laser therapy for activated Achilles tendinitis with microdialysis measurement of peritendinous prostaglandin E_2 concentrations

J M Bjordal, R A B Lopes-Martins, V V Iversen

Br J Sports Med 2006;40:76-80. doi: 10.1136/bjsm.2005.080842

Rat Achilles Tendinitis

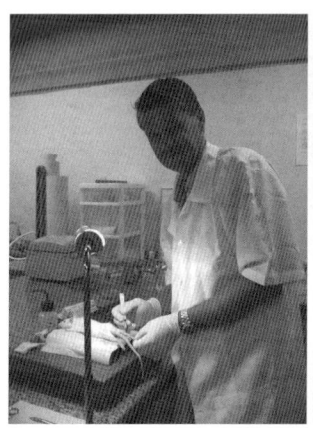 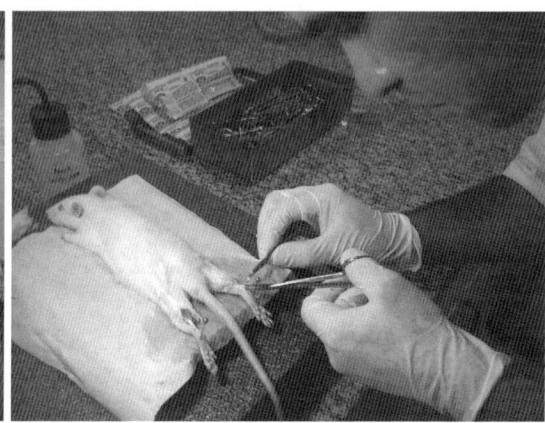

Experimental set-up Achilles tendinitis

All animals subjected to this procedure
(controls and placebo too)

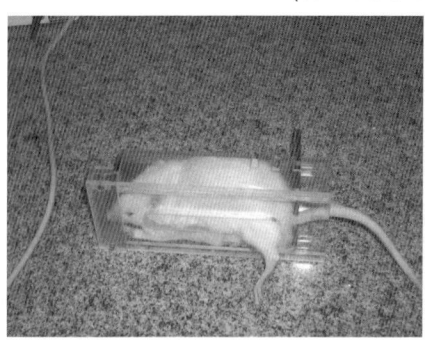 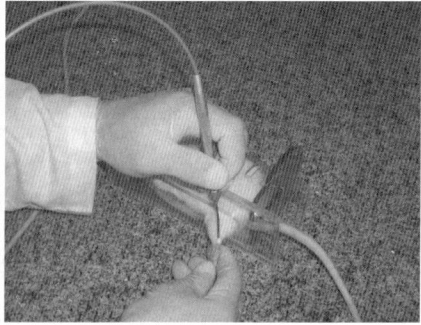

Rat Achilles Tendon Edema induced by Collagenase Injection

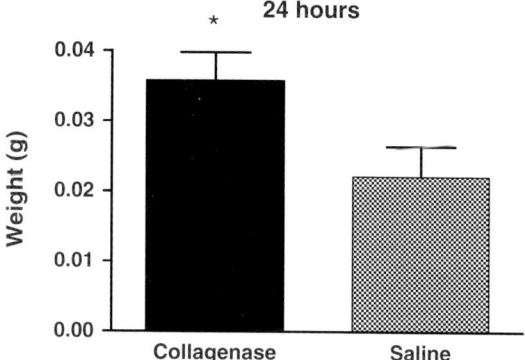

Rat Achilles Tendon Edema induced by Collagenase Injection

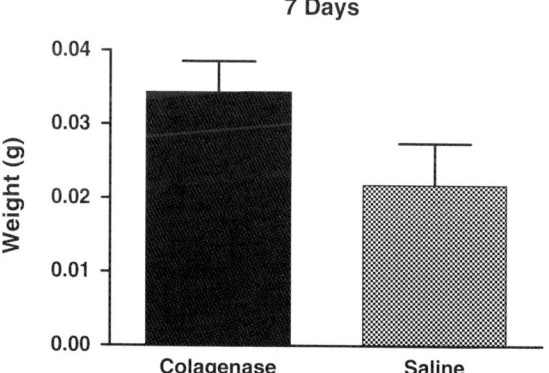

Peritendineous Tissue of Collagenase-induced Rat Achilles Tendinitis

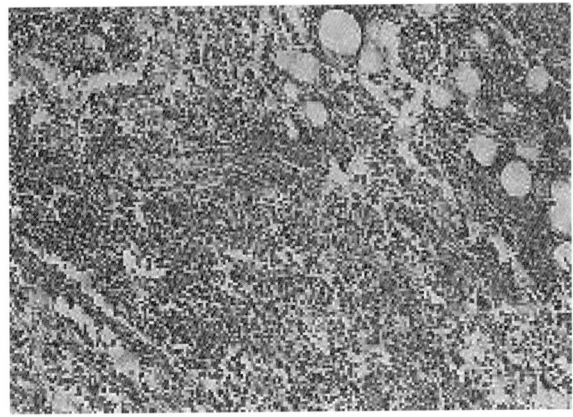

Non-irradiated

Irradiation 808 nm (100 mW) 6 Joules of Energy

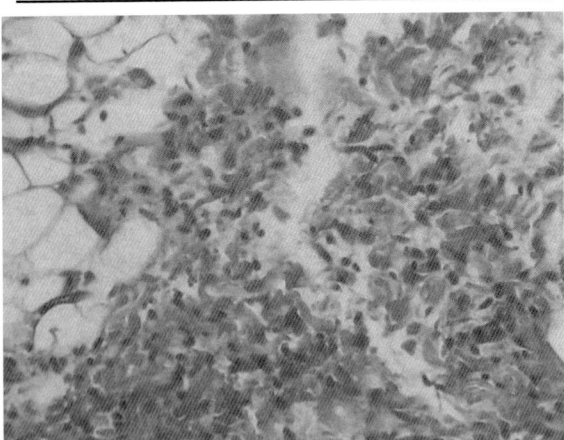

Time-Course of Protein Nitration in Collagenase-induced Rat Achilles Tendinites

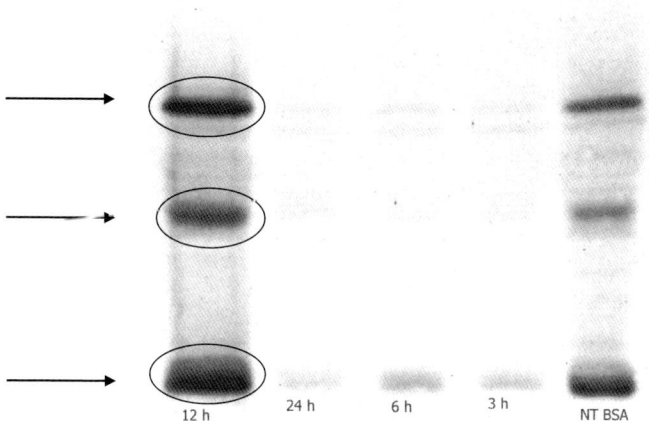

Western Blotting with Nitrothyrosine. The greatest expression was found 12 hours after injection of collagenase in rat Achilles tendons.

The Effect of LLLT on Nitrotyrosine Residues in Collagenase-induced Rat Achilles Tendinitis

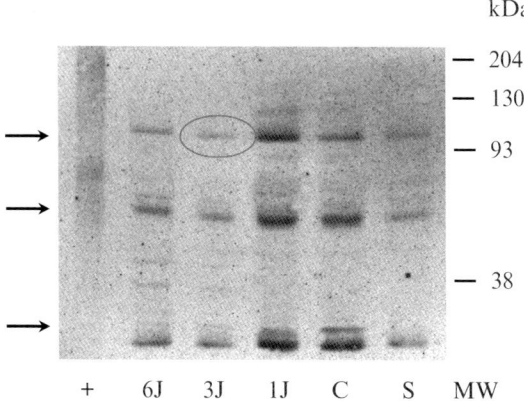

COX-1 Expression in Rat Achilles Tendon after Collagenase Injection and Laser Treatment

Western Blotting

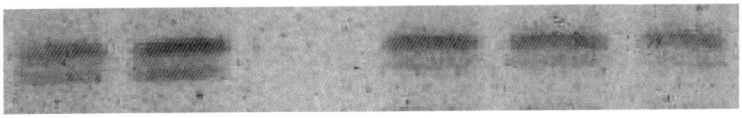

Colagenase   Colagenase   1Joule   3 Joules   6 Joules

Western Blotting da Enzima Ciclooxigenase-1 demonstrando a expressão da enzima em animais injetados com Colagenase somente, e colagenase + irradiação com Laser de 810 nm nas energias de 1, 3 e 6 Joules

COX-2 Expression in Rat Achilles Tendon after Collagenase Injection and Laser Treatment

Western Blotting

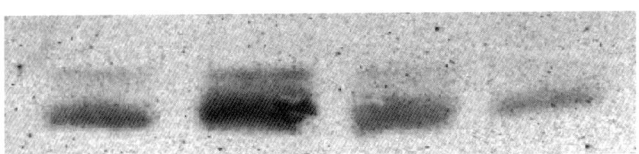

6 Joules   3 Joules   1 Joule   Colagenase

Western Blotting da Enzima Ciclooxigenase- 2 demonstrando a expressão da enzima em animais injetados com Colagenase somente, e colagenase + irradiação com Laser de 810 nm nas energias de 1, 3 e 6 Joules

COX-1 Derived PGE2 Production in Rat Achilles Tendon After Collagenase Injection

COX-1

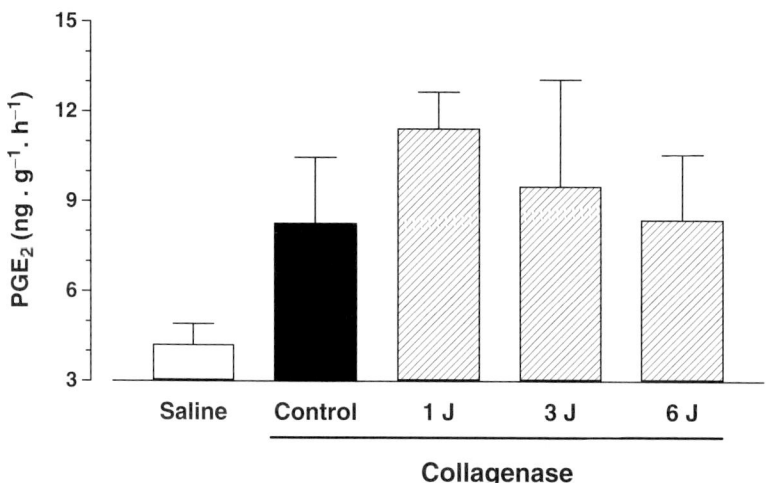

COX-2 derived PGE2 Production in Rat Achilles Tendon After Collagenase Injection

COX-2

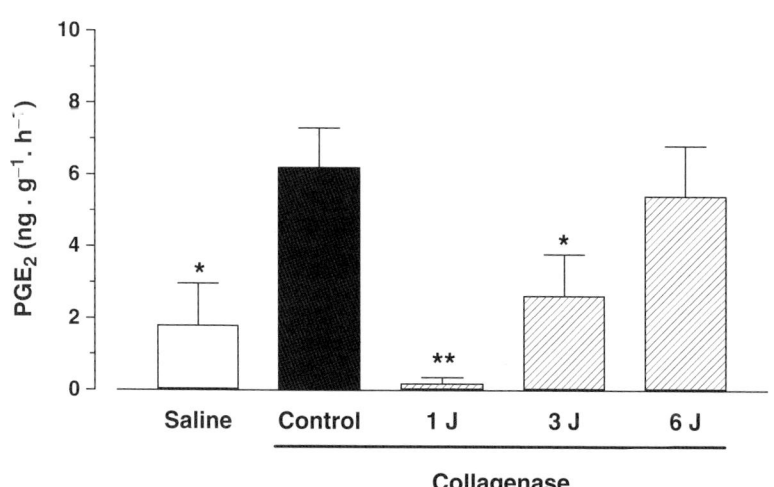

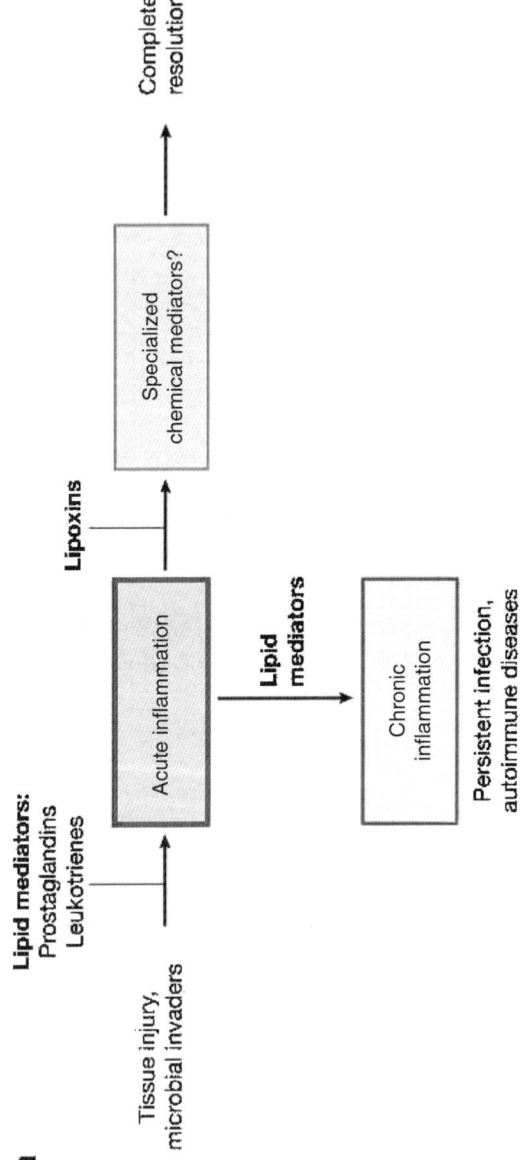

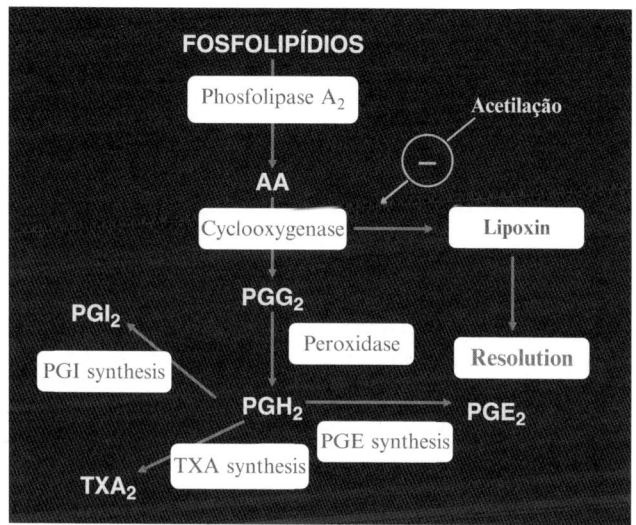

Therapeutic windows

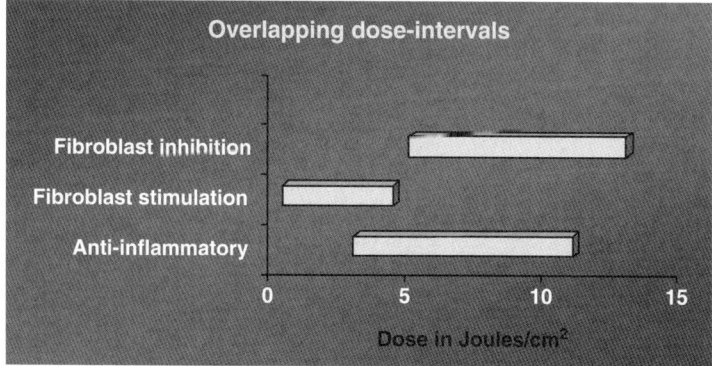

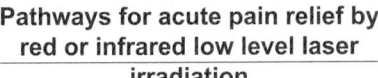

Pathways for acute pain relief by red or infrared low level laser irradiation

Local LLLT effects after first irradiation, enhanced effect by repeated irradiation

Effects on biochemical inflammatory markers

( ) Number of controlled laboratory trials verifying results

| Reduced PGE$_2$ levels (5) | Reduced TNF ? levels (2) | Reduced IL1? levels (3) | Reduced COX-2 mRNA levels (2) | Reduced plasminogen activator levels (1) |

Found in 21 out of 24 controlled laboratory trials

Effect not due to:
Endorphin and opioid receptors (2)

Effects on cells and soft tissue

| Reduced oedema formation (7) | Reduced hemorrhagic formation (5) | Reduced neutrophil cell influx (4) | Enhanced local microcirculation (4) |

Bjordal et al. 2006
Photomed Laser Surg(

Conclusions

- Laser Therapy is quite effective in reducing inflammation comparable to classical NSAIDS such as Sodium Diclofenac and Indomethacin;
- Laser Therapy:
  - reduces edema formation;
  - inhibits leukocyte (neutrophil) migration;
  - Decreases Protein Nitration (oxidative stress);
  - Inhibits PGE2 production;

Appendix 6

A Systematic Review of Post-operative Pain Relief by Low Level Laser Therapy (LLLT) After Third Molar and Endodontic Surgery (Slides Only)

Jan M. Bjordal

J.M. Bjordal
Bergen University College & University of Bergen

R. Waynant and D.B. Tata (eds.), *Proceedings of Light-Activated Tissue Regeneration and Therapy Conference.*
© Springer Science + Business Media, LLC 2008

EDITORIALS

COX-2 Inhibitors, Other NSAIDs, and Cardiovascular Risk
The Seduction of Common Sense

David J. Graham, MD, MPH

" . . . (A) long habit of not thinking a thing wrong, gives it a superficial appearance of being right, and raises at first a formidable outcry in defence of custom. But the tumult soon subsides. Time makes more converts than reason."

Thomas Paine, Common Sense, 1776

Analysis of Cochrane reviews of randomized placebo - controlled trials with pain therapy

Physical agents patients n = 5482
Drug patients n = 6284

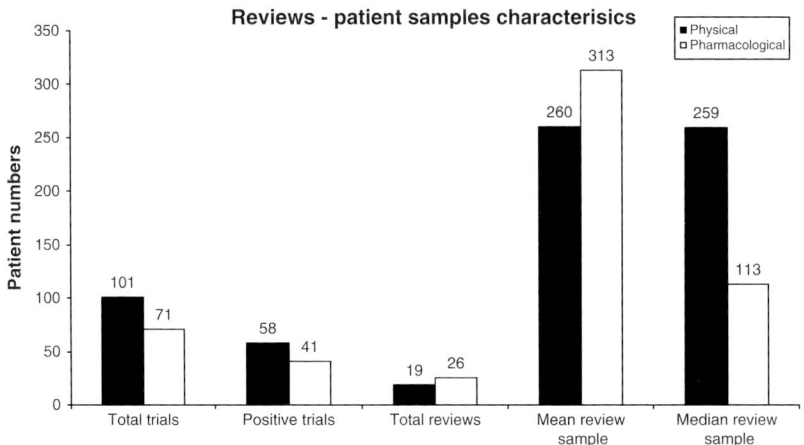

Analysis of Cochrane reviews of randomized placebo-controlled trials with pain therapy

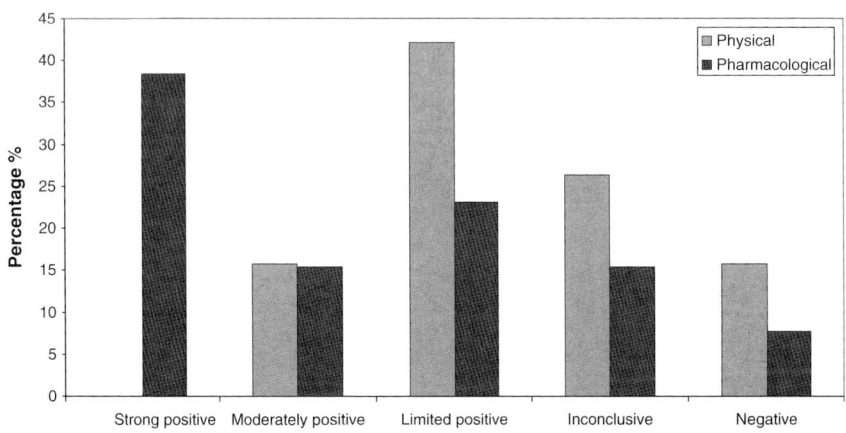

Possible bias by funding sources

	For-profit funding addressed in reviews	For-profit funding found in trials
Drugs	7/26	22/24**
Physical agents	3/19	5/28

The odds ratio for drug trials being industry-funded
was significantly higher than for trials with physical
agents (OR = 44 [95%CI: 8 to 252])

Why are funding sources important for trial conclusions?

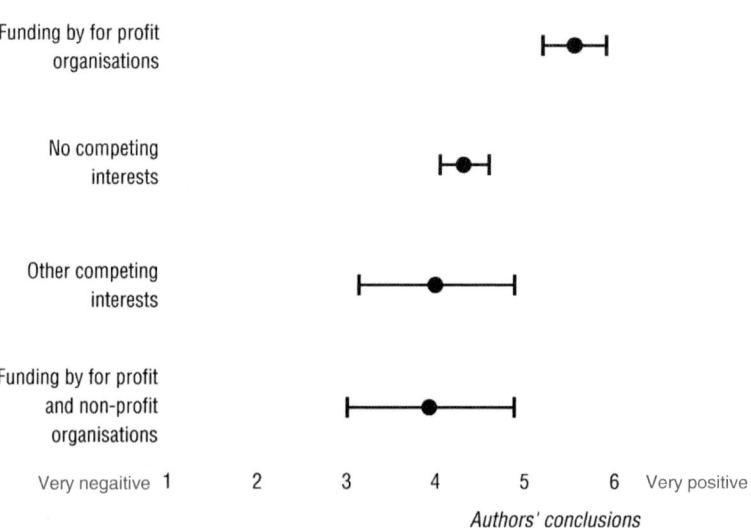

Some basics about pain therapy and meta-analyses

1. Chronic pain patients report that "important pain relief" is on average 30% or 15-20 mm on 100 mm VAS (Tubach et al, 2004, Farrar et al. 2001)

2. The least perceptible pain relief for knee osteoarthritis is 10 mm on 100 mm VAS. Ehrich et al. 1999

3. In this presentation, plots to the right of the middle line means a positive effect, plots crossing the middle line means a non-significant difference over placebo, plots to the left are significantly poorer than placebo

Patient selection criteria inflates effect estimate in many oral NSAID trials

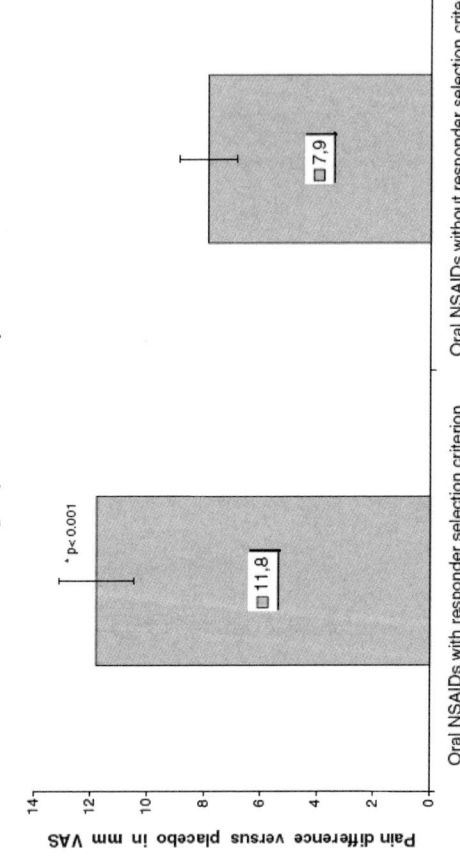

Efficacy within 4 weeks in oral NSAIDs trials according to patient selection procedure

Fig. 3. The result of the sub-group analysis for oral NSAIDs is shown. Trials were sub-grouped in one subgroup which recruited patients from regular NSAID-users and only included those who experienced an increase in pain intensity of at least 15 mm on VAS after discontinuing medication in the pre-trial wash-out period and had baseline pain over 40 mm on VAS. We call this the responder selection criterion, as this patient selection procedure ensures that only known NSAID-responders are allowed to participate. The second subgroup only used the inclusion criterion of a minimum baseline pain intensity of 40 mm on VAS.

Pharmacological interventions for knee osteoarthritis

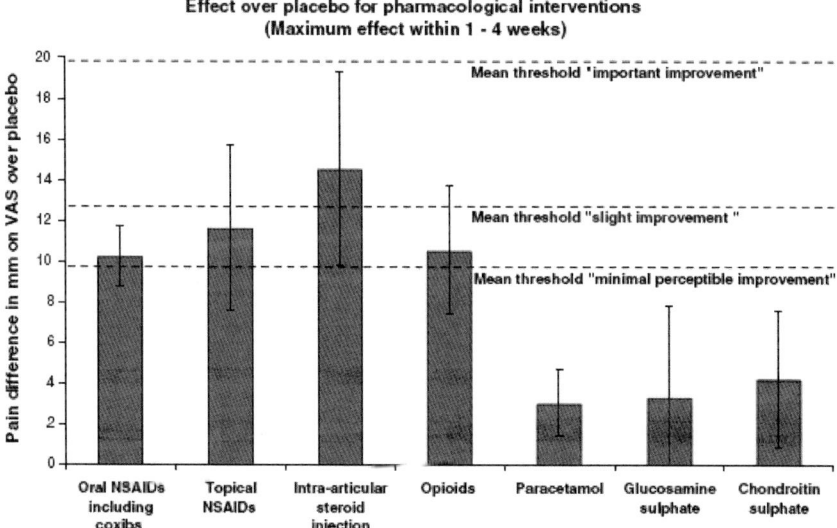

**Effect over placebo for pharmacological interventions
(Maximum effect within 1 - 4 weeks)**

☐ **1:** <u>BMC Musculoskelet Disord.</u> 2007 Jun 22;8(1):51 [Epub ahead

Short-term efficacy of physical interventions in osteoarthritic knee pain. A systematic review and meta-analysis of randomised placebo-controlled trials.

Bjordal JM, Johnson MI, Lopes-Martins RA, Bogen B, Chow R, Ljunggren AE.

ABSTRACT: BACKGROUND: Treatment efficacy of physical agents in osteoarthritis of the knee (OAK) pain has been largely unknown, and this systematic review was aimed at assessing their short-term efficacies for pain relief. Aims and methods: Systematic review with meta-analysis of efficacy within 1-4 weeks and 5-12 weeks. RESULTS: 36 randomised placebo-controlled trials (RCTs) were identified with 2434 patients where 1391 patients received active treatment. 33 trials satisfied three or more out of five methodological criteria (Jadad scale). The patient sample had a mean age of 65.1 years and mean baseline pain of 62.9 mm on a 100 mm visual analogue scale (VAS). Within 4 weeks of the commencement of treatment manual acupuncture, static magnets and ultrasound therapies did not offer statistically significant short-term pain relief over placebo. Pulsed electromagnetic fields offered a small reduction in pain of 6.9 mm [95% CI: 2.2 to 11.6] (n=487). Transcutaneous electrical nerve stimulation (TENS, including interferential currents), electro-acupuncture (EA) and low level laser therapy (LLLT) offered clinically relevant pain relieving effects of 18.8 mm [95% CI: 9.6 to 28.1] (n=414), 21.9 mm [95% CI: 17.3 to 26.5] (n=73) and 17.7 mm[95% CI: 8.1 to 27.3] (n=343) on VAS respectively versus placebo control. In a subgroup analysis of trials with assumed optimal doses, short-term efficacy increased to 22.2 mm [95% CI: 18.1 to 26.3] for TENS, and 24.2 mm [95% CI: 17.3 to 31.3] for LLLT on VAS. Follow-up data up to 12 weeks were sparse, but positive effects seemed to persist for at least 4 weeks after the course of LLLT, EA and TENS treatment was stopped. CONCLUSION: TENS, EA and LLLT administered with optimal doses in an intensive 2-4 week treatment regimen, seem to offer clinically relevant short-term pain relief for OAK.

PMID: 17587446 [PubMed - as supplied by publisher]

Short-term effcacy of physical interventions in knee OA

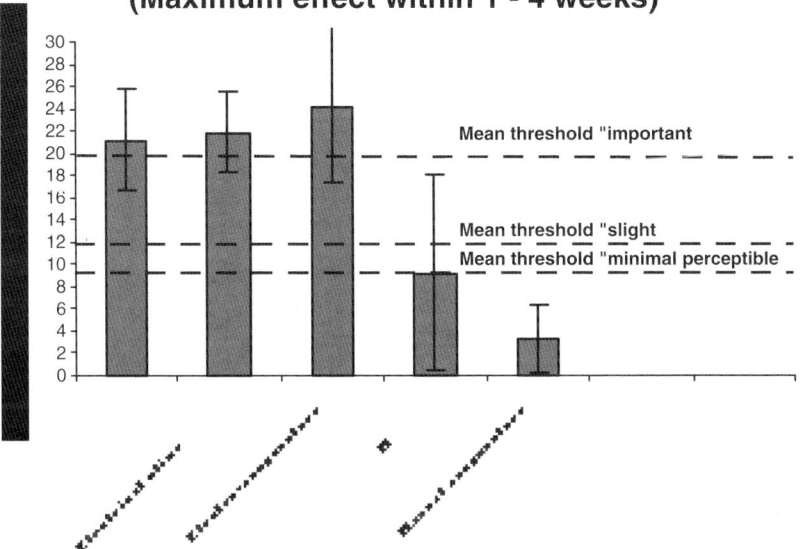

**Effect over placebo for physical interventions
(Maximum effect within 1 - 4 weeks)**

Mean threshold "important

Mean threshold "slight
Mean threshold "minimal perceptible

Interventions for knee osteoarthritis

Pain relieving effect over placebo after 4 weeks

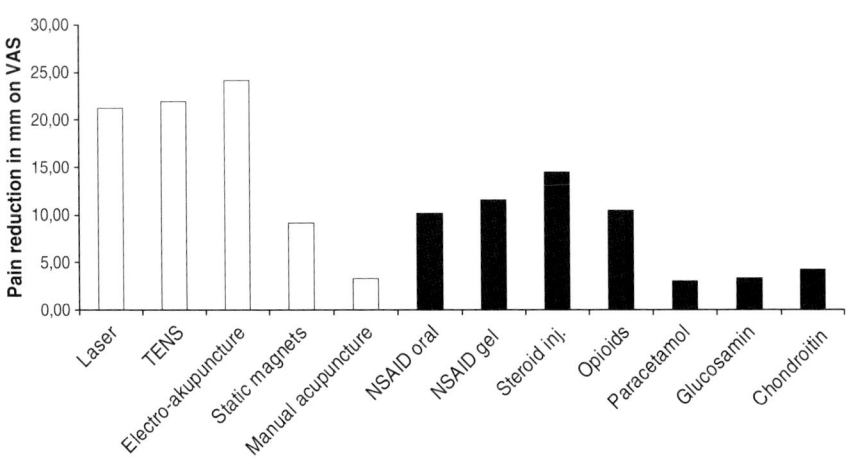

Interventions for knee osteoarthritis

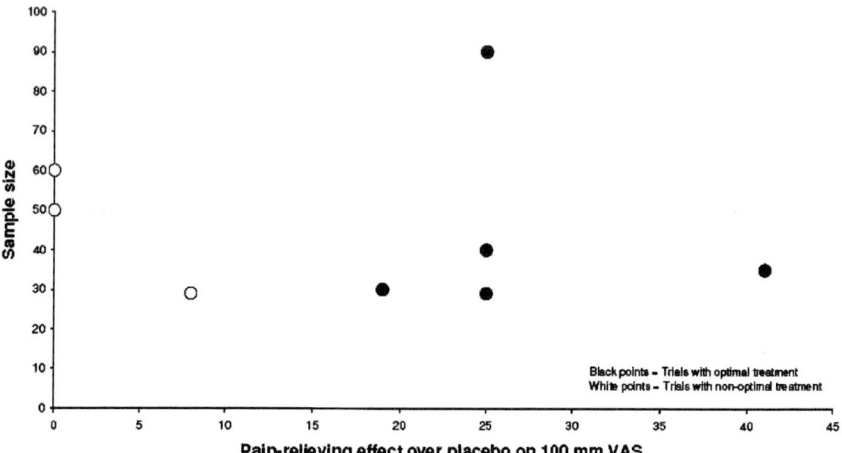

Interventions for knee osteoarthritis

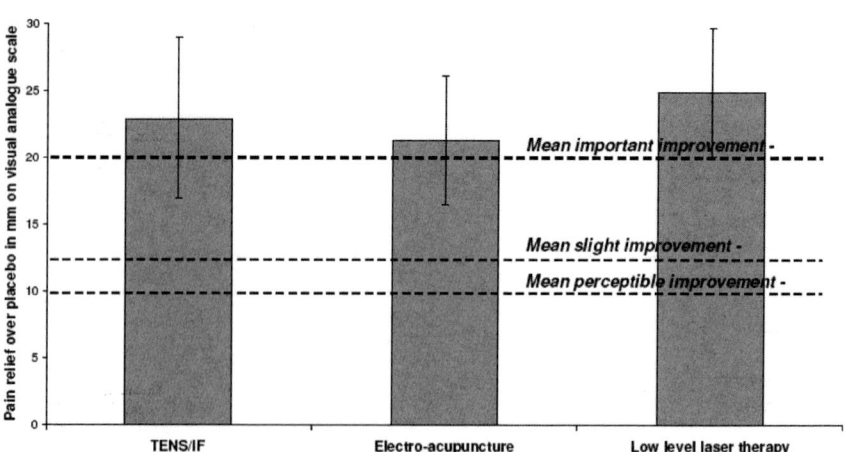

Interventions for low back pain

NSAID

LLLT

n = 535

n = 318

6 RCTs with positive results
7 RCT not significantly different from placebo

4 RCTs with consistently positive results
1 RCT not significantly different from exercise therapy
1 RCT with non-significant results used a dose which is 3%
of the minimal recommended dose listed by World
Association for Laser Therapy

Pain VAS post therapy	0.54 SMD (3 RCTs) n.s.
Disability post therapy	No data
Improvement post therapy	1.2 RR (6 RCTs)
Follow up VAS, 1 month	No data
Improvement follow-up 1 month	No data
Improvement 6 months	No data

Pain VAS post therapy	11.7 WMD (2 RCTs)
Disability post therapy	0.58 SMD (2 RCTs)
Improvement rate	2.2 RR (3 RCTs)
Follow up VAS, 1 month	15.0 WMD (1 RCT)
Improvement follow-up 1 month	2.6 RR (3 RCTs)
Improvement 6 months	3.2 RR (2 RCTs)

EBM and electrophysical agents research

Process of obtained approvals for clinical use of drug and EPA therapy

Drug therapy

Identify chemical agent and test if desired biological effect can be achieved in cell cultures

Test for adverse effects and beneficial biological effect can be achieved in animals.

Narrow optimal dose range determined for animals through multiple controlled trials

Test clinical effects on humans through several controlled clinical trials

Test of dose for optimal effect on the most suitable diagnosis in randomised placebo-controlled trials on humans

Approval for clinical use obtained

Narrow optimal dose range determined

EPA therapy

Safe dose range determined, uncontrolled testing on patients

Scattered clinical evidence of effect in controlled studies
Authorative persons' recommendations

Approval for clinical use obtained

Optimal dose range uncertain

The most pressing challenge is to identify and establish consensus for optimal laser doses

Available laser trials are different from drug trials in the sense that drug trials are performed with known optimal doses while several laser trials use doses which are unlikely to cause any biological effects. In systematic reviews without assessment of dose validity, these non-optimal dose trials will confuse the results and possibly erase positive effects from optimal laser dose studies. Even the Cochrane Library have failed to address this important issue in their laser reviews.

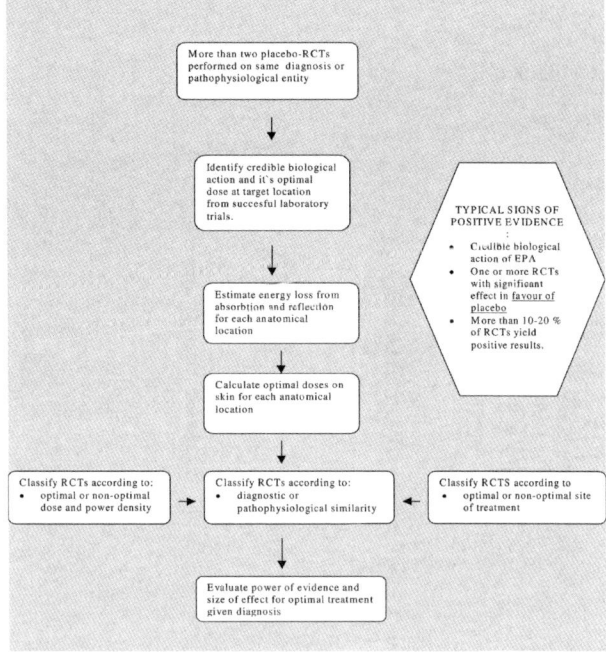

**Our model for
investigating
dose response
patterns for
empirical
therapies**

More than two placebo-RCTs performed on same diagnosis or pathophysiological entity

Identify credible biological action and it's optimal dose at target location from succesful laboratory trials.

Estimate energy loss from absorbtion and reflection for each anatomical location

Calculate optimal doses on skin for each anatomical location

Classify RCTs according to:
- optimal or non-optimal dose and power density

Classify RCTs according to:
- diagnostic or pathophysiological similarity

Classify RCTS according to
- optimal or non-optimal site of treatment

Evaluate power of evidence and size of effect for optimal treatment given diagnosis

TYPICAL SIGNS OF POSITIVE EVIDENCE :
- Credible biological action of EPA
- One or more RCTs with significant effect in favour of placebo
- More than 10-20 % of RCTs yield positive results.

Laser in Achilles inflammation

Dose: 141 Joules/cm^2
Energy Dose: 200 mW for 30 s = 6 Joules
Power density: 4.7 W/cm^2

Region of the main collagen fibers of the tendon.
No signs of inflammation nor oedema. 400 X

Dose: 282 Joules/cm^2
Energy Dose: 200 mW for 60 s = 12 Joules
Power density: 4.7 W/cm^2

Region of the main collagen fibers of the tendon.
Discrete signs of inflammation and oedema. 400 X

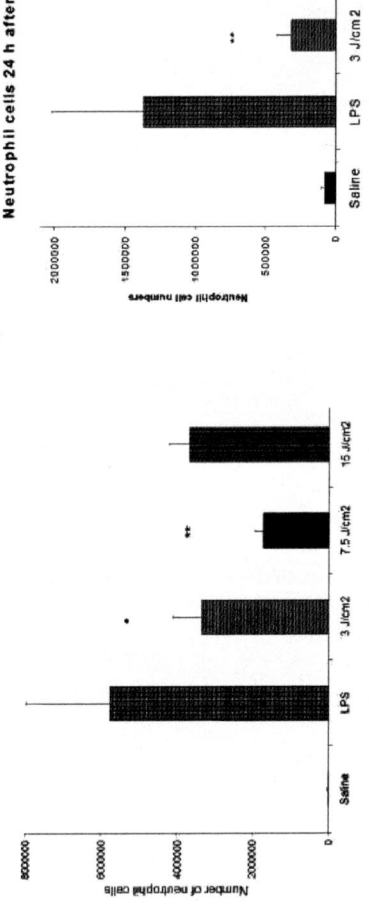

Laser therapy mechanisms

LLLT reduces neutrophil cell influx in peritonitis

LPS - from *E.coli* bacteria

6 hours

24 hours

$\lambda = 904$ nm, 4 mW, 50 mW/cm^2, 30 -152 s

Laser in Achilles inflammation

GAPDH and COX expression reduced by 30-45 s laser irradiation

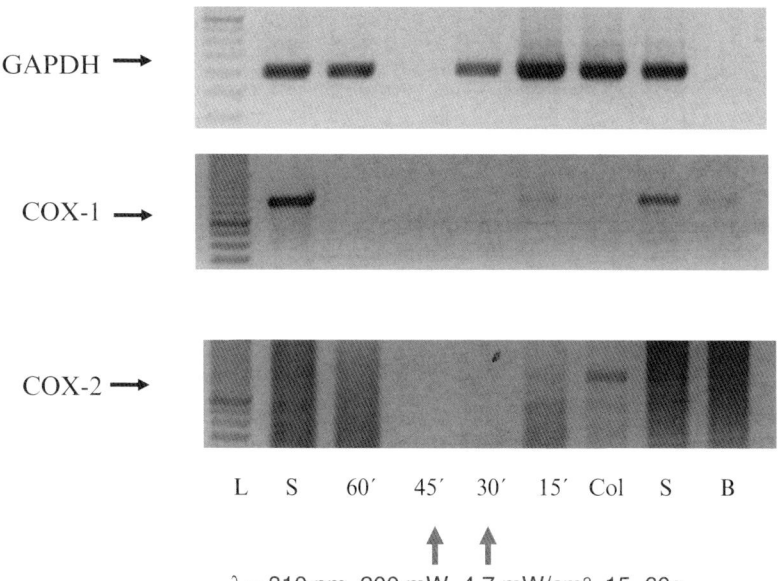

$\lambda = 810$ nm, 200 mW, 4.7 mW/cm², 15-60s

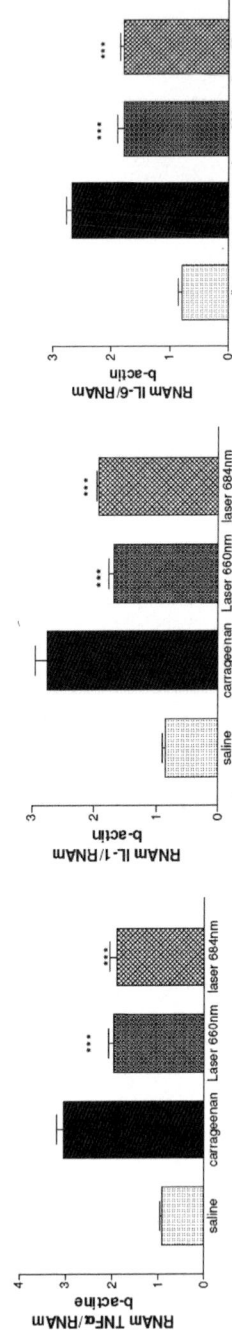

Carrageenan-induced rat paw edema

mRNA-expression measured at 3 hours post-carrageenan injections

λ = 660-684 nm, 30 mW, 38 mW/cm^2, 196 s

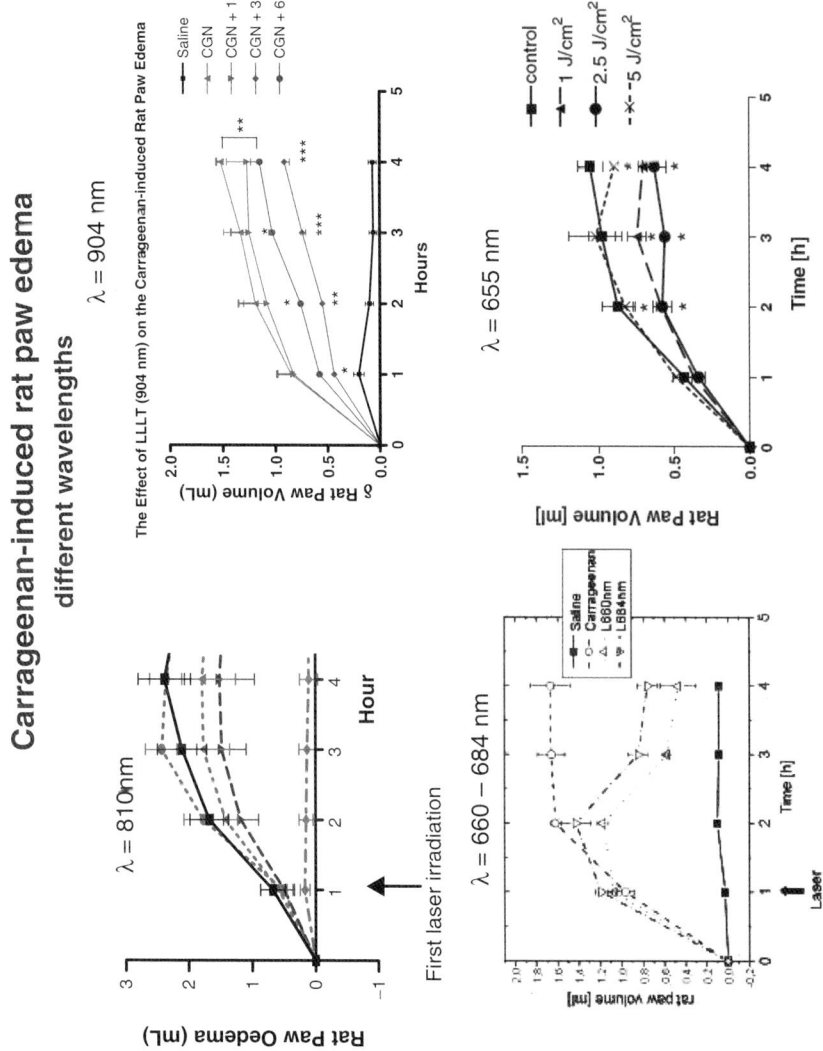

Carrageenan-induced rat paw edema
different wavelengths

Time points of LLLT effects in the inflammatory process

	Carrageenan	LPS	LPS	LPS/Ovalbumin
				Leukocyte cells
				Neutrophil cells
Injury	Oedema	Neutrophil cells	Neutrophil cells	Myeloperoxidase
PGE^2	Neutrophil cells			Hemorhagic index

| 1 | 4 | 6 | 12 | 24 | HOURS |

$\lambda = 655, 680, 810$ nm, 904 nm ; $2.5 - 200$ mW ; $31 - 4,700$ mW/cm2, $15 - 150$s

Laser therapy in acute inflammation

First author, year, model	Inflammatory agent	Laser type, mean output power (mW)	Power density (mW/cm²)	Dose (Joules/cm²)
Campana 1993 arthritis animal	Urate crystals	633nm, 5mW	6	0.72
Honmura 1992rat paw edema	Carrageenan	830nm, 60mW	32	9.6
Honmura 1993 rat paw edema	Carrageenan	830nm, 60mW	32	9.6
Shimizu 1995 ligament cells	Mechanically stretched	830nm, 30mW	12	2.3 – 7.4
Ozawa 1997 ligament cells	Mechanically stretched	830 nm, 700mW	6 – 13	3.9
Sattayut 1999 myofibroblast cells	Carrageenan	820nm, 200mW	22	4-19
Campana 1999arthritis animal	Urate crystals	633nm, 30mW	30	8
Sakurai 2001 fibroblast cells	Lipoly-saccharide	830nm, 700mW	21	1.9 – 6.3
Nomura 2001 fibroblast cells	Lipoly-saccharide	830nm, 50 mW	6-13	4 – 7.9
Dourado 2004 mice	Snake venom	904 nm 50 mW	90	2.8
Albertini 2004 rat paw edema	Carrageenan	660 nm 2.5 mW	31	7.5
Ferreira 2004 Rat paw edema	Carrageenan PGE₂	633 nm 12 mW	171	7.5
Pessoa 2004 rat skin wound	Excised skin flap 0.5 cm²	904 nm 2.8 mW	5	0.66
Lopes-Martin 2005 mice pleurisy	Carrageenan	660 nm 2.5 mW	31	7.5
Aimbire 2005 Airway hyperreactivity	Lipoly-saccharide	660 nm 2.5 mW	31	7.5
Aimbire 2005 rat lung injury	Bovine serum	660 nm 2.5 mW	31	7.5
Median results		Median 830 nm [633-904nm]	Median 31mW/cm² [5-50]	Median 7.5 J/cm² [0.7- 19]

Anti-inflammatory effects

Pathways for acute pain relief by red or infrared low level laser irradiation

Local LLLT effects after first irradiation, enhanced effect by repeated irradiation

Effects on biochemical inflammatory markers

( ) Number of controlled laboratory trials verifying results

| Reduced PGE₂ levels (5) | Reduced TNF α levels (2) | Reduced IL1β levels (3) | Reduced COX-2 mRNA levels (2) | Reduced plasminogen activator levels (1) |

Effect not due to: Endorphin and opioid receptors (2)

Effects on cells and soft tissue

| Reduced oedema formation (7) | Reduced hemorrhagic formation (5) | Reduced neutrophil cell influx (4) | Enhanced local microcirculation (4) |

Found in 21 out of 24 controlled laboratory trials

Therapeutic windows

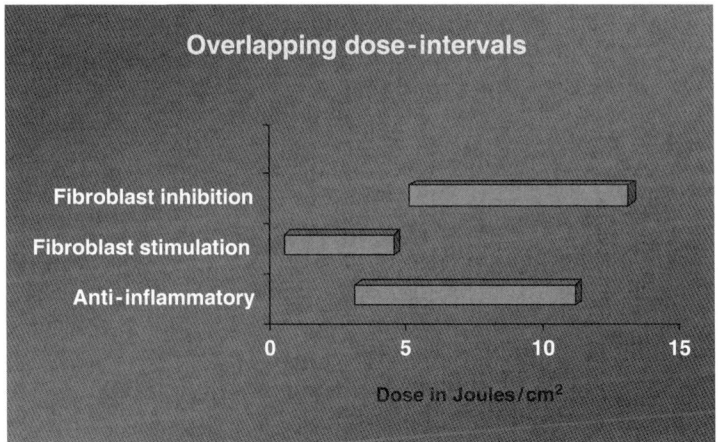

Lessons from *animal studies*

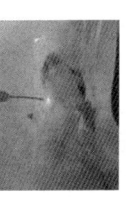

1. LLLT is a COX-inhibitor which can reduce formation of oedema, hematoma, leucocyte cell influx and levels of inflammatory markers

2. The anti-inflammatory effect is not wavelength-specific within the red and infra-red spectres (600 -1000nm)

3. The anti-inflammatory effect appears to be dose-dependent and most of the pathological area needs to be covered

4. Treatment can be commenced as early as 5-10 minutes after incisions are made

5. The time profile shows optimal effect at 1-6 hours, whereafter the effect subsides. A second treatment session could be commenced from 6-12 hours post-surgery.

6. Assumed optimal treatment for 820 nm after third molar extraction 3-6 Joules applied in one buccal point, one lingual point and 1 points over the incision, Total dose 9-18 Joules

Randomized placebo-controlled LLLT trials in acute post-operative dental pain

Systematic review

1. *Literature search for:* Randomised controlled trials

2. *Primary outcome measure:* Pain relief within first 24 hours after surgery

3. *Treatment:* LLLT irradiation with wavelengths 600-1000 nm and direct exposure of pathology

Randomized placebo-controlled LLLT trials in acute post-operative dental pain

Results

9 randomised placebo-controlled trials (RCTs)found

751 patients included

1 RCT excluded for not giving LLLT before day 1 (Payer et al 2005)

1 RCT did not give pain scores on VAS or NRS (Necker et al 2001)

2 comparisons made with NSAID in control groups (Masse et al 1994, Markovic et al. 2006)

Methodological quality acceptable (mean 3.1 points on Jadad 5 point scale)

150 sec, 50mW, Dose 7.5 J

LLLT in third molar surgery

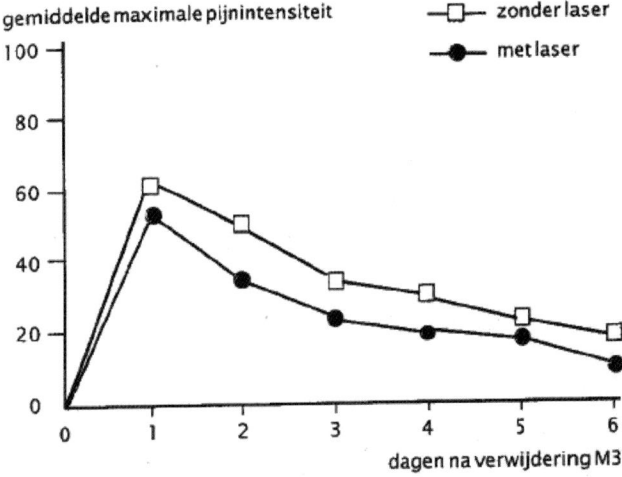

gemiddelde maximale pijnintensiteit   ⎯□⎯ zonder laser
                                           ⎯●⎯ met laser

dagen na verwijdering M3

830 nm, 30mW, Dose 0.9 J, 33 sec   Braams et al.
Paracetamol 0.5 g as needed   Ned Tijdschr Tandheelkd 101 (1994) maart

Randomized placebo-controlled LLLT trials in acute post-operative dental pain

First author year	Patient numbers	Wavelength nm	Laser output (mW)	Spot size (cm2)	Dose (J)	Irradiation time (s)	Pain relief (%)
Braams 94	43	830	30	0.06	6	200	18
Carillo 90	100	632	8	0.03	0,24	30	3
Fernando 93	64	830	30	0.06	4	132	8
Kreisler 04	52	810	50	0.5	7,5	150	43
Markovic 06	90	637	50	0.5	4	?	28
Masse 93	28	632, 904	4	?	0,36	150	5
Necker 01	210	810	36	0.45	5	150	37
Ong 01	48	830	30	0.06	6	200	45
Payer 05	49	680	75	0.5	3	?	Not measured
Roynesdal 93	50	830	40	0.06	6	150	20
Taube 90	17	632	8	?	0,96	120	16

Randomized placebo-controlled LLLT
trials in acute post-operative dental pain

Pain on VAS all included trials

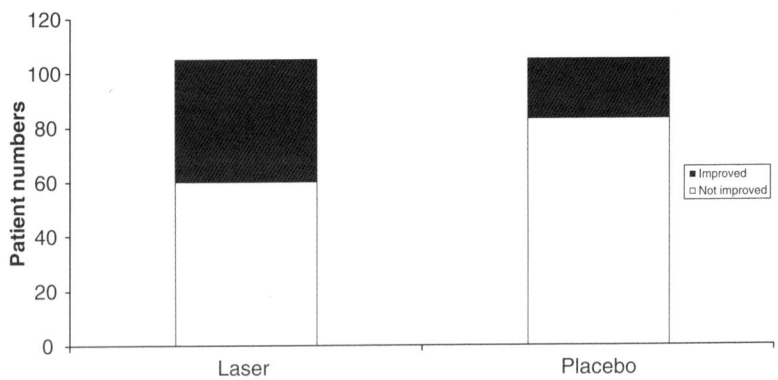

Review:	Laser acute pain
Comparison:	01 Pain on VAS day 1
Outcome:	01 Pain on VAS

Study or sub-category	N	Control Mean (SD)	N	Laser Mean (SD)	WMD (fixed) 95% CI	Weight %	WMD (fixed) 95% CI
Carillo	50	26.00 (22.00)	50	24.00 (22.00)		13.40	2.00 [-6.62, 10.62]
Taube	17	30.10 (27.00)	17	26.90 (27.50)		2.97	3.20 [-15.12, 21.52]
Masse	28	12.00 (26.00)	28	14.00 (26.00)		5.37	-2.00 [-15.62, 11.62]
Roynesdal	26	37.00 (20.60)	26	31.00 (18.60)		8.42	6.00 [-4.88, 16.88]
Neckel	105	36.40 (18.30)	105	26.90 (18.70)		39.79	9.50 [4.50, 14.50]
Ong	48	31.00 (26.60)	48	17.00 (26.60)		8.80	14.00 [3.36, 24.64]
Kreisler	26	23.00 (23.60)	26	10.00 (23.60)		6.05	13.00 [0.17, 25.83]
Markic	30	29.40 (16.00)	30	21.20 (16.00)		15.20	8.20 [0.10, 16.30]
Total (95% CI)	**329**		**329**			**100.00**	**7.81 [4.65, 10.96]**

Test for heterogeneity: Chi² = 6.46, df = 7 (P = 0.49), I² = 0%
Test for overall effect: Z = 4.85 (P < 0.00001)

-100   -50   0   50   100
Favours control   Favours treatment

Randomized placebo-controlled LLLT trials
in acute post-operative dental pain

Patients improved after LLLT

■ Improved
□ Not improved

Relative Risk for improvement: 2.1 (95% CI 1.3 to 3.2) p=0.001

Randomized placebo-controlled LLLT trials in acute post-operative dental pain

Low dose < 1 Joules

Review: Laser acute pain
Comparison: 01 Pain on VAS day 1
Outcome: 03 Pain on VAS day 1 low dose

Study or sub-category	Control N	Mean (SD)	Laser N	Mean (SD)	WMD (fixed) 95% CI	Weight %	WMD (fixed) 95% CI
Carillo	50	26.00(22.00)	50	24.00(22.00)		61.63	2.00 [-6.62, 10.62]
Taube	17	30.10(27.00)	17	26.90(27.60)		13.66	3.20 [-15.12, 21.52]
Masse	28	12.00(26.00)	28	14.00(26.00)		24.71	-2.00 [-15.62, 11.62]
Total (95% CI)	95		95			100.00	1.18 [-5.59, 7.95]

Test for heterogeneity: Chi² = 0.29, df = 2 (P = 0.86), I² = 0%
Test for overall effect: Z = 0.34 (P = 0.73)

-10   -5   0   5   10
Favours Control   Favours Laser

Not significantly different from placebo

Randomized placebo-controlled LLLT trials in acute post-operative dental pain

Pain on VAS with optimal doses 2-8 Joules

Review: Laser acute pain
Comparison: 01 Pain on VAS day 1
Outcome: 02 Pain on VAS day 1 high dose

Study or sub-category	Control N	Mean (SD)	Laser N	Mean (SD)	WMD (fixed) 95% CI	Weight %	WMD (fixed) 95% CI
Reynesdal	25	37.00(18.60)	25	31.00(18.30)		12.00	6.00 [-4.23, 16.23]
Neckel	105	36.40(18.30)	105	26.90(18.70)		50.13	9.50 [4.50, 14.50]
Ong	48	31.00(26.60)	48	17.00(26.60)		11.09	14.00 [3.96, 24.64]
Kreisler	26	23.00(23.60)	26	10.00(23.60)		7.63	13.00 [0.17, 25.83]
Markic	30	29.40(16.00)	30	21.20(16.00)		19.15	8.20 [0.10, 16.30]
Total (95% CI)	234		234			100.00	9.60 [6.05, 13.14]

Test for heterogeneity: Chi² = 1.52, df = 4 (P = 0.82), I² = 0%
Test for overall effect: Z = 5.31 (P < 0.00001)

-100   -50   0   50   100
Favours control   Favours treatment

LLLT significantly better than placebo

Randomized placebo-controlled LLLT
trials in acute post-operative dental pain

Dose and pain relief (%) plot

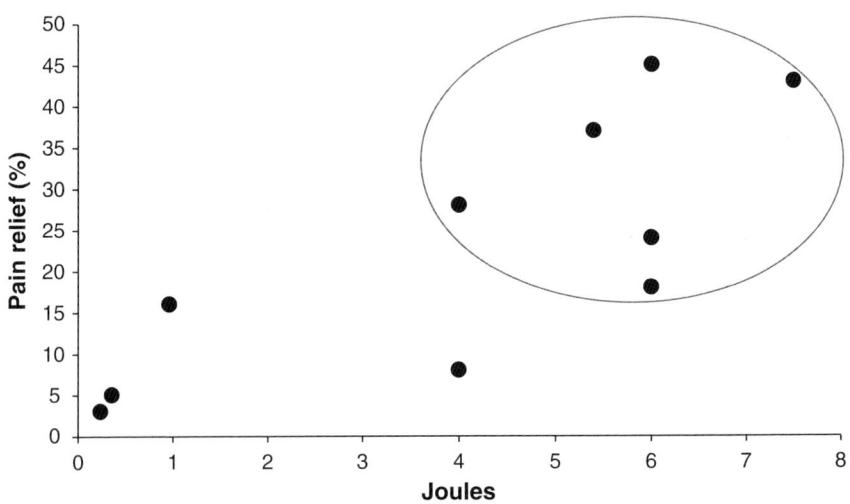

LLLT near-optimal doses 4-8 Joules (810-830 nm)

Randomized placebo-controlled LLLT trials in acute post-operative dental pain

Conclusions

1. There is moderate evidence that LLLT is more effective than placebo control in relieving pain after third molar and endodontic surgery

2. There is weak evidence that adequate doses of LLLT offers equal effectiveness as NSAIDs.

3. The potential of LLLT effectiveness is probably not fully explored as the published trials appear to be under-dosed when compared to the assumed optimal LLLT dose and application regimen

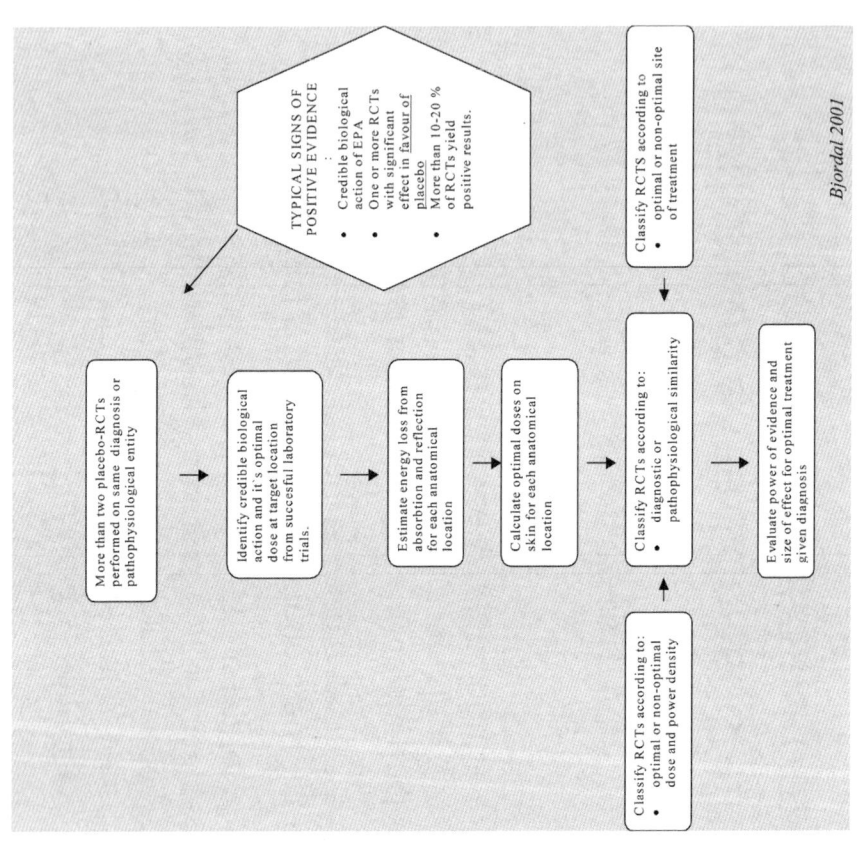

Identification of dose-response patterns

Identification of success factors

Identification of failure factors

More than two placebo-RCTs performed on same diagnosis or pathophysiological entity

Identify credible biological action and it's optimal dose at target location from succesful laboratory trials.

Estimate energy loss from absorbtion and reflection for each anatomical location

Calculate optimal doses on skin for each anatomical location

Classify RCTs according to:
- diagnostic or pathophysiological similarity

Evaluate power of evidence and size of effect for optimal treatment given diagnosis

Classify RCTs according to:
- optimal or non-optimal dose and power density

Classify RCTS according to:
- optimal or non-optimal site of treatment

TYPICAL SIGNS OF POSITIVE EVIDENCE
- Credible biological action of EPA
- One or more RCTs with significant effect in favour of placebo
- More than 10-20 % of RCTs yield positive results.

Bjordal 2001

Laser dosimetry

DISTANCE TO TARGET TENDONS

Results

Carpal tunnel

The distance to the supraspinatus tendon is shortest with the hand-in-back position

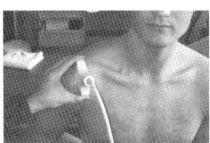

The distance to the Achilles and common finger & wrist extensor tendons is 5 times shorter than to the supraspinatus tendon

The distance to the patellar tendon is less than half the distance to the supraspinatus and carpal tunnel tendons

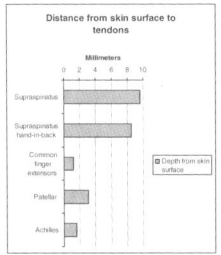

Therapeutic windows

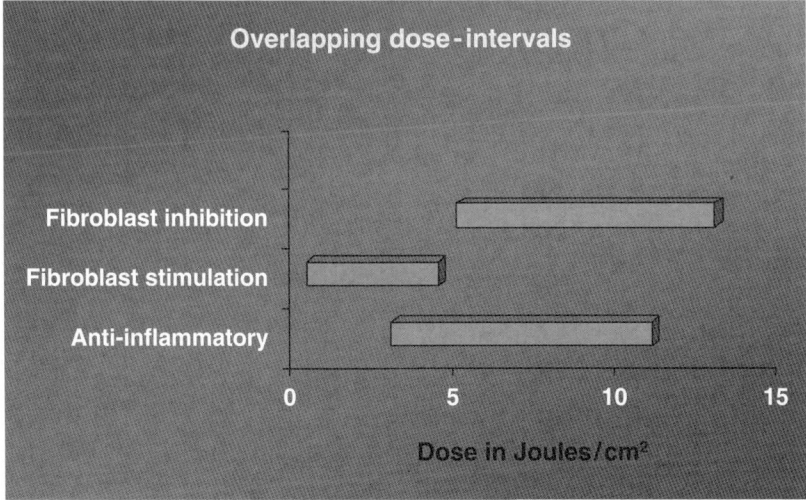

LLLT in tissue repair

Dose-dependent increase in collagen production after laser exposure of fibroblast cells

Peak 0.4 - 2 J/cm^2

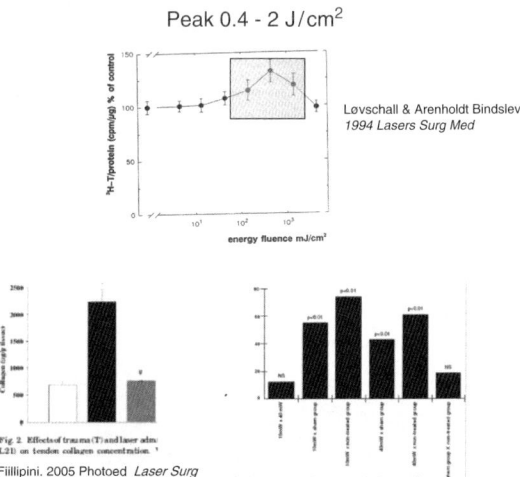

Løvschall & Arenholdt Bindslev. *1994 Lasers Surg Med*

Dose-dependent inhibition of collagen production and angiogenesis after laser exposure (> 5J & 100mW/cm2)*

Fiillipini. 2005 Photoed *Laser Surg*

Salate. 2005 Photomed *Laser Surg*

Optimal dose range for common tendon disorders

Tendon	IR 820 – 830 nm		IR 904 nm (pulsed)		HeNe 632 nm	
	Power density (mW/cm^2)	Dose (J/cm^2)	Power density (mW/cm^2)	Dose (J/cm^2)	Power density (mW/cm^2)	Dose (J/cm^2)
Plantar fasciitis	10 – 200	1.4 - 14	5 – 200	0.6 - 6	20 – 400	4.2 - 42
Achilles	5 – 200	0.7 - 14	2 – 200	0.3 – 6	10 – 400	1.4 - 14
Patellar	5 – 200	0.7 – 14	2 – 200	0.3 – 6	10 – 400	1.4 – 14
Epicondylitis	5 – 100	0.7 - 7	2 – 100	0.3 - 3	10 – 200	1.4 - 14
Rotatorcuff *	30 – 600	4.2 - 42	15 – 600	0.6 - 12	60 – 1.200	8.4 - 84

Threaths to validity in tendinopathy trials

LLLT is no more effective in the reduction of symptoms of CTS than is sham treatment.

Muscle Nerve 30: 182–187, 2004

DOUBLE-BLIND RANDOMIZED CONTROLLED TRIAL OF LOW-LEVEL LASER THERAPY IN CARPAL TUNNEL SYNDROME

JAMIE IRVINE, MD,[1] SU L. CHONG, BSc,[2] NASIM AMRJANI, MD,[2] and K. MING CHAN, MD[1,2]

[1] Division of Physical Medicine and Rehabilitation, University of Alberta, Edmonton, Alberta, Canada
[2] Centre for Neuroscience, University of Alberta, Edmonton, Alberta, Canada

Interventions and Masking. *Treatment Device.* The Eriel TOP 250 (Coradon Rehabilitation, Calgary, Alberta, Canada), which emits a low-level galium/aluminum/arsenide (GaAlAs) laser beam with an 860-nm wavelength, was used. A single-probe diode emitting a 60-mW beam with an intensity of 3 J/cm² per second over an area of 0.01 cm² delivered a total dose of 6 J/cm² in 15 s. A sham probe, identical in

To standardize the total dosage that each subject received, a thin clear plastic template with 1-cm × 1-cm grids was placed over the wrist and palm. Each

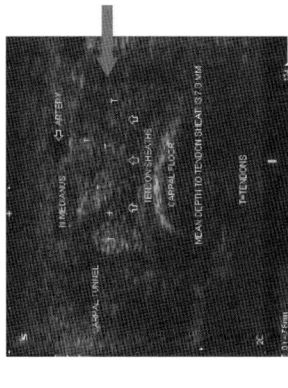

Treat the tendon — not the nerve !

0.06 W (Joules/second) x 15 second

= 0.9 Joules per point and per cm²

Diagnoses	Points or cm2	Joules 780 - 6.0nm	Notes
Tendinopathies			Energy dose delivered to the skin over the target tendon or syncvia
Carpal-tunnel	2-3	12	Minimum 6 Joules per point

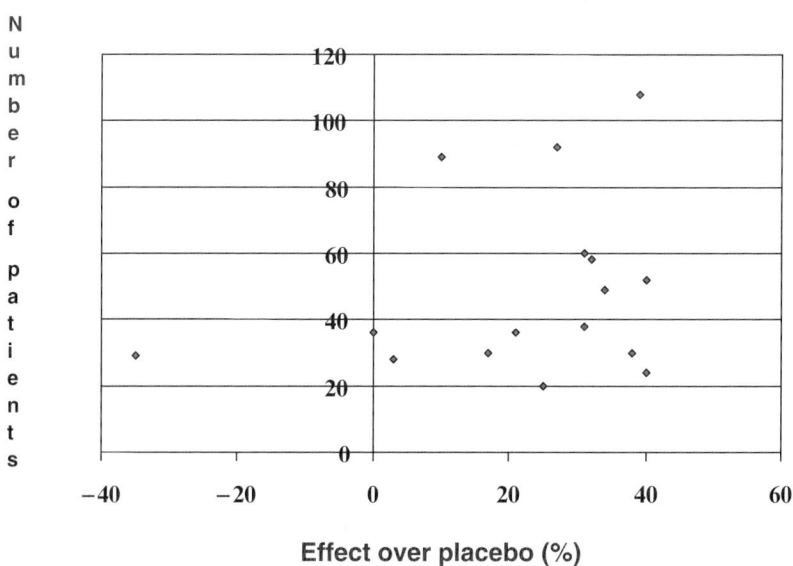

Effect vs sample size

(All trials)

Effect over placebo (%)

Review of laser trials for tendinitis

Investigating the outliers

Author/ laser type (single diode point lasers)	No. of patients	Diagnosis	Results (% better than placebo)	Power density at skin (mW/cm^2)	Dose at skin surface (J/cm^2)
Darre 830 nm IR	89	Achilles	10 %	150	20
Krashemminkof 830nm IR	36	Lat.epicond.	0 %	110	13.2
Papadopoulos 820nm IR	29	Lat.epicond.	- 35 %	714	30
Basford 830nm IR	28	Plantar fasc.	3 %	955	31.5

Bjordal & Couppe -00

Review of laser trials for tendinitis

Investigating the outliers

- In this trial the author (Siebert et al. -87) concluded that LLLT was ineffective because there were no significant differences between groups with regards to pain reduction

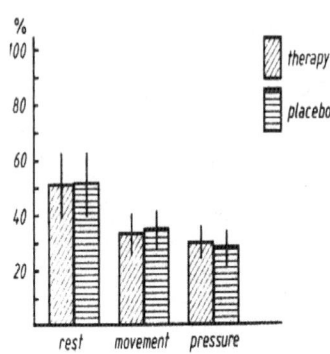

Fig. 4. Comparison of two groups with respect to percentage of pain reduction

Bjordal et al -01, PhysTherRev

Review of laser trials for tendinitis

Investigating the outliers

- Pain reduction in both groups were highly significant after treatment

- The "placebo effect " was far larger (35-50 %) than any other placebo-controlled LLLT-trial have described for similar patient samples (typically 10-20 %) and duration of treatment (2-4 weeks).

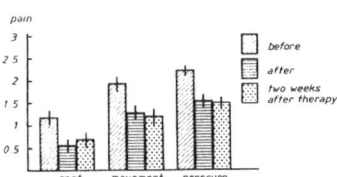

Fig. 2. Reduction of pain in therapy group according to scale

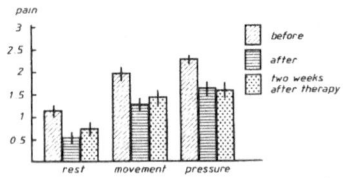

Fig. 3. Reduction of pain in placebo group according to scale

Bjordal et al. 2001 PhysTherRev

Review of laser trials for tendinitis

Investigating the outliers

- The highest power density was used in a trial on epicondylopathy (Papadopoulos et al. -96, dosage at skin surface : 714 mW/cm^2, 30 J/cm^2)

- Note that pain reduction was significantly less (!) in <u>active</u> LLLT group

VAS scores during the study

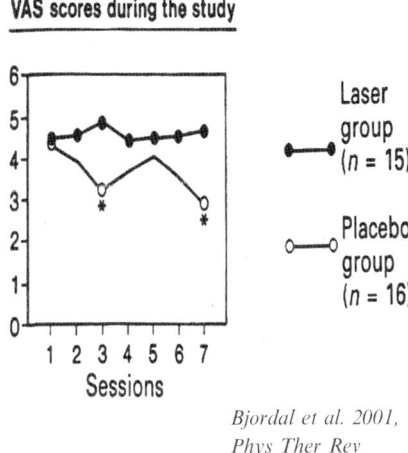

Laser group (n = 15)

Placebo group (n = 16)

Bjordal et al. 2001,
Phys Ther Rev

Review of laser trials for tendinitis

Investigating the outliers

- In one trial on lateral epicondylopathy (Haker & Lundeberg -91) the authors violated the recommended application procedure by placing the laser source in skin contact.

Fig 3—Application of the laser probe.

Bjordal et al. 2001
Phys Ther Rev

Review of laser trials for tendinitis

Investigating the outliers

Laser exposure area according to figure 5 on page 10
in technical manual SPACE MIX 5-UP

- The faulty treatment procedure left a large blind spot (about 3 cm^2) in and around the small area (< 0.3 cm^2) where the tendinitis was located.

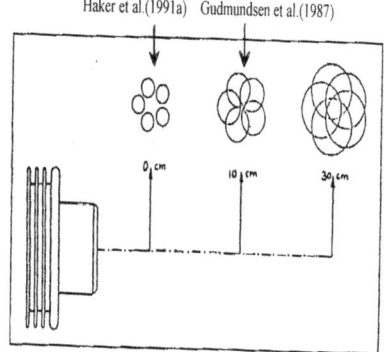

Haker et al.(1991a)   Gudmundsen et al.(1987)

Bjordal et al. -01

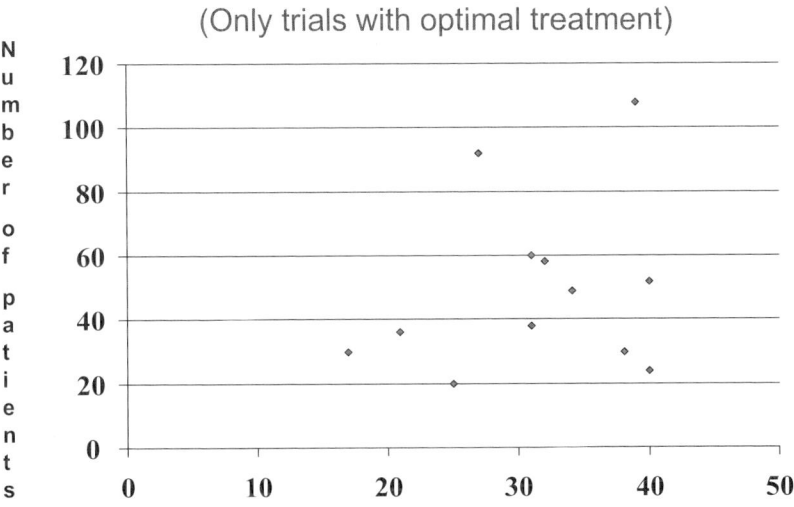

Effect vs sample size

(Only trials with optimal treatment)

Number of patients

Effect over placebo (%)

Mean effect over placebo : 32.0 %

Bjordal et al -01 Phys Ther rev

Patient selection differences in LLLT
trials versus Celecoxib (Celebrex)

## Celecoxib	## LLLT

Celecoxib

Patients. Patients (age ≥ 18 yrs) who had experienced an acute episode of tendinitis and/or subacromial bursitis within 7 days before the first dose of study medication were eligible for inclusion. Inclusion required a patient's Maximum Pain Intensity at Rest measure of moderate to severe [i.e., at least 50 mm on a 100 mm visual analog scale (VAS)]. In addition, at least 2 of the following had to be present in the affected shoulder to be included in the study: (1) painful abduction at any degree of motion with a VAS score ≥ 50 mm; (2) painful arc of movement from 45° to 120°, again with a VAS score ≥ 50 mm; and (3) tenderness over insertion of the supraspinatus tendon or over the subacromial bursa.

Patients with the following conditions were excluded from the trial: history of uncontrolled chronic disease, surgery to the affected shoulder, history of inflammatory arthritis, significant degenerative joint disease of the shoulder, evidence of rotator cuff tear (weakness of arm elevation/not due to pain, positive "drop arm sign," or high-riding humerus) visible on shoulder radiograph, a current fracture, or chronic calcific tendinitis with a

LLLT

Table 1: Affected Arm, Cause, and Previous Treatments

	Laser group (n)	Placebo group (n)
Total patients	29	29
Affected arm - right	23	25
Dominant arm - right	28	29
Work	12	11
Sport	8	13
Other activities	11	8
Unknown	3	1
Steroids*	13	8
NSAID +	5	5
Ultrasound	4	5
Other treatments	11	13
Untreated	12	13

*Steroids from local injection; NSAID + = nonsteroid anti-inflammatory drug

- Only acute tendinitis (< 7 days symptom duration)
- Short duration 14 days
- Long list of exclusion criteria
- No previous treatment

Petri et al. J Rheumatol 2004

- 7 months long symptoms duration
- Few exclusion criteria
- Majority had previous treatment failures

Haker et al. Arch Phys Med 1991

Oral NSAID and LLLT in tendinopathies

Type	No. of placebo-controlled RCTs	No. Patients	Mean symptom duration (weeks)	Number of trials not including patients with prior treatment failures	Trials allowing prior (≤ 3 months) or concomitant use of steroids
NSAID	9	1097	7	All	None
LLLT	28	1465	31	4/26	20/26

LLLT in neck pain

Pain on VAS for all trials regardless of dose

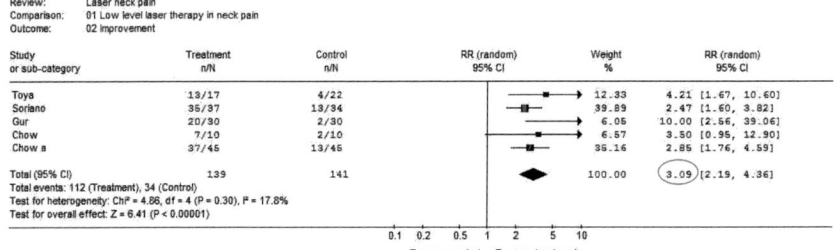

Review: Laser neck pain
Comparison: 01 Low level laser therapy in neck pain
Outcome: 01 Pain reduction on 100 mm VAS

Study or sub-category	Treatment N	Mean (SD)	Control N	Mean (SD)	WMD (random) 95% CI	Weight %	WMD (random) 95% CI
Özdemir	30	53.00(18.40)	30	5.00(14.30)		13.22	48.00 [39.66, 56.34]
Gur	30	42.80(32.30)	30	10.80(36.80)		11.48	32.00 [14.48, 49.52]
Hakguder	30	41.30(22.80)	30	12.10(22.40)		12.72	29.20 [17.76, 40.64]
Chow	10	27.00(19.00)	10	7.00(16.80)		11.96	20.00 [4.68, 35.32]
Ilbuldu	20	43.50(24.00)	20	21.00(27.40)		11.82	22.50 [6.54, 38.46]
Altan	23	27.20(6.90)	25	23.20(5.30)		13.72	4.00 [0.50, 7.50]
Chow a	45	27.00(21.00)	45	-3.00(21.00)		13.17	30.00 [21.32, 38.68]
Dundar	32	9.00(31.40)	32	10.00(31.80)		11.92	-1.00 [-16.48, 14.48]
Total (95% CI)	220		222			100.00	23.16 [8.78, 37.54]

Test for heterogeneity: Chi² = 124.26, df = 7 (P < 0.00001), I² = 94.4%
Test for overall effect: Z = 3.16 (P = 0.002)

-100   -50   0   50   100
Favours control   Favours treatment

LLLT in neck pain

Global improvement - all trials regardless of dose

Review: Laser neck pain
Comparison: 01 Low level laser therapy in neck pain
Outcome: 02 Improvement

Study or sub-category	Treatment n/N	Control n/N	RR (random) 95% CI	Weight %	RR (random) 95% CI
Toya	13/17	4/22		12.33	4.21 [1.67, 10.60]
Soriano	35/37	13/34		39.89	2.47 [1.60, 3.82]
Gur	20/30	2/30		6.05	10.00 [2.56, 39.06]
Chow	7/10	2/10		6.57	3.50 [0.95, 12.90]
Chow a	37/45	13/45		35.16	2.85 [1.76, 4.59]
Total (95% CI)	139	141		100.00	3.09 [2.19, 4.36]

Total events: 112 (Treatment), 34 (Control)
Test for heterogeneity: Chi² = 4.86, df = 4 (P = 0.30), I² = 17.8%
Test for overall effect: Z = 6.41 (P < 0.00001)

0.1   0.2   0.5   1   2   5   10
Favours control   Favours treatment

LLLT in neck pain

Effect/dose per point plot

LLLT in neck pain

Effect size/wavelength plot

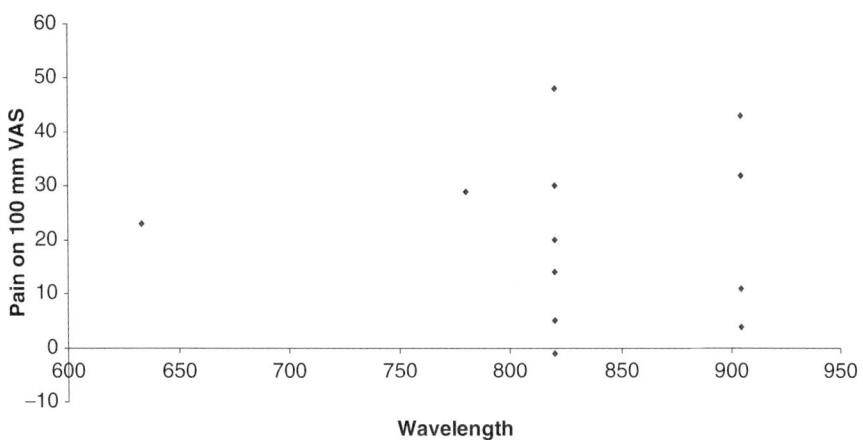

LLLT in neck pain

Effect/points plot

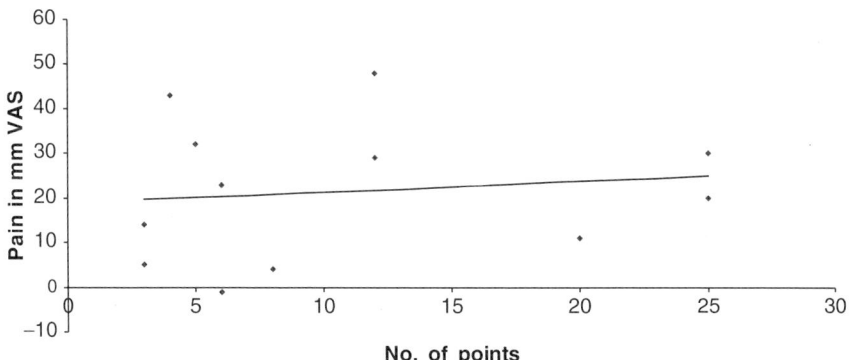

Clinical dosage recommendations

http://www.walt.nu

Recommended anti-inflammatory dosage for Low Level Laser Therap

Laser classes 3 or 3 B, 780 - 860nm GaAlAs Lasers. Continuous or pulse output less than 0.5 W

Energy dose delivered to the skin over the target tendon or synovia

Diagnoses	Points or cm2	Joules 780 - 820nm	Notes
Tendinopathies			
Carpal tunnel	2-3	12	Minimum 6 Joules per point
Lateral epicondylitis	1-2	4	Maximum 100mW/cm2
Biceps humeri c.l.	1-2	6	
Supraspinatus	2-3	10	Minimum 5 Joules per point
Infraspinatus	2-3	10	Minimum 5 Joules per point
Trochanter major	2-4	10	
Patellartendon	2-3	6	
Tract. Iliotibialis	2-3	5	Maximum 100mW/cm2
Achilles tendon	2-3	8	Maximum 100mW/cm2
Plantar fascitis	2-3	12	Minimum 6 Joules per point

Arthritis	Points or cm2	Joules	
Finger PIP or MCP	1-2	6	
Wrist	2-4	10	
Humeroradial joint	1-2	4	
Elbow	1-4	10	
Glenohumeral joint	2-4	15	Minimum 6 Joules per point
Acromioclavicular	1-2	4	
Temporomandibular	1-2	6	
Cervical spine	2-4	15	Minimum 6 Joules per point
Lumbar spine	2-4	40	Minimum 8 Joules per point
Hip	2-4	40	Minimum 8 Joules per point
Knee medial	3-6	20	Minimum 5 Joules per point
Ankle	2-4	15	

Daily treatment for 2 weeks or treatment every other day for 3-4 weeks is recommended

Irradiation should cover most of the pathological tissue in the tendon/synovia.
Tendons
Start with energy dose in table, then reduce by 30%, when inflammation is under control
(Does not apply for carpal tunnel tenosynovitis)

Therapeutic windows range from typically +/- 50% of given values
Recommended doses are based on ultrasonographic measurements
of depths from skin surface and typical volume of pathological tissue
and estimated optical penetration for the different laser types in caucasians

Disclaimer
The list may be subject to change at any time when more research trials
are being published. World Association of Laser Therapy is not responsible
for the application of laser therapy in patients, which should be
performed at the therapist/doctor's discretion and responsibility

Revised August 2005

Recommended anti-inflammatory dosage for Low Level Laser Therapy

Laser classes 3 or 3B, 904 nm GaAs Lasers (Peak pulse output more than 1 Watt)

Energy dose delivered to the skin over the target tendon or synovia

Diagnoses	Points or cm2	Joules 904nm	Notes
Tendinopathies			
Carpal tunnel	2-3	4	Minimum 2 Joules per point
Lateral epicondylitis	1-2	2	Maximum 100mW/cm2
Biceps humeri cap.long.	1-2	2	
Supraspinatus	2-3	3	Minimum 2 Joules per point
Infraspinatus	2-3	3	Minimum 2 Joules per point
Trochanter major	2-3	2	
Patellartendon	2-3	2	
Tract. Iliotibialis	2-3	2	Maximum 100mW/cm2
Achilles tendon	2-3	2	Maximum 100mW/cm2
Plantar fascitis	2-3	3	Minimum 2 Joules per point

Arthritis	Points or cm2	Joules 904nm	
Finger PIP or MCP	1-2	2	
Wrist	2-3	3	
Humeroradial joint	1-2	2	
Elbow	2-3	3	
Glenohumeral joint	2-3	6	Minimum 2 Joules per point
Acromioclavicular	1-2	2	
Temporomandibular	1-2	2	
Cervical spine	2-3	6	Minimum 4 Joules per point
Lumbar spine	2-3	10	Minimum 4 Joules per point
Hip	2-3	10	Minimum 4 Joules per point
Knee anteromedial	2-4	6	Minimum 2 Joules per point
Ankle	2-4	3	

Daily treatment for 2 weeks or treatment every other day for 3-4 weeks is recommended

Irradiation should cover most of the pathological tissue in the tendon/synovia.
Tendons
Start with energy dose in table, then reduce by 30%, when inflammation is under control
(Does not apply for carpal tunnel tenosynovitis)

Therapeutic windows range from typically +/- 50% of given values
Recommended doses are based on ultrasonographic measurements
of depths from skin surface and typical volume of pathological tissue
and estimated optical penetration for the different laser types in caucasians

Disclaimer
The list may be subject to change at any time when more research trials
are being published. World Association of Laser Therapy is not responsible
for the application of laser therapy in patients, which should be
performed at the therapist/doctor's discretion and responsibility

Revised August 2005

Bulk of evidence from LLLT trials

Type	No. of placebo-controlled RCTs	No. Patients
LLLT Low Back Pain	6	313
LLLT Tendinopathies	28	1465
LLLT Neck Pain	14	732
LLLT Osteoarthritis	11	431
Total LLLT	59	2941

Magnitude of effects

Magnitude of effect (mean mm on VAS)	LLLT regardless of dose	Oral NSAID published data	LLLT optimal dose and treatment procedure	Oral NSAID after correction for patient selection bias
Low Back Pain	11	12	14 (only 1 trial with VAS data)	9
Tendinopathies	13	6	19	6
Neck Pain	23	-	25	-
Osteoarthritis	16	10	22	7
Median	13	10	19	7

Take home messages

1. When doing systematic reviews, look behind the evidence and assess the procedures and doses used.

2. Scientific evidence can be graded by how likely it is that <u>future research will change</u> the conclusions.

3. Guidelines are the results of professional discussion and is graded by the <u>level of agreement</u>

4. Guidelines are notoriously confounded by stakeholders and industrial interests.

Bergen - Norway

Index

Accelerated repair, 357
Acellular dermal network, 357
Action and absorption spectra,
    xxviii–xxix
Acupuncture, 273
AIDs, 233
ALA, 82–86
Alpha waves, 322–324
Aneurysm progression, 53
Angiogenesis, 151, 152, 154, 155
Antigen presentation, 101, 102, 106
Anti-tumor immunity, 99, 100, 105, 108,
    110
Apoptosis, 67, 71, 72, 76–77
Arndt-Schultz Law, 327
ATP solution, 297

Bioresonance, 216
Biphasic response model, 327
Blue light, 33, 34, 36
Bone, 181–185
Broadband light sources, 59–63

Cardioprotection, 151, 154
Catalase, 3, 7, 8, 11, 13, 16, 17
Cell cultures, 12
Cell therapy, 248
Cellulite, 332–336, 338, 340
Clausius-Mosotti equation, 299
Clinical, 24, 26
Colostrum, 233–237
Complex regional pain syndrome, 284
Craniofacial pain, 273

Cross sectional tomograms, 173, 174
Crushing, 348
Cytochrome c oxidase, xxvii–xxix, xxxi,
    xxxiii, 67, 71, 72, 77
Cytokines, 53–56
Cytotoxic T-cells, 101, 103, 105–108

Dendritic cells, 101, 102, 107
Dental implants, 191, 195
Dentistry, 173, 175, 273, 274
Dependence on biological response
Diabetes laser therapy, 215, 218–220
Diabetic ulcers, 23, 24, 29
Diabetic wound healing, 222, 229
Dose, 3, 5–7
Dosimetry, 115–117, 119, 122, 123

Effective field, 331
Er:YAG laser, 182, 184–188

Far infrared, 348
510(k), 269
Fluence, 327
4 J/cm2, 357

Glioblastoma cells, 309, 311, 312
Glycate haemoglobin, 215, 217

Heart muscle, 151, 154
High density lipoproteins, 127, 135–136
High resolution, 173, 174, 178
Human brain cancer (glioblastoma), 11, 13,
    15–18

Hydrogen peroxide, 3–8
Hypoglycaemia, 215–219

IDE, 269
Implant failure, 159
Inflammation, 280
Inflammation complex regional pain
        syndrome, 274
Infrared thermography, 283
Inhibition, 3, 4, 6–8, 12, 277–280
Intensity, 327, 328

Laser, 68–70, 75, 76
Laser bioinhibition, 327
Laser biostimulation, 327
Laser dosage, 216
Laser induced biostimulation, 259
Laser induced stimulation, 222
Laser light dose, 221–224
Laser medicine, 207, 208, 210
Laser phototherapy, 247–251, 253,
        254, 257
Laser reconstruction, 259, 264–266
Latent-TGF-beta1 activation, 207
LED, 68, 69, 194–198, 201–203
Light therapy, 3–8, 283
LILT. See Low intensity light therapy
Lorentz-Lorenz law, 301, 303
Loretzian curve fitting, xxvii, xxx
Low density lipoproteins, 127, 128, 130–135
Low intensity light therapy, 227, 273, 331,
        334, 337, 338, 340
Low level laser irradiation, 151–155
Low level laser therapy, 273
Low-level light therapy (LLLT), 67–78, 327
Low power laser therapy, 207
Lysine residues, 135

Magnetic fields, 215, 217
Microstructure, 173–175, 179
Mid-IR results, 295
Mitochondria, 67, 72–73, 77, 82–84
Mitochondrial dysfunction, 39, 40, 47
Mobile phones, 315, 316, 318, 322–324
Molecular beacons, 141, 146
MRSA, 33–34, 36

Nanoparticles, 127, 128, 130, 135, 145
Near infrared, 15
Near infrared laser therapy, 259
Nerve injury, 247–250, 257
Nerve regeneration, 247, 248
Nerves, 278, 279
Nitric oxide, 53–56
Noninvasive, 173, 174, 179
Novel light-activated cellular signaling,
        xxix–xxxiii

Optical biopsy, 173, 174, 179
Optical coherence tomography (OCT), 159,
        161, 163, 168, 173–179
Optimization, 94
Oral wounds, 209
Oxidative stress, 39, 40, 44, 46, 47

Pain relief, 274, 275
PBR, 83, 84, 86
Periodontal diseases, 181, 182, 189
Periodontal pocket, 181–183, 189
Periodontal tissue regeneration, 173,
        174, 179
Periodontitis, 181, 182, 184, 189
Photobiomodulation, 39–41, 46, 48, 49,
        59–60, 67, 68, 77, 170, 171,
        181–185, 188, 189, 191, 194, 195,
        198, 201–203, 207, 208, 210, 273
Photodynamic therapy (PDT), 81–86,
        89–96, 99–105, 107–110,
        115–119, 121–123
Photomology, 335, 338, 340
Phototherapy, 273
PMA, 269
Polarization, 297–299, 302, 304
Pre-IDE, 269
Protective device, 316, 322
Protoporphyrin IX (PpIX), 89–95

Quantitative electroencephalogram, 315

Rats, 357
Real time diagnostic imaging, 173, 175
Receptors, 127, 128, 131, 133–137, 141, 146
Redox sensing, 210

Reflex sympathetic dystrophy, 284
Refractive index extension, 295
Refractive index, 297, 299, 303, 304
Relieve of NO block, xxvii
Retinal degenerative disease, 39, 40, 43, 47, 49
Retinal injury, 39–41
Retrograde mitochondrial signaling, xxxiii
rf irradiation, 307
rf studies, 295

Sciatic nerve, 348, 351, 352, 354
780 nm radiation, 53, 54, 56
Singlet oxygen, 81, 82
685 nm, 357
660 and 880 nm wavelengths of laser light used together, 23, 29
Skin, 90–94, 96

Smooth shapes, 338–340
Spectroscopy, 119–122
Spine disc healing, 260
Spine treatment, 260, 265
Staphylococcus aureus, 59, 61, 62
Stimulation, 3, 4

TGF-b1, 171
Tissue optics, 116–119
Toll-like receptors, 99, 101, 102, 104
T regulatory cells, 99, 108

Ultrasonic evaluation, 193, 195, 197

Vasomotor changes, 283

Wound healing, 207–209

Printed in the United States of America